AF497382

A FARM JOURNAL FARM

Our Folks—by which is meant the four million subscribers and readers of "The Farm Journal"—live in homes such as the one pictured above. They are the cream of the agricultural people of America. They have solid, substantial homes, big barns with tight roofs, money in bank, fertile land, the best stock, the biggest apples, the richest milk. It is for their benefit that "The Farm Journal" is published, and to them this volume "How To Do Things" is offered.

AND HERE WE HAVE—

PETER TUMBLEDOWN'S FARM

No one can read "The Farm Journal" and be a Peter Tumbledown too. Many have tried, but they have to give up one or the other.

HOW TO DO THINGS

A Compendium of New and Practical Farm and Household Devices, Helps, Hints, Recipes, Formulas and Useful Information from The Farm Journal

WILMER ATKINSON COMPANY

Publishers of the Farm Journal

Washington Square

Philadelphia

Foreword

In introducing this volume to the millions of readers of THE FARM JOURNAL —the great family of rural Americans whom we often call with respect and affection "Our Folks"—only a very few words are necessary.

At intervals for the last forty years we have had friends of THE FARM JOURNAL come to the Editors and say, "Why don't you make a book of the best things out of the paper? It is a shame," they say, "to have this wealth of information lost, or to force readers to make scrap-books to preserve special articles that interest them."

The Editors have always agreed that such a book would be a good thing. From their daily contact with subscribers all over the country, the countless letters and inquiries they receive and answer, they know very closely what farm folks are interested in and want to know about. On this contact they are accustomed to base the choice of what goes into THE FARM JOURNAL, and they have always believed that a book containing these important things would be appreciated by many.

It was more a question of finding the time and energy to do the job. Other matters pressed, the family of Our Folks kept getting bigger and bigger, and demanding more and more information and advice, the Great War came on, and so on and so on. However, the time came at last, and here finally is the long-desired book, full to overflowing with interesting and important matters.

The Editor's Desk

The "Contents" pages which follow a little further on, and, even more, the thirty-two pages of "Index" which come next, will sufficiently show what is in the book, and where it may be found.

To many other millions of farm folks to whose eyes this may come and who are not familiar with THE FARM JOURNAL, we believe some special explanation may be due. In spite of its more than forty years' existence, and its enormous circulation in every State and in every civilized country of the world, THE FARM JOURNAL still remains unknown, or only vaguely known, in many American farm homes.

We believe it may be interesting and may help to account for the publishing of this book, if we devote a few pages to a brief history of THE FARM JOURNAL,

THE FARM JOURNAL is just the paper for stock breeders.
Its teachings are sensible and practical—
Cream not skim-milk.

Foreword

The old familiar heading of the biggest little paper in the world.

its foundation, growth and present position as the foremost farm and household magazine. We shall not feel badly if impatient readers skip these few pages, or only pause to look at the illustrations, but in that case we shall hope that they will return later, and learn something of the periodical from which the rest of the book is reprinted.

THE FARM JOURNAL came into existence in the month of March, 1877, and is therefore near middle age.

The Editor was reared at the plow handles, and all his ancestors on both sides were farmers since the time of William Penn.

The first number was an eight-page paper, and 25,000 copies were printed; now we often print a hundred and twenty or more pages and every edition exceeds a million copies.

The Editor did not then aspire to other than a local circulation; now this magazine goes to every State in the Union, to Canada, and to many distant countries.

At first two small rooms held the entire plant, and all the work was done by less than a dozen assistants. The Editor successfully canvassed the county fairs personally for subscribers.

The subscription price was 25 cents a year, cash in advance, and "stop when the year is up." John Wanamaker was just starting his great department store, and for a time was the principal advertiser.

The growth in circulation was not very rapid, but constant; when we obtained a subscriber we held him until he died, and then his children took his place. It often required several copies for the children and grandchildren.

The first office of the paper was at 726 Sansom Street; the second, 914 Arch Street; the third, 144 North Seventh Street; the fourth, 125 North Ninth Street; the fifth, 1024 Race Street, where we were for seventeen years; thence we moved to our present location on Washington Square. All the other offices could be placed inside of this one.

Poor little chaps in a big, cold world! THE FARM JOURNAL will help you keep them healthy and happy

Foreword

This is the guarantee that appears every month on Page 1. Since its first insertion in 1880, nearly every first-class farm paper in the country has followed our lead and makes some such promise

From the first, unreliable and shady advertisers were excluded; but the Editor felt that something more was necessary to safeguard the interests of subscriber and advertiser alike. So in the issue for October, 1880, we put up the "Fair Play" notice, substantially as it may now be seen on the first page, and we have held firmly to it ever since. THE FARM JOURNAL was the first periodical in the world to print such a notice. Many others have since adopted the plan. Any subscriber dealt with dishonestly by an advertiser is protected from loss by the publisher.

In July, 1879, Jacob Biggle, "a retired city merchant," began his letters from Elmwood Farm, and Aunt Harriet, his wife, took a hand at times. Old Peter Tumbledown also began to figure about this time.

In the February issue, 1880, the Editor stated that "you can count on the fingers of one hand all the farm papers of the country that have more subscribers than THE FARM JOURNAL." By this time it had outgrown the local field, and subscriptions poured in from all the States.

It required six tons of paper for the big edition of December, 1880, which was pretty good for a youngster of four summers.

In 1881, the Editor announced that he wanted one hundred thousand subscribers and was bound to get them. By 1884 he did get them.

THE FARM JOURNAL announced that it was "a tub that stood on its own bottom," and this has always been its policy. It never has had any side issues of any kind. "Do one thing and do it as well as we can," is our motto.

The best crop on the farm.

THE FARM JOURNAL entered its sixth year "healthy, happy and spunky."

In 1882 the Editor established the custom of taking a journey through the country every summer, visiting subscribers and points of interest. These trips were and continue to be eye-openers.

It was during the following winter that the 100,000 mark was reached,

Foreword

Farm and village people up and down these roads and all about your neighborhood and all over this broad land read THE FARM JOURNAL and speak in praise of it.

and the claim was made that no other farm weekly or other monthly had so many. In proof of this, our subscription books were offered for examination. They have been open ever since for that purpose.

The next year THE FARM JOURNAL started in for 200,000 subscribers and announced that "we are bound to get them." "Our paper does not wear a long face," the Editor said, "it is hopeful, cheerful, looking on the brightest and best side of everything, and it likes fun—a grin better than a groan."

In March, 1884, the Editor said: "THE FARM JOURNAL now enters its eighth year," and was "never happier in its life." By December of this year the 125,000 mark was reached.

In 1887 we began to take subscriptions for a longer term than one year—two years at first; then, after finding the system to be right, we extended the time to three years; now we seek only four-year subscriptions. We have many subscribers who are paid ahead for twenty years; some for thirty, and one until 1957.

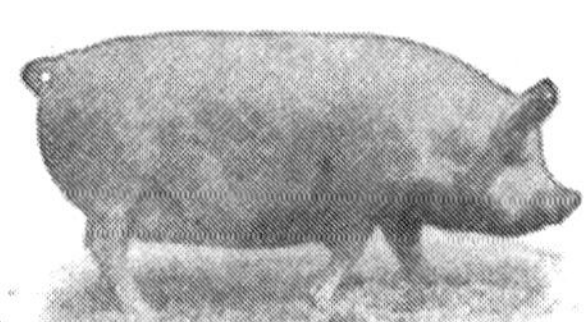

The rent-payer and mortgage-lifter. THE FARM JOURNAL tells the How of Hogs from beginning to end.

Foreword

The hen turns grass into greenbacks, grain into gold, and even coins silver out of sand:
THE FARM JOURNAL is full of hen lore and chicken wisdom.

It is often hard to get the one-year idea out of the heads of new people who do not know THE FARM JOURNAL. But when they have read it for awhile, and understand why we take only four-year orders, and the saving it makes, it is very easy to get their renewals for as long a time as we choose to ask for.

The number of subscriptions kept creeping up steadily to 1890, until 200,000 was almost in sight, and a clerical force of seventy or more was needed to handle the business in the busy season.

In September, 1891, the Editor announced that he was after 1,000,000 subscribers, a mark which he has now reached, after eighteen years of steady effort.

In 1892 THE FARM JOURNAL appeared for the first time stitched and bound like a book. New type was bought and better quality of paper used. In March, 1893, a thirty-two page paper was printed for the first time.

Scientific fertilization means profitable crops. THE FARM JOURNAL constantly teaches that on the farm, as elsewhere, knowledge makes wealth.

During the winter of 1893-94 the plan was adopted of having all subscriptions begin with January and end with December.

In August, 1894, appeared the first half-tone illustrations (engravings made direct from photographs of the objects).

Foreword

By the following spring the editions had passed the quarter-million mark.

From 1895 to 1898 the paper ran along smoothly with no important events to chronicle. Its circulation was growing right along, and its advertising patronage increasing every year. In March of the latter year the paper became of age, having completed twenty-one years of life.

In 1899 the publishers concluded that some better way must be found for printing the paper. Heretofore the work had been done by outside printers under contract, and often as many as ten presses were at work printing an edition, with as many as sixty men and girls at work folding and binding it.

Not having close enough control over this work, the deliveries of the paper were often delayed. It was therefore decided to put in our own press, and this was done in the fall of 1899, in time to print a part of the October number. For six years this press turned out all editions of the paper at 40,000 copies a day of a forty-page magazine. In 1905 it was replaced by a new press of the same size but equipped with several important improvements.

In 1902, for the first time, the circulation of the paper went permanently above the 500,000 mark, and the claim was made that two and a half million people read every issue.

The big twin presses on which THE FARM JOURNAL is printed at the rate of two hundred a minute. Driven and controlled by electricity throughout. Come see them when in Philadelphia.

About 1903 THE FARM JOURNAL began its great fight against the San Jose scale, and redoubled its efforts to secure parcels-post and postal savings banks for the farmers of the country. Both have now been secured.

In the summer of 1905 our second big press was installed. In February and March, 1907, it was necessary to print sixty-page magazines for the first time in our history. About this time, too, a complete new outfit of type was secured.

In 1909 the big printing press again became inadequate for the demands of our growing family of subscribers, and the whole building (1024 Race Street) was overcrowded. After some trouble, an abandoned church property which stood back to back with us on Cherry Street was bought, and this was

Foreword

remodeled to house the printing and mailing departments. At the same time we installed press No. 3, a huge machine as large as both the old presses put together.

This was followed by press No. 4, a duplicate of No. 3; and these wonderful pieces of machinery, each driven by a 30 horse-power electric motor, now print THE FARM JOURNAL. Each of them will print, trim, fold, wire-stitch, count and deliver 45,000 eighty-page FARM JOURNALS a day, or about 1½ copies per second.

In the early years of the paper subscribers' papers were addressed with pencil, but this was so slow and inaccurate that it was necessary to put all names in type by type-setting machines and have the addressing done by mailing machines, each capable of 10,000 per day. In 1908 this system was in turn replaced by an improved method using light aluminum plates instead of the heavy lead type. While more expensive, the clear-cut addresses given by this system, insuring prompt receipt of each issue by the subscriber, more than justify its cost.

Stamping subscribers' names on aluminum plates for the mailing list.
The machine works something like a typewriter.

Even with the Cherry Street church and the property next door to it, which was bought in 1911, the magazine continued to outgrow its home, and it was apparent that we must enlarge again. A large lot on Washington Square was bought and erection of a new building was begun in the summer of 1911. This was completed and occupied fourteen months later, and there we are now "at home" to our friends.

Here we have three large, airy, well-lighted floors, each one of them nearly as large as any building we ever had before. Two more floors above are rented for a term of years, and we can get them for our own use at the end of that time. If necessary, the concrete foundations and framework are heavy enough so that we can put still another floor on top.

In 1913 THE FARM JOURNAL first offered to subscribers a straight guarantee that subscriptions might be discontinued "at any time, for any reason, or for NO reason," and the subscription money would be refunded. This in-

Foreword

Sheep are the most profitable livestock kept on the farm, next to the boys and girls, and there are few farms for which they are not adapted. They live on the least expense, breed and grow two crops at the same time—wool and mutton. THE FARM JOURNAL will help you on the sheep question.

sured a list of subscribers, all of whom read and want the paper, since those who would rather have their money back can get it without fuss or delay. This "Guarantee to Subscribers" is now printed in every issue of THE FARM JOURNAL. No other periodical has done this.

The vast multitude of well-to-do persons who read THE FARM JOURNAL each month and have confidence in its pages have insured a gratifying advertising patronage; yet we have endeavored to see that the quantity and quality of reading matter should more than keep pace with the advertisements.

On the editorial page of THE FARM JOURNAL you may read under the title the words, "Unlike Any Other Paper." This states a truth that is obvious to those who form the habit of reading it.

In the great oats and wheat States, where the threshing crews pile their glittering straw mountains, THE FARM JOURNAL is best known and best loved.

Foreword

For one thing it says a thing and stops after it has said it.

For another it is cheerful and hopeful, and wants everybody to have a good time. It is young in spirit if not in years.

For another it is full of snap and ginger, and hits the nail squarely on the head.

It never has printed and never will print a quack medical advertisement, or other advertisement of doubtful character. (We could easily obtain $50,000 worth of such advertisements per year were we so inclined.)

Hence, its columns are perfectly clean and pure; it does not have to be hid from the children, nor carried out of the house with the tongs.

For another, it prints no long-winded, tiresome essays, and it knows what to leave out as well as what to put in.

It is at home in every State, in all latitudes and longitudes, and welcome from the rising to the setting of the sun.

It treats every branch of farming—gardening, stock breeding, dairying, poultry raising, bee culture, sheep and swine husbandry, fruit growing and trucking.

It thinks the humans of the farm are the best stock on it, yet it teaches the kindest, wisest care for every species of farm animal. Its teachings are practical and therefore profitable.

It is a woman's paper of the purest and best type, and all the young folks of the farm find the keenest delight in it.

Its purpose has always been, and now is, not how much profit will accrue, but how much good it can do. If our business were a mere matter of dollars and cents, we would quit today.

Responding to our desire to encourage and benefit Our Folks, we have the assurance of countless messages day by day, brought to us by every mail, that four million readers wish us well and rejoice in our prosperity, and this cheers us and gives us unspeakable comfort and happiness.

If we should sum up in a few words our feelings about THE FARM JOURNAL and "How To Do Things," we would say something like this: "A favorite description of THE FARM JOURNAL is to say that it is 'Cream, not Skim Milk.' The contents of this book have been selected as the very best of THE FARM JOURNAL contents. We therefore present it to Our Folks, old and new, as the 'Cream of the Cream.'"

WILMER ATKINSON COMPANY,
Publishers.

September, 1920
Philadelphia, Pa.

CONTENTS

CONTENTS (Continued)

Free Information Service

Readers of THE FARM JOURNAL and this book are familiar with the great variety of facts and information presented in their pages.

It often happens, however, that subscribers need special information and aid on some point that is not covered, requiring special correspondence and often special investigation by the Editors of THE FARM JOURNAL.

With this in mind, we have included, as a special feature of "HOW TO DO THINGS," the sheet of Free Information Coupons which follows this page. When writing to the Editors clip off a coupon and attach to each letter you send. This will show that you are one of "Our Folks" and therefore entitled to the best service, and most prompt attention, that the Editors can give. If you run short of coupons, use the last one to ask for a new sheet, which will be sent if your subscription is paid ahead.

All Kinds of Information

The inquiries received and answered by our Editors cover every conceivable subject. They are asked about crops and fertilizers, about garden truck and flowers, about horses and cattle and hogs and poultry, what to do for every kind of ailment, animal and human. "Our Folks" ask about buying automobiles and tractors, buying farms, borrowing money, income tax, about the standing of different firms, how to collect debts, questions of law, where to sell farm products, how to form co-operative organizations, etc., etc.

To all these inquiries, the Editors return answers, as promptly as they can, or refer subscribers to the proper source of information. All of this service is free, except that inquiries of the Law Department, requiring immediate answer, must be accompanied with remittance of one dollar. The free service rendered by the Veterinary and Family Doctor Editors alone is made use of by thousands of "Our Folks."

In using these Free Inquiry Coupons, be sure to follow instructions. Do not fail to attach the label from your last copy of THE FARM JOURNAL, and sign your name legibly.

WILMER ATKINSON COMPANY,
Publishers The Farm Journal.

Washington Square, Philadelphia.
Mallers Building, Chicago.

The Farm Journal
Philadelphia, Pa.

Coupon No. 6

FREE INFORMATION COUPON

Attach here the address label from your last issue of The Farm Journal in order to insure prompt attention to inquiries.

Subscriber's
Signature_______________________

The Farm Journal
Philadelphia, Pa.

Coupon No. 1

FREE INFORMATION COUPON

Attach here the address label from your last issue of The Farm Journal in order to insure prompt attention to inquiries.

Subscriber's
Signature_______________________

The Farm Journal
Philadelphia, Pa.

Coupon No. 7

FREE INFORMATION COUPON

Attach here the address label from your last issue of The Farm Journal in order to insure prompt attention to inquiries.

Subscriber's
Signature_______________________

The Farm Journal
Philadelphia, Pa.

Coupon No. 2

FREE INFORMATION COUPON

Attach here the address label from your last issue of The Farm Journal in order to insure prompt attention to inquiries.

Subscriber's
Signature_______________________

The Farm Journal
Philadelphia, Pa.

Coupon No. 8

FREE INFORMATION COUPON

Attach here the address label from your last issue of The Farm Journal in order to insure prompt attention to inquiries.

Subscriber's
Signature_______________________

The Farm Journal
Philadelphia, Pa.

Coupon No. 3

FREE INFORMATION COUPON

Attach here the address label from your last issue of The Farm Journal in order to insure prompt attention to inquiries.

Subscriber's
Signature_______________________

The Farm Journal
Philadelphia, Pa.

Coupon No. 9

FREE INFORMATION COUPON

Attach here the address label from your last issue of The Farm Journal in order to insure prompt attention to inquiries.

Subscriber's
Signature_______________________

The Farm Journal
Philadelphia, Pa.

Coupon No. 4

FREE INFORMATION COUPON

Attach here the address label from your last issue of The Farm Journal in order to insure prompt attention to inquiries.

Subscriber's
Signature_______________________

The Farm Journal
Philadelphia, Pa.

Coupon No. 10

FREE INFORMATION COUPON

Attach here the address label from your last issue of The Farm Journal in order to insure prompt attention to inquiries.

Subscriber's
Signature_______________________

(Use this Coupon to write for new Coupon Sheet)

The Farm Journal
Philadelphia, Pa.

Coupon No. 5

FREE INFORMATION COUPON

Attach here the address label from your last issue of The Farm Journal in order to insure prompt attention to inquiries.

Subscriber's
Signature_______________________

Index

Index

Index

Index

Index

Index

Index

Index

Index

Index

Index

Index

Index

Index

Index

Index

Index

Index

Index

Index

Index

Index

How To Do Things

Balanced Dairy Rations

Most feeders of dairy cows can produce protein more cheaply than they can buy it. Wise is the man who has a good supply of legume forage on hand from his alfalfa, clover, cow-peas, or soy-bean fields, for the dairy cow cannot do her best on corn, timothy hay or fodder, even with good succulent silage to help maintain summer pasture conditions.

Bran at $1 a hundred furnishes protein at a cost of about eight and a half cents a pound, while cottonseed. at $1.90 a hundred, furnishes it at a cost of a little more than five cents a pound. The cost of producing the protein on the farm in alfalfa, clover, or other leguminous crops, varies with local conditions, so that no such definite figures can be given; but almost any farmer should be able to supply himself at much less than five cents a pound. If he has neglected to do so, however, he must purchase protein in some form to supplement the abundance of silage and corn which he doubtless has on hand.

More Protein, More Milk

Feeding two pounds of cottonseed-meal a day to a cow that has been giving milk on such a ration as corn, fodder and timothy hay, will increase the flow to a surprising degree. In spite of the many things that have combined to raise the price of cotton-seed-meal this year, it is probably the cheapest concentrate to buy.

Some good dairy rations:

1. Corn silage, 25 lbs.; clover hay, 10 lbs.; corn, 4 lbs.; bran, 4 lbs.

2. Corn silage, 30 lbs.; alfalfa or cow-pea hay, 10 lbs.; corn, 6 lbs.; bran, 2 lbs.

3. Clover hay, 20 lbs.; corn, 4 to 6 lbs.; bran or oats, 2 to 4 lbs.

4. Clover hay, 2 lbs.; corn-and-cob meal, 5 to 7 lbs.; gluten or cottonseed-meal, 2 lbs.

5. Alfalfa or cow-pea hay, 10 lbs.; corn fodder, 10 lbs.; corn, 5 to 7 lbs.; bran, 2 lbs.

6. Alfalfa or cow-pea hay, 15 to 20 lbs.; corn, 8 to 10 lbs.

Each of these rations includes a whole day's feed for the ordinary cow, half to be given in the morning and the other half in the evening, but they are not intended for the cow of unusual dairy capacity, or one that is being fed for heavy production, or to make a record. The amounts specified are those to be fed to a cow giving from twenty to twenty-five pounds of milk a day, and cows giving more than this should receive more feed, especially more grain, while those giving less milk should have the grain cut down. The general plan followed is to give each cow all the roughness she will eat, and about one pound of grain for each three pounds of milk produced.

Cows Differ

Where it can be practiced, individual feeding for results is worth while. Cows are not mere uniform machines; they differ greatly, and the only way to get the best response at the milk-pail from various combinations of rations is to try them out on each cow, to ascertain the limit of responsiveness.

Churning and Temperature

The butterfat in milk or cream is in the form of very small globules that are held apart by a thin coating of milk serum around each globule of fat. The object in churning is to throw the fat globules together with enough force to break through the serum layer and make the fat globules stick together. The ease with which this can be done will depend upon the temperature of the fat globules, and this temperature is controlled by the three factors—period of lactation, feed and weather.

How Long Since Calving

Period of lactation is the term used to show the length of time since the cow calved. As a cow's milking period advances the fat globules in the milk become smaller and harder. This will not be noticed if there are a number of cows in the herd, because it is probable that there will be about an equal number fresh and dry at one time. But where only one cow is milked, the temperature of cream for churning will have to be increased as the cow's milking period advances.

Some feeds, such as hay, cottonseed-meal, and hulls, harden the butterfat

Other feeds, as green grass, bran and corn-meal, make a softer butterfat. Thus, the kind of feed given will have a bearing on the temperature at which to churn.

Butterfat Changes Slowly

The time that cream is held at any temperature will have quite an influence. Milk serum changes temperature very much faster than fat. Therefore, if cream is held at a low temperature for only a short time, the fat of the cream has not become so cold as the milk serum, and the churning temperature may be set lower than when cream has been held at a low temperature for a long time. If the day is very cold the cream may chill as it is being churned, and this often is the reason why butter takes so long to come. If the temperature is taken during such a churning, it will be found to be lower than when the churning began.

There is no fixed rule for churning temperatures. Only practice can determine this. Use a thermometer, determine the best temperature for different conditions, and save patience, time and money.

How Much Fat in Cream?

It Depends on Temperature

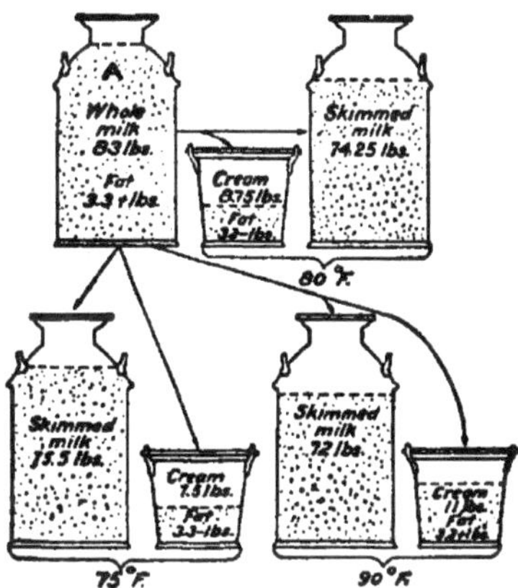

The amount of fat in the cream was about the same, no matter whether separated at 70°, 80° or 90°; but the weight of cream was greater when separated at 90°

Temperature of the whole milk has a direct effect on the percentage of fat in the cream and the skimmed milk. The illustration shows the average results of tests with five types of separators. The tests were made to show the effect of variations in temperature on the amount of fat in cream that was separated. These should interest every man owning a cream separator, as they will aid him in securing a greater profit with the same effort.

The temperature of milk being separated should be such that the milk will flow easily, facilitating rapid and thorough separation of the cream and the skimmed milk. It is a wise plan to separate the milk as soon as possible after it comes from the cow. In that case, the temperature is high enough that a thorough separation is effected. If the milk is allowed to cool after being drawn, the temperature needs to be raised to about 85 degrees or 90 degrees to secure the best results when separated.

Many dairymen think that there is an advantage in having the whole milk at a low temperature, because the cream possesses a higher percentage of fat when the temperature is low. However, in these tests the loss of fat in the skimmed milk was greater. It should be noticed that the weight of fat in the whole milk and in all three pails of cream was approximately the same, but that there was a distinct variation in the weight of the cream, and this is the cash end of the dairy business.

The richness of cream or the percentage of fat, derived from whole milk by use of a separator may be regulated by either the cream screw or skimmed milk screw.

The Hand-Fed Calf

The average man makes a mighty poor "wet nurse" for a calf. Too often he seems to think that the calf should be as hungry as possible at feeding time so that it will take a hearty meal. For that reason he lets it have but two big drinks of milk, night and morning.

That is a wrong and unnatural way of feeding a young calf. Watch nature's plan. The youngster takes a suck or two of milk at frequent intervals, changing teats now and then and often kneading the udder with its head. Note that it sucks the milk instead of drinking it, and sucking is a comparatively slow process. On the contrary, when the calf is fed by hand, the man who arrives with the pail of milk is usually in a big hurry, and so makes the calf hurry up the drinking process as much as possible; then he wonders why the calf suffers from indigestion, or at least fails to thrive.

Why Nature's Way is Best

What a difference there is in the appearance of the hand-fed calf, raised on the plan we have outlined, and the suckling calf at six weeks of age! The difference surely is an argument in favor of the natural plan of feeding. The natural plan gives the better results for the reason that the milk is sucked in small quantities, so slowly that it mixes at every swallow with the saliva poured into the mouth during the operation. The saliva renders the milk alkaline and properly prepares it for the fourth stomach which contains the ferment (rennin, of rennet) that curdles the milk. When a calf drinks fast the milk is swallowed without addition of the necessary amount of saliva and, entering the fourth stomach, is quickly coagulated into a tough curd which can not be easily digested by the calf. Indigestion of one form or another is the natural consequence. The calf may scour acutely and die, or suffer from less acute chronic scouring and fail to thrive. Or it may suffer from a sudden acute attack of indigestion characterized by a fatal fit or convulsion; or it may show no severe symptoms, but fail to grow and look as well as it should do or would do if properly fed.

Better Let Wife Do It

The lesson from all this is that it would be better, if possible, to have a woman do the calf feeding. If that can not be managed, then have it done by a mature, careful person, as young children do not understand the importance of careful feeding. Then try to feed milk at least three times a day, instead of twice, and from a self-feeder which will make sucking necessary. If that can not be done, then make the calf drink slowly and keep it stanchioned for some time after feeding. This alone will prevent a great many cases of severe indigestion, and most of the chronic indigestion so common among hand-fed calves.

Next remember that the fourth stomach, or true digestive stomach (abomasum), is the only one ready in the young calf to digest or care for food. The first stomach (paunch or rumen) is undeveloped at first and can not care for roughage. It therefore is a serious mistake to make the calf eat coarse fodder too early, in place of feeding sufficient milk and easily digested meals which go almost directly into the fourth stomach.

The calf fed on roughage becomes pot-bellied and stunted, and the growth which should be made during calf-hood, if not made then, can not be attained later. The proper feeding and developing of the young heifer calf really is more important work than the feeding of the adult dairy cow.

Just What Calves Shall Be Raised

Everybody says at once, "Mostly heifers." Of course this is wise. But if we are going to raise cows, and three-quarters of them sell for less money than they actually have cost us, what is the use? The fact is we need to raise something besides merely heifers. That a heifer at milking age has cost nearly or quite $75 is an undisputed fact today. It has been demonstrated time and again. And this doesn't include the time spent on the calf nor the interest on capital invested and in use.

Consider the Mother

The remedy is to follow common-sense rules like the ensuing, and not be led by sentiment even then. Certainly raise full-bloods if it is possible, and it is possible more often than some folks can realize. Remember, the daughter of a cow that is irritable, and subject to irregularity in milking every day or two because unduly excitable, will very rarely make a good dairy animal. Discard her. A man we know has just given away a two-months-old heifer because her dam has developed a tendency to become excited over trifles that make

her milk record fall every few days. He doesn't want another like her.

Physical Qualifications

Select a calf with a broad muzzle, one that is a hungry feeder and having a docile disposition if you wish to mature a cow that will produce the largest amount of milk or fat from a given quantity of food. The calf must be lengthy, full of vitality, and possessed of four good teats, evenly set at four corners equidistant. We know a heifer about to come in that has seven teats. If a real udder shows in a calf, so much the better. Good markings are worth while, but are not of supreme importance.

Pedigree is excellent, but it is not of so much value as high performance in the calf's direct ancestors. specially should the calf's sire and that sire's dam be high-brows. The calf's dam should be a choice cow, but her sire cannot be too well bred to insure her future. Also, if the calf's granddam and great-granddam on the sire's side were not cows of great natural individual ability, do not waste good money and time on the calf.

Lice on Farm Animals

Owing to the closer confinement of animals during the winter season, lousiness is usually more prevalent than in summer. Lice are of two general classes or kinds; namely, those with wide mouth parts adapted for piercing the skin—"blood-sucking" lice—and those adapted for feeding upon the hair and other products of the skin—"biting" lice.

Treatment is usually very satisfactory in ridding the animals of lice, but the difficulty comes in keeping them free. It is necessary, therefore, not only to destroy the adult lice and those hatching from the nits on the animals, but also to free the quarters in which the animals are kept.

Disinfectants to Use

Frequent spraying of the quarters with some efficient disinfectant will do much to prevent or limit the infestation. For this purpose creolin or other coal-tar disinfectants in five per cent. solutions, kerosene or kerosene emulsion, corrosive sublimate, 1-1000, or whitewash containing disinfectants, may be used.

Applications to the infested animals should be made two or three times at four or five day intervals, so as to destroy any lice hatching from nits since the previous application.

Creolin and other coal-tar disinfectants in two per cent. solution, or kerosene emulsion may be safely used. Kerosene emulsion is made as follows: Dissolve one-half pound of common soap in one gallon of boiling water; while hot, pour in two gallons of kerosene, care being taken to remove the vessel from the stove before adding kerosene, and stir for ten minutes; when ready to use add one part of this solution to nine parts of water and thoroughly mix. Smaller quantities can be prepared in like proportions.

The applications can be made by a cloth or sponge to the affected parts, they can be sprayed on from a sprinkling can, or, better, by a spray pump. If the hair or wool is long, better results will be obtained by clipping the animals before making the applications.

Baby Beef

A creep should be provided in the pasture so that calves may have access to grain without being disturbed by cows, as it is very important to start beef calves on grain before they are weaned. They may be kept in a separate lot into which the cows are turned twice a day, if this method is preferred. In this case there will, of course, be no need for creeps or anything else to keep cows from the grain, which may be fed at such times that the cows will not disturb the calves.

The calves may be started on a mixture of two parts shelled corn to one part of oats by weight. The oats may be gradually reduced until none are being fed at the end of eight weeks; but while this is being done a little old-process linseed oil-meal or cottonseed-meal should be added and the quantity gradually increased until it makes up about a seventh of the weight of the ration. On full feed calves should eat about two pounds of grain for every 100 pounds live weight, in addition to good roughage. Well-bred calves handled in this way should be in prime condition at the end of about ten or twelve months.

Water Helps Calves Grow

Every well-informed mother knows how important it is that her babe shall have water. No matter if he has plenty of milk, he must have water also, and the more milk he uses the more water is required by his little system. We know these things but forget that the same principle applies to babies in the animal world. I confess, myself, last year, to having been among the thoughtless. We attempted to raise a dozen calves. They had an abundance of skimmed milk, good feed, a good grain ration and care, but for weeks they did not do well. They seemed to be crying day after day for something more. Finally they began to cough, one and all of them.

Alarmed, I stated the case to one of the experiment stations, and was told that tuberculosis might be feared. But we had scarcely received the reply when my wife began to make inquiries. Almost the first question she asked was, "Do the calves get all the good drinking water they need?" It requires a woman to bring up young stock and make good animals of them. Manlike, I took exception to the question at once, and argued that "this talk about calves needing water is all moonshine. They need it just as sheep do. Did you ever carry a sheep a pail of water and see it turn away in disgust?" "But do these calves have water? I really want to know." This was her stand.

I had to admit finally that they did not. We were awfully busy and had forgotten that part of it. Then we began to supply water as well as milk, and it would have done you good to see them drink it and gain. Almost immediately the coughing began to decrease and growth to increase. I had learned one more lesson.

This summer every calf gets water twice a day, regardless of its age, sex or complexion. If he is a little baby, kept in a cool shed away from heat and flies, water is carried to him. And it is quite surprising the amount each one drinks. Even young calves will rarely refuse water, cold or warm, and right after warm milk. I am disposed to believe that want of plenty of pure water encourages scours and prevents benefit from good food, and so dwarfs the young animal.

Marking Systems for Animals

Did you ever hear of friends becoming enemies because one man's stock crept through the fence into his neighbor's field, or into the road, and the ownership could not be definitely established because it was not marked? Or did you ever know a man who spent his life in building up a good herd, only to have his efforts largely wasted at his death because his animals were not marked in such a way that they could be identified in the records he left? Such things often happen.

A good system of marking stock will prevent such things. Earmarking is used more than any other system. The tattoo method, Fig. 1, is the best for most purposes. It is simple, effective, permanent, painless, and does not disfigure the animals. An instrument something like a hog-ringer is used to tattoo numbers or letters in colored ink beneath the skin of the ear. It can be used on cattle, sheep and swine. Some breeders do not like this method for animals with dark skin, but red ink can be used for those animals, and the marks show plainly on examination. A set of figures which can be changed makes it possible to tattoo any number beneath the skin.

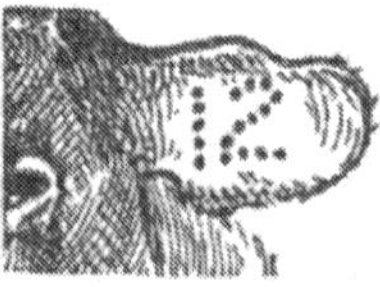

Fig. 1

Buttons, Fig. 2, are much used for swine, each button bearing a number and some symbol to establish the

Fig. 2   Fig. 3

animal's ownership. These metal buttons can be read from a greater distance than tattoo marks, and the animals do not have to be caught. Another style of metal marker is seen in Fig. 3. This is used for cattle, sheep and swine. All metal markers must be properly put in, else they will catch and tear out, leaving a disfigured ear.

Some cattle breeders use the tattoo method, and also a metal tag, Fig. 4, fastened on a strap around the neck. This is much used with the dark-skinned animals. The tag can be read at a distance and if lost, the tattoo mark is still there and can be read on close examination. It is unnecessary to catch and hold those animals and read the earmark unless the tag is lost.

What to put on the buttons, or what to tattoo beneath the skin, is an important question. With regis-

Fig. 4

tered stock, the herd book number may be used. Then the registry association can tell all about the animal in case the owner's records are lost or destroyed by fire. Where animals are not registered, the number on the ear must refer to the animal's entry on the breeder's record, where the sire and dam can be determined, as well as other points of interest.

A plan used with horned animals is to brand a number on the horn. This practice disfigures the horn, and when a horn is broken off, the number is gone.

Live Stock Shipping Association

The main purpose of these associations is to enable their members to ship in car-load lots to the central markets, instead of being more or less at the mercy of local buyers in disposing of a few animals from time to time. The fact that no capital is required for the organization of such an association makes these associations possible in communities in which more complicated forms of co-operation would not succeed. Such associations are scarcely practicable, however, in regions where there is so much live stock that it is generally marketed in car-load lots under any circumstances, or where there is so little that the association has practically nothing with which to work.

How to Organize

To organize such an association it is only necessary for the farmers of the community to meet together, adopt a simple constitution and by-laws, to elect officers, and, in turn, for them to appoint a manager.

It is recommended, although it is not absolutely necessary, that the organization incorporate. This can be done at a nominal cost—usually not more than $10. For this small expenditure of trouble and money, the association usually enables the farmer to market his stock when it is ready, instead of compelling him to wait until the local shipper is ready to buy it. He obtains for himself the benefits of the cheaper car-load transportation, and the shipments of the association realize for the owner the market price for his stock less the actual cost of marketing. In this particular it has been found that when thin stock, calves or lambs are sold in small numbers, the local price is usually very low. It is on this class of stock that the associations have been able to save their members the most money.

Stock Should Be Marked

In order to avoid misunderstanding, it is important that all stock be marked at the shipping point. This precaution prevents disputes in regard to shrinkage and dockage, and assists in making adjustments in case of loss or damage in transit. There are three common methods of marking. Numbers or other characters may be clipped in some conspicuous part of the animal, paint may be employed, or numbered ear tags used. The last method is the less frequent because it is somewhat difficult at the stockyards to get close enough to the animal to see the number on the tag. If the second method is adopted, ordinary paint is undesirable, especially for hogs, as it does not dry readily enough to prevent smearing. This difficulty may be overcome by using paint containing about one-fourth varnish. In the case of sheep, however, painting is objectionable because the marks will not scour out and wool manufacturers object to them, and branding fluid therefore is preferable.

Since no payments are made for stock shipped until returns from the central market are obtained, these co-operative associations may be formed without capital. All that is necessary is for the farmers to comply with their engagement to furnish the stock to the manager when, where and in such quantities as they say they will. In some associations a fixed sum of money is exacted from a shipper for failure to deliver stock to the manager as agreed. In every case the amount to be exacted should be reasonable, and should fairly represent the actual loss which it is estimated the association will suffer as the result of non-delivery. Fuller details in regard to the organization and management of such associations are contained in Farmers' Bulletin No. 718. Address Department of Agriculture, Washington, D. C.

Scientific Hand Milking

Its Importance Often Overlooked

Too little attention is paid to the subject of scientific hand milking. A poor milker may easily do enough harm in a herd of cows in one year to equal in loss the amount of his wages. In other words, it would pay to hand him his year's salary in a lump sum and "buy him off" instead of allowing him to milk poorly ten or twelve cows each night and morning. Such a milker, if he is rough, cross, noisy, unclean, irregular or imperfect in his milking, may quickly or gradually dry off the cows.

A School Would Be Good

We know of one case in which a beginner, in two months, completely dried off the milk secretion in the cow upon which he was allowed to practise. In another case a new milker by his roughness and harshness so reduced the milk flow that the owner had

Cattle and Dairy   **Hand Milking**

to fire him in self-defense. It probably is a fact that in every herd where the milk is not weighed night and morning and close tally kept, one or another of the milkers is doing indifferent or disastrous work. In Great Britain girls who are taking up farm work are learning to milk by practising on dummy cows until they become sufficiently expert safely to tackle the living animal. It would be well were our boys and would-be hired hands put through such a course of training to make them proficient without spoiling or injuring a cow or two in the process.

The Milker Often to Blame

Seeking the cause for many mysterious cases of intermittent garget experienced in some dairies, it must be suspected that the milker often is to blame. We think that incomplete milking is a possible cause, but one that is little suspected. The way a milker feels at milking time will in many instances determine the amount of milk he ob-

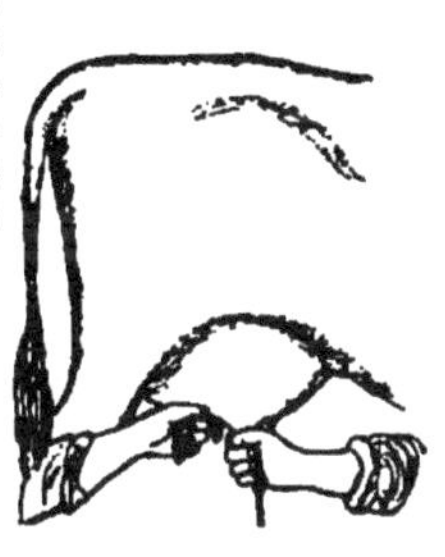

tains. If he quickly extracts it all, it will be well for the cow and the employer. If he is in a hurry, indifferent, tired or feeling sick, and does not strip the cow clean, slight, unexplained garget may result. If such work continues, the cow will soon show a serious shrink in milk, prove profitless or dry off entirely and have to be discarded.

It would be a good plan for every dairyman, especially in herds where slight cases of garget are prevalent, to have an expert strip the cows ten minutes after the milkers have finished. By this means some very rich milk will be obtained for use on the farmer's table and at the same time a check will be kept on the work of each milker, and some cases of garget possibly prevented. Knowing that the cows are going to be stripped the milker will, if conscientious and anxious to please, milk just as well and completely as he knows how, and so all concerned will be benefited. If he is the other sort of a worker he will be detected and discharged before he has done permanent damage.

Fast Milking Pays

The man who can make the milk fairly boil in the pail and raise a lot of foam usually is getting the maximum flow of milk from each cow; while the slow milker no matter how particular and faithful he may be, often fails to get all that the cow would let down to the fast milking expert. A change of milkers may have a good or bad effect. In one experiment, two equally proficient milkers changed cows and at once there was an increase in milk yield from each lot of cows. A change of milkers however, more commonly results in a decrease in milk production, and this sometimes is so noticeable that the accustomed milker has to resume his work with affected cows.

Feeding Corn Ensilage

Some farmers express disappointment at the results they have had from feeding corn ensilage to milch cows, and in most cases their disappointment is due to misunderstanding the properties of their feed.

They seem to have begun with the idea that corn silage is a complete feed and that cows need nothing else. For the benefit of the dissatisfied, here are a few tested and satisfactory formulas:

Corn silage, thirty pounds; clover hay, eight pounds; gluten-feed, dry (not meal), four and two-thirds pounds; cornmeal, three and one-third pounds per day.

Or, alfalfa hay, six and one-quarter pounds; or cow-pea hay, five pounds; or soy-bean hay, five and one-half pounds, in place of the clover hay, but with the silage, gluten and meal the same.

Or, to meet the needs of many others, use corn stover (cut), eight pounds per day, with the silage, gluten and meal; but add three pounds good wheat bran.

Or, if you use gluten-meal in place of gluten-feed, take about three to four pounds (depending on the grade), and add a couple of pounds more corn-meal.

Or, if preferred, in place of the gluten first mentioned, use two or three pounds cottonseed-meal; or, one and one-half pounds wheat bran.

The proportions mentioned for each of these suggested combinations make an average cow's daily portion. Some cows require less, some more.

Making the Babcock Test

When a man begins to think of testing his cows and keeping a record of them, he is getting on higher ground. Without recording the length of time a cow is in milk, her total milk production and its fat contents, no man is able to build up a great and a paying herd. The use of the Babcock milk-testing machine may be learned by anybody. It is a centrifugal machine which holds annealed glass bottles that are carefully gauged and with measurements marked on their necks. The process was invented by Prof. S. M. Babcock, who gave it to the world without patenting it to make money for himself, and it has made millions of dollars for dairymen.

To test milk, first carefully stir it from the bottom up, or pour it from pail to pail, but do not churn it. This is to mix it well and so get a true sample. As soon as it is quiet, suck up into the milk pipette more than enough to cover the mark, 17.5 cubic centimeters (c.c), cap the end with the finger and slowly let the milk drop out until its upper level agrees w i t h the mark. Then pipe it into one of the bottles o f t h e machine, where it will be safe from change, if needful, for a week. If the test is to be made at once, pipe in a similar amount of sulphuric acid, taking care not to get it on the hands or clothes, as it is a p o w e r f u l acid. W h e n putting it into the milk, let it flow down from the inside of the bottle and not run directly into t h e milk, as this will blacken or burn the curd and prevent a clear reading. Acid and milk should be at 60 degrees temperature to produce clear readings. Buy acid with a specific gravity of about 1.82. As soon as the acid is added, take the bottle by the neck and gently swirl the contents until they are thoroughly mixed. The curd must be fully dissolved. Then close the machine and whirl the samples for five minutes at a speed of about 700 to 1,200 revo-

Pipette ——→

17.5 C.C.

Whole milk bottle.

17.5 C.C.

Acid jar.

lutions per minute. Next fill each bottle to the base of the neck with hot water and whirl for two minutes more. Then fill to about the seven per cent. mark and repeat the whirling for two minutes. The measuring of the fat must be made while the sample is hot. Measure from the top of the curved upper level. If the fat extends from 0 to 4 in the neck there is just four per cent. fat, or four pounds of fat in 100 pounds of milk. If it should run from two to seven, the amount is five per cent. The scale is graduated so that tenths of pounds are as easily read as full pounds. A little practice with the machine will readily make any boy an expert in its use.

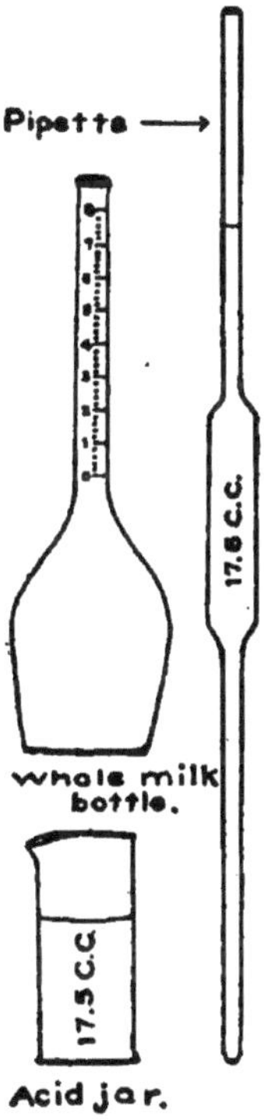

When testing milk it must not be forgotten that the fat contents do not measure the exact butter production. For instance, if milk is four per cent. fat, it should make about four and one-half pounds of butter, because in all butter there is some water, salt and minute parts of other things, like ash. If there was no loss in churning, and the overrun were just sixteen per cent. (the law forbids it to be more), the amount would be four and sixty-four one-hundredths pounds. The buttermaker who is getting but 109 or 110 pounds of butter from 100 pounds of fats is not doing so well as he should. The loss of fats in churning should never exceed one and one-half per cent. in the buttermilk, and may be less.

Any dairyman who does not own and operate a good Babcock milk tester and keep records of all his individual cows, should not complain if his purse tells him that "farming doesn't pay," for in all untested herds are cows that eat up the profits which should go to the owner.

Love of Horses and Cows

I have frequently read of the importance of being kind to cows and horses; to speak quietly and avoid exciting them.

From practical experience with cows, calves and horses, I have come to the conclusion that they recognize what is required of them by the tone of the voice alone until they associate certain words, short ones, as meaning a definite action on their part. The fewer words the better.

Curing a Kicking Cow

I can illustrate the method I have found to be most satisfactory by giving instances. For thirty-four months I have had entire charge and care of two cows that were calved on the premises. One was cross and turbulent. She was broken to milking by placing a cord around her so as to catch her flank, and that would pinch her at any attempt at kicking. She soon learned that the pinch was her own doing and stood fairly quiet; but if the cording work was forgotten, her milk did not reach the separator.

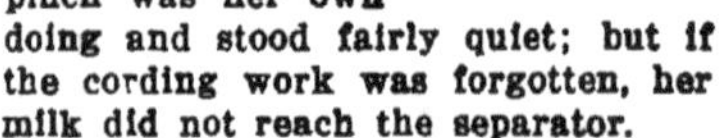

The other cow I broke to milking myself, but she had a long reach and I used a device I saw in use on the farm I lived on in Yorkshire, England. It consists of a soft rope, having a loop at one end and a cross-piece of wood to fit the loop when twisted around the hocks.

This cow expects the movement of putting the cord on and places her feet together as a part of the operation.

Making Milking Comfortable

Bathing the udder with some warm water, taking care to dry it perfectly, and in cold weather using a small quantity of vaseline to prevent chapping, and making the action of tripping more agreeable to both man and cattle, are what I call kindness to animals.

A Remedy for Bot-Worms

More important, perhaps, is to mention a safe and sure remedy for bot-worms in horses. You may never have heard of the following method: Obtain the young leaves of English walnut trees—they are plentiful in Pennsylvania; dry them carefully in a moderately hot oven so as to scorch them; pulverize them, keep them in an airtight jar and give a teaspoonful in the grain feed once a day— morning feed. In a few days if there are bots they will pass out in clumps and the horse will get the good of all his feed. There is no reaction from this medicine. I have used it for years in America and have known it in use in England for seventy years; probably in use long before my time.

How to Find the Robber Cows

The farmers and dairymen of the United States are milking 14,000,000 cows twice a day, spending good money to feed them, devoting many hours to their care, and yet are not getting one cent of profit out of them.

Allowing ten minutes for the milking and care of each of these cows 700 times a year, it is found that on the average every farmer in the United States is spending 72.2 days in work for which he gets not one cent of profit.

There is just one way to account for the fact that we keep on feeding and milking these 14,000,000 unprofitable cows—which is, we do not realize that we are doing it.

We are milking good and poor cows alike, giving them the same degree of care, not knowing that part or all of the profit earned from part of the herd is being wasted because some of the remainder are not even paying for their feed.

There is only one way to find out which cow is a profitmaker and which should be fattened and butchered without delay, and it is not hard to learn and practise. That way is by the Babcock test, the scales and the milk sheet.

By this method the poor cow can be detected very quickly, and at the same time the milk and butter production of each cow is known at the end of each month.

Not only are poor cows found, but the good ones as well, so that the owner knows from which cows to save his heifer calves. When this is accomplished he has taken a big step toward doubling the production of his herd.

Weigh the Milk

For the quick and convenient weighing of milk there are many kinds of scales on the market which will do if they weigh correctly. The handiest kind is shown in the illustration.

It costs about $2.50, will weigh up to sixty pounds, and is substantial enough to keep accurate for many years. To weigh each cow's milk is the work of a few seconds. It takes but a second or two more to record the weight on the milk sheet.

The sheet is so ruled that each cow's name appears at the top of her own column, and each day's milk yield, both morning and evening, can be put down. At the bottom of the page are spaces for the month's total of pounds, the average butterfat test and the total pounds of butterfat, which is found by multiplying the total of milk by the percentage of butterfat shown in the average test for the month. By the use of this sheet each cow's exact work is known. There is no guesswork about which cow is the good cow and which is less profitable. No sane man can afford to keep cows and not find out by weighing the milk which are the good and which are the poor ones.

The Scale is the Foot-Rule of Stock Raising

Something like ten years ago we put in our platform scale. After we had had it for a short time we were only sorry we had been without it so long. There is a pleasure in dealing with certainties. I do not believe with the Psalmist, in his haste, "that all men are liars." Neither do I believe that all men are dishonest. The laws are not for those who obey, but for those who have not been educated up to the point of obedience. Scales make for honesty. "To err is human," and we often err when such a thing never once entered the mind.

Don't Guess—Weigh It!

It is one thing to say that it is a ton or that an animal weighs 100 pounds. You may say so and then proceed to check your weight. It may be right, it may not. The scale will tell you. Definiteness gives a man poise. Resolving to be accurate resolves itself into dollars and cents, and that is the acid test today! How much is there in it for me? asks the man you want to hire.

Weigh the Rations

The successful farmer, as well as the successful man in business, is first of all methodical. He has a system worked out that he follows, even though you may not know it. The Babcock tester has helped him weed out the cows that are not 100 per cent. efficient. The weighing of milk has stopped leaks and the man who works with these things and is all the time studying, begins to do a little experimenting in feeding. And in doing so he comes to know about "balanced rations." To feed these he needs to get the right proportions, and here a reliable scale comes in. Or the man may have a poor field. He gets the soil tested and the analysis reveals its needs, in order to grow a certain crop. He needs a scale to get the right proportions of the elements in the fertilizer that are required. If he has a scale, he can feed certain animals certain kinds of feeds, and weighing before and at the end of the experiment see if they have gained or lost weight.

Buy a Good Scale

My scale is large enough for all kinds of weighing. The pit was made of brick, but concrete does well. The scale was located handy to the various buildings, in a spot that was high and dry. Care was taken to have the foundation level and firm. The approaches were plank. At its installation it was so delicately adjusted that a silver dollar thrown on the platform affected the weighing beam. It has never cost a cent for repairs.

Ten years is a long, long time and many changes and improvements have come, and with it a pitless platform scale. All that is necessary is to place the scale on firm ground and supply inclined approaches and it is ready.

Cattle and Dairy   **Use of Scales**

Poor Scale Worse Than None

A worthless scale is worse than none. Unless you can accurately weigh and know you are right, better not weigh at all. That means to get a reliable scale, well made and finely adjusted so that it will stay in repair. Most scales will weigh pretty well at first, but unless they are made to last they lose their accuracy and are a snare and delusion to their owner. A farm scale is subjected to hard wear and tear in the natural run of affairs, and it must be built to stand hard knocks.

Now we don't know any more about a scale than we do about a diamond, because it isn't in our line. So the only way to do when we want to buy, is to go to a firm with a reputation and character, and depend upon them and let them put it in. They have made a specialty of scales just as you have of certain farm lines, and know how. A firm stakes its reputation on the character of its goods. A reliable firm will not betray your trust, for it knows that a pleased customer is the best advertisement.

Box Stalls and Slatted Mangers for Horses

No Animal Likes to Be Tied Up

One of the blessings of modern barn architecture is that more and more box stalls for tired horses are being erected. Light, airy, slatted box stalls are a joy and comfort both to owner and animal.

When we bring our animals in tired and sweated, instead of almost criminally tying them up in a close, dark, partition stall, we turn them loose in knee-deep bedding, remove head-stall, collar and harness, and then they eat their noonday meal or rest at night in cool comfort. A stuffy tie-up stall never was intended for anything but punishment to the hard-working brute creation. Out with it!

We have the bottom of our horse hay-mangers slatted, so that all dust, dirt and waste filters through and works into manure with the bedding. A few days ago we found in a newly-built barn tight-bottom mangers with a trap door which lifts up in order to sweep all dust and dirt out. This is a pretty good idea when we know what disagreeable tasks we get into trying to plow and harrow with heaving, short-winded animals fed in dusty, dirt-ridden mangers.

The Stable Floor and Other Matters

How is that stable floor? Has it been pawed out in front and is there a space under the manger? A horse, the most ambitious animal when up, is the most helpless when down, Many a fine animal has lain down naturally enough, and in some manner, when trying to rise, has forced himself forward under the manger and been found dead or badly injured in the morning.

All our folks know about this danger, but in the mutiplicity of things to be looked after it is sometimes forgotten. Let us consider the manger. To begin with, every horse ought to eat off the ground. If the bottom of the manger is on, or within two inches of the earth or floor, no horse can get under it, and where the knee touches the manger when the animal paws it will rarely be continued. Standing upon an earth floor is good for the feet and necessary for some horses, but such a floor must be leveled often.

Watch the Coat

A horse's coat is a good indication of his condition at this season of the year. If it "stares," or looks rough and unkempt regardless of the daily brushing, he is not fully nourished and needs a change of feed. A molasses addition to the ration, of say a gill or half-pint twice daily, or a small handful of oil-meal gradually increased to a pint twice a day, or two quarts of potatoes or apples twice daily, will presently work wonders in his appearance and spirits. A warm bran mash once a week is also good.

Beware of Icy Spots

Do not run risks on icy spots. If the road is mostly bare with perhaps but one or two places of ice, it is better to carry half a peck of sand and make a gritty path for the horse rather than force him forward with dull shoes.

Dentist Work

I have seen two horses of late that plainly said they had ground their teeth to a sharp edge and now were suffering with sore cheeks due to laceration. It is easy to smooth down the sharp and rough outer edges of the molars when they get in this condition. Gently draw out the tongue, hold it first on one side and look carefully at the teeth next the cheek and then next the tongue. Then hold the tongue on the other side and repeat the inspection. Let the sun shine into the mouth so you can see plainly. Next, after giving the horse a little rest, take a sharp file by the handle and rasp off the troublesome sharpness. If you doubt your ability to do this, employ a veterinarian.

Systems in Harnessing

Harnessing and unharnessing necessarily take up much time on every farm. But, on some, time is wasted thus that might be saved. To stop this extra labor let a carefully planned system be followed by all who handle the teams, in both harnessing and unharnessing, that every man shall know exactly where to find each strap and snap. Speaking of snaps, these useful little things need looking after once in a while, to see that they have not gotten out of order and so are ready to fail and perhaps make trouble and expense.

The Horse in Spring

In March we find thousands of farm work horses in poorer condition than usual. Much poor corn was fed to the horses, for oats have been high in price, and such feed has not balanced the ration with timothy hay. Lucky is the man that has had good clover, mixed clover or bright alfalfa hay to feed his idle horses. That kind of hay is rich in protein, so that less grain is needed with it, and it makes a balanced ration with corn of good quality.

With roots and sound oat straw as a "side dish," the clover and alfalfa hay-fed horses have wintered well and will go into spring work in fit condition; but the "hay belly" horse that has been distended with marsh or slough hay, which should have been used for packing iron castings or crockery instead of bluffing the belly of the beast, is in sorry shape for work. At once he should be put upon a ration of whole oats and one-ninth part of wheat bran, fed wet, but not in the form of a mash. If a few ears of sound, old corn can be added at noon, so much the better, and the hay should be the best obtainable. Prairie hay is suitable, or fine timothy hay will do well for spring feeding.

Special Menu for Thin Horses

To the very thin, hide-bound, weak horse that shows a tendency to collar and saddle sores, we should feed blackstrap molasses, along with cornmeal and wheat bran in the way set forth in "Horse Secrets." The horse may have to be starved to the molasses feed at first, but soon will take to it with relish.

In getting the horse into shape for April and May work feed small quantities of feed often. If molasses can not be had, and coarse brown sugar can be had cheap, gradually add some of it to the ration. The horse likes it, and it not only is digestible but helps to drive worms out of the body. Raw potatoes have the same effect if fed to young colts, but we do not believe in feeding them to the horse that has to work hard.

Treatment for Worms

If it is known that worms are present, and they are seen to be doing harm, do not trust to feed for their removal. Mix together equal parts of salt, sulphur and dried sulphate of iron (copperas), and of this give the

Buying Horses | **Horses**

adult horse a tablespoonful night and morning in his feed for a week; then skip ten days and repeat. Omit iron for a pregnant mare and increase salt and sulphur. Have the teeth put in order by a veterinarian, if that was not done last fall.

Things to Watch in Buying Horses

In buying horses for spring work look at the shoulders closely. If no sore spots or lumps or scars are seen, handle the muscles thoroughly. Reject the horse at once if a hard mass (fibroid tumor) is found where the collar will bear. Such a tumor swells when the collar irritates it in spring and the only remedy is to have it cut out. That means expense and retiring the horse from work when his help is most required.

Also reject a horse that has soft baggy swellings on the shoulder, under the collar. These lumps are infected by the organism known as botryomyces (a fungus) and the only cure is to have them cut out. A small red sore forms in the center of such growths when the collar irritates, and the horse consequently is unable to work well. Avoid, too, the horse that will not let you handle his neck, where the collar rides, without rearing and biting or kicking. It is certain he will have a sore neck and have to be laid up when he is most needed in summer.

Give the Colt's Hoofs a Chance

"No foot no horse." This is an old proverb, but not less true because of its age. It is not very hard to put the middle on a horse after he is grown, but it is difficult to put good

feet on him if they have been neglected until that time. If left alone, the feet of colts will not always develop as they should. The toes or heels may need to be lowered in order to straighten the axis of each foot. Note that in the illustrations there is a break in the axes of the first two feet shown, while in the third foot the line is straight from the fetlock joint to the base.

Hot Weather Care of Horses

A horse sweats profusely. His skin is far more richly supplied with sweat-glands than is that of man. The flow of sweat calls for much water in the system. If it is not provided the sweating tends to stop suddenly. The horse "peters out," says the driver, or the "sweat stops on him," he adds; and when that occurs the horse is in great danger of his life. The skin becomes dry and hot, the horse lags or stops, pants, shows red injection of the membranes of the eyelids, nostrils and mouth, staggers, has extremely high fever, often bloats high up in the right flank, and soon falls in a stupor or becomes delirious and dies.

In such cases indigestion practically always is present. That should have been noticed by an observant driver, and the manure would have given the hint, if looked at carefully, as always should be done morning, noon and night, more particularly in hot weather.

Watch the Manure

The state of the manure indicates with absolute fidelity the condition of the digestive apparatus. The manure should come in bright golden-yellow balls, if the horse is fed on oats, and should have a clean, earthy and not unpleasant smell. If the manure is clay-colored, slimy and foul-smelling, that indicates derangement of the liver and digestive organs; so does scouring or constipation and dark color of the manure, or foul odor, or failure of the manure to form into balls, or if it is very hot, steaming and mucus-covered. The slightest change from the normal condition of the manure should put the feeder on his guard, and usually it is necessary to reduce the rich feed or to withhold it for a time.

What to Feed

Do not feed heavily on corn during hot weather. Allow sound old oats and feed them whole with the addition of a ninth to a sixth part of wheat bran, by weight, according to the state of the bowels. Dampen the feed. Remove all that is not eaten up clean. Give the horse a pound or so of hay when cooling off at noon. Then allow drinking water and then the grain feed. Give only a sip or two of water as the horse comes into the stable from hard work in hot weather, and do not allow him to go to the trough on his way to work when his stomach is distended with feed. Take drinking water to the field and allow a little of it often during working hours. Many humane farmers are doing this. They put a barrel of cold water on a low wagon or stone boat, put in a floating top, cover up with a thick blanket, and water the horse at frequent intervals.

In Case of Heat Exhaustion

If a horse shows any symptom of heat exhaustion, stop work, take off the harness, stand the horse where there is shade and a breeze, sprinkle his body with cold water from a sprinkling can held a foot or two from the skin, put cold wet swabs (not ice packs) on the poll of the head; or, if gravitation water can be used, tie the hose to the brow-band of the bridle and let a stream of cold water trickle over the head. Give stimulants internally. Coffee will do, but two or three ounces of a mixture of two parts alcohol and one part each of aromatic spirits of ammonia and nitrous sweet ether (sweet spirits of nitre), in a quart of cold water, will be better. Repeat the stimulant in half dose, once an hour until the horse revives.

Care of New-Born Colts

Many colts are arriving at this season of the year, and we wish to remind our readers of the necessity of proper care for both mare and foal. Thousands of foals die during the first two weeks from infection of the navel, causing formation of pus abscesses at that point, internally and in the joints. The infection is usually derived from dirty bedding and floors, or from a string tied around the navel.

Have the colt born on grass by preference. If that cannot be arranged, prepare a special box stall by cleansing, disinfecting with four ounces of formaldehyde to each gallon of water, and then applying fresh whitewash to floor, walls and ceiling. Have the stall well lighted and ventilated. Bed it with fresh, clean straw or planing-mill shavings.

When the foal is dropped, cleanse the navel, squeeze out any fluid present and at once immerse in tincture of iodine contained in a wide-mouthed bottle. Now dust with finely powdered screened slaked lime, repeating the application twice daily. Use iodine again if the navel cord is slow in drying.

Do Not Use String

If possible, do not tie a string around the navel cord. Sever it with an emasculator or ecraseur. If a cord must be used, soak it in a five per cent. solution of coal-tar disinfectant or a 1-1000 solution of bichloride of mercury. Tie it an inch or so from the body of the foal and place a second string an inch lower. Scrape through the cord between the strings, immerse the stump in tincture of iodine, and remove the string just as soon as it is seen that the blood will not flow. To that end it is well to tie the string with a bow-knot, which may be loosened to see if the blood will flow, and removed if it does not. Saturate the stump of the navel with tincture of iodine after squeezing out the fluid, and dust with slaked lime as already advised.

Disinfection is Essential

Leaking of urine from the navel commonly follows application of a string and failure to remove it promptly. If leakage starts apply a 1-500 solution of bichloride of mer-

Horses   **Harnessing**

cury once daily and then dust with starch powder. If this does not suffice, the leaking orifice should be cauterized with a lunar caustic pencil, or a red-hot knitting-needle. Some veterinarians prefer to inject a few drops of turpentine. Infection generally follows leaking of urine, so this should be promplty stopped and the disinfectant used.

The foal's bowels should move promptly after birth. The first milk of the mare has a purgative effect for the purposes of removing the first feces (meconium) from the bowels. If the movement does not take place promptly, inject warm sweet-oil, warm cream, molasses and water, or a few ounces of warm water containing one tablespoonful of glycerine. The latter mixture is particularly effective. If injections do not bring away the feces, the latter may be found blocking the rectum in a firm mass. This must be broken down with the fingers after injecting warm sweet-oil, or forceps may have to be used to bring it away. Afterward give one to three tablespoonfuls of castor-oil shaken up in milk, if the bowels do not move naturally. This must not be delayed.

Harnessing 3-Horse and 4-Horse Teams

One of the best ways to make more money and get around the shortage of labor, is to use more horses per man in seeding and plowing. Instead of using two two-horse teams, use one four-horse team and one driver. In every case where this plan is followed and colts are used, take every precaution that the lines are properly strung and all the tie-straps securely fastened.

A three-horse team works to good advantage on a fourteen-inch plow in heavy stubble or sod; the extra horse lessens the pull and the horses stand up better under the work. Three-horse teams are much used in harrowing, too, so some young drivers may want to know how to string the lines. There are two ways of doing it. In Fig. 1 two lines are used and the outside horses are tied to the middle horse's hame by use of the hitch strap. If three lines are used, string the lines on two horses just the same as if that team were to be used alone. Then tie the third horse to the outside hame on left horse. Snap the outer check of the third horse's line, but allow the inside check to hang

Harnessing   **Horses**

loose. Take this line in the left hand, making two lines for the left hand and one for the right.

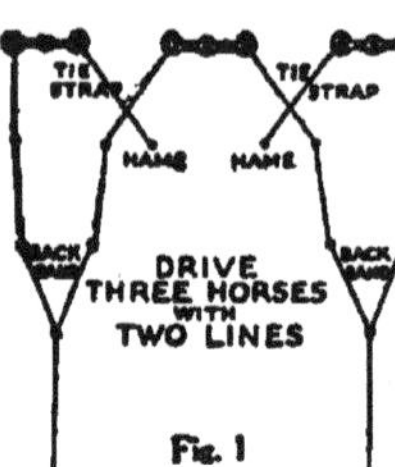

Fig. 1

For a four-horse team either two or four lines may be used. Fig. 2 shows how to arrange four lines, two lines for each hand. The four-horse team is used most on the disk and harrow. It may be advisable to use spreaders for the lines at the hames. The outer horses are tied to the inner ones with tie-straps.

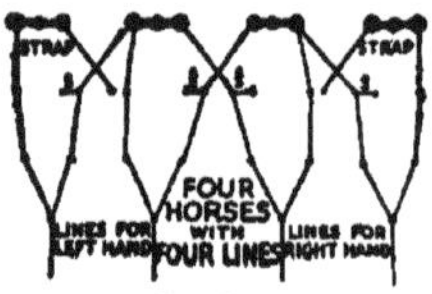

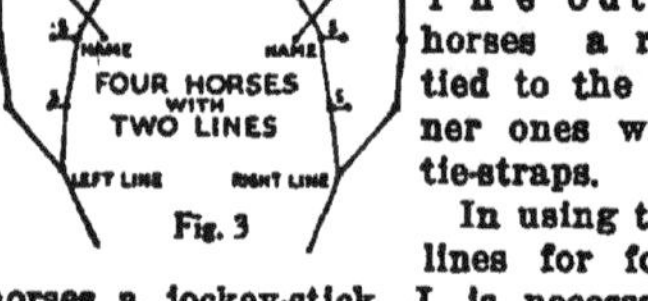

Fig. 2

Fig. 3

In using two lines for four horses a jockey-stick, J, is necessary between the middle horses. This can be made of a broomstick about three feet long, with a snap on each end to snap into the bits. Spreaders, S, are used on the hames and back-bands, as in Fig 3. Three-horse and four-horse eveners will be necessary for these hitches, and can be bought at most implement stores.

The length of the tie-straps may have to be changed after they are tried. The only way to get them right is to use judgment, and then shorten or lengthen the tie, whichever is necessary. If a horse is nervous and tries to go ahead of the others, tie him shorter. Put him where the lines will give the best control of him, so he will not forge ahead and wear himself out. With a four-horse team control with two lines is never quite so good as with four, and if colts are used four lines are advisable. The disk or harrow is a poor place to use an unbroken team; always have two steady, well-broken horses if the others must be colts.

The use of more horses per man not only solves one part of the labor problem, but it helps to cheapen the cost of horse labor by putting all the horses to work. The horse that works 1000 hours a year and costs $100 to keep is cheaper than one which costs $75 to keep and works but 500 hours. The former costs ten cents an hour worked, and the latter fifteen cents an hour.

Hogging Down Corn

When corn climbs toward the two dollar mark, many hog raisers are tempted to cut down the amount of grain, even though it means a poorer finish on the spring pigs. A more logical solution of the problem would be to give a full feed, but to give it in an economical way that brings good results. Many experiments, together with the testimonies of practical farmers, show that such a solution is to be had by "hogging down" the crop.

Advantages

What are the advantages of this practice? Unquestionably the saving of labor connected with husking and handling the grain is one of the greatest. The scheme takes the "backbone" out of husking and puts it into feeding. Husking and cribbing are dispensed with. Shoveling into the crib, back again to the wagon when the corn is hauled to the feed lots, and a third time to the feeding floors is eliminated. The same saving is effected in the handling of manure, there being no need of hauling and scattering it. The fertility of the soil is built up. Storage charges on the grain are reduced to a minimum and the rats are cheated out of their customary toll.

Quality Just as Good

The quality of pork is just as good and the gains more economical than when the corn is hand fed, although best results are secured when the hogs have forage along with the grain, preferably some crop seeded in the corn at the time of the last cultiva-

tion. The animals are healthier because of the exercise, than when the shotes are fed in a dry lot. Many weeds are cleaned up by the hogs. Fall plowing may be done, for the hogs will gather the corn quickly and tramp the stalks down.

Every good thing has its bad features, to be sure. It is true of "hogging down" corn. Here are the things that can be said against the plan:

Disadvantages

It is claimed that the soil will be packed by having the hogs running on it. This objection may have some weight in very wet weather, but in dry weather it is doubtful whether the hogs do any more damage than the other animals usually turned in after the corn is husked. The loss of stover is another objection. Generally a certain amount of stover is left in the field, anyway. If the corn in those fields is harvested by hogs, the stalks will decay much sooner. The slight waste occasioned when heavy hogs are turned into a field is cared for by turning in some shotes.

Now you have both sides of the question. Does it appeal to you? A man out in Iowa said, "No" to that question several years ago, but he was persuaded to try the plan. After a trial he observed that his ground soon produced better corn in those fields where the hogs were doing the harvesting.

If the plan looks good enough to try, fence the field hog tight. The larger fields need to be divided so that the shotes will not have too large a field to harvest at one time. Eight spring shotes to the acre is the customary rule, and it is well to have a field cleaned up in ten or twelve days. Fences of woven wire, if fastened to solid corner or end posts, may be attached to the cornstalks through the field.

Lice on Hogs

The farmer should frequently examine his hogs about the ears, flanks, and inside of the legs to see if they are lousy. Lice are common pests among swine, and vigorous and persistent treatment is required to eradicate them. They may be readily seen traveling among the bristles, particularly in the parts just mentioned. The eggs, or "nits," are small white oval bodies attached to the bristles. Dipping does not as a rule destroy the vitality of these eggs. Swine should be dipped frequently in order to kill the lice that hatch out of the eggs after the previous dipping. These lice are blood-sucking parasites, and by biting the hog and sucking blood they cause a great deal of skin irritation. Furthermore, they act as a drain on the vitality of the hog, through the loss of blood which they abstract. When lousy the hog is usually restless and rubs on posts and other convenient objects. The coat looks rough and harsh. This pest is transmitted from one animal to another by direct contact, or by contact with infected bedding or quarters.

Dipping Vigorously and Persistently

To free hogs from lice they should be dipped two or more times at intervals of about two weeks. Several dippings may be required before complete eradication is accomplished. Do not fail at the same time to clean and disinfect thoroughly the sleeping quarters. Cresol compound (U. S. P.) may be used for dipping and disinfecting. For dipping mix in the proportion of two gallons to 100 gallons of water; for disinfecting, in the proportion of three gallons to 100 gallons of water. Although not always so effective as might be desired, coaltar products of the kind ordinarily sold as stock dips are commonly used to treat hogs for lice. For use they are diluted with water in accordance with directions supplied by the manufacturers. Cresol compound and coaltar dips may be purchased at the drug store.

Dipping vats are made of various materials, but the most durable is cement.

All-the-Year Hog Pasture

Pasture is a pig's paradise. The longer the pasture lasts, the bigger will be the paradise. If the pasture lasts all the year, the pigs will have an everlasting paradise until they become pork. Here is the way to have a lasting paradise for sixty to seventy-five head of hogs:

Divide the hog pasture into four parts having the relative size of the four lots shown, and put a watering trough in the middle. Handle the lots as follows:

Lot 1: Sow with rye late in August or early in September, about one and one-half to two bushels of seed to the acre. This will do to pasture from December to May. In the spring plow the rye under and sow soy-beans for pasture the following September and October.

Lot 2: This lot needs no attention until early next spring, when oats and rape should be seeded on it to furnish pasture for the hogs as soon as they are removed from the rye on Lot 1 in May. This will keep them supplied with pasture till the last of June. It can be pastured again in August until the soy-beans are ready on Lot 1. Then when the hogs are placed on the soy-beans, Lot 2 can be plowed under and seeded with rye which will do to pasture in December, and until the following May.

Lot 3: This lot should be seeded with rape about the first of May. It can be pastured from the last of June until the middle of August when the hogs are put back on Lot 2. It can be pastured again when the hogs are through with the soy-beans on Lot 1, until the rye is ready on Lot 2.

Lot 4: This is a permanent pasture of alfalfa, and should be seeded the latter part of August. This lot is used as a sort of a reserve pasture for times when other lots might not be ready for use, or could not be used very conveniently.

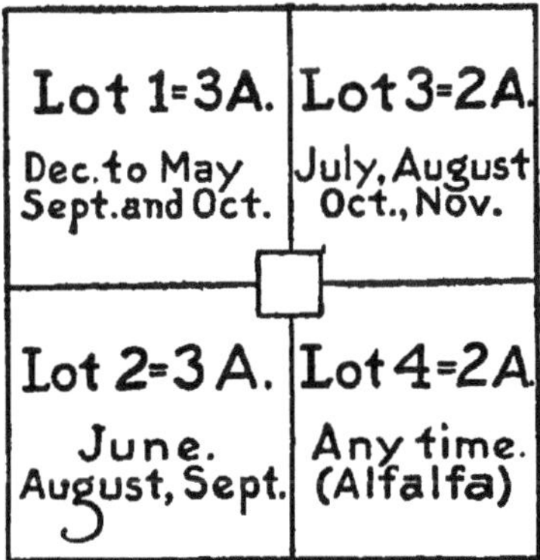

Individual Hog-House

The new man from the Simpson place drove past his neighbor's farm with five sows in the wagon one morning in early April. His neighbor was fixing the fence at the foot of the hill, just where the new man stopped to rest his team.

"Got them from the Jones' farm," came from the driver in answer to his neighbor's question. "I don't know where I'm going to put them, though."

"Well, what you ought to do is to make some movable sheds, one for each sow, and pull them out into the lot so the sows can farrow undisturbed. I use them almost entirely, they're cheap. Drive on up the hill, tie your team and come and see them."

"Now, here's one in use," he continued as they walked about the hog lot. "This sow farrowed out here three days ago, all alone. Saved every pig."

"Looks as if those would be easy to make," observed the new man.

"Easiest thing in the world. You can make one in an hour or two. See, you take two pieces of 4x4 eight feet long for runners and build a floor on them, using two-inch planks for floor. Then use a 2x4 nine feet long for the ridge, and 2x4's eight feet long for the slanting studs, making an A-shaped frame. Put the roof boards up and down, and hinge them on one side for two doors which can be opened toward the sun on bright days. Between the upright pieces of 2x4 you can make a door at one end. I've made mine to slide up and down."

"What would five of those cost me?"

"That depends on what grade of lumber you use. I'll give you a list of the lumber for one of them, and you can take it to the lumber man for his prices."

After figuring on the roof of one of the houses for a few minutes, he produced this list, which he copied on the back of an envelope and handed the new neighbor.

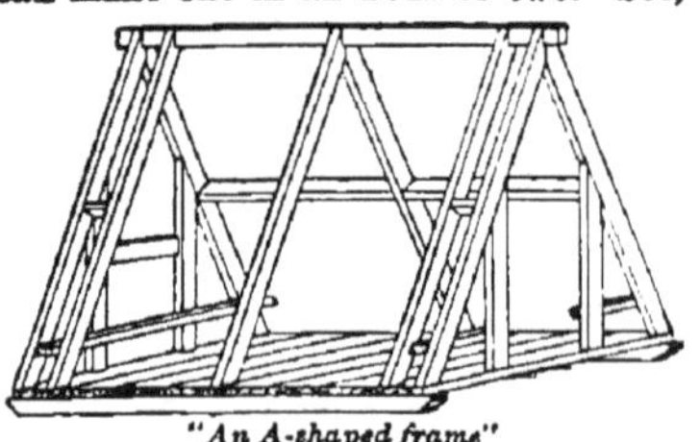


```
 2 pieces 4 x  4,  8 feet, for runners
     "    2 x  4,  9  "    "  ridge
 2   "    2 x  4,  8  "    "  studs
 4   "    2 x  4,  4  "    "  end studs
 1   "    1 x  4,  2  "    "  above end door
 1   "    2 x  4, 10  "    "  fenders
 9   "    2 x 12,  6  "    "  floor
11   "    1 x 10, 16  "    "  shiplap for roof
 4   "    1 x 10, 12  "    "  shiplap for ends
 4   "    1 x  4,  8  "    "  door battens
 3 pairs  6-inch strap hinges
 1 pair   4-inch strap hinges; nails.
```

Self-Feeder for Pigs

"It's too wet to work in the field this forenoon," said John Smith's hired man as he pushed his chair back from the breakfast table. "How about those self-feeders for the pigs?"

"Fine," came from the head of the table. "There are some short pieces of lumber overhead in the double crib. The pigs are old enough to wean now, and if we use self-feeders it will take less time to put them on the market."

"What difference does a little time make to a hog, uncle?" This time it was a cousin from the city who spoke.

"Maybe the hired man will explain that while you help him check up on the lumber. Take this list and help him to find all the pieces."

Together the hired man and the city boy went over the list:

```
2 pieces 1 x 6,  3 feet for end-crosspieces
2   "    1 x 12, 2  "  9 inches for end-uprights
11  "    1 x 3,  3  "  shiplap, top and sides
1   "    1 x 12, 2  "  10-in. for front sloping board
1   "    1 x 10, 2  "  10-in. for front sliding board
1   "    1 x 3,  2  "  10 in. for back sloping board
1   "    1 x 3,  2  "  10 in. for front of trough
1   "    1 x 4, 12  "  for cleats
1 pair hinges, 2 wing-nut bolts.
```

"There's enough lumber for three self-feeders," said the hired man, "but the hinges and bolts are missing. What kind of feeders do you want?" John had just come into the driveway.

"The one-way kind; nothing elaborate. Here is a drawing, Fig. 1, the county agent gave me, showing a cross - section of one, with all the dimen-

Fig. 2

sions you'll need except the floor, which is fourteen inches from front to back, inside. Fig 2 is the finished feeder. The top is hinged on at H, Fig. 1. The feeder is three feet long. First cut the end cross - pieces and nail a cleat on the inside of each to support the floor. Put the floor boards in. Cut the top of end-uprights to slope one inch toward the front, so water will drain off the top, then mark the position of deflecting boards on the end-uprights. Nail on the back, put in the sloping boards and nail on the front. Fasten the top together by 1x4 cleats and hinge it. Make quarter-inch slits in the front sloping board for the wingnut bolts. Two 1x2 cleats at each end will hold the slide snug. Finally, nail on the eight-inch board at front of trough, and we'll have our cousin from the city paint them. Then they'll look something like the manufactured feeders offered for sale."

"Each one will hold about five bushels," commented the hired man, as he nailed the floor in place. He put a handful of nails into his mouth. "Make-less work-at ch-chore time," came from between the nails.

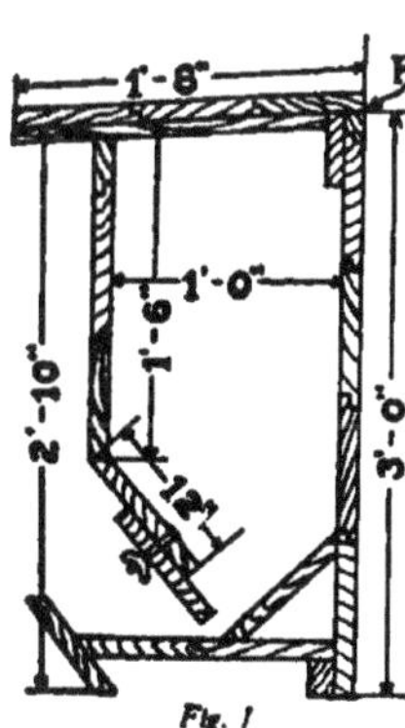
Fig. 1

Winter Pigs Without Milk

The average farmer has faced the problem of how to raise winter pigs without milk. A satisfactory ration, when milk is not to be had, is a thick mash of the consistency of cream, composed of one-half cornmeal; the other half may be wheat middlings, equal parts of finely ground oats and barley, or oats or barley with middlings. Mix with hot water and feed warm three times a day until the pigs are four months

of age. After that twice daily, at about eight in the morning and five at night, is enough. Feed just what they will clean up greedily and no more. A cheap trough for feeding the mash is shown in the accompanying illustration. Give a drink of water occasionally.

When pigs are four months old the slop may be thinned. Water may not be necessary in addition to that in the slop. Give a little alfalfa hay, or better still, ground alfalfa hay may take the place of the middlings or oats after the pigs are five months old. Scatter soaked corn and whole oats on the floor to induce exercise, and always keep in the pen a pile of hardwood ashes on which a handful of salt has been scattered. An acre or two of winter rye pasture will be found very desirable in this connection. Dry bedding and lots of it should be provided at all times and changed at least once a week.

Shearing Sheep

A sheep's fleece pays for its keep; the mutton is clear profit. When each fleece represents several dollars, there is pretty good reason for care at shearing time.

About the first of May is the time to shear sheep in most sections. When the weather is warm enough that the shorn s h e e p will not suffer, the work can be safely done. Sheep that carry their fleeces into the summer lose more in body weight than the slight i n c r e a s e in weight of fleece will repay.

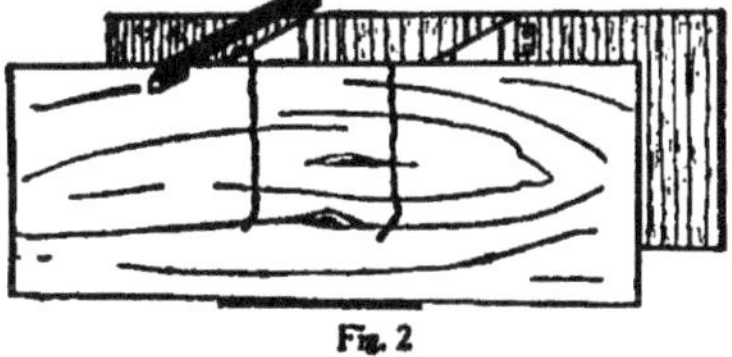
Fig. 1

If a clean floor is not available on which to do the work, make one of some boards, so the fleeces will not be

Fig. 2

filled with dirt. For most purposes it is advisable to have a shearing-machine, but where there are but a few head on the farm, and no neigh-

bors to share the cost of buying a machine, hand shears may be used. On quite a few farms the shearing is done with hand shears. An expert can work about as rapidly with hand shears as with a shearing-machine, but an amateur can not do such good work with the hand shears.

The fleece must not be torn apart when it is removed. It should be rolled into a compact bundle and tied. A handy board for tying is shown in Fig. 1; the boards are hinged The dotted lines are twine. After a fleece is placed on the boards, they are closed as shown in Fig. 2, closing parts A first, then B, of Fig. 1. A notched stick holds the boards together. Then the fleece is tied. Binder twine is objectionable for tying because the fibre from the twine gets into the wool and makes it bring a lower price. Bags should be packed separately, else the fleece will sell for a lower price.

It is a good plan for neighbors to go together to do their shearing. Likewise they may sell all their wool together, and buy their supplies together. Enough growers should unite so that several thousand pounds of each grade of wool can be marketed at a time.

Dipping sheep for the control of ticks is usually done after they are sheared. Any reliable stock dip is all right for this purpose.

A Rack for Feeding Sheep

Will Prevent Waste of Hay

Sheep need plenty of good alfalfa or clover hay. If they have it, they require less grain. That does not mean they should have access to the hayracks. Neither does it endorse the practice of scattering hay about on the ground. The plenty is best provided by feeding the ricks in racks where not a bit of the hay is wasted. This rack shown here is easy to build. It will save both time and feed.

A hayrack like this, sixteen feet long, will have about $7 worth of lumber in it, if it is substantially put together. Four by four inch timbers are used for posts. It will take two at each end and two for the center.

Nail cross supports to each post about ten inches from the ground; then nail on the floor boards which will form a bottom for the rack. A 2 x 4 inch piece is nailed along the outside of the posts, making a fender for the trough.

It pays to make everything solid. The four-inch strips for the rack are spaced about three inches apart and supported at the top by a frame made of 2 x 6 inch material. Board the end up solid for a brace.

These materials are used: Six 4 x 4 inch x 6 foot posts; three 2 x 4 inch

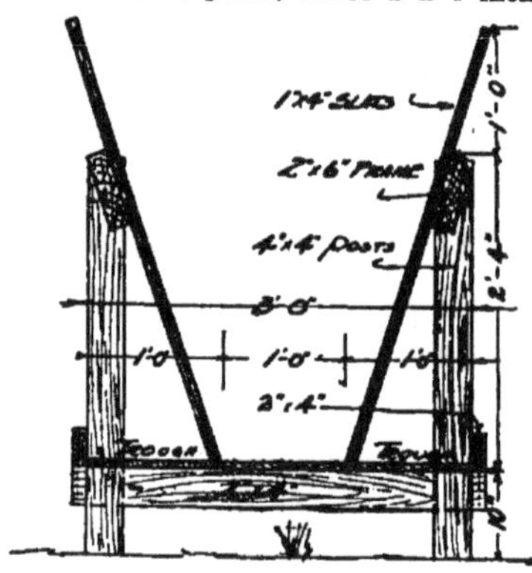

x 3 foot cross supports; two 2 x 4 inch x 16 foot trough fenders; seventy 1 x 4 inch x 3 foot slats; five 1 x 8 inch x 16 foot floor boards.

Treatment for Sheep Parasites

Sheep raising must be made safe, for there is a serious shortage of wool. Two arch enemies which cause great loss of life and vitality in sheep are the lung-worm and stomach-worm.

Modern munitions have been developed for fighting these foes. The old method of fighting lung-worms was to "gas" them by fumigating the animals with burning sulphur, or by sticking each animal's head into a sack containing a hot brick from which iodine was evaporated by the heat. The new method is to inject chloroform directly into the sheep's nostrils.

The injection may be made with a medicine dropper, fountain-pen filler or small syringe. The dose is from thirty to sixty drops, but we scarcely can advise any one other than a trained veterinarian to give the treatment. If it must be done by the layman, one lamb should be treated with a half dose and the effects watched; then others may be experimentally treated with increasing amounts until the safe dose is found. The chloroform stupefies the threadlike worms in the windpipe and air-passages of the lungs, and they are coughed up and swallowed by the sheep. This being true, it is well to give them a full dose of Epsom salts shortly after the chloroform has been administered. The dose for an adult sheep is four ounces dissolved in warm water. This is the best purgative for sheep.

Preventives

More important than medicinal treatment to keep lambs free from lung-worms is to pasture them upon new grass each spring, never allowing them to graze bare-bitten, sheep-tainted pastures. It also is imperative to keep the lambs thriving at all times by supplying plenty of nutritious feed. A mixture of oats and bran may be fed in addition to grass, if the pastures should become short; and other green feed should be supplied as a soiling crop.

The old method of fighting stomach-worms was to give three doses of gasoline on three successive days, the gasoline being mixed with new milk and raw linseed-oil to make an emulsion. This treatment did not always kill the worms, and sometimes killed the sheep. The new plan recommended by Drs. Hall and Winthrop, zoologists of the Department of Agriculture, conserves time and man power, and those who have tried it say it is much more effective than the gasoline treatment.

A one per cent. solution of pure sulphate of copper (bluestone) is made by adding one and one-quarter ounces of the bluest crystals to one gallon of hot water; of this the dose is one ounce for a lamb of comparatively small size and one and three-quarter ounces for a large, strong lamb or sheep. Only one dose is needed and no physic need be given after this drug. The solution may be measured in a glass graduate and administered by means of a small rubber tube and funnel inserted in the sheep's mouth, or it may be given from a long-necked bottle.

Good Clover is Better for Lambs Than Poor Alfalfa

When the same grain ration is fed, no matter whether the grain consists of corn alone or of clover and linseed-meal, the lambs fed clover or alfalfa make larger gains, require less feed per pound of gain and produce gains at a lower cost for feed than do those given either oat straw or corn stover.

The value of leguminous roughages is generally appreciated by lamb feeders; and most of them know that so far as efficiency is concerned clover and alfalfa hay are unsurpassed for finishing lambs. On most farms, however, there are always such roughages as oat straw or corn stover that it is desirable to utilize in the feeding operations. There are years when legumes are a partial or total failure, or not enough legumes may be raised to feed as many lambs as desired. In such cases timothy hay, oat straw or corn stover are fed.

Protein Must Be Supplied

These non-leguminous feeds, however, do not prove satisfactory as sole roughages for fattening lambs. They may be used in the earlier part of a long feeding period, provided the lambs are finished on a more efficient roughage like clover or alfalfa, or they may be used as a part of the roughages.

Such feeds as timothy hay, oat straw and corn stover are low in protein, and should be supplemented with a high-protein feed like linseed-meal. The addition of linseed-meal to corn and either oat straw or corn stover results in larger and cheaper gains and higher finish on the lambs. Still, the gains made by these lambs are more costly than those produced by clover or alfalfa.

When both roughages are of equal quality, clover and alfalfa have about equal values for fattening lambs. Alfalfa usually is harvested in better condition than clover, and commands a higher price on the market. The results of these feeding tests justify a warning to feeders not to overvalue alfalfa to the extent of feeding an inferior grade of this hay when good clover may be had at a lower price.

Care of Horses' Teeth

Always examine the horse's teeth when the digestive organs are out of order. Attention by an expert dentist may be necessary. Chewing is made difficult and sometimes painful when the teeth are "cutting" through the gums, especially when milk (temporary) teeth are being displaced by second (permanent) teeth. Roots of milk teeth are absorbed and the remaining part, cap or crown, is forced off by the incoming second teeth. Crowns often lodge between the teeth and cheek, or fail to come off promptly, hence keep the second teeth back, or cause them to come in crooked.

Teething Troubles

When colts under five years of age "quid" their hay, or do not properly chew grain, examine the mouth and remove crowns with forceps. Between the ages of four and five years lancing swollen gums over teeth about to cut through often gives relief, especially as regards corner incisors (nippers or front lower teeth) and tushes or bridle teeth. One stroke of a rough file over the incoming tooth may take the place of lancing.

Wolf or blind teeth in the upper jaw, just in front of the first back teeth (pre-molars), seldom do harm, do not cause eye disease (moon blindness) or weak eyes, but should be pulled if they interfere with the bit.

Using the File or Rasp

Filing, rasping or floating back teeth (molars) is necessary when sharp points tend to cut the cheeks or tongue. Those points are found on the outer edge of the upper back teeth and inner edge of the lower back teeth. Rough grinding surfaces of teeth never should be filed smooth. Sharp tushes should be shortened and filed blunt when they cut the tongue or interfere with the bit; long points of back teeth and hooks of the two last upper and lower molars are to be treated in the same way.

Pull Decayed Molars

"Quidding" of hay, or pain (toothache), shown by holding the head to one side when drinking cold water or chewing feed, slobbering, or foul smell from the mouth, usually indicates a split or decayed back tooth. It should be pulled with forceps or punched down into the mouth through a hole made in the bone of the face above the root of the tooth by means of a bone augur (trephine). Persistent discharge from one nostril, with or without bulging of the bones of the face below the eye, often is due to a diseased molar tooth which must be removed.

When the front teeth greatly overlap the lower teeth, causing "parrot mouth," the horse cannot graze properly. Correct this in curable cases by notching deeply with a sharp triangle file across the front of the upper teeth at the proper height; then nip off the overlapping portions by means of strong, sharp pincers and file smooth.

Bits Sometimes Make Trouble

The lower jaw, on the floor of the mouth (bar) between the first incisor and the front molar tooth, may be injured by the bit in younger or hard-mouthed horses. The jawbone may be chipped or splintered by the bit. A raw sore forms and gathers feed and decaying saliva, which soon give off a foul odor. Probing of the wound discovers exposed diseased bone which flakes off (exfoliates) and should be removed. This is best done by means of a bone scoop or bone forceps in ten to fourteen days from the time of discovery.

A discharging sore (fistula) may form under the jaw on the edge of the bone. The diseased bone, due to fracture, must be removed, else healing will not take place. Cleanse the sore in the mouth and the external sore and pipe (sinus) by syringing with a two per cent. solution of permanganate of potash twice daily, and swab with tincture of iodine on alternate days until healed. Allow soft feed. Use a rubber or leather-covered straight bar bit. Do not use any overhead check. Similar discharging sinuses, farther back, are often due to diseased molar teeth which must be extracted. Fistula of the salivary duct is located nearby.

Stock Medicines and a Medicine Chest

All the medicine in the world will neither prevent nor cure a disease unless the remedy is at hand when it should be used. That is why so many animals die while the owner is making a trip to the drug store for remedies which are so necessary and so frequently used that they should be kept in a definite place in the stock barn.

A small cupboard about two feet wide and three feet high, with a depth of eight or ten inches, is a good thing to have in the feed way or in some convenient corner of every stock barn. On the several shelves of such a cupboard remedies of various kinds, and instruments for different purposes, can be kept for use. They will be handy when a sick animal is discovered, and it will not be necessary to go to the druggist, or even to the house, for medicines. There may be two doors on the cupboard, as shown in the illustration.

Instruments and Implements

In such a medicine chest, which is best made of grooved lumber to keep out dust and cockroaches, there may be hooks for hanging up syringes, scissors, trocars, clippers, knives, etc. These are necessary parts of every stockman's medicine kit, for their use is not complicated. The lower shelf in the cupboard may be placed twelve or fourteen inches high to make room 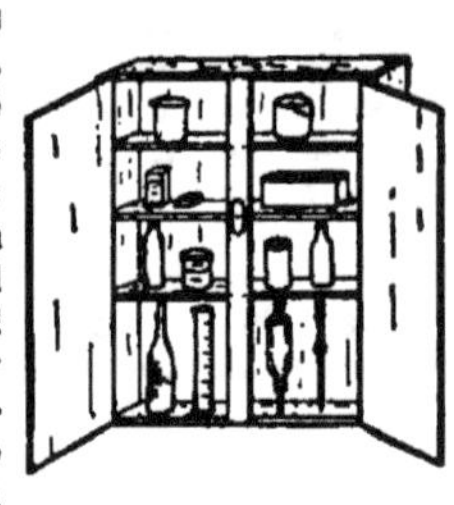 beneath it for the instruments, a long-necked bottle to be used in drenching animals and for any tall pieces of glassware. Smaller pieces of glassware may be placed on the upper shelves. A small strip of wood placed on the upper surface of the shelves at the outer edges will prevent the bottles being forced off the shelves.

The Most Necessary Drugs

It is neither possible nor economical to keep in such a chest a remedy for every trouble. There are a few remedies, however, whose uses are so varied that they need to be included in the list. One of these is tincture of iodine, used to swab out deep cuts and wounds. Ordinarily such wounds heal in a couple of weeks; if they fail to do so, a snag or an abscess may be present, necessitating the services of a veterinarian.

Aloes and linseed oil are almost a necessity, particularly for use in purging a horse. Epsom salts answers well as a purgavtive for cows and sheep, and castor oil for calves infected with scours. Baking soda can be kept for colic, and some powdered tobacco for worms in horses. A box of salve, for use on sore teats, is worth having on hand. Just as necessary is a bottle of linement for use on sprains and bruises.

Not Much, But Some

Copper sulphate, commonly called blue vitriol, has a wide use as a wash and is also used in controlling proud flesh in wounds. Silver nitrate is entitled to a place on the shelf, because of its use as an eye wash. Red iodide of mercury is used for blistering, making it into an ointment for use on spavins and splints. Carbolic acid is effective as a wash and disinfectant. Sulphur chloride of lime, powdered chalk, sulphate of iron and lead acetate are a few other articles worth a place in the cupboard. The amount of these kept on hand need not be large; it is important to have them on hand, rather than in any large amount. Soap is an important article to have. Powders for dusting on sores may be kept in cans with perforated tops, so the dust can be sprinkled on easily.

Labels on bottles containing medicine need to be renewed as often as they become faded or discolored. A bunch of labels may be kept in the cupboard to be used as needed. It is imperative that all bottles or boxes containing poison be labeled "Poison" in large letters All such bottles might be kept on the top shelf, where they will be out of reach of any of the children who might wander into the barn. Better still, keep the door of the chest locked.

Dry Stables, Sound Feet

Every practical horseman knows that the disease known as thrush is caused by allowing the horse to stand with his feet constantly in wet and filth; but few, comparatively, understand that canker of the frog and sole is caused in the same way.

Thrush is characterized by inflam-

Foot Troubles   Veterinary

mation of the fine skin between the toes in cattle; pus forms and tends to underrun the horny wall of the foot. In horses the frog (see F, in illustration) is the part affected, and its cleft, normally shallow, becomes deep and exudes a thin, foul-smelling l i q u i d. Gradually the frog becomes rotten and

A sound foot. F, frog; S, sole

loose and the disease may spread to the surrounding parts; lameness is rare.

Canker of the frog, F, and sole, S, differs from thrush in that the horn of the sole becomes soft or spongy and readily bleeds when cut. In canker, the sensitive tissue (pododerm) of the sole, which ordinarily is covered with solid horny tissue, seems to have taken the place of the solid material. The sole is made up of sprouting fungous tissue and is extremely sensitive and vascular. If it be cut away it may grow again in a single night and the entire affected part is covered with a stinking fluid.

Prevention is all important in these diseases. Stable management should be such that no horse is allowed to stand for any length of time in wet and filth. In horse stables where the manure is removed "now and then," the "nows" and the "thens" sometimes coming months apart, so that the horse has to jump into bed over a high barrier of manure, it is little wonder that the animal contracts thrush or canker.

Treatment of Thrush

Treatment of thrush consists in removing the cause, cleansing the af-

fected foot thoroughly, then cutting away all loose, rotten and underrun horn of the frog and on each side of it, and packing the cleft of the frog full of calomel, or a mixture of calomel, subnitrate of bismuth and slaked lime. This is to be covered with oakum, upon which pine tar has been spread, and the dressing is to be renewed at intervals of three or four days. The stall floor should be kept clean, sprinkled with slaked lime or gypsum (land-plaster) and bedded with sawdust or planing-mill shavings.

Treatment of Canker

Canker is best treated by the trained and experienced veterinarian, but there is no specific remedy. Before succeeding with a bad case, it usually is necessary to alternate remedies and try a great number. The first step in all cases should be to cut down the sprouting growth level with the walls of the foot; then it is usual to cauterize the sole with a red-hot iron or some strong caustic. We usually employ terchloride of antimony, or full strength formaldehyde to start with, and if that does not suffice, change to chromic acid, or strong nitric acid. After applying the caustic, oakum saturated with tincture of iron, or a solution of two ounces of sulphate of copper (bluestone) to the pint of hot water, is bound upon the sole in such a way as to cause firm pressure, for pressure is absolutely necessary. The dressing is changed or renewed every twenty-four hours. Dry dressing powders also are useful, such as a mixture of equal quantities of calomel, subnitrate of bismuth and tannic acid or burnt alum, or a cheaper mixture of slaked lime, alum, sulphur and charcoal. Naphthalin sometimes is added.

The Terrible Foot-and-Mouth Disease

It is sincerely to be hoped that contagious foot-and-mouth disease is about stamped out of this country; but that remains to be seen. As the disease has caused terrible losses thus far, and may remain as a menace for some time to come, every owner of farm animals should perfectly understand its nature and characteristic symptoms.

The disease is technically known as aphthous fever, or epizootic aphtha. It is extremely contagious or communicable; indeed, it is the most contagious disease of farm animals. It is peculiar to cloven-footed animals, but may affect persons who happen to drink the milk of an affected cow. It usually attacks from forty to fifty per cent. of the cattle of a district when it gets a start; but may, in some instances, be immediately stopped in its progress.

An Organism Too Small to See

It is caused by an organism too small to be seen by the strongest microscope and which passes through a fine porcelain filter. It is therefore called an ultra-microscopic or filterable virus. The virus is present in the liquid filling the vesicles characteristic of the disease, in the saliva, urine and manure, and to a less degree in the milk. It also is present in the blood during the stage of fever, but not after the vesicles have appeared. An outbreak of the disease always is due to a previous outbreak. It is impossible for it to appear spontaneously. The virus is easily destroyed by an effective disinfectant. A temperature of 165 degrees Fahrenheit will destroy it, but it withstands freezing.

Infection, Direct or Indirect

The disease is contracted by way of the mouth in contaminated feed, or may enter by abrasions or wounds of the skin, or by infected dust getting into the lungs. Saliva of an affected animal rubbed into the eye of another will produce the disease. The fluid of the vesicles readily causes the disease if it gets into a cut or broken place on the skin. Direct infection comes from the secretions and excretions, saliva and milk. It affects the new-born calf of an affected cow.

Indirectly infection takes place by means of any object or substance contaminated by the disease, such as milk from a creamery, dust, manure, cars, chutes, pens, yards, stock-yards, auctions, shows, water troughs, feed, hides, horns, hoofs, wool, hair, men, dogs, cats, birds, rats or mice, and an apparently well animal may carry it or it may be spread by contaminated hog cholera serum or vaccine against smallpox. It is suspected that cattle recovered from the disease may possibly act as "disease carriers," in the manner in which typhoid is sometimes carried and communicated by a recovered person. There is no definite immunity from an attack of the disease. Immunity may last months or years.

After exposure to the infection the disease appears in three to six days in cattle, or a little longer in some cases; hogs, one or two days; sheep, one to six days. The mouths of cattle are most severely affected, and the feet of hogs and sheep.

Symptoms Easily Recognizable

In three to six days the cow has a chill, which is at once followed by a fever running 104 degrees to 106 degrees, and she droops, loses appetite and shrinks in milk. In one to two or three days more vesicles (blisters) appear upon the lining membrane of the mouth, inside of the cheeks, on the gums, on the tongue near the tip and along the borders of the hard palate (dental pad). The vesicles are from the size of a pea to that of a silver dollar, and are filled with colorless or yellowish fluid. The animal slobbers profusely and smacks the lips loudly as soon as the vesicles appear. The vesicles rupture in one to three days more, quickly scab over and heal in ten days to two weeks.

Similar vesicles appear upon the skin between the toes and along the upper edges of the hoofs; also upon the teats. The animal is lame and stiff when the feet become affected and the vesicles on the teats make milking painful and lead to a shrinkage of fifty to seventy-five per cent. in the milk flow. When the vesicles rupture the remaining ulcers are very easily infected with filth, and the animal may die of such infection or be slow in recovering.

Ordinarily the attack kills only one to three per cent. of the affected animals, but it may leave many injured, and always proves seriously hurtful. The disease is controlled by immediate slaughter, burial in quicklime and maintenance of strict quarantine.

Mouth Troubles of Stock

One would think that any man who found that one of his animals was slobbering and could not properly masticate feed, would open the beast's mouth and have a look at the teeth, gums, palate, tongue and throat, but such is not the case. We have had a number of letters within the past two months describing symptoms such as we have mentioned, and in not a single instance had the owner or attendant made the necessary examination. We had to tell him to do so, and then the cause was readily discovered. In every case where there is evident difficulty in chewing, or swallowing, or if the feed forms "quids" and is expelled or dropped from the mouth, or if saliva flows unnaturally, examine the mouth or have some expert do so.

"Under Tongue" and Lump Jaws

Called to determine why a cow would not eat properly and why saliva dropped from her mouth, we asked the owner if he had examined the tongue, and he answered in the negative. We then told him, on general principles, that he should have done so, as actinomycosis of the tongue was a common disease of cattle and was known as "wooden tongue." In this case that disease was found present and curable. It is caused by the ray fungus (actinomyces), which also causes lump jaw, getting into the tissues of the tongue through an abrasure, or is carried on a sharp spear of straw, fodder or hay, or on a grain hull. Barley awns are also a frequent cause of the disease.

The tongue enlarges, becomes hard and shows sores on its surface. Scarification, painting with tincture of iodine and administration of iodide of potash internally in the drinking water, generally cures the disease, and that was the effect of treatment in quesion.

Watch the Animals—They Can Not Speak

Observation was necessary for its detection. Observation is needed in diagnosing every ailment of animals, for the beasts cannot speak for themselves. Let every owner learn to observe the animal in normal health so that he will understand its signs and symptoms, then he will the more readily detect departures from the normal which indicate sickness. By making this his practice he often will find it unnecessary to employ a veterinarian, and, better still, he will be able to promptly save an animal from suffering.

Not as Bad as it Seemed

A few cases may be quoted where prompt examination and attention would have saved trouble. A cattleman became convinced that his animals were affected with contagious foot-and-mouth disease, because they drooled and acted sick. He sent a long distance for a veterinarian and when the doctor arrived he found the lips and tongues of the cattle filled with sharp needle-grass to which they had recently found access. The offending needles were extracted, so far as possible, and a simple mouth wash of borax prescribed, and soon the animals recovered.

In another instance a man called a veterinarian a distance of seven miles to determine what caused a big, hard bunch on the face of his weanling colt. He thought the doctor should cut out the bunch or split it open, and so telephoned him to bring along the necessary instruments. But a forefinger did the work. A mass of fodder had lodged between the molar teeth and cheek and was easily pried loose. The owner looked "mad," but could not complain when he had to pay the veterinarian a pretty stiff fee for the long run and "operation."

The Deadly Rubber Band

In two other cases dogs showed horrible gaping wounds of the neck, and were brought to the veterinarian to be chloroformed. The owners were asked if they could tell what caused the awful wounds and answered, "No"; then the veterinarian made the needed examination, and in each case he found a small rubber band at the bottom of the wound, having eaten its way in as far as it could go, and stopped when in contact with windpipe and vertebrae. A lady put a nice warm flaxseed poultice upon the wounded foot of her big St. Bernard dog, and held it in place by means of a small rubber band placed around the cloth and leg. Naturally the dog promptly ate the poultice, and nobody remembered the rubber band, or imagined that it was the cause of a great wound above the foot and several discharging holes in the terribly swollen toes; but the veterinarian discovered that too late to save the foot. Assuredly here is need to observe closely and correctly, and by so doing prevent suffering and loss.

Wire Cuts and Other Cuts

Horses running loose in pasture or barn-yard are quite apt to run into, or be crowded into, the fence, especially if they are running with a number of others. Usually slight cuts heal quickly and are unnoticed; but, nevertheless, every cut and scratch, however trifling in itself, should be attended to, as serious cases of blood-poisoning come from small beginnings.

In a farm paper, lately, a man asked what could be done for one of his horses. Its leg, he said, was swollen and stiff, apparently caused by a cut on the shoulder. Before the leg began to swell, he should have cleansed the wound with a *weak* solution of carbolic acid and peroxide of hydrogen, and salved it with oxide of zinc ointment. As it is, his horse is laid up at the time he most needs its services. Better be safe than sorry.

Start in Time

Get after those insignificant wire-cuts while they *are* insignificant. If your horse is working, tie a bandage over the cut to keep it from being chafed and rubbed. Liniments are more apt to irritate the raw flesh than to benefit. Some horses get a liniment bath whenever they complain, but if they could they would protest against it. One man who put liniment on his horse and started to the store, wondered what made his horse "act up."

A horse was turned out in a pasture to graze. Later, a loud rattle and commotion was heard in that direction, and investigation disclosed the fact that the horse had slipped or rolled under the fence. The tight wire prevented his getting up, he became panic-stricken and kicked and struggled till he was covered with a hundred cuts and gashes. The men folks cut the wire and helped the animal to stand, then heated water and got bandages to bind the wounds and stop the bleeding.

On examination it was found that one vein was severed in the left hind foot, and his chest and sides were covered with deep gashes that were sewed up with a needle and fine silk thread, after the cuts had been bathed with peroxide of hydrogen and salved with oxide of zinc ointment. The bandages were tied tightly to hold the stitches in place, and the horse was tied in his stall where he could be looked after. The cuts were dressed every day until they healed, and there were no scars left.

The Heifer's First Calf

It is certainly alarming to most dairymen, but especially to the beginner, when a heifer that has been raised with much work and trouble approaches the calving with a big, hard, caked udder, and after calving does not quickly get over the condition. In such cases the udder will be found to contain watery fluid (serum), when stripped before calving, and after the calf comes this fluid changes rather slowly to true milk and is liable to be stained with blood. In the most severe cases there is also a dropsical swelling of the belly, around the navel, and in some instances the swelling is hard, but more often it "pits" under pressure like dough or putty. It is found necessary to strip away the fluid to relieve the congestion before calving, and milking three times a day is imperative after calving.

Circulation is Not Sufficient

The condition we have described constitutes "congestion," by which term is meant a rush of blood to the part and failure of the circulation to quickly and perfectly remove the blood. The copious flow of blood to the udder is natural and necessary toward calving time and after calving; for it is from blood that the milk ingredients come, and a large flow of blood indicates the probability of a correspondingly large secretion of milk as soon as the udder takes up its work in a normal and perfect way.

The serum mentioned above is forced out of the capillary blood-vessels and invades the connective tissues. As soon as the circulation of the blood, through the veins, becomes normal, the surplus serum is reabsorbed. The lymphatic vessels also join in this work of removal. The condition largely is due to lack of circulation away from the udder through the "milk veins," which are seen as prominent vessels upon the abdomen of the adult cow. The enlargement of these veins is a gradual process and not complete until the adult age. The vessels are undeveloped in the heifer, hence the young animal is most liable to have congestion of the udder. This congestion is almost absolutely certain as an indication of large milk producing capacity, and therefore is not alarming, but the reverse, provided the udder is given proper treatment.

Increased Exercise and Careful Feeding

Abundant daily exercise is necessary to stimulate perfect circulation of the blood. Such exercise should be given before calving and then daily when the cow has sufficiently recovered from calving. It also is important to keep the excretory organs active before and after calving, and for that reason, constipating feed, such as much corn, should not be allowed. The heifer, on the contrary, should have laxative feed, such as bran mashes with some flaxseed-meal added.

A little sound silage or roots and clover or alfalfa hay will add the needed protein material.

When congestion is acute before calving, stripping away of the serum may become necessary and should be done twice daily; but the chief necessity of the case is to increase exercise, without causing chill, and to massage the udder patiently two or three times a day, also rubbing the dropsical swelling of the navel and abdomen toward the heart. It is well to use warm sweet-oil, melted lard or lanolin upon the palms of the hands when doing the rubbing. If the congestion persists, two drams of saltpeter may be given twice daily in water or soft feed, and a teaspoonful of turpentine may be added to each ounce of lard or oil used for the rubbing of the udder.

After Calving

When congestion persists or is aggravated after calving, let the calf suck for a few days, but also strip away some of the milk or serum twice daily, if the calf does not take it all.

Increase the saltpeter to a tablespoonful dose twice daily, adding an equal dose of powdered poke-root, and to the lard or oil used to rub the udder two or three times a day, add one teaspoonful each of turpentine and fluid extract of pokeroot and belladonna. Do not use camphor in such cases, or much belladonna, as both tend to dry up milk secretions. In extra severe cases the treatment may well start with a full dose of saline purgatives, such as one pound of Glauber or Epsom salts, and half a cupful of blackstrap molasses in three pints of warm water as one dose to be administered very slowly and carefully from a long-necked bottle. A very large cow takes from one and one-quarter to one and one-half pounds of salts as a purgative dose. The purgative never should be repeated sooner than twenty-four hours, and seldom is found necessary after one good dose has been given. If the milk is bloody reduce the rich feed, milk gently and be careful that the udder is not bruised or chilled upon a hard, cold floor, such as one of cement.

Symptoms, Prevention and Treatment of Milk Fever

While milk-fever of the cow may attack the animal at any time of the year, it seems most prevalent in late winter or early spring. That at least is the experience of the writer; but locality and climate may alter the case in some circumstances. We think that it indicates lack of resistant power, and the disease will therefore be most liable to strike when resistant powers are at their lowest. That time often is when cows have been tied up in hot, badly ventilated stables for many months, and during that time have been heavily fed a ration rich in the protein necessary for milk production. It seems certain, at least, that pampered cows, of dairy breed, that have had two or more calves are most susceptible to attack, and it is such cows that are pampered for the maximum milk production.

Having such a cow the owner hates to dry her off for six weeks before calving or to withhold rich feed. He wishes to have the cow give as much milk as possible and calve when in high condition, that she may again give a big "mess" of milk. That system of management is wrong and often disastrous.

A Disease of Pampered Cows

Milk-fever never attacks a cow that is in natural condition. It is unknown on the range where cattle run out. It is a disease of domestication, overfeeding for great production, and of weakened condition, the result of pampering, lack of exercise, prolonged lactation, early breeding, stimulating feed and warm stables.

To prevent milk-fever every dairy cow should be "dried off" for at least six weeks before calving, and should have enough exercise every day throughout pregnancy to keep her muscular and to regulate her bowels. The pampered cow becomes soft, sluggish and constipated. Effete matters of the blood are not eliminated by the liver and kidneys under such circumstances, and the system of the cow becomes poisoned as a result. Such a cow is subject to milk fever, or any other disease, and when attacked is liable to suffer or quickly succumb.

Treatment Before Calving

In addition to properly feeding, exercising and stabling the adult cow that is nearing the calving time as preventives of milk-fever, it is important to treat her properly at the time of calving. If she is fat and constipated she should have bran mashes containing flaxseed-meal toward the time of freshening, and at calving a full dose of Epsom salts (one pound or more in warm water as a drench, or smaller doses, say four ounces a day, for a week before that time. It is best that the bowels should be active at calving time; exercise helps induce such a condition.

When the calf comes the udder should not at once be milked dry. That is a common and serious error in management. Let the calf suck for three or four days and milk-fever will not be likely to occur, or strip away only a little of the milk three or four times a day at first, for the disease strikes in its worst form a few hours to two or three days after calving.

Remember that it may also attack a cow that is going on lush spring grass. Keep the big milking adult cow off rich grass just before freshening, and feed her dry hay and light, laxative mashes, as is done with the cow in winter. If the udder is greatly congested and distended before calving, massage it well two or three times daily, and even strip away some of the serum which will be found present. Also give a physic and reduce rich feed.

Symptoms and Treatment

The first symptom of milk-fever usually is restlessness, as manifested by stepping up and down with the hind feet, thrusting out the tongue; then follows weakness of the hind parts, suppression of milk and feces, and finally paralysis and unconsciousness.

It seems almost unnecessary to indicate treatment, so widely is that understood nowadays; but for those who may not have heard of it, it may be said that the udder is stripped clean, washed and then inflated with air, one teat at a time, by means of a special inflating apparatus or cleansed bicycle pump, fitted to a small rubber tube and large sterilized milk tube through which air is pumped to the udder. The cow is kept upon her chest. If she is allowed to lie upon her side she will be likely to die of pneumonia.

Large quantities of medicine must not be administered by way of the mouth, for the cow cannot swallow and will choke. The veterinarian may give hypodermic treatment with strychnine, but usually the inflation suffices. When a quarter has been pumped full of air, a wide tape is lightly tied around the teat, and the other quarters are similarly treated; then the udder may be gently massaged. The treatment is repeated if the cow does not soon recover.

Barrenness—Causes and Treatment

The causes of barrenness in cows are many and different, so that there can be no one cure for all cases. Indeed, no specific remedy for barrenness has been found.

First, we may set down the fact that when a heifer is heavily fed and not bred by the time she is sixteen months old she tends to become barren, her generative organs suffer fatty degeneration. This is most likely to affect the heifers of beef breeds, but pampered dairy heifers are similarly affected. Heavy feeding, especially of corn, may also induce an acid condition of the secretions of the organs, which prevents conception. It is important, therefore, to raise the heifer in a natural manner and to breed her before the detrimental fattening stage is reached. The precocious fat heifer comes in heat very early, and if not bred may soon stop coming in heat.

When periods of heat cease, apart from the cause just stated, the cause may be the presence of a mummified fetus in the womb, or a diseased condition of the ovaries, or germ infection, notably that of contagious abortion.

Abnormalities

Local abnormalities also may cause barrenness. Of these the most common is persistence of the web of tissue closing the vaginal passage. Tumors in the vagina, or twisting or abnormal closure of the mouth of the womb, cause barrenness. Other causes are change of climate, feed and water, semi-starvation, debility from any cause, retention of the afterbirth leading to a diseased condition of the womb, excessive nervousness, and effects of prolonged in-and-in breeding.

To remedy barrenness the cause, so far as possible, should be ascertained and removed. When a heifer that is not a freemartin (twinned with a bulb) fails to come in heat, an examination of the vagina should be made for removal of causes of barrenness already mentioned; then the veterinarian should massage the ovaries two or three times at intervals of a week, to break cysts, which sometimes cause barrenness. The massaging is done by way of the rectum, and can be properly done only by a specially trained and experienced expert, who also may have to open the womb with special instruments. This most commonly is necessary in the cow that has become barren after giving birth to one calf, or several. In such cases the mouth of the womb may be found blocked with dried catarrhal discharge substance; or adhesions may have resulted from laceration caused in difficult calving; or pus or a mummified fetus may be present in the womb. These foreign matters must be removed before conception can take place.

When a heifer or cow fails to come in heat there is a chance that she may do so if fed a quart of stove-dried whole oats the first thing each morning. Along with this she must live a natural out-door life, so far as possible; and a vaginal injection consisting of at least two quarts of lukewarm water containing a dram of coal-tar disinfectant to the quart should be given every other day. A solution containing a dram of pure phosphate of soda to the quart should be used if a cow comes in heat regularly but fails to conceive.

Castration of Young Animals

The castration of quite young animals is commonly undertaken by the farmer, but older animals should have the attention of a trained surgeon or experienced expert. No matter who does the work, certain general principles necessary to success should be carefully put in practice. A few of these principles will here be set down:

The colt should be castrated on a fine, dry day, when flies are not about. Castrate when the colt is sufficiently developed, but not so old that his neck has become coarse and thick and the general appearance that of a stallion. In most instances the colt is sufficiently well developed at one year of age. If he is too effeminate at that age, better let him grow for another year. In our opinion it is a mistake to castrate the colt much under one year old, as often advised by writers on the subject. The geldings in such cases lack masculinity and fail to make well-grown horses. If possible operate with the colt cast on green grass. Avoid doing the work in a dirty place where infection is almost certain to be contracted. Avoid casting the colt upon a manure heap or bare ground. Tetanus (lockjaw) germs abound in such places.

Before the operation starve the colt for at least twelve hours and withhold drinking water. Lay him down gently. The standing operation is popular with some, but is somewhat dangerous to colt and operator. Take great care to use only clean instruments and with clean hands. Wash the seat of the operation with warm water containing

A standing operation is not advised. It is better to throw the colt gently in a clean place. By means of a web casting harness like the above this is easily done.

one teaspoonful of coal-tar disinfectant to the pint. Wash the hands with a similar solution and immerse the instruments in a like solution. After removal of the testicles by means of an ecraseur or emasculator, cleanse again and let the colt up. The cuts in the scrotum should have been made large and in such a position as to allow free drainage. No pockets should be allowed to remain.

After the operation it is important to enforce exercise every day to keep down swellings, promote free circulation of blood and so induce healing. The day after the operation rip the scrotal wounds open with cleansed fingers, to allow drainage, and at the same time break down any adhesions that may have formed between the severed cord and wall of the scrotum. Repeat this once daily until white pus starts flowing. At that stage there will be no danger of tetanus or infection.

To castrate a calf, have an attendant set the calf on its rump, hold its fore legs firmly, the holder's back being in a corner. Now double a rope about two yards long and tie the free ends together to form a loop. Place one loop over one hind pastern, then twist the rope and place the other loop upon the other pastern. The operator should stand upon the rope, with his feet pressed against the calf's feet and so keep the calf's legs spread apart and prevent kicking. The operator now stoops down, cuts the end off the scrotum, or makes two free incisions upon the testicles, pulls these out and scrapes through the cord and artery as these are drawn away from the body.

The Most Money in the Early Chicks

The best time for hatching future layers or breeding stock is between the fifteenth of March and the fifteenth of May. This gives the youngsters a good start, and before the hot weather of July and August strikes them they will have matured sufficiently to be able to withstand the depression. Late-hatched chicks are likely to become stunted when hot weather comes. This causes a setback and there will be few if any eggs before the latter part of February or March. Early-hatched pullets, those that lay at or before eight months of age, are the ones that lay when eggs are scarce and prices high.

Early, But Not Too Early

The eggs from early chicks will be of good size. On the other hand, late-hatched pullets at times lay eggs that are so small they are practically unsalable even for table use. Early-hatched pullets are more steady layers, and their yield can be regularly counted on. However, it is not advisable to get pullets out before March 15, for the reason that such birds will go into molt in the fall, and thus pass over a valuable season without laying any eggs.

Better for Meat

Aside from laying there is a big value in the early chicks from a meat standpoint. Such birds will develop large carcasses and come into the broiler or the soft roaster age in fine condition, and just at a time when there is a strong demand and prices are high.

From a breeding standpoint, we get better vigor, better laying, better fertility, and better size from hatches made between March 15th and May 15th than we do from later hatches; in many ways, too, they are better than hatches brought out during February and the early part of March.

Good Feed and Care Essential

The value of early-hatched pullets and cockerels does not depend alone upon the time of the year they were hatched. Instead, good care and good feed go hand in hand with the date of incubation. In other words, they must be kept growing from the very beginning. Any setbacks will show themselves at once. There must be good, nourishing food, and it must be given so that the chicks do not stall at it. It must be growing feed—developing feed. I never took kindly to the old advice of feeding every two hours. However, the chicks should be fed three times a day—morning, noon and night. Besides, I believe in having a trough of dry bran or some commercial chick feed where the little ones may help themselves between meals. Part of their ration must be mash food, and part cracked grains or commercial chick feed.

Equally important with the kind of feed is regularity in giving it. There must be a regular hour for feeding and at that hour the food must be given. It is almost incredible how well the chicks, with their ravenous appetites, know when feeding time has come, and every moment's delay has a telling effect upon them.

With wise feeding comes exercise, which is induced by scattering the grain among the litter on the house floor. This exercise not only sharpens the appetite, but it puts the pullets in a good, vigorous condition. Early pullets, well hatched, well fed and well cared for, will mature rapidly and go to laying at from six to eight months of age, according to breed.

Hatching and Brooding by Hen Power

The broody hen is in demand despite the thousands of incubators and brooders now in use all over the United States. There are many folks who still cling to hen power for hatching and brooding, maybe as a matter of choice, but more likely because they are not operating on an extensive scale.

I use both hens and machines, and about the only difference between the two that I can see, is that with incubators I can get out a greater number of chicks at one time. From March 15 to May 15 I set all the broody hens I have. They are set outdoors in barrels laid on their sides, with a small lath pen in front of each barrel. Both food and water are always at hand, and Biddy is at liberty at any time to get off her nest to eat or drink, or to dust herself. The nesting material is composed entirely of tobacco stems, and in consequence I am never troubled with vermin on either the old or young.

Generally I set no fewer than three hens at a time, and divide the young between two of the hens, giving the third hen her liberty so that she may rest up, get into good condition and soon return to laying.

Eleven to Thirteen Eggs

I never give a hen more eggs than she can comfortably cover—eleven early in the season, and thirteen when the weather becomes more settled. Nor do I disturb a hen while she is incubating, and when the hatch is starting I darken the nest and do not look at it for twenty-four hours.

Whole corn, sharp grit and fresh water are Biddy's daily diet during her hatching period, and she does not need anything else. Corn digests slowly, furnishes heat to the body, and results in better satisfaction than when mixed grains are given.

When hens first become broody I remove them to their barrel nests at night, and place a china egg under each one. In two or three days if I find they are hugging the nest and show a desire to sit, I remove the nest egg and give each a setting. Some hens, after being removed from the hen house, become restless and want to get back. They should not be risked with eggs, as the broody fever is not strong enough.

Do Not Feed Newly-Hatched Chicks

When a chick is born it still has in its crop a portion of an undigested yolk of the egg. It will be at least twelve hours, and possibly double that time, before the yolk is digested. Up until then the chick will not need any food. So many beginners are apt to think otherwise, and will force the little ones to eat. That is a sad mistake which any one will discover after being in the poultry business a few years.

The feed of young chicks, at least for the first few weeks of their lives, should be of a dry nature. There is no ration better fitted for them than rolled oats and finely cracked wheat and corn. If equal parts of this, by weight, are thoroughly mixed and kept before the youngsters, they will get a balanced ration that will carry them through for three or four weeks. After that it will be safe to give them, in addition, a moist ration in which there is a certain percentage of animal food. This mash must be governed by the purpose for which the young are intended. If raising for future breeding stock, there must be more nitrogenous material like wheat and oats; if for table poultry, then the ration should be of a carbonaceous nature, like corn, meat and starchy material. Grit, charcoal and fresh water must be constantly within reach.

The hen can be allowed with her young as long as she treats them well, for the more motherly care they receive the better they will thrive. If it is noticed that she rebels and will not hover her young at night, or if she is discovered driving them, the hen must be removed, for the youngsters will do far better alone. When a hen begins laying she is very apt to be indifferent in the care of her young.

Running the Incubator

Use soft water in the pipes if the machine is a hot-water affair. Hard water will leave a deposit of lime, etc., the same as it does in a teakettle.

A spirit-level is not necessary. A small pan of water will do just as well for leveling up an incubator. Try it on all four corners of the machine. It is important that the incubator stands perfectly level.

Keeping Up Heat

One frequently has difficulty, during a cold spell of weather, in keeping up the temperature in the egg chamber. Try setting a lamp on the floor near the incubator. A lamp with a squatty bowl can be set directly under the machine. I have used a lantern with good results. Anything which will warm the air of the room will answer the purpose.

Moisture Often Necessary

If the incubator is in a dry location and if a cloth is wrung out of hot water and laid over the bottom of the machine about the eighteenth or nineteenth day, it will materially assist the hatching process.

Spread newspapers over the nursery floor before the hatch begins. This will do away with the necessity of washing after the chicks are removed. Cover the papers with a cloth to help the chicks in getting on their feet.

Open Door Quickly if at All

The instruction book will probably advise keeping the doors closed tightly during the hatch. But if a watch is kept through the inner glass, some eggs will be seen with empty shells attached. Sometimes an egg rolls into an empty shell, and thus prevents the chick from breaking through. Remove the empty shells, but do it quickly, leaving the doors open only a few seconds at a time, so as not to lower the temperature inside the machine.

Don't be in a hurry about taking out the chicks. Leave them in the nursery for several hours after they are thoroughly dry. Quarter them in several small flocks rather than in one large flock; give dry feed, clean water, exercise and warmth.

The Use of Trap Nests

Trap nests are the best guideposts to success. They point out the hens that are doing work, and expose the drones. They not only tell you how many eggs each individual hen lays in a year, but they also point out the color of the shell and the shape of the egg. Other systems make mistakes; the trap nest makes no mistakes.

The time is near at hand when hens will be sold on their egg records, and prices governed accordingly.

It is an accepted fact that the only way to build up a laying strain of hens is to breed from those giving the best records. By annually picking out the best of the flock, it is possible each year to increase the average of the flock.

In line with the introduction of trap nesting came the question of the laying hens giving a better percentage of fertile eggs, as well as receiving an extra allowance of feed. It is more difficult to overfatten a hen that

is doing steady laying than it is one that is not laying.

The excellent system devised by Albert Angele, Jr., consists of a house and yard divided into two unequal parts, and the smaller side is for the cock. In the house are trap nests with two openings. Every night the hens are put in the larger house. The cock stays permanently in his own quarters. Every hen that lays an egg or enters a trap nest goes out into the apartment with the male. When night comes the laying hens are all with the male and they then are returned to their own side of the house and yard, to go through the same process.

How to Prevent Crowding the Chicks

Avoid all angles in the brooder. If there are square corners, pile them full of litter. Clean straw cut into two-inch lengths makes ideal material; so does chaff, chopped alfalfa, or fine clover tops and leaves from the barn floor. The first lesson the chicks must learn is where to go to get warm. The temperature of the hen's body is about 105 degrees, and it is her habit, when brooding young chicks, to hover them for ten to twenty minutes at a time, as frequently during the day as they seem to need it. Her bodily heat is sufficient to warm them quickly, so they do not need to be hovered long at a time.

Warmer Rather Than Cooler

Many of the instructions put out by brooder manufacturers say to start the heat at 100 degrees and gradually reduce it to 90 degrees at the end of the first week. If the weather is cool and the heat is run like this, the chicks to get warm have to stay too long, and they quickly learn that crowding helps. I believe that 100 degrees is not too warm for the first four weeks in early spring. Have heat enough so that after they have learned their way about there is a pretty steady stream of chicks going in from exercising outside, or coming out from brief warm-ups.

Choice of Temperatures

Chicks must be given a choice of temperatures. That is, the whole brooder is to be kept at 100 degrees only under the hover. As soon as they can be taught the way back and forth they should have a small outside run and pure outside air. In teaching them to go in and out from the outside runs, make the new spaces very small at first, and enlarge these as they get their bearings. The first day or two they will have to be attended to closely, and frequently pushed under the hovers for a warming-up, and again brought out to exercise and eat. They quickly learn these things, and there is usually no further trouble.

Give the Last Meal Late

If the chicks are fed the last meal of the day too early they are apt to get discontented and begin crowding. The last feed should always be dry finely-cracked grains, etc., all they want to eat, with enough left in the litter or scattered there later, for the early morning workers. This prevents crowding in the early morning, or in getting out of the brooder.

If in any way they ever do begin to crowd, draw their attention by giving them some unusual treat, like an onion chopped into fine bits and scattered over their backs, or some bread-crumbs, or bits of finely minced fresh beef. By the time they have finished scrambling for these, they will have forgotten the crowding. Meanwhile, you can have been increasing the heat, or remedying the cause of the trouble whatever it was.

Be careful not to overheat them. This can be detected by their panting or discontent, or by pushing about the drinking cups too much. Give all the clean tepid water they want, from the very first. If the heat is right, the air pure and the food rightly selected, they will never hurt themselves by drinking.

The Danger Signal of Disease

Colds are usually contracted at night by crowding on the roost and "sweating," or by poor ventilation, or drafts. It is important that these matters be looked into. Proper precaution is better than medicine. The most common ailment in the fall is a cold. Sneezing, heavy breathing, slight coughs, and even a closed eye, are warnings that call for immediate attention, or that dreaded disease, roup, will follow.

Two Causes of Colds

The two common causes for colds are an unequal heating of the body and the rapid reduction of the temperature.

The temperature of a hen is not reduced by perspiration on the surface, as is the case with man, for the hen has no sweat-glands in the skin. In the hen's case, the moisture is carried out through the breath; so, for this reason, if a hen is very warm she will have her mouth open, breathing the air in and out to take out the moisture and not to get an extra supply of oxygen into her lungs.

Hens Are Different from Humans

Fowls crowded at night while on the roost become too warm. The temperature goes high. When they get off the roost they meet a temperature ten or more degrees cooler, especially on a cold morning, and the breathing organs become so chilled that a cold is the result. The temperature of a hen's body is about 105 degrees. Nature has provided her with a coat of feathers for protection. Therefore, hens should not be expected to live under conditions which are comfortable for man. More birds are injured by housing too closely and crowding, than by the opposite course.

Almost any disease can be prevented. Fowls that have a sound constitution and are active and vigorous, rarely become sick. Such a condition can best be secured by breeding only from well-matured stock birds selected for their vigor and sound constitution and, if possible, whose ancestors were physically sound.

Won't Stand Neglect

But even if the fowls are in the pink of health they will not stand neglect. Many a sound constitution has been ruined by unsanitary surroundings and indiscretion in feeding. It is well, also, to look into the condition of the drinking vessels. Green scum will pollute water and invite disease. Many a disease germ lurks in filthy drinking vessels. They can not be kept too clean.

To have healthy fowls, bad air, filthy water, damp houses, lice, undue exposure to cold winds and rain, overcrowding, accumulations of manure, must all be avoided.

Feeding a hot mash in the morning is indirectly the cause of colds and kindred ailments. The writer moistens his mash with hot water, but by the time it is ready for the fowls it is only slightly warm. The same may be said of the drinking water. It is well enough to take the chill off the water during cold weather, but to give hot water is a mistake.

Big-Four Egg Formula

The essential thing is to get the hens to eat enough of any combination to bring results. While the following proportions may not produce the highest results, yet they will certainly raise the general average of egg production wherever put to practice. There are many groups that might form a big-four combination. In this instance the four most common feeds have been selected. The feeding standard of 1:4:5 has been adopted, as it will meet average conditions of temperature. The nutritive ratio, as given, will be found reliable for autumn and spring feeding. But for the winter period twenty-five per cent. more of cracked corn may be fed to advantage. For summer, wheat or first-grade screenings should be substituted for the corn.

    Cracked corn .... 49 pounds.
    Mangel-wurzels 29   "
    Beef scrap 13   "
    Wheat bran 31   "

Mix Feed With Exercise

A scratching place should be pre-pared, and supplied with clean straw to the depth of a foot. Scatter all cracked corn in this, several times daily. This exercise will not only do good, but will add to the egg basket. Place the bran and beef scrap in an open hopper. Do not mix them. Reduce the beets or place in a clean spot for the hens to pick at. They may be hung up, just enough to force Biddy to jump for them. The quantities, as given, may be increased or decreased to accommodate the size of the flock.

Other Conditions to Watch For

The most careful, systematic feeding in the world will not produce eggs satisfactorily unless other conditions are favorable. Your hens must have a warm, dry place for roosting and exercise. Especially at night they should be kept free from direct drafts of air. Cleanliness and absence of vermin are essential to success in egg production. This all means work, but it pays. Given the same attention as other things, the egg industry can be rendered as profitable as any other department of farming.

Marketing Undrawn Fowl

Poultry should be marketed undrawn, and with heads and feet still on. When the carcass is full drawn, and the head and feet removed, it decomposes most rapidly.

Undrawn poultry keeps much better, for the reason that when the body is opened for cleaning, the delicate tissues in it are open to the bacteria of the air, which multiply very rapidly and soon destroy the flavor of the chicken, even if they do not bring about actual putrefaction.

The head and feet are good indications of age and condition, and help the honest poultryman in making sales. The legs and feet of young chickens are clean and smooth, while those of old birds are rough and scaly and show buttons or spurs. The head of a chicken that is not fresh shows a greenish color below the bill, sunken eyes, and a darkening or discoloration on the neck, all of which indicate decay.

The skin of a dry-picked chicken is flexible, translucent, with the feather papillae plainly visible, and contains short hairs which have to be removed by singeing.

The skin of a scalded chicken is hard, thick, close to the muscles underneath, and almost free from these hairs.

The wetting of a chicken, and especially scalding, lessens or destroys the delicate flavor of the meat.

The entrails of a good chicken should be almost empty, round, firm in texture, and show little red veins here and there.

Caponizing the Cockerels

The Most Desirable Fowls on the Market

There are millions on this new and progressive North American continent who have not so much as heard the name of capon, to say nothing of the question of how to caponize and what the benefits are. Naturally, there are reasons for this state of affairs.

A Gentle Bird

If properly understood and utilized the noble, graceful, gentle, dignified and affectionate capon will reduce the feed bills, death rate, fighting, noise, annoyances, and correspondingly increase the profits, efficiency and pleasure, in poultry raising.

The improvement in surgical instruments during the past decade has been marvelous, but when it comes to caponizing tools much can not be said regarding them, as it is only in recent years that there has been real improvement in design.

A Question of Instruments

Owing to the internal location of the generative organs of cockerels, and the early age at which the operation should be performed, instruments specially designed are absolutely necessary. In this, as in anything else, a practical knowledge of the operation is essential in designing and manufacturing such tools.

About twenty-five years ago caponizing in this country was widely advertised, and poultrymen in large numbers were induced to take it up; but their efforts were not successful, owing to the faulty and primitive design of the tools, and the imperfect workmanship of impractical mechanics who had no real knowledge of the art. The instructions sent out with these instruments were also very meager and confusing, and failure was inevitable. The effect was detrimental, rather than beneficial, and has done more to discourage the practice than any other cause.

The design and workmanship of caponizing instruments must be considered very carefully. The instructions accompanying them must be clear and explicit, and reasonable care and judgment must be exerted in following directions.

Any design or type of tool that brings the hand in the way, thus causing shadows and uncertainty, or that is controlled or actuated by powerful or fickle springs or is imperfectly adjusted, will give trouble and cause failure, loss and discouragement.

The Proper Age

The following advice and details as to proper size or age to caponize apply to the English and American breeds and their crosses (the Mediterranean breeds must be worked when younger, or weighing around twelve to eighteen ounces).

As soon as cockerels are distinguishable and weigh from one to one and a half pounds they are just right for best results. At this age their organs are usually just beginning to develop, and the attachments are so small there is practically no danger when they are removed or extracted; but if left until the development has progressed too far, the bird will die of hemorrhage upon their removal. A little experience will enable one to determine the proper stage of development.

A "Slip" is Worse Than Nothing

A "slip" is the result of an unsuccessful attempt to caponize. If the slightest particle of the testicle, or of the cords and attachments, is left in the bird the operation is a failure and a "slip" is inevitable, which is neither capon nor cockerel, and is a regular nuisance. "Slips" are no better than cockerels as food.

Capons are sweet, tender and juicy, equalling and excelling the flesh of young hen turkeys.

Capons Improve Until Maturity

It is a well-known fact that for every chick hatched and raised to four weeks of age at least three to four eggs are necessary, allowing for the rejected, the infertiles, the dead in the shells and the deaths after hatching. About one-half of those raised will be cockerels.

Cockerels must be sold before they get too large and too old. Roosters are always a drug on the market.

A castrated cockerel, like a castrated lamb, pig or calf, gets better every day until he reaches full maturity at least, and I personally know of capons four years old that were as tender as chicken when roasted.

Feeding Stock on Range

The very best way of growing poultry, especially when intended for future breeding, is to give them free range and provide them with food just when they need it.

An Outdoor Feed Hopper

The outdoor feed hopper for dry mash here given is so plain that it will not be difficult to understand making the hopper. A roof can be constructed over it to keep out the rain and hot sun. The hoppers are to be kept well filled with dry mash, and the youngsters, after taking vigorous exercise, are able to sharpen their appetites so that they eat greedily of the meal set for them. The hoppers can be made any size.

Selecting the Breeders
The Foundation of Success in Good Breeding Stock

There are two ways of selecting good breeders; one is by trapnesting and the other by observation and study. That trapnests are of the greatest importance in this particular has been demonstrated by the State experiment stations and by others who make poultry raising a business.

In trapnesting for the selection of breeders, attention must be given to other factors besides ability to lay. For instance, if a hen lays 200 eggs in a year, but has had some contagious disease earlier in life, she should not be permitted in the breeding pen, because the disease is liable to be transmitted to her offspring. A hen with a trapnest record of 200 eggs must necessarily be a healthy fowl, and with ordinary precaution one cannot go far astray in selecting her for the breeding pen.

In selecting by observation, health and vigor must be the main factors.

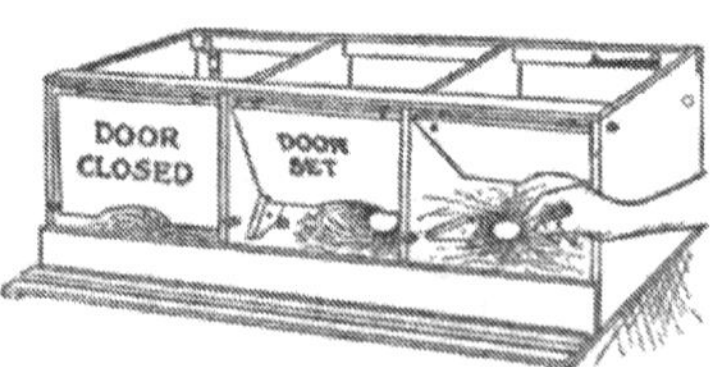

In several of the largest egg-laying competitions, this trapnest is being used. It is a simple affair, and can be made of ordinary box material.

Chickens   **Breeding**

The individual selected must be active and carry her body in an erect and proud fashion. The comb should be bright red in color, soft and velvety; the eyes should be steady and clear. A fowl that stands moping around or roosts in the daytime, is either weak or sick, and should never be selected, no matter what her record may be.

Although feathers are only a covering for a bird, some attention must be paid to them also. Good feathered birds not only look better but sell better. Size and shape are also important in a dual-purpose hen; good, heavy layers, of large size and uniform shape, with vigorous constitutions, are the ideal stock to breed from.

A good layer can invariably be distinguished by her actions and willingness to scratch for food. If examined on the roost at night, it will be found that she has a full crop. This, too, is a good indication of health. A hen may be compared to a small factory—food is the raw material and eggs are the finished product. The output is greater when the factory runs efficiently and consumes large quantities of raw material.

A soft, red comb, lying to one side (in single-comb varieties), a short curved beak, lack of color in shanks, worn-off toenails, are all indications of laying ability. It is said that hens lose color in the shanks because they lay it out of them; and the toenails are short and worn as a result of much scratching.

Observed from the side, a good layer has a small head, rather round, and the general appearance of the body is decidedly wedge-shaped because of extreme fullness in its back. Large-headed birds with oval-shaped bodies are never good layers. If a good layer is picked up she will be found to possess considerable weight for her size. Examination will show the distance between the pelvic bones to be from two to three inches. Examining the pelvic bones for egg-laying ability is a good method, but the inexperienced are likely to have difficulty at times in estimating the distance, especially if the hen is an old one and with much fat. The distance between the pelvic bones seems more than it is, for the lower bone is forced down by superfluous fat.

In selecting a male bird, find one which has good size and color and holds his body erect. Pick a hearty eater if possible, the tendency of males being to allow the hens to eat everything and have nothing for themselves. He should be of good mating qualities, not quarrelsome, and yet possess a fighting spirit, and be continually with the hens.

Breeders do not require different housing from that of laying hens. They must at all times have fresh air and plenty of it. I believe that the correct type of house is the fresh-air one. More eggs are lost than gained, considering the whole country, by keeping the fowls confined too much in warm, mild winter weather.

There must be a different method employed in feeding breeders than is used in feeding layers, the object being to produce eggs of quality rather than many of low fertility. Mention should have been made before that either pullets or hens are good as breeders, but each must be handled differently. The old hens should have no corn except in very cold weather. Too much animal food causes the production of more eggs than can be properly fertilized.

How to Tell What is Killing the Chickens

We may pretty well determine the character of the animal that visited our hen house by the condition of the fowls as found. Should an opossum get into the coop he will kill but one or two on his visit. He eats the head and neck of the victim, and doesn't seem to care for the rest of the carcass.

A mink is more deadly. He will slaughter a dozen or more birds in a night, biting them in the neck and sucking the blood. Both the mink and the opossum leave the carcasses in the coop or house where they found them.

Rats drag their prey into the holes or runways. Rats, however, very seldom attack a half-grown chicken or a fowl. Their appetite is more for the youngsters, so the front of each coop should be closed with a wire-covered frame, which keeps out the rats and permits ventilation.

Cats and foxes carry their victims away with them; the cat, like the rat, cares only for the baby chicks, seldom doing damage to birds that weigh more than a pound.

The skunk seems to select poultry for his diet only as a last resort. He prefers refuse meat or scrap. If any of the latter is found he will fill up with it and then retire to his den. The next night he will return, and in case the refuse meat or scrap is insufficient to satisfy his appetite, he will top off on poultry.

The weasel crawls on the roost, selects his victim, taps a vein and sucks the blood. The weasel is a regular contortionist, and is able to so contract his body that he can wedge through the smallest opening.

Where there are foxes, opossums, minks, weasels and skunks will also abound. None of the above pests should be allowed to harbor near the poultry yard. Piles of rail or stones, stone walls, brier patches, make a safe harbor for these vermin. A good active dog will do much to keep them away.

Hawks Are Deadly

Hawks are deadly enemies. They have the habit of perching nearby and surveying the territory, and having once laid plans, descend upon the young chicks. Crows, while not as a rule poultry enemies, will attack chicks when very hungry, and especially in the spring when there are young crows in the nest to feed. Little chickens that run around the coops in which their hens are penned up, are particularly apt to be victims of the crows. If the hen is running at large, she can usually protect them. The owl is a night-bird, and frequently feasts upon large poultry roosting in trees.

Foxes

The fox has a cunning way of hiding in some woodland nearby the range of the fowls. He keeps a close watch and waits until some reckless hen wanders near him, then suddenly grabs his prey and carries it to his haunts. If unable to capture any poultry during the day, he will visit the hennery at night.

In sections where foxes are known to exist, the foundations of the poultry houses should begin a foot or two under the ground, as the sly old reynard has a knack of making a tunnel under the sill to gain entrance to the house Care should also be taken that tne nouses are so closed at night that it is impossible for any of the night-prowling enemies named above to enter; and it is a wise poultryman who keeps a padlock on the door, for there are other poultry enemies besides those mentioned—the two-legged kind. There can be nothing more discouraging than to find, on opening up the pens in the morning, that some "midnight poultry raiser" has cleaned up your entire stock of chickens.

Leghorn Egg Machines

The Leghorn is an Italian—chock-full of business and plumb-full of "pep."

The demand for large, white-shelled eggs has made Leghorns and Minorcas very fashionable breeds. However, there are some markets that prefer brown eggs, and in such sections the Leghorns and Minorcas are not extensively kept.

Of the Leghorn family I have had experience with the Single Comb Brown and Single Comb White varieties. I was always able to get more eggs from the Brown Leghorns than I could from the White, but eggs from the Brown variety were so much smaller in size that I was getting more pounds of eggs from the White Leghorns than from the Brown. While size eggs may not have much to do with the general market, it does have great influence in holding private trade.

I found the Leghorns excellent layers during the spring, summer and fall, but rather indifferent in very cold weather, unless extra protection was given them. They are lightly clad and quickly affected by changes of weather in winter; yet they are very hardy and suitable for almost any climate.

Single Comb White Leghorns lay good-sized eggs, generally of good shape and uniform; but occasionally some hens will persist in laying eggs of creamy color instead of pure white. As table poultry they are not to be recommended, as their activity toughens the meat.

My first objection to Leghorns is their tendency to fly. They are able to go over the highest fence with ap-

The Leghorn Breed | Chickens

parent ease unless one wing is clipped. Second, they are noisy. This is very objectionable, especially to a nervous person.

The Single Comb Black Minorca is a larger fowl than the Leghorn, and lays a much larger egg that is perfectly white in color. There is more meat in the carcass of the fowl, and consequently this breed is better adapted for table purposes. The Minorca is not in demand for that purpose, however, because of the dark pin-feathers and whit skin.

As a rule, none of the Mediterranean birds become broody. Occasionally one will show the trait. When they can be induced to hatch out a

brood, the hens prove to be very proud and faithful mothers.

The Black Minorca was formerly known as the Red-Faced Spanish. It is a noble breed worthy of more popularity.

Profits in Ducks and Geese

A Desirable and Lucrative Business

Green goose culture is profitable, but the industry is not carried on so extensively as that of green ducks. The demand for geese is not constant, although there is a limited demand the year around. The trade is best during the winter months, and especially at the holiday seasons.

Geese are natural grazers; ten geese will eat as much grass as a cow. Geese live to a great age. The females are reliable and productive for many years.

Goose feathers are usually worth about thirty-five cents a pound. It takes the feathers of four geese to make a pound. White goose feathers are more valuable than colored ones.

Toulouse the Preferred Goose

It is claimed that the Toulouse is the most profitable breed to raise. It not only grows larger, but it matures quicker than any other breed. Toulouse geese are not so much inclined to roam, and they grow more rapidly and accumulate fat faster than other breeds.

Ducks More Prolific

Ducks are more prolific than geese, and artificial methods are resorted to in hatching and rearing the young. At eight weeks of age a pair of ducklings will average about nine pounds. They are then known as green ducks, and are in demand. At ten weeks old they will weigh from nine to fourteen pounds to the pair. The best prices are obtained about May 1, but from then until July the price gradually grows less. From July to September prices for green ducks remain unchanged. From September to November they again command good prices.

Ducklings are rapid growers. When a duckling comes out of the shell it weighs two and a quarter ounces.

Possible Profits

The Weber Brothers, in Farm Journal's booklet, "Duck Dollars," claim they make a net profit of fifty cents on each duckling marketed. As they market from 40,000 to 45,000 ducklings a year, their annual income is $20,000. When grain was low they made that profit. When grain is high, as it is now, they get more for their ducklings, so the average net profit of fifty cents a duckling has remained the same.

One way of making money with ducks is to have a lot ready to sell when the season opens. The first essential is to start with good breeding stock. Birds that have been inbred until their constitutions are weakened are not fit for use as breeding stock.

White duck feathers sell for thirty-seven to thirty-nine cents a pound, while colored feathers run from seventeen to twenty-three cents.

Water or No Water

While a body of water affords much needed exercise for both geese and ducks, it is possible to profitably keep water-fowl exclusively on dry land.

Breeding fowls will keep down surplus fat and give better fertility to their eggs when allowed to bathe regularly each day; but stock being grown for market should be given as little exercise as possible for the best and quickest results.

A point in favor of water-fowl is that they do not require very elaborate houses, but these should be dry and well bedded.

Six Breeds of Money-Making Geese

The leading market breeds of geese are the Toulouse, Embden, Wild or Canadian, African, Brown Chinese and White Chinese. The first two are probably the most popular. The Toulouse,

An adult Toulouse gander weighs twenty-six pounds, the adult goose, twenty pounds; the young gander, twenty, and the young goose sixteen pounds. An adult Embden gander

the African and the Canadian are gray in color; the Embden and White Chinese have white plumage; the other breed is brown.

weighs twenty pounds; adult goose, eighteen; young gander, eighteen; young goose, sixteen pounds. The weights of Wild or Canadian adult gander, adult goose, young gander, and young goose, are twelve, ten, ten and eight pounds, respectively. The adult birds of the African breed have the same weight as the Embden adult birds; but the African young gander and young goose weigh sixteen and fourteen pounds, respectively. The Chinese geese have the same weights as the Canadian. The African and Chinese breeds show distinctive knobs or protuberances on their heads, while the heads of the other breeds are plain.

Toulouse geese have massive bodies of medium length, broad and very deep, almost touching the ground. They are good layers, producing from twenty to thirty-five eggs a year; they are docile, grow rapidly, and are excellent birds for the market.

Embden geese are considered only fair layers, the egg yield varying greatly among individuals. As market birds, they mature early, grow rapidly, and on account of the white pinfeathers are popular with marketmen. They furnish more attractive carcasses than Toulouse.

Wild or Canadian geese are rather poor layers and are often difficult to breed successfully in captivity. They are crossed with other breeds to produce the so-called mongrel type, which is much prized for market purposes, but is usually sterile.

African geese are good layers as well as good market geese. They grow rapidly and mature early.

There are two standard varieties of the Chinese goose (the Brown and White) and both varieties mature early. As layers, they are prolific, and as market poultry are rapid growers.

The Chinese geese are naturally shy, and therefore rather difficult to handle.

Geese can be raised in small numbers where there is low, rough pasture land, with a natural supply of water. Grass makes up the bulk of their feed. While they can be kept without the use of water to swim in, it will be all the better, especially during the breeding season, if a body of water is available, as it affords exercise which the geese could not obtain otherwise.

The period of incubation of goose eggs varies from twenty-eight to thirty days. The first eggs are usually set under hens, while the last eggs which the goose lays may be hatched either under hens or under the goose, if she becomes broody.

The broody goose plucks off more or less down from her breast, with which to line the nest and cover the eggs whenever she leaves them. A goose can conveniently cover eleven eggs, but a hen should not be given more than five.

Year-old geese are not mature enough for breeders. The females lay fewer eggs of smaller size, and usually more of them are infertile than is the case with females two or three years old.

The bill of the goose is provided with sharp, interlocking, serrated edges, designed to cut and divide vegetable tissues easily. The tongue, at the tip, is covered with hard, hairlike projections pointing toward the throat, which serve to convey the bits of grass and leaves into the throat quickly and surely.

Keeping the Pigeons Healthy

The various diseases of the pigeon can in almost every instance be traced to some neglect on the part of the owner. There are, however, some cases which can not be accounted for and which sometimes appear in the best managed lofts; but such cases are not common.

Serious illness can, as a rule, be averted if prompt treatment is given the birds the moment the ailment is discovered. It is delay at this time that often renders the disease serious.

Neglect the Chief Cause

Neglects are costly. For example, the night is damp and chilly and a stiff breeze is stirring. In the hurry to close the lofts for the night, the windows facing the side from which the wind is coming are left open, and those birds roosting nearby are then in contact with a most dangerous draft. It is needless to add that such a neglect will at once invite colds and other kindred ailments.

It must not, however, be inferred that fresh air must be kept out of the loft. On the contrary it is required. Some of the most successful squab raisers have either open windows facing the south, or have muslin tacked over the windows in place of glass. This admits both sunshine and fresh air.

Cleanliness is Necessary

There seems to be a good argument pro and con about cleanliness in the pigeon loft. The "pros" believe that the lofts should have a thorough cleaning every week, pointing to the fact that filth is a disease and a vermin breeder.

The "cons" say that frequent cleaning of the loft is unnecessary, for the reason that pigeon manure is different from hen manure, and does not produce disease and vermin. They further argue that frequent cleaning of the loft does great damage in scaring the breeders.

The writer has had some years of experience in pigeon keeping, and is a firm believer in cleanliness for best results, from both a health and productive standpoint. He regularly cleans out the nest boxes as they are deserted, and every week gathers up the loose manure on the floor. By moving about quietly, the pigeons soon understand that no danger will come to them, and there is consequently no alarm. It is wonderful how easily pigeons are tamed.

A reckless, careless fellow—one who goes pell-mell about the place—is more dangerous than useful, and should never be allowed to enter the lofts. He will keep the stock in constant fear whenever he approaches them. Tame birds are not only attractive but they are productive. They live in happiness and contentment.

Pigeons, like chickens, are naturally hardy. Disease is practically given them by man in unsanitary lofts and surroundings, in damaged foods, in impure water, in poorly constructed houses and in indifferent care. No live stock respond quicker to good care, or more readily show the effects of bad treatment, than pigeons.

Fixtures for the Pigeon Loft

Next to the proper construction of the loft, comes the necessity of having proper loft requisites, and of this list one of the most important is the "bathtubs," if they may be thus styled. These bathtubs, or pans, can be made any size desired, of lumber, lined with zinc or galvanized iron to make them waterproof.

It is not advisable to have them more than four inches in depth.

Pigeons are naturally clean in their habits, and one of their great enjoyments is to bathe. This bathing cleans the feathers of considerable dirt, and also destroys what vermin may be on them. As soon as the birds are through bathing, the tubs must be emptied so as to give the birds no chance to drink any of the water, which is of an oily nature, closely resembling milk.

The Bath

During the summer, bathing should be allowed at least every other day. In winter, however, once a week is sufficient, and then only on bright, clear days. The general practice is to fill the pans just about an hour before noon, and allow them to remain about two hours, which gives ample time for all birds to take advantage of the bath.

Nests

Although in many cases a part of the construction of the house, nests really belong to the list of fixtures. Some pigeon men prefer small boxes tacked to the wall, but this practice is not generally followed, and certainly does not make as presentable an appearance as does the regulation style. Other breeders have the nests built solid, with no cleat in front, so as to be able to scrape out the manure. Others have the front and partitions movable, and still others use a type of nest with movable front and bottom. The latter affords a much better chance for thorough cleaning.

Feed Hoppers

Self-feeding hoppers are advocated by some authorities. In these hoppers sufficient food is supplied for three or four days' feeding, and the birds are allowed to help themselves at will. But this system of feeding is not carried out on the large plants, the argument being used that where birds are thus fed they raise inferior squabs mainly on account of a lack of a regular feeding hour. When hoppers are used, not only will the parent birds lose their appetite, and the squabs in consequence suffer, but the grain (particularly the corn) can not be kept fresh, resulting in sickness. The feed trough is preferable.

For drinking vessels, the two-gallon galvanized-iron fountains are best. That size is sufficient for a loft of fifty pairs of breeders.

Use of Nappies

Going back to the question of nests, some writers advocate the use of nappies, using the argument that they are better proof against the breeding of lice than the ordinary nest box. But other writers claim that they are a disadvantage, as only a small percentage of birds will use them unless forced to do so. They illustrate that if a nappy is placed in one nest, and the adjoining nest is empty, in nine cases out of ten the empty nest will be used first. Furthermore, when nappies are used, the death rate among the squabs is greatly increased, as the birds frequently crawl over the edges and become lodged behind, and as they have no way to extricate themselvcs, they finally perish. Another reason given is that where nappies are used there is a large percentage of weak-legged squabs. Lastly, they are very difficult to clean.

The Carneaux—a Coming Breed

On account of its wonderful qualities as a squab producer, the Carneaux is coming into more consideration. An interest being created, the breed has been carefully tested and proved to be good for the purpose intended.

As a squab producer the Carneaux has no superior, having the quality and capacity necessary to convert every ounce of grain possible into fat, juicy, solid flesh.

The skin color of the Carneaux squab is a beautiful yellow, several degrees lighter than the skin of Homer squabs. This makes it an ideal squab for market purposes, and it presents a handsome appearance when dressed.

When four weeks old the Carneaux will average, each, three to seven ounces heavier than Homer squabs of the same age, and when you stop to consider that the parents are only a trifle larger than Homers, it will give you an idea of this breed's adaptability for producing squabs.

A Coming Breed

The Carneaux is gentle and peaceful, not so nervous as the Homer. The birds are very active, full of life and fire, and can dance and strut the neatest of all pigeons. They are hardy and prolific and can stand a whole lot of abuse without a protest, and are not affected by captivity. If you prefer to let them fly at large, they will stay home even if in a new one.

From a fancier's standpoint they offer great possibilities, as the Belgian standard calls for a clear red, red-and-white splash, and red with white rosette on wings. To hold the beautiful shade of red, a sort of rufous red, not seen in any other bird, is a trick that can be made very interesting, especially so when taking into consideration the proud carriage, grace of movement, show of strength when on the wing, and slick, handsome appearance when bred right.

The origin of the Carneaux is very remote. Whoever originated the breed has done a good work for the squab producer. Being naturally of a domestic nature, and ready and continuous breeders, these birds may be called blessed by the squab-eating public. They are gradually taking the place that belongs to them of right, and in the near future may contend with the Homer for first place in public favor.

Hints on Handling

Some pigeons are more devoted to the young than others. It has been remarked that those birds with short beaks are generally least devoted.

All pigeons have in each wing ten strong feathers, counting from and including the outside ones, which are called "primaries," from being the first that shows in the wing of a bird, and called by pigeon fanciers "flight feathers." They are the strongest quilled and webbed, and do the greater part of the work in flying. Immediately following these are others not so strong in either quill or web, called "secondary" feathers.

Wing diseases may be cured, if taken in time, by plucking the flights and anointing the joint on the under side with turpentine. Wing disease in the pinion joints is difficult to cure.

Pop-corn is said to be better for pigeons than any other variety of corn.

Pigeons will not eat food that has been fouled, and they must have plenty of sand and gravel in the loft, as well as ground oyster shells.

To avoid shy pigeons you must be kind and careful of them, and enter the loft whistling a call of some kind to let them know you are coming; and always have a cup of grain to throw to them, that they may expect something each time you visit the loft, and thus get acquainted with your movements.

Never catch any bird in daytime unless absolutely necessary; this alone will make tame birds wild. If you wish a particular bird, wait until evening when they are roosting, when it is an easy matter to catch the one you want and place in a pen until the next morning, at which time you can make your examination and let it fly back with the rest of the birds.

Pigeons are often kept and do well in a box nailed on the side of a house or barn, and others have houses placed on poles. If either of these should be adopted, they should have a southern exposure, and the roof of the box should project over as far as possible, so as to shelter it from the rain and snow.

The pigeon loft should be well lighted. A dark room is not healthful, especially if the birds are not allowed to fly out. The loft should be well ventilated, but not by a direct draft through it.

Points on Squab Raising

"The more one studies his pigeons, the more money will he get in return," says Edison W. Kalm, in the Squab Magazine. "Go up and look into the breeding pen. Don't scare the birds. Look! See that big bird working away at his nest! Doesn't he look industrious? Watch him and see him quit and start to drive the first female that alights near him. Make a note of it—it means about six pairs of squabs at the end of the year. See that little bird working over in the corner? He stops for nothing. If he drops a stem he goes back for another. He doesn't appear to notice anyone. Put him down for ten pairs."

"Solid colored birds are not particularly well adapted for squab-raising purposes," writes D. Russ Wood. This may appear peculiar to a beginner, nevertheless it is true. Size has been sacrificed for color in the breeding of solid-colored birds.

The Market Age

The weight of squabs varies from six to eighteen pounds to the dozen; nine pounds is a fair average. It requires from four to six weeks to bring squabs to marketable size. At this time the down disappears from the head and they are fully feathered under the wings. They should then be plump and heavy. When this period is passed their fat increases, the once-tender flesh becomes hard, and the birds, learning the use of their wings, will leave the nest.

Pigeons are at the most productive age between two and six years, but it is not impossible to have some do good work up until ten years old. Where it is intended to hold squabs as breeders they should be leg-banded before they are able to leave the nest and a record kept of their breeding. When it is possible to determine the sex the males should be banded on the right leg and the females on the left.

Do Not Feed Before Killing

Squabs intended for market should be caught before they are fed their morning meal, so that the crops will be empty. The method of killing, plucking and cooling is practically the same as employed with poultry. Never save poor, inferior squabs for breeding, as they will reduce the quality of the stock. Dispose of weak or inferior breeders, especially if they are males, as one always has a surplus of the latter.

The most precarious period is when the birds are from four to eight weeks old. This is the time of the first molt.

When the birds look droopy and seem to lack in appetite, give them a physic. Put a tablespoonful of Epsom salts in their drinking water. Do this at night so they will get the full benefit in the morning.

Two parts of corn to one part of cowpeas is a mixture that will make plump, fat squabs. One who has tried it says that his birds have never been in a healthier condition, nor produced better, than since he began feeding this mixture. This is one way of saving wheat, and apparently with as good results.

When Bees are Put Outdoors

March the Beginning of the Bee Calendar

Beekeepers who wintered their bees out-of-doors in hives properly protected, will have little to do with them in March, unless they live in the Southern States. Those who wintered their colonies in special repositories or cellars, however, will find the month of March to be an exceedingly busy one; for, generally speaking, March is the time to put the bees out-of-doors. Some progressive beekeepers, in places where the temperature during the winter gets below zero, wait until the pussy-willows are in bloom before setting colonies out.

Wrapping the Hives

In taking the colonies out of their winter quarters it is best to give them some added protection in the form of telescope cases, after first wrapping the hives with old newspapers. Where cases are not at hand, the hives can be wrapped with old papers and the whole then wrapped with waterproof paper, tied in place, leaving the hive entrance open for air, etc.

Spring Dwindling

Reports from many beekeepers tell me that spring dwindling is one of their greatest handicaps, that the colonies become so weak that they are not strong enough to secure much surplus from the early flows, that it takes practically all spring and summer for the bees to recover from spring dwindling, and that they become strong only in time for the late flow from buckwheat, goldenrod and aster.

I have in mind one beekeeper in New York State with colonies running into the hundreds, whose whole surplus is practically from buckwheat. He has told me on several occasions that the reason he secures so little from the clover and basswood is because of spring dwindling, which so reduces the number of bees in each colony that it takes the colonies all summer to build up. The added protection given the colonies when placed outdoors will in a large measure overcome the spring dwindling.

Starting the Bees in the Spring

April is ordinarily a busy time among the bees, as experience proves; they gather much pollen and often considerable nectar.

Naturally where the colonies have ample stores with the added nectar, brood rearing is going on at an astonishing rate, and much brood is maturing.

If the colonies have not sufficient stores on hand, it is a good thing to remedy matters at once; and, personally, I haven't much faith in the idea of stimulative feeding for brood rearing—that is, giving the colony a small amount of syrup every day. I believe the better plan is to give them what they need at one feeding, say from ten to twenty pounds of syrup.

Feed if Necessary

This feed is made by mixing equal parts of hot water and granulated sugar and stirring it until the sugar is completely dissolved.

Care should be exercised in feeding, to prevent robbing by other bees, and not to chill the sensitive brood.

If colonies are rearing much brood and there comes a sudden shortage of food for same, then there is great danger of the entire colony of bees and brood perishing from starvation.

This need not be feared if the colony went into winter quarters with ample stores; but if the colony is not amply supplied at this time, then the condition must be remedied at once to avoid disaster.

The most desirable way to feed is in some overhead feeder, with all the feed given at one feeding, seeing that the hive is warmly packed, especially the upper story hive body in which the feeder is placed, and all packed with cloths or other material that will conserve the heat of the colony.

Transfer Brood from Strong Colonies

Weak colonies should be strengthened by giving them frames of sealed brood and bees from the strong colonies, care being exercised that the queen from the strong hive shall not be upon the frame of brood given the weak one, or else there will be a queen lost to the strong colony and brood rearing will be seriously retarded in that hive.

The combs taken from the strong colony must be replaced with a like number of combs from the weak one. I would not advise taking more than two or three combs of brood and bees from any strong colony.

Watch for Bee-Moth

The larvæ of the bee-moth will begin to make its appearance, and should be destroyed, as one moth destroyed now will prevent the appearance of thousands in the fall. While there isn't much danger to a strong colony from the bee-moth, yet it is decidedly annoying to have perfectly good combs destroyed later on. That is, combs stored in the honey house. Swatting the bee-moth now will largely prevent its appearance in the fall.

Some colonies will possibly be found to be queenless, and this is a condition which should be remedied as soon as possible by securing queens from the Southern queen rearers. If this is not possible, then two or three frames of young larvae, not more than two to three days old, should be given the queenless colonies, that they may rear a queen at the earliest possible time.

Rear Your Own Queens

It is a mighty good thing to rear one's own queens, and requeen every colony later with a young queen of the season's raising, as these young queens will be splendid and prolific layers, and colonies headed by young queens are not so likely to cast swarms as those headed by older ones.

Queen rearing is neither a difficult nor an intricate thing, and where the bee-keeper has a good strain of bees it is a decided advantage in many respects always to rear his own queens.

From many parts of the country come complaints of foul brood being introduced into apiaries as the result of securing queens from apiaries that are more or less affected; but where the beekeeper raises his own queens from perfectly healthy colonies, he has nothing to fear from outside contamination, so far as introducing queens is concerned.

If the supers are not ready for surplus honey in April, it is high time they were all assembled, as May is the time for the colonies to work in the supers.

Unite such colonies as may be very weak, and in the uniting save the surplus queens, especially if they are young and vigorous, as they can be used for making increase in May.

Swarming Time Among the Bees

In the northern section of the country, June is more prolific in swarms than May, but the experienced beekeepers look with positive aversion upon natural swarms, with the possibility of their emergence and loss in the absence of the beekeeper from the yard.

For this reason the expert beeman in no sense depends upon natural swarming for increase of colonies, but at this time resorts to artificial increase made at his and not the bees' convenience.

Artificial Swarms

The method is both simple and easy: From strong colonies take two or three frames of hatching brood with adhering bees, but without the queen, and in addition a frame heavy with honey, and place all in an empty hive, filling out the hive with combs or frames of foundation and with a queen in a cake inserted between two of the frames of hatching brood.

Close the entrance with a strip of wood, and remove the hive to its permanent location and leave it severely alone for four or five days until the bees become accustomed to their new quarters and have released the queen.

Then pull back the strip at the hive entrance and give the bees an opening of about an inch, increasing it as the bees become more numerous; and before the time comes for winter quarters the nucleus will build up into a booming colony if there is a reasonable flow of nectar in the fields.

In the hands of an expert, even one frame of hatching brood can be made into a strong colony; but for the novice two to four are better.

What to Do With Natural Swarms

Should natural swarms emerge, then after they are hived in a new hive all the section supers with bees should be removed from the nearest parent colony and placed upon the hive containing the swarm, as the swarm is made up largely of the working force of the colony, and to leave the sections on the old hive is a loss of time and honey.

E v e r y weak colony should be strengthened by frames of hatching brood from the strong ones. Thus equalize things in the apiary.

Select Your Breeding Stock

By using the eggs, or rather the larvæ from some queen whose offspring possess good qualities—such as gentleness, good wintering and energetic honey-gathering—the quality of the bees of the apiary may be enhanced.

A beekeeper should be just as anxious to build up a good strain as a stock man or poultryman; and a little care will spell success.

All queens are not equally desirable. Here and there we may find a queen whose bees are a pleasure to work with, and it is from these queens that we should use the larvæ to propagate and make new queens.

A concrete stand for each hive is a thing to be desired, as it not only lifts the hive above the mud, weeds and natural enemies (such as toads and snakes), but makes it much easier for the field bees to leave and return to their homes.

All hives should be painted with a good quality of white paint, as this not only preserves the hives but reflects rather than absorbs the heat of the sun.

The Busy Bee in Midsummer

The swarming season is practically over, though some swarms will emerge during July.

The bees are on the home-stretch now, gathering much nectar from the clovers; and about the middle of July is the proper time to harvest the crop and keep it separate from the later fall flow of darker and not so richly flavored honey.

In the matter of giving the bees additional storage room and harvesting the clover honey, one can not go entirely by the calendar, but must be governed by the conditions afield and the condition of the colony, as each colony will require individual treatment.

Watch the Nectar Supply

If the clovers show a tendency to stop secreting nectar, and here and there patches of it are beginning to die, then we should give no further supply of empty section supers, but rather permit the bees to concentrate all their energies upon what sections they have, as it is folly to end the season with a large number of unfinished sections on hand.

If the season has been late, then the wise thing to do is to let the supers remain on even beyond the twentieth of the month, and be governed by the conditions prevailing afield.

Uncapping the Combs for Extracting

In harvesting the honey from the extracting combs, a good sharp honey uncapping knife is a necessity; and two are better than one, as one can remain in a pan of hot water on the stove while the other is in use, and the knives changed with every two or three combs uncapped. Hot water not only cleans the knives of honey and wax, but by keeping the knife hot makes it cut more readily.

In any case the knives should be very sharp.

Holding the knife at just the right angle, it is a very easy matter when it is heated to slice off the caps of the combs in the extracting frames, and make them ready for extracting.

The honey and the wax that adhere to the knife can be scraped off into a can, and the honey separated from the wax by heating the honey until the wax melts and rises to the surface; when the mass is cooled the wax can be taken off without trouble.

Storage

In harvesting the comb-honey greater care must be exercised, and under no condition should the section combs be stored in a cold or damp place.

The cellar is the worst place imaginable for it, and an ice-box is certain to make it sweat and run over the combs and spoil its appearance.

The warm attic is an excellent place, but be careful to have it stored so that no robber bees can gain access to it.

By all means use a bee escape board in taking off the supers containing the sections.

Late Summer With The Bees

It is a serious mistake to believe that little if any honey can be gathered after the early clovers are exhausted, for in many sections of the country where buckwheat is not grown the fall flow from the asters and goldenrod even exceeds that from the earlier bloom.

Extracting Honey
Bees

While the honey from the late bloom is lacking in the light color and delicate aroma so characteristic of the clovers, yet there is a steady demand for it, and with the increased cost of living its price in many instances equals that paid for the clovers a few years ago.

Naturally there will be a short period of scarcity between the clover flow and the late fall one, and the bees will be inclined to rob, and careful attention should be given the apiary to prevent this.

Watch Out for Robbers

It may be wise to contract the entrances of the hives for a short period to prevent robbing, and in the case of weak colonies this is absolutely necessary, for as a rule strong colonies are amply able to repel robbers, while the weaker colonies are the ones which suffer.

In sections of the country where buckwheat is abundant and constitutes the main flow, the beekeeper will be kept busy, and swarming may be resumed on a large scale.

Buckwheat Honey

While the buckwheat honey has a somewhat rank flavor, very distasteful to some people, still there are others who prefer it to all others, as it has a characteristic bee flavor, and makes a nice combination with buckwheat cakes and country sausage in the winter.

Perhaps the largest producer of buckwheat honey in the country is Frank Alexander, of Delanson, N. Y., and on one occasion his apiary produced as much as 70,000 pounds of extracted buckwheat honey.

Extracting and Artificial Ripening

Mr. Alexander's methods of harvesting these bumper crops vary from the established rules. He extracts from the super combs before the cells are capped over and while the nectar is very watery, and allows it to ripen in huge vats, into which the nectar runs right from the extractor.

The apiary has about 800 colonies right in one yard, the largest individual apiary in the country, and only possible because Mr. Alexander is located right in the heart of the buckwheat region of New York State.

On one occasion I sampled some of this artificially ripened buckwheat honey, and found it equal to that left in the hive and ripened and capped by the bees.

Strengthen the Weak

All queenless colonies should be given queens at once; and, if weak, strengthened by frames of hatching brood from strong colonies able to spare them.

It is a wise thing to examine every colony, know its exact condition, and remedy anything which needs attention before the season is too far advanced.

Keep the weeds down. Look out for the bee-moth. Watch out for robber bees.

Preparing Bees for Winter

Merits of Outdoor and Indoor Wintering

For many years I have been an ardent advocate of wintering bees out-of-doors, right on their summer stands, for at best it is a troublesome business to have to cart the hives into a cellar each fall, and cart them out again each spring.

Where bees can be kept successfully during the honey season, they can likewise be wintered outdoors successfully if proper protection be given each hive.

When wintered outdoors the bees have a decided advantage over those wintered indoors, in that they may take occasional cleansing flights on warm days, and by thus voiding their feces and emptying their bowels, are not so prone to dysentery as bees compelled to retain their feces until they are set outdoors in the spring.

Outdoor Wintering Best

One of the most successful beekeepers of Canada, who for many years stored his bees in an up-to-date and expensive concrete repository, has recently become an advocate of outdoor wintering, and in hives properly protected, has been as successful under the new method as under the old, and feels that it has been a decided saving of work.

Another advantage lies in the fact that it simplifies the matter of taking care of the outyards, and does not compel the carting of bees from the yards many miles to the home cellar.

In the Northern States it would be suicidal to leave the bees out-of-doors in single-walled hives. Where they are to be left outdoors for the winter they should be in double-walled chaff hives with the hive entrance considerably contracted.

Protect from Cold and Slush

Each hive should be elevated upon either a wood or concrete stand, to keep it above the snow and slush which will accumulate on the ground; and while it is true that an entire apiary covered with snow and out of sight will in no sense suffer, but on the contrary be more comfortable for the extra blanket of snow, yet when the snow melts it is liable to run into the entrance of the hive to the discomfort of the bees. A hive stand will prevent this.

Wintering Indoors

If, however, it is your purpose to carry the bees into a cellar for the winter, about the middle of November is a good time, being governed, of course, by the weather conditions.

There are no hard-and-fast rules, but a good principle to adopt is to wait until there has been a warm spell of a couple of days after a cold spell, and after the bees have taken advantage of the two or three warm days to cleanse themselves; then it is a good time to carry them in.

The cellar should be a dark one, not damp, but such as would be suitable for keeping vegetables during the winter.

If there are windows in the cellar they should be covered with pieces of tar paper, until the place is absolutely dark even in the daytime.

The temperature should not be permitted to go above 60 degrees at any time, and this can be controlled by leaving the cellar windows or door open for several hours at night if a warm spell should come on after the bees have been stored.

Don't open the doors before it is dark outside, and be sure to close them before dawn. Having them open a couple of hours, say from 8 P. M. to 10 P. M., will usually be sufficient.

The hives with bees in them may be tiered in rows in the cellar without removing either the hive top or its bottom board.

Orchard Planting

The Staking-Board a Great Aid

When the young tree is dug up for the nursery the balance is broken, because so many of the small roots are several and the large roots are cut off and left in the ground. Therefore, a large top induces too much evaporation, and consequently the living protoplasm, which requires so much moisture to sustain it, dies. This accounts for many trees dying after being transplanted. Cut back carefully, so that there may be less branch than root.

With apples and pears it is advisable to cut back only the wood which is a year old. Peaches may be cut back to a "whip" with no side branches. This is also true with ash and elm trees.

Dig No Holes in Advance

Don't think to expedite matters by digging the holes for the trees beforehand. They dry out and are not fit to receive the tender roots of the trees. Dig the holes large enough to receive the roots without cramping them. Put them in a natural position. Put fine, mellow soil about the roots. Plant the trees a little deeper than they grew in the nursery, to allow for the settling of the soil. Tramp to make the ground firm about them.

Using the Staking Board

Here is an excellent method of staking and planting: Begin by using a small rope or wire, stretching this from the base-line to the corresponding line on the opposite side; then put in small pegs along its entire length, the distance apart at which it is intended to plant the trees. After the whole orchard has been so laid out, take a double staking board having three V-shaped nicks in it. This board may be about four inches long and perhaps four inches wide; any thickness desired. Start at the first peg and place the board so that the stake fits in the center notch; then remove this stake and place it in the notch made at one end of the board, and put another stake in the notch in the opposite end of the board; and continue until the whole orchard is double-staked in this way.

A hole can now be dug between each two stakes. To do the planting the staking-board is again brought into requisition and placed over the hole, so that the two stakes fit into the notches at the end. Then the young tree is held so that its trunk fits into the center notch—just where the single stake stood before the double-staking took place. In either double-staking or planting always work from one end of the row to the other, and always keep the center notch of the board facing away from you.

If one is careful in performing this work it is sure to turn out well.

Not Complicated

This sounds like a complicated operation perhaps, but really it is very simple and easy in practice. The picture plainly shows how the board is made and used, and it is hardly necessary to add that each tree is to be held exactly in place in the notch until most of the soil is shoveled into the hole and firmly secured in place.

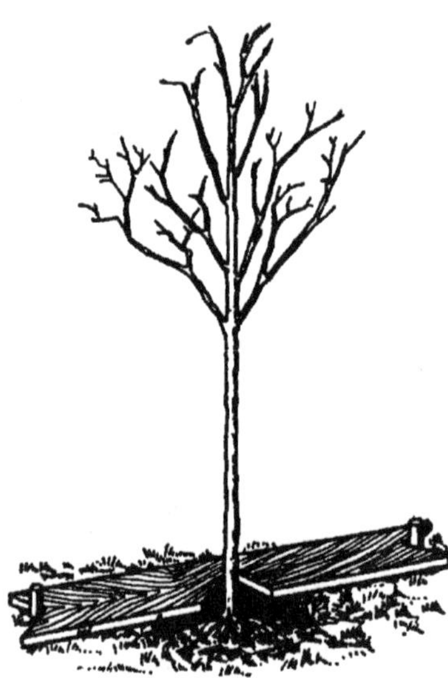

Showing Method of Using the Staking-Board. Note Position of Tree in Notch.

On windy, exposed fields, incline the newly-set trees slightly toward the northwest; the trees will straighten as they grow.

Do not let tree roots lie around in sun and wind, unprotected; as fast as an armful of trees is dug from the heeling-in place, wrap the roots in a blanket until all are set. Look out for crown or root gall, or San Jose; better burn infested trees.

Squares or Hexagons?

Trees may be set either in squares or hexagons; the latter system has some advantages—more trees to the acre and absolute uniformity between trees. The one objection to planting in squares is that it does not cover the ground uniformly with trees; for instance, A is farther from D, and B from C, than A from B or C, or B from D or A—making a waste of space in the middle of the square. This is sometimes utilized by planting a tree there, such as a peach or some quick-bearing or short-lived tree, temporarily to occupy the ground; but this results in crowding in a very few years. Apples should not be closer than thirty-five or forty feet apart; pears, twenty or twenty-five feet; peaches and plums about twenty feet; cherries (sour), sixteen to eighteen feet; cherries (sweet), twenty to twenty-five feet; quinces, twelve feet.

Tree-Pruning Hints

Spring is a good time to prune trees, unless you prefer to wait until June. The rule is that spring pruning induces wood growth and June pruning induces fruit growth. Of course, on young trees you should want only wood growth until they are good-sized and fully able to endure the strain of fruit bearing. Some growers do part of their pruning in March and part in June.

Don't prune mature trees too severely. A tree must have some place upon which to produce its fruit; otherwise it will produce watersprouts instead of fruit.

Don't prune off a single branch unless you know just why you are removing it and why you are removing that particular branch in preference to some other.

Don't neglect to paint all large wounds. Painting will improve the appearance, prevent decay, prevent evaporation of the tree's supply of moisture, and facilitate healing.

Above all, don't allow any man to prune your trees if his chief recommendation is his ability to handle an axe and a saw.

Don't prune your trees because some one else thinks they need pruning. He may not know any more about them than you do.

Don't prune your trees unless you can tell the difference between a dead and a living branch, between a bearing and a non-bearing branch, between a fruit-spur and a water-sprout, and between a fruit-bud and a leaf-bud.

Don't prune off the large limbs when equally good results can be had by removing a few of the smaller limbs. The large ones form the framework of the tree and are needed to support the bearing branches.

It is sometimes stated that the fruitgrowers of the Pacific slope, who produce some of the finest fruit in the world, prune away "nearly half of the tops of their trees" every year. They do nothing of the kind. They remove from one-quarter to two-thirds of the annual growth of the previous season. But they give their trees culture that causes the trees to make a terminal growth of from two to three, and often four, feet. The average eastern farmer gives his trees only enough care to permit the growth of four or five inches of terminal growth; and so his tree tops do not need the same treatment that a larger growth would require.

Trim fruit trees a little every year, rather than much in any one year. Peach trees require more pruning than most trees; at least one-half of the length of the new growth should be removed each season. Cherry trees require the least pruning; merely cut out dead, broken or "crossed" limbs. Other trees need a judicious thinning-out and, sometimes, cutting-back. Avoid cutting so as to leave "stubs"; make neat cuts close to union.

The harder you prune the more suckers you will have; don't overdo a good thing.

Protecting the Orchard from Frost Damage

The heavy demand for fruit at profitable prices, and the fact that Europe must depend upon our orchards for her supply, due to the depletion of her famous groves, should awaken fruit growers to the importance of making arrangements to insure their fruit growth against frosts in the early spring.

The most practical and economical method yet devised for protection of large areas is the direct addition of heat by means of numerous small fires properly distributed over the area to be protected.

When to Start the Fires

An orchard alarm thermometer obviates the inconvenience of remaining on watch. This instrument is so delicate that an alarm bell sounds in the owner's bedroom when the orchard temperature approaches the danger point. The instrument is placed at what is considered the coldest or lowest part of the orchard, and is set so as to ring when the temperature is a few degrees above the danger zone. This allows time in which to light fires and get the heat up to the protection point before the frost becomes dangerous.

When a severe drop in the temperature is accompanied by a heavy wind, it is almost impossible to accomplish anything in the way of heating the orchard sufficiently to ward off the cold. But, fortunately, ordinary frosts occur during still weather.

Fuel

Several kinds of fuel are adapted for use as orchard heat. Wood is very good where it is plentiful enough. Coal, coke and fuel oil all have given satisfaction, depending on the worth of the crop. J. G. Gore, a successful fruit grower in Oregon, saved his crop, which sold for $1,000 per acre, for several years by the use of old rails and stumps for fuel. A little dash of kerosene or crude oil, and the application of a torch, starts the fire. An iron rod about four feet long, wound at one end with cotton waste or rags saturated with oil, makes a good torch.

Anywhere from twenty-five to fifty fires per acre are needed, according to the size of the trees and the degree of frost. Fires need not be built large, as these tend to create currents that draw in cold air from outside, and thus defeat the purpose. Carefully prepare the piles of wood beforehand, every other one being set off first, and those in between held in reserve for further use if the first are insufficient. In this manner it is possible to hold the temperature of the orchard from six to ten degrees above the temperature outside. When the frost is severe a dense smudge should be made about sunrise, which serves to retard the process of thawing until the sun has warmed the atmosphere.

Oil Now the Favorite

Oil has been more satisfactory with commercial orchardists who have a large area to fire. Oil is more quickly handled, either crude or distilled being acceptable. The crude is cheaper, but is less satisfactory because it contains a higher percentage of water, which tends to extinguish the flame and cause the fire-pots to boil over. It is more difficult to handle in cold weather, and in burning gives off large quantities of soot. Distillate, having no water content, ignites freely, burns readily, and creates little residue.

Fire-pots for burning oil may be had of any orchard supply house at reasonable terms, in quantities. From fifty to 100 are used per acre. Oil should be on hand so that it may be used quickly, as the success of this system depends on being ready for business when there is a frost indication.

Proper Props for Trees

Apple growers of the Northwest employ some novel and efficient methods in propping the limbs of bearing trees so as to insure their not breaking under a heavy load.

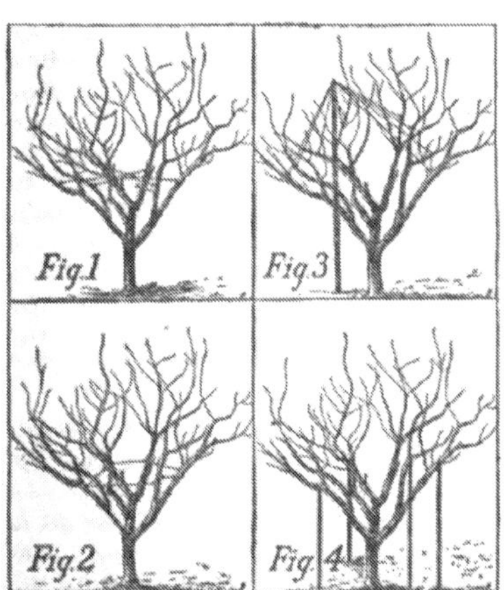

Propping is accomplished any time throughout the growing season when the weight of the fruit trees bears the limbs down so that there is danger of their breaking. Four methods of propping are employed by the orchardists: The center-ring-and-wire, Fig. 1; the cross-wire, Fig. 2; the center-pole-and-wire, Fig. 3; and the single-pole-prop, Fig. 4.

In the center-ring-and-wire method, screw-eyes are placed in the main limbs at some distance above the crotch of the tree. Wires are attached to the screw-eyes and brought to a ring placed approximately in the center of the tree. This holds the tree in shape and prevents the breaking of the limbs at a time when the crop is on.

In the cross-wire method screw-eyes are placed in the main limbs at some distance above the crotch. From each screw-eye a wire extends and is attached to a limb opposite or nearly so. This answers the same purpose as the former method.

In the center-pole-and-wire method, screw-eyes are placed in the main limbs, to which are attached long strands of wire. At the end of each strand is a loop which is placed over a nail driven in the end of a pole.

Cleaning Up   **Orchard**

This pole is raised to a position nearly parallel to the trunk of the tree. This draws the wires tight, holds the trees in shape, and prevents the limbs from breaking.

The single-pole-prop method is most common. This usually consists of 1 x 2 inch or 1 x 4 inch pine strips varying in length as conditions demand; eight to twelve foot lengths are most commonly used. These props are usually sharpened at one end so as to make it easy to place them in the ground. The end which is to hold the limbs is V-shaped, or small lath strips are nailed on each side of the prop, practically forming a notch.

In large commercial apple orchards there are three methods of single-pole propping in common use. First, one crew may haul and scatter the props through the orchard while another crew set them up; second, a crew may haul the props out and set them as they go; and third, the props may be hauled out and set up as needed. It has been found that the average cost is $6.36 per acre for propping, which is oftentimes saved on a single tree's crop, to say nothing of conserving the tree for future development and production.

Clean Orchard in the Fall

The dried peaches and plums that hang on the tree as "mummies" in the fall are those that rotted by the brown rot or ripe rot last summer. They have the germs of decay in them at the present time, and if allowed to remain until spring may scatter such germs through the orchard, filling it so full of the spores of the rot that we would expect a serious outbreak of this disease to injure the fruit again next summer, if climatic conditions are favorable.

The best thing to do is to take off the dried or mummied fruit from the peach and plum trees at any time during the fall or winter, and burn it. It is not enough merely to drop them on the ground, because the spores may possibly spread from there. This removal of the dried or mummied fruits can be done almost any time, but it should be before growth starts in the spring.

It is as bad to leave these germ-laden rotten fruits in the orchard, as it would be to leave in the poultry yard a number of fowls that have died from roup or cholera. It would make little difference what other sanitary precautions were taken if this were done, as the disease germs would spread. It is true that brown rot germs also pass the winter as brown blisters in the bark of the twigs. As they are beneath the surface of the bark they cannot be reached by spray liquids.

The best possible means of preventing damage by brown rot or ripe rot to plums or peaches, is to remove and burn, or bury deeply all of the mummied fruits, and prepare to spray next summer with the self-boiled lime-sulphur solution, made according to the formula of eight pounds of flowers of sulphur and eight pounds of quicklime in fifty gallons of water.

The Orchard Cover Crop

August the Time to Consider It

In August it is time to be thinking about the cover crop you are going to sow in the orchard. You ought to have it in not a bit later than August 1st or 15th.

What's it going to be—clover or rye? I believe they use soy-beans and cow-peas in the South, but I know nothing about those cover crops. I'm talking with the Northern farmer. I myself would recommend red clover, though some farmers like rye or buckwheat. Clover is—what's that big word? — oh, yes, "leguminous." It makes nearly as much humus as rye, and it adds nitrogen to the soil, which is important to good fruit crops.

About ten pounds—no less—of red clover to the acre, sown broadcast and smoothed over.

It is taken for granted that early in the spring you got busy with the disk harrow and chopped up the soil, at the same time working in the manure you spread on it last winter; and that after you were done with the disk you used the spring-tooth harrow, set shallow, to keep down the weeds and keep up a surface mulch. The condition of the soil when you sow is a vital point, you know.

Picking Apples

Eleven Rules That Will Insure Results

Some good apple-picking rules: 1. Pick lower limbs first. 2. See that the ladder is pushed into the tree gently so as not to knock off or bruise the fruit. 3. Hang the basket so as to be able to pick with both hands. 4. Lay the apples in; not drop or throw them. 5. Pick no specked apples. 6. Pick no small, green ones. 7. Do not take much time picking a few little apples out of reach—let them go. 8. In emptying, pour gently, as you would eggs. 9. Do not set one basket or crate on another so that the apples below will be bruised. 10. Lift and set down gently all filled crates. 11. Use spring wagon in hauling, avoid rough ground, and go slow except on smooth road.

Surgery is Valuable Aid in Preserving Trees

It is only in the last few years that a systematic study of tree doctoring has been practiced for the preservation and restoration of valuable fruit or shade trees.

The cavity to be treated must be thoroughly cleaned out. Every speck of rot must be removed.

It must be shaped so that it widens inward instead of outward—which is the simple method used to hold the filling in place. An oval opening should be made, for an oval wound grows over more quickly.

Reinforcing for Large Cavities

When the cavity is large it is necessary to brace it by bolts placed from twelve to eighteen inches apart through the wood, holes of the same size being drilled to take the bolts. A square hole is then cut through the bark for the nut, just deep enough so that it goes entirely under the bark, which soon covers the whole thing. This system of bolts gives the filling something to cling to and prevents it from cracking away from the sides of the cavity.

Disinfecting Necessary

The cavity and the bolts are next creosoted, and then fumigated with bisulphide of carbon to kill any borers that may be in the sound wood. Simply saturate a cloth with the bisulphide, place it in the cavity, and cover with paper over night.

Then the place is ready for filling. Concrete filling is the one most used.

"Trimming" a Tree

"Tinning" is a method used when the cavity is small. A ledge is first made about three-quarters of an inch wide and a quarter of an inch below the cambium (this is the layer next to the bark). This ledge should extend all around the opening. Then a piece of tin, zinc or copper is used the exact size of the hole and fitted into place after painting the inside of the cavity. The metal is also treated in the same manner and fastened to this ledge by galvanized nails.

How to Brace

Sometimes trees have to be braced when the decay has occurred in the fork. The old method was to put a band around the limb, and in a young tree it is not unusual to see the band grown into the bark and the tree frightfully disfigured and often crippled. Now the tree surgeon braces the limbs by cables fastened to eyebolts.

Protect Your Trees from Mice and Rabbits

November is the time to protect orchard trees against injury from mice and rabbits. For field-mice, soil mounding is recommended. The earth should be mounded to a height of from six to eight inches and from twelve to sixteen inches in diameter around the base of each tree and well tamped down. All grass and litter should be cleaned away from the trees.

Cylinders of one-fourth-inch mesh galvanized iron wire, are a good protection against either mice or rabbits. Pieces of such wire 12 x 24 inches make cylinders of a convenient size for small trees. The cylinder, after its edges have been fastened together, should be

Wrong—sinfully wrong!

slightly imbedded in the ground to secure it.

Many protective washes have been suggested from time to time, but most of them have not proved satisfactory. Extensive experiments have shown the ordinary lime-sulphur mixture to be quite satisfactory. It is used at the ordinary strength, as for scale insects. The trunks of the trees should be sprayed or painted close to the ground and to a height of two feet above it.

A wash recommended by Ohio fruit growers is made of one peck of fresh stone-lime slaked with old soap-suds, and the mixture thinned to the consistency of whitewash. To one peck of lime, one-half gallon of crude carbolic acid, four pounds of sulphur and one gallon of soft soap are added. The trunks of the trees should be painted with this wash in late autumn.

In may be that some suckers have started around the base of your trees since the last trimming. Cut every one of these out before snow comes.

The right way

Late in the fall plow a furrow down through the orchard between every two rows of trees if the ground is apt to be wet. The trees will do a great deal better for this surface drainage. Also perhaps some tile drains are needed underground.

County demonstration orchards are showing good results. The cost of pruning, spraying and managing the Nicholas orchard of twenty-nine trees has been forty-three cents per tree.

Grafting and Budding

Plain Grafting

When in the spring the sap begins to move in the stock, be ready; this occurs early in the plum and cherry, and later in the pear and apple. Do the grafting, if possible, on a mild day during showery weather. The necessary tools are a chisel, or a thick-bladed knife or a grafting-iron (with which to split open the stock after it is sawed off smoothly with a fine-tooth saw), a hammer or mallet to aid the splitting process, a very sharp knife to trim the scions, and a supply of good grafting wax. Saw off a branch at the desired point, split the stock a little way down, and insert a scion at each outer edge—taking care that the inner bark of the

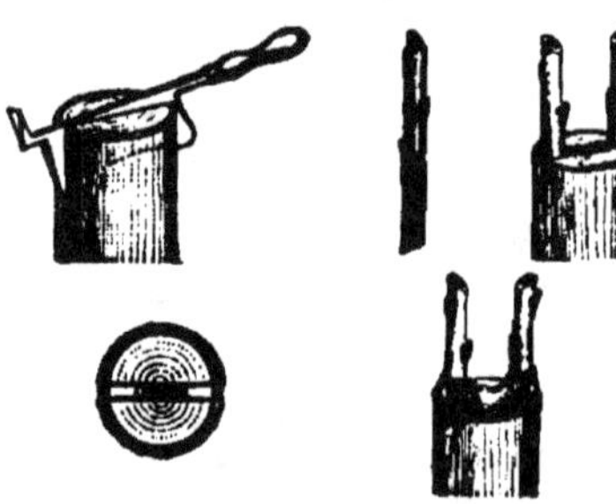

scion fits snugly and exactly against the inner bark of the stock. This—together with the exclusion of air and moisture until a union results—constitutes the secret of success.

Trim the scions to a long edge, as shown in the picture; insert them accurately; the wedge should be a trifle thicker on the side which comes in contact with the stock's bark. Lastly, apply the grafting wax with a brush. Each scion should be long enough to have two or three buds, with the lower one placed as shown. The "spring" of the cleft holds the scion in place, and therefore tying should be unnecessary. If both scions in a cleft grow one may later be cut away.

When grafting large trees it is best not to cut away too much of the tree at once; therefore a few secondary branches should be left untouched and these after the scions are thriftily growing, can gradually be cut away the following years. Or, part of a tree can be thus top-grafted one year and the remainder the next. Many a worthless tree has thus been entirely changed.

You can't graft a pear or an apple on a cherry or plum tree, or vice versa. The stone fruits and the pomaceous fruits are separate families and refuse to intermarry.

Judge Biggle likes to make his grafting wax this way: One pound of resin, one-half pound of beeswax, and one-quarter pound of tallow, melted together. Keep in an iron pot; heat for use when wanted. He says: "It is best to use scions which were cut early this spring or last fall; they can be kept in moist sawdust or sand."

Common putty may be used for grafting wax, and is much cheaper; put it on good and thick and fill all cavities smoothly. Then take cloth, tear it in strips, wind it around the putty and tie it with string. Many fruit men say they have better luck with putty than with wax.

Bridge Grafting

Rabbits have seriously injured fruit trees in many orchards by girdling. When the girdle is only three or four inches wide the tree may be saved by bridge grafting. Trees with large patches of bark removed entirely around the trunk cannot be successfully treated, though those not too badly injured may be saved by special treatment.

Bridge grafting should be done in early spring, scions from healthy trees being selected from twigs produced last season. The torn edges of the wound should be cut off smoothly, and all badly loosened bark removed. The scion should be cut half or three-quarters of an inch longer than the wound and the ends of the scions pointed.

The scion may then be inserted under the edge of the bark, care being taken to have the cut on the scion made rather slanting, to give considerable space for it to unite with the bark of the tree. Several of these scions should be put in around the tree at intervals of not more than one and one-half inches. (See illustration). On small trees, three or four scions will be sufficient.

It is a good practice to paint over the wound areas with white lead, and they may further be protected by binding with cloth. Care should be taken, however, to see that the twine that holds the cloth is not so tight as to girdle the newly-set scions. After the scions have become firmly established, the cloth may be removed.

The scions will continue to increase in size, and as they approach each other the union of one scion to the other may be accomplished by shaving the sides of the scions. In time the entire girdle area may be entirely healed over in this way.

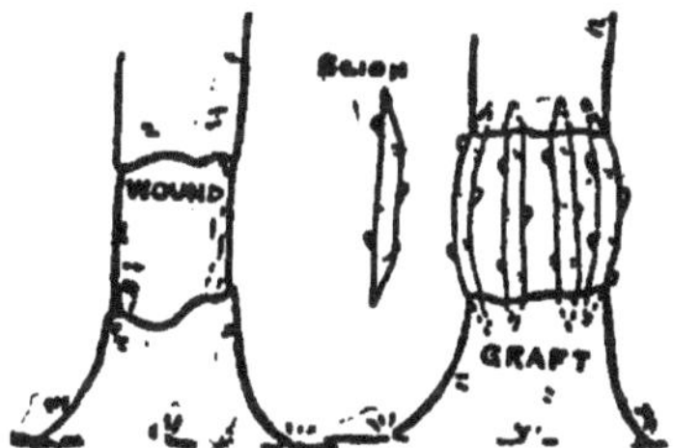

In some cases, bridge grafting will not be necessary. If the inner bark has not been removed by the rabbits, the tree may be saved by immediately protecting the girdled area before it has had time to dry out, by wrapping with cloth which has been treated with grafting wax. The inner bark will then form an outer bark, without serious injury. Where it can be used this method is better than bridge grafting. Trees on which the bark has been removed along the sides and not entirely around the trunk, will be benefited by painting the wound. Before this is done, however, the rough edges of the bark should be removed so as to facilitate healing.

After all is said, the fact remains that it is much safer and better to prevent injury than to cure it. As we have often stated, mice and rabbits can be kept off by wrapping the

Budding | **Orchard**

tree trunks with strips of wood veneer, laths building paper or wire screening. Of course, however, such wrappings do no good after the injury is done.

Budding

The art of budding consists in taking a bud from one tree and inserting it under the bark of some other tree. The union of the two, the bud and the stock, takes place at the edges of the bark of the inserted bud. For this reason, the bud should be inserted as soon as cut from its twig and before it has had time to dry out. The bud should also be full, plump and well matured, and cut from wood of the current season's growth. The stock should be in active growth so that the bark will slip easily.

In cutting the bud a sharp knife is required, as a clean, smooth cut is desirable. The knife is inserted a half inch below and brought out the same distance above, shaving out a small wedge of wood under the bud along with the bark. This wedge is no hindrance to the union and should not be removed. The leaf is always clipped off.

To insert bud, make a T-shaped incision just through the bark of the stock, as shown in the illustration.

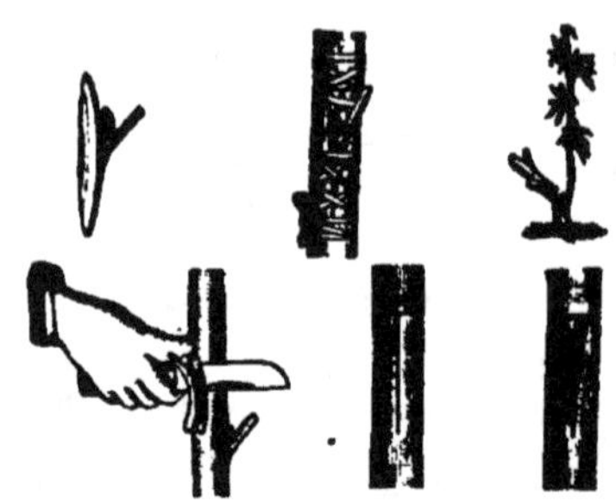

Raise the bark carefully without breaking it and insert the bud. Practice will give ease and dispatch to the operator. The bud must be held firmly to the stock by a bandage wound about the stock, both above and below it, being careful to leave the eye of the bud uncovered. Raffia, bast, candlewick, or waxed cloth may be used for tying. In about ten days, if the bud "takes," the bandage must be removed or the stock will be strangled and its growth hindered. The work of budding is usually performed in July or August in the North, and in June in the South. When the bark peels easily and the weather is dry and clear, is the ideal time.

The Codling Moth

Next to San Jose scale the codling-moth is the chief enemy to fight in the average apple orchard, for this moth is the cause of wormy apples—and farmers don't like apples that way.

The falling of the apple blossoms is the signal to begin spraying; the closing of the calyx lobes a week or two later is the signal to stop spraying.

Orchard   **Codling Moth**

Begin in time but never spray until most of the blossoms have fallen from the tree. Use the regulation Bordeaux arsenical mixture. One thorough application may answer, provided that rains do not wash off the poison during two weeks. Fig. 1 shows an apple from which the petals have recently fallen. Note the wide-open nature of the calyx lobes—the "blossom end." Here the conditions are just right for spraying. The apple stands up straight on its

Fig. 1

stem, the cup-like calyx is held upright and open—and consequently, a worm and a drop of poison may find an easy lodging together in the cup.

Fig. 2

Fig. 2 shows the same apple about two weeks later. Note that the calyx lobes are drawn nearly together, and that the cup is no longer a cup; 'twould be difficult for any poison to enter now.

The Right Way to Saw a Limb

To save time and labor in trimming trees, try sawing the underside of the limb about one-third of the way through or till the saw begins to pinch, and then saw on top about one-half inch in front of the under cut, and when you have sawed down almost to the under cut, the limb will break off and not peel down the side of the tree. One cut will do the same work as two cuts.

Showing Wrong and Right Ways of Sawing Tree Limbs

SPRAYING CALENDAR

PLANT	FIRST APPLICATION	SECOND APPLICATION
APPLE .... (*Scab, rot, rust, codling moth, bud moth, tent caterpillar, conker worm, curculio, etc.*)	When buds are swelling, but before they open, Bordeaux.	If canker worms are abundant just before blossoms o p e n, Bordeaux-arsenical mixture.
ASPARAGUS .... (*Rust, beetles.*)	Cut off all shoots below surface regularly until about July 1st.	After cutting ceases, let the shoots grow and spray them with Bordeaux-arsenical mixture.
BEAN .... (*Anthracnose, leaf blight, weevil, etc.*)	Treat the seed before planting with bisulphide of carbon. (See remarks.) When third leaf expands, Bordeaux.	10 days later, Bordeaux.
CABBAGE .... (*Worms, lice, maggots, etc.*)	Pyrethrum or insect powder.	7-10 days later, repeat.
CELERY .... (*Blight, rot, leaf spot, rust, caterpillars.*)	Half strength Bordeaux on young plants in hotbed or seedbed.	Bordeaux, after plants are transplanted to field. (Pyrethrum for caterpillars if necessary.)
CHERRY .... (*Rot, aphis, slug, curculio, black knot, leaf blight, or spot, etc.*)	As buds are breaking, Bordeaux; when aphis appear, tobacco solution or kerosene emulsion.	When blossoms drop, Bordeaux-arsenical mixture.
CURRANT .... GOOSEBERRY .... (*Worms, leaf blight.*)	At first appearance of worms, hellebore.	10 days later, hellebore. Bordeaux if leaf blight is feared.
GRAPE .... (*Fungous diseases, Rose bugs, lice, flea, beetle, leaf hopper, etc.*)	In spring when buds swell, Bordeaux.	Just before flowers unfold, Bordeaux-arsenical mixture.
MELONS .... CUCUMBERS .... (*Mildew, rot, blight, striped bugs, lice, flea, beetle, etc.*)	Bordeaux, when vines begin to run.	10-14 days repeat. (Note: Always use half strength Bordeaux on watermelon vines.)
PEACH .... (*Rot, mildew, leaf curl, curculio, etc.*)	As the buds swell, Bordeaux.	When fruit has set, repeat. Jar trees for curculio.
PEAR AND QUINCE...... (*Leaf blight, scab, psylla, codling moth, blister mite, slugs, etc.*)	As buds are swelling, Bordeaux.	Just before blossoms open, Bordeaux. Kerosene emulsion when leaves open for psylla, if needed.
PLUM .... (*Curculio, black knot, leaf blight, brown rot, etc.*)	When buds are swelling, Bordeaux.	When blossoms have fallen, Bordeaux-arsenical mixture. Begin to jar trees for curculio.
POTATO .... (*Flea beetle, Colorado beetle, brown rot, etc.*)	Spray with Paris green and Bordeaux when about 4 in. high.	Repeat before insects become numerous.
TOMATO .... (*Rot, blight, etc.*)	When plants are 6 in. high, Bordeaux.	Repeat in 10-14 days. (Fruit can be wiped if disfigured by Bordeaux.)

NOTE.—For San Jose scale on trees and shrubs, spray with the lime-sulphur mixture in autumn after leaves fall, or (preferably) in early spring, before buds start. The lime-sulphur

Third Application	Fourth Application	Remarks
When blossoms have fallen, Bordeaux-arsenical mixture.	8-12 days later, Bordeaux - arsenical mixture.	For aphis (lice) use one of the lice remedies mentioned elsewhere. Dig out borers from tree trunks with knife and wire. For oyster-shell scale, use whale-oil soap spray in June.
2-3 weeks later, Bordeaux-arsenical mixture.	Repeat in 2-3 weeks.	Mow vines close to ground when they are killed by frost, burn them, and apply a mulch of stable manure.
14 days later, Bordeaux.	14 days later, Bordeaux.	For weevils: Put seed in tight box, put a cloth over seed, pour bisulphide of carbon on it, put lid on and keep closed for 48 hours. Use 1 oz. to 4 bus. of seed.
7-10 days later, repeat.	Repeat every 10-14 days until crop is gathered.	Root maggots: Pour carbolic acid emulsion around stem of plants. Club root: Rotate crops; apply lime to soil; burn refuse; treat seed with formalin before planting.
14 days later, repeat.	14 days later, repeat.	Rot or rust is often caused by hilling up with earth in hot weather. Use boards for summer crop. Pithy stalks are due to poor seed; or lack of moisture.
10-14 days, Bordeaux.	Hellebore, if a second brood of slugs appear.	Black knot: Dark fungous-looking bunches or knots on limbs. Cut off and burn whenever seen.
10-14 days, repeat, if necessary.	2 to 4 weeks later, repeat.	Cane-borers may be kept in check by cutting out and burning infested canes.
When fruit has set, Bordeaux - arsenical mixture.	2 to 4 weeks later, Bordeaux.	For lice, use any of the lice remedies. For rose bugs, use 10 pounds of arsenate of lead and one gallon of molasses in 50 gallons of water, as a spray. Or knock the bugs into pans of kerosene every day.
10-14 days, repeat.	10-14 days, repeat.	Use lice remedies for lice. For striped bugs, protect young plants with a cover of mosquito netting over each hill. Or keep vines well dusted with a mixture of air-slaked lime, tobacco dust and a little Paris green.
When fruit is one-half grown, Bordeaux.	Note:—It is safer always to use half-strength Bordeaux on peach foliage.	Dig out borers. Cut down and burn trees affected with "yellows."
After blossoms have fallen, Bordeaux-arsenical mixture.	8-12 days later, repeat.	Look out for "fire blight." Cut out and burn blighted branches whenever seen.
10-14 days later, repeat.	10-20 days later, Bordeaux.	Cut out black knot whenever seen.
Repeat for blight, rot and insects.	Repeat.	To prevent scabby tubers, treat seed with formalin before planting.
Repeat in 10-14 days.		Hand-pick tomato worms.

mixture is a fungicide as well as a scale cure, and if it is used the *first* early Bordeaux spray may be omitted.

Spraying Formulas

Fungicides

BORDEAUX MIXTURE is made by taking three pounds of sulphate of copper, four pounds of quicklime, fifty gallons of water. To dissolve the copper sulphate, put it into a coarse cloth bag and suspend the bag in a receptacle partly filled with water. Next, slake the lime in a tub, and strain the milk of lime thus obtained into another receptacle. Now get some one to help you, and with buckets, simultaneously pour the two liquids into the spraying barrel or tank. Lastly, add sufficient water to make fifty gallons. It is safe to use this full-strength Bordeaux on almost all foliage—except, perhaps, on extra tender things, such as watermelon vines, peach trees, etc. For these it is wiser to use a half-strength mixture.

FORMALIN.—This is also called formaldehyde, and may be purchased at drug stores. Its principal use is to treat seed potatoes to prevent "scab." Soak the whole seed for two hours in a mixture of one-half pint formalin and fifteen gallons of cold water; dry the seed, cut, and plant in ground that has not recently grown potatoes.

BORDEAUX COMBINED WITH INSECT POISON.—By adding one-quarter pound of Paris green to each fifty gallons of Bordeaux, the mixture becomes a combined fungicide and insecticide. Or, instead of Paris green, add about two pounds of arsenate of lead. The advantages of arsenate of lead over Paris green are, first, it is not apt to burn foliage even if used in rather excessive quantities; and, second, it "sticks" to the foliage, etc., better and longer.

Insecticides

ARSENATE OF LEAD.—This is the best insecticide for chewing insects, and is for sale by seedsmen. Use about two pounds in fifty gallons of water.

WHITE HELLEBORE.—This, if fresh, may be used instead of Paris green in some cases—worms on currant and gooseberry bushes, for instance. (It is not such a powerful poison as the arsenates, and would not do so well for tough insects such as potato-bugs.) Steep two onions in one gallon of hot water, and use as a spray.

For Sucking Insects

Now we come to another class of insecticides, suited to insects which suck a plant's juice but do not chew. Arsenic will not kill such pests; therefore we must resort to solutions which kill by contact.

KEROSENE EMULSION.—One-half pound of hard or one quart of soft soap; kerosene, two gallons; boiling soft water, one gallon. If hard soap is used, slice it and dissolve it in water by boiling; add the boiling solution (away from the fire) to the kerosene, and stir or violently churn for from five to eight minutes, until the mixture assumes a creamy consistency. If a spray pump is at hand, pump the mixture back upon itself with considerable force for about five minutes. Keep this as a stock. It must be further diluted with water before using. One part of emulsion to fifteen parts of water, is about right for lice.

CARBOLIC ACID EMULSION.—Made by dissolving one pound of hard soap or one quart of soft soap in a gallon of boiling water, to which one pint of crude carbolic acid is added, the whole being stirred into an emulsion. One part of this is added to about thirty-five parts of water, and poured around the bases of the plants, and about four ounces per plant at each application, beginning when the plants are set out and repeated every week or ten days until the last of May. Used to fight maggots.

WHALE-OIL SOAP SOLUTION.—Dissolve one pound of whale-oil soap in a gallon of hot water, and dilute with about six gallons of cold water. This is a good application for aphis (lice) on trees or plants. For oyster-shell or scurvy scale use this spray in May or June, or when the tiny scale lice are moving about on the bark.

TOBACCO FLEA.—Place five pounds of tobacco stems in a water-tight vessel, and cover them with three gallons of hot water. Allow to stand several hours; dilute the liquor by adding about seven gallons of water. Strain and apply. Good for lice.

LIME-SULPHUR MIXTURE.—S l a k e twenty-two pounds of fresh lump lime in the vessel in which the mixture is to be boiled, using only enough water to cover the lime. Add seventeen pounds of sulphur (flowers or powdered), having previously mixed it in a paste with water. Then boil the mixture for about an hour in about ten gallons of water, using an iron but not a copper vessel. Next add enough water to make, in all, fifty gallons. Strain through wire sieve or netting, and apply while mixture is still warm. A good, high-pressure pump is essential to satisfactory work. Coat every particle of the tree. This is the standard San Jose scale remedy, although some orchardists prefer to use the soluble oil sprays now on the market.

PYRETHRUM, OR PERSIAN INSECT POWDER.—It may be dusted on with a powder bellows when the plants are wet; or one ounce of it may be steeped in one gallon of hot water, and sprayed on the plants at any time. It is often used on flowers, in greenhouses, on vegetables, etc.

BISULPHIDE OF CARBON.—This is used to kill weevils in beans and peas, etc. It comes in liquid form and may be had of druggists. When exposed to the air is quickly vaporizes into a poisonous and explosive gas which is heavier than air and which will destroy all insect life. (Caution.—Do not inhale the vapor, and allow no lights near.)

Tobacco stems, tobacco dust, kainit, soot, freshly-slaked lime, dust, etc., are often used as insect preventives—in the soil around plants to keep away grubs, worms and maggots, or dusted on to discourage the visits of cucumber bugs, etc. (Note.—The first four are excellent fertilizers as well as insect preventives.

Crows and blackbirds frequently pull up planted corn. The best preventive is to tar the seed, as follows: Put the seed in a pail and pour on enough warm water to cover it. Add a teaspoonful of coal-tar to a peck, and stir well. Throw the seed out on a sieve or in a basket to drain, and then stir in a few handfuls of land plaster (gypsum), or air-slaked lime.

A New Fungicide

Some orchardists are now using the following self-boiled lime-sulphur spray, claiming that it is less liable to spot or burn fruit and foliage: Put eight pounds of unslaked lump lime in a barrel; add enough water to cover. When the lime begins to heat, throw in eight pounds of flowers of sulphur. Constantly stir and gradually pour on more water until the lime is all slaked; then add the rest of the water to cool the mixture. About fifty gallons of water, in all, are required. Strain. Two pounds of arsenate of lead may be added, if desired, to the finished mixture, which then becomes a combined fungicide and insecticide, and may be used in the same manner as advised for Bordeaux-arsenate of lead. (Special note. —The self-boiled mixture is not the same as the lime-sulphur advised for San Jose scale, which is too strong for trees in foliage.)

If you do not care to bother with making spraying mixtures at home, they can be purchased, already prepared, of seedsmen. For only a few trees or plants, the extra cost of these factory mixtures is not great.

Testing the Seed Corn

What Every Farmer Should Do

The Rag Doll Method

Every ear of corn, whether old or new, should be tested. Now is the time to make the tests before the rush of spring work comes on. The "rag doll" method is the cheapest, simplest way of testing.

Take strips of heavy, unbleached muslin, 12 x 54 inches. Mark down the middle lengthwise with a lead-pencil, and then crosswise every three inches, beginning twelve inches from one end and making eleven lines. Number the twenty divisions and at the same time number twenty ears of corn to be tested. Take six grains from ear No. 1 (two from near tip, two from middle and two from near butt), no two kernels from same row, and place them on division No. 1 on the cloth, with tips of all kernels pointing the same way, crosswise of the cloth. Place kernels from No. 2 on space No. 2, and so on for all the ears.

Next place a handful of moist sawdust on a piece of blotting paper on one end of the cloth and roll the rag around it carefully so the kernels will not be displaced; roll fairly compact but not too tight. Tie the "rag doll" at both ends. Soak it in lukewarm water over night, drain for half an hour, and stand it on end in a pail lined with a wet cloth—tips of kernels pointing down. A few pieces of brick in the bottom of the pail will afford air circulation and drainage. Fold the pail cloth-lining over the top, put a fairly heavy dry cloth over the pail, set it in a warm place, and moisten the cloths with warm water every day. In seven days, when the sprouts will be about two inches long, take the doll out and unroll carefully. Any ear whose kernels have not grown vigorously should be thrown out. Be careful to throw away the right ear.

Make six or eight "dolls"—a pailful—at the same time. To prevent mold, scald all the cloths used.

Testing the Seed-Corn

The Earth Box Method

Here's how to make the test: Any kind of a shallow box filled with soil and divided into two-inch squares will do for a tester. The easiest way to divide the box into squares is to string wire or cord back and forth and crosswise as shown in the lower picture.

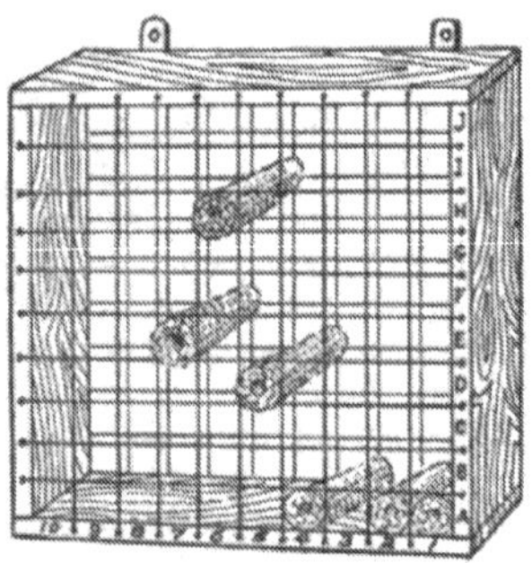

In each square place six kernels from different parts of one ear,—and so on until the box is full. Each square then represents an ear, and the ears are arranged elsewhere to correspond in position with the squares. Slightly press the kernels into the soil, and stretch a wet cloth over all. On this lay another cloth, and cover the latter with soil about half an inch deep, packing it down carefully all around to exclude air. For the sake of cleanliness put another cloth on top, and sprinkle water on this,—about two quarts daily. Keep the box in a warm place in the house, and in about a week the kernels will have sprouted sufficiently to show which ears are best. Discard all ears that do not show vigor as regards germination.

The upper illustration shows a home-made ear-rack numbered and lettered to correspond with the squares in the tester. As an example, to make the idea perfectly clear, the upper ear of corn shown is number "H-6"; and in the "H-6" test-box square below you can plainly see that only four kernels from that ear have sprouted. Discard it.

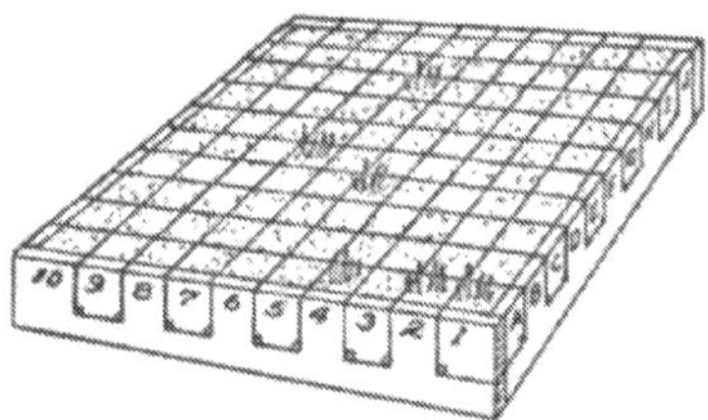

The outfit illustrated will test 100 ears at a time; the size can be increased or reduced to fit your needs, or several outfits can be made and used at once.

Safe Dates for Seeding Wheat

The map below shows the dates when winter wheat can be seeded in the different States in the winter wheat area and still be safe from the Hessian fly menace. Wheat should not be seeded earlier than the date indicated for any section. September 10 is the earliest date, and applies to the Northern States, where frost comes sooner than in sections farther south. The seeding dates indicated will give wheat a chance to get a good start before cold weather.

Raising Beans

Why do we import annually such a vast quantity of beans to supply the home market? Why haven't we awakened to the fact of abundant profit from easy crops of beans raised at home? Farmers all over the country should become aroused to the importance of this hardy, nourishing vegetable, its importance as a staple article of diet—wholesome and nourishing—and its value to the farmer as a safe and profitable crop. Almost anywhere in the Union, beans of one variety or another can be grown profitably.

Beans Mature Anywhere

They will mature successfully in all Northern States, and many of the Southern States have made excellent showings in the way of good bean production.

More people are eating beans. It is no longer the laboring classes, those who are forced to economize, who select beans as an article of steady diet. People are learning the remarkable nutriment of well-cooked beans.

American canners by vigorous exploitation have created a new demand for beans. They have brought people to a knowledge of the delicious flavor of beans when rightly prepared, as well as their rich food value. And, because of the trade they have thus built up, these industries consume annually quantities of beans so vast that they are almost unbelievable.

Watch the Market Quotations

The market page, with its reports of high and steadily increasing prices, is proving a decided encouragement to the farmer—is opening his eyes to the really tremendous possibilities in bean culture.

Simple white "navy" beans are gaining in popular favor at an extraordinary rate. Wherever wheat and corn will grow you can raise these beans.

Of Great Food Value

From every point of view there is wisdom in the old saying about the man who "knows beans." Everybody ought to know them—from the consumer whose bone and tissue, brain and muscle, are all enriched and improved by a diet of good old-fashioned beans, to the farmer, who in beans has an opportunity of raising a safe and heavy crop and of finding a ready market at top prices.

Of all vegetables, beans come nearest to taking the place of meat as a food. No family having access to a bushel of beans need have meat on the table more than six times a week, and hardly that.

Potatoes

Potatoes do best in a loose, well-drained sandy loam, well provided with humus. A clover sod, plowed under in the fall, makes an ideal field. Owing to scab and other potato peculiarities, the potato grower needs to practice a systematic rotation of crops, and to use commercial fertilizer rather than stable manure.

Varieties: Each locality has its favorites; study your market's requirements. Among the best early varieties we might mention: Early Rose, Early Michigan, Early Ohio, Early Northern, Early Bovee, Early White Ohio, etc. Among the best late or main crop varieties are: Rural New-Yorker, Carman No. 3, Sir Walter Raleigh, Great Divide, Vermont Gold Coin, Nebraska, Mammoth Pearl, Rose Seedling, Burbank, Uncle Sam, State of Maine, etc.

Medium-sized seed is best; avoid "jumboes" or "littles." Treat the seed with formalin, as follows: To prevent scabby potatoes, soak the seed for two hours in a solution of one-half pint of formalin (formaldehyde) in fifteen gallons of water. Then dry and cut the tubers for seed. Cellar-sprouted tubers are not so good for seed as those which are unsprouted. Tubers sprouted a little in sunlight just previous to planting are desirable when extra early crops are wanted.

Just how to cut the seed: Leave two strong eyes on each seed piece,

Vegetables   **Potatoes**

and discard the "seed end" (a cluster of tiny eyes) of each tuber.

An important point is to have the soil in perfect condition before planting. Use the harrow thoroughly. Rows for horse cultivation should be two and one-half feet apart. Drop seed pieces fifteen inches apart; cover four inches deep. There are several good machine potato-planters now on the market; but except on large areas it is customary to open and close the furrows with a plow or horse-hoe and drop the seed by hand.

The cut shows an excellent plank clod-crushing drag for smoothing land after it is harrowed. Any farmer can easily make this drag at home.

Cultivation should begin soon after the seed is planted. Go diagonally over the field with a weeder or a light peg-tooth harrow, to break up the soil crust and to kill any weeds which may start. Go over the field again within a week, the other way, diagonally. These early harrowings greatly lessen the after work of keeping the field clean. Some folks are afraid to harrow a planted field, but it is wonderful how the little plants seem to dodge the harrow teeth and come up smiling, uninjured.

When the potatoes are several inches high, a cultivator should be

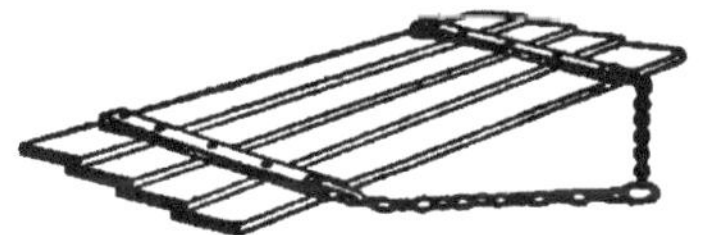

used between rows. If the ground is well drained, hilling-up is unnecessary; although a little soil may be thrown toward the rows at the last cultivation, if preferred. One hand-hoeing during the season is desirable.

Every few weeks the vines should be sprayed with a mixture of Bordeaux and arsenate of lead. Spraying should begin when the plants are about five inches high.

Hints to Potato Growers

1. A loose, rich, gravelly or sandy loam soil is desirable for potatoes.

2. Manure should be applied to the crop that precedes rather than to the potato crop.

3. A clover, alfalfa, cow-pea or soybean sod, plowed under in the fall, will make a good potato seed-bed. Measure depth of furrow to see that it is eight inches or more deep.

4. Like produces like. Hill-selected seed potatoes should be more productive than those from unselected plants.

5. If potatoes are sprouted in the light before planting it will hasten growth. Sprouts should be one-fourth inch long.

6. Treat all seed potatoes for scab before planting. Here is the most approved method of treating them to prevent a scabby crop: Soak the whole seed for two hours in a mixture of one-half pint of formalin (often called formaldehyde) and fifteen gallons of cold water; dry the seed, cut, and plant in ground that has not recently grown potatoes.

7. Do not plant late potatoes too early. Large potatoes planted early are checked during the dry summer and fail to mature a full crop.

8. Never follow potatoes with potatoes. Rotate crops.

9. A well-prepared seed-bed is firm and in good tilth. Preparation before planting is half the battle.

10. Spray plants with Bordeaux mixture at least four times at two-week intervals after the potatoes are well up.

11. Arsenate of lead added to the mixture will destroy bugs.

12. Don't wait until the bugs begin work. Get the arsenate of lead on the plants first.

13. A crop of 200 bushels of potatoes requires 650 tons of water—equivalent to six inches of rainfall.

14 Destroy the weeds. Harrow the soil before the plants appear above ground. Such harrowing kills millions of sprouting weeds, and prevents much future work. The best harrow to use for this purpose is a spike-tooth instrument, for with it there is practically no danger of harming the potato sprouts.

15. Make it a business to push the potatoes. Do not allow the potatoes to push you. Cultivate them six or seven times during the season.

16. When growing potatoes on irrigated land the following things are essential: A carefully leveled piece of ground with a fall of not more than one and a half feet to the hundred, plenty of water, good drainage for surplus water, and a careful study of methods suggested by the State college of agriculture.

17. In growing potatoes in the Great Plains or dry-land section, every method should be used to store up moisture and conserve the supply.

Fresh Vegetables All Winter

A large proportion of the vegetables sold in Chicago in winter are preserved in outdoor pits. It is estimated that not less than 10,000 of these pits are made use of inside the city limits of Chicago, and there are thousands of others in Cook county. Some of the pits are thirty feet long and from ten to twelve feet wide, and hold more than 150 bushels of potatoes, roots, etc. Some are even larger and hold a carload of vegetables, keeping them out of reach of Jack Frost. This is practically the same system, but on a larger scale, as that used by our grandfathers to winter a small stock of garden vegetables.

Not all of the cabbage, carrots, beets, radishes, parsnips and other vegetables stored in these pits are sold in Chicago. Carloads are shipped to other points in Illinois, and even out of the State.

How the Pits Are Used

The pits are sunk in well-drained soil to the depth of fifteen or eighteen inches, and then the vegetables are carefully piled up to a cone, after which there is a thick covering of soil twelve inches thick all over,

Then comes a top covering of straw or strawy manure.

When removing the vegetables to be hauled to market, one end of a pit is opened and the desired quantity removed. The stuff is sacked and then piled in a wagon-box which has been thickly lined inside with gunny sacks or old comforters and blankets. Some of the sacks are kept in a cellar over night and hauled to market the next morning and sold to consumers. Vegetables so preserved are crisp and fresh the winter through.

Strawberry Culture

We wish we could prevail upon everybody who has even a patch of land, to set out a bed of strawberries this spring, take good care of them through the season, and revel in this delicious fruit next summer. Surely we cannot do our people a better service than to persuade them to give immediate attention to this important matter.

It is not a difficult thing for any person with even a small garden, to grow strawberries in such abundance that every member of the family shall have enough for at least three weeks of the summer, for it is an easy fruit to grow, and yields certainly and profusely in response to intelligent effort.

The strawberry bed should be started as early in the spring as the soil can be got into mellow condition. In the latitude of Philadelphia, this occurs usually early in April—farther south, earlier, farther north, later. Let it be understood that it will not do to delay if the best results are to be attained. It will do to plant in early May, but not in early June nor late May.

Select a piece of ground that, from its lay, is well drained, for the strawberry does not like wet feet (neither does it like dry ones); an old sod is not suitable, because it may harbor the white grub, which is very destructive to the roots of strawberry plants. A patch of ground that is likely to be as free as possible from weed seeds is, of course, best, and one out of the way of the chickens.

If you are a beginner and want to grow for market, a quarter-acre is enough to begin with; if only for family use, a bed 20x40 is large enough to supply the family lavishly for nearly a month.

For the quarter-acre 1,500 plants are enough; for the 20x40 garden bed, 100 plants are sufficient. This number allows the rows to be four feet apart and the plants two feet apart in the rows.

In garden culture some growers set the plants a foot apart in the bed and cut off all runners, and this is a good plan, and insures larger berries than can be grown in close-matted rows.

Take horse manure and throw in a heap until it heats enough to sprout and kill the weed and grass seeds that it may contain; then give your ground a big dose, bigger than you ever applied before to any land; plow down and harrow thoroughly.

In setting the plant, make a fan-shaped hole in the ground with a mason's trowel and insert the roots, close the soil around the crown, and with the toe of your boot press your whole weight down, to insure a firm setting. First puddle the roots in mud, so they will begin growing at once; never allow the roots to become the least bit dry at any time.

is well to cut off the tip ends of the roots before planting. Discard all feeble plants; set none but those with plenty of roots.

Under this treatment, if the plants are good ones, you will soon be delighted to see that growth has started and the foundation of a future crop is well laid.

The plants received, puddled, that is stood in mud to keep the roots from drying and planted in the manner described, you have made a fine start; if you keep on at that pace you will come out ahead of the race.

Now, no weed must get a week old in that patch or bed; the soil must be kept as mellow as mellow can be; runners must never be allowed to take root within eight inches of the mother plant, nor of each other; or, if they do take root, they must sometime be pulled up or cut off; and an occasional dusting of wood ashes and bone meal along the rows will be a fine thing. Next fall you will see the rainbow of promise, in the form of matted rows two feet across, filled solidly with ro-

Fig. 1

Fig. 2

Fig. 3

Fig. 4

Fig. 1 shows a plant set too deeply. One thus set will have the crown covered with soil, which will injure it. Fig. 2 shows too shallow setting, which will allow of the roots dying. In Fig. 3 is shown the roots bunched improperly, and Fig. 4 indicates how a sensible man will do this job. It

bust, healthy plants eight inches apart, provided, however, you keep the old hen away and do not allow leaf rust to prevail. You must watch for rusty brown and reddish spots on the leaves, and use Bordeaux mixture when needed. Then, when winter fully sets in, and the ground becomes solidly frozen, cover the rows all up out of sight with a stable manure (free from weed seeds), three or four inches deep, and leave it on until freezing weather is over, and the plants are ready to wake up; then rake some of the manure off into the path between the rows, but only enough to let the little plants see the sun. Let no weeds grow until fruiting time, but pull, not hoe them out, for the ground should not be disturbed. Next summer be sure and have a good supply of cream on hand and send us an invitation to visit you. We will do the rest.

Our Neglected Bush Fruits

Two very important bush fruits that are not receiving the attention they deserve, are the currant and gooseberry. Known and recognized as two of the most hardy varieties of fruits, growing and bearing fruit under the most adverse circumstances, this hardiness is their own worst enemy. Too frequently they are relegated to unused portions of the garden, usually along the fence, where they are permitted to grow without care and to be choked with grasses, briers and weeds. No efforts are made to check their only serious pest—the imported currant-worm — and they cease to thrive, the fruit grows smaller and poorer in quality, and after a time the grower remarks: "Beats all how they run out!" Then he wields the scythe or grubbing-hoe and these noble fruits are no more.

Excellent Flavors When Ripe

Both the currant and gooseberry, while having distinctive flavors not appreciated by all people, have most deserving but little understood qualities. Much of the prejudice against the gooseberry arises from the fact that it usually goes to market in a green and immature state; and, to the person unfamiliar with the fruit, one bite of a green gooseberry is sufficient to make an everlasting enemy. The gooseberry is at its best when fully ripe. A green plum, a green blackberry or a green strawberry has no appealing qualities. Neither has a green gooseberry. One has only to know the toothsome quality of gooseberry preserves and gooseberry butter to appreciate it; and as a fruit to be eaten out of hand, when ripe, it has few superiors.

Market Never Overcrowded

The market for currants is never overcrowded when clean, attractive and well-ripened fruit is offered. Currants stand at the head of most fruits as a jelly foundation, possessing a pleasing flavor and beauty of coloring. Some of the more recent varieties are so mild as to be greatly relished as a dessert, and are most highly palatable when eaten with cream and sugar. A special advantage of the gooseberry and currant is that they can remain on the bush for a much longer period after ripening than most other fruits. The crop is not one that easily spoils. White Grape currants may be kept on the bushes for three weeks after they have ripened, and the flavor and quality will be unimpaired at the end of that time. The White Grape currant, by the way, makes a pink jelly of remarkable beauty.

Varieties to Grow

Almost any variety of the currant or gooseberry, if given good care and cultivation, will prove satisfactory. Among the varieties commonly grown are Fay's Prolific, Red Cross, Perfection and Wilder, among the red sorts; White Grape is the only white variety worth bothering with. It no longer pays to grow the Red and White Dutch; they are pioneer sorts that have been greatly improved upon.

Among gooseberries, popular sorts are the Downing, Chautauqua, Smith's Improved and Red Jacket; as well as English varieties like the Industry, Crown Bob and Lancashire. English varieties are more susceptible to mildew than the native sorts, but where they can be grown successfully they are highly appreciated on account of their size and coloring.

In my experience I have found no variety of gooseberry that equals the old Houghton. I have had three-year-old-plants bear an average of two quarts each; mature bushes running as high as ten and even fifteen quarts. I have never known the Houghton to fail as a cropper, and the plants are hardy and quick-growers. The most serious pest of both the currant and gooseberry is the imported currant-worm, which may be easily controlled by spraying with hellebore at the first appearance. Once a plantation becomes thoroughly infested, the fight is a hard one and the worms usually conquer.

Berry Bushes and Grape Vines

Cut back one-third or more of the length of last year's growth on currants and gooseberries, and cut out surplus or unthrifty shoots entirely. Very old shoots are likely to be infested with borers, and should gradually be replaced with younger growth.

Grape or currant cuttings can be made from the largest and best developed wood of the past year's growth. This should be cut into pieces eight inches long, leaving at least two buds, and packed in damp sand or moss in the cellar until planting time in spring. Make cuttings as early as possible.

When you prune raspberries and blackberries, cut out all canes that fruited last season (should have been done last fall); also all winter-killed canes. Shorten remaining canes to three or four feet, and cut off at least a third of each long side-shoot. Rake up and burn all brush promptly, and thus get rid of many insect pests and disease germs.

How to graft old grape-vines: Saw off the vine about three inches below the ground level, let it bleed twenty-four hours, then split down one side of the stock and insert the scion as shown in the drawing. No grafting wax is needed—simply cover with two inches of moist soil, as indicated by the lower dotted line, and drive a stake on which to tie the new vine when it grows. When the stake is in place, or within a few hours, mound up the soil as shown by the upper dotted line. The mound should not be disturbed by hoe or cultivator until the union is well formed. If the mound forms a hard crust, it should be carefully broken with the fingers. Remove suckers which grow from the stock, but be careful not to injure the scion.

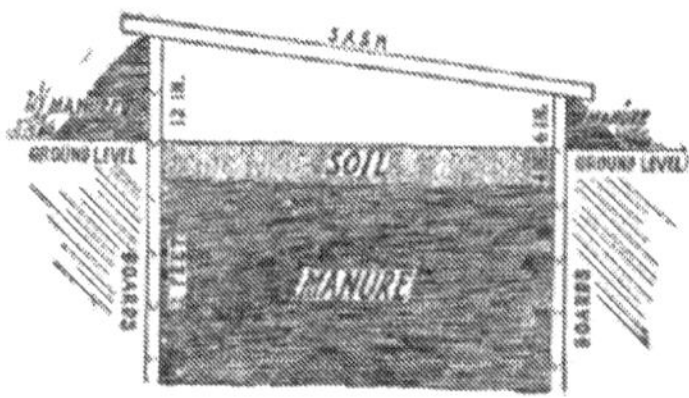

One shoot from each scion bud is enough; surplus shoots should be cut out—leave only one or two, or three at most.

The Standard Hotbed

Did you ever make a hotbed? It's easy. Follow the directions here given and you cannot fail.

A hotbed is nothing more than a board-edged pit, in which there is fermenting horse manure covered with several inches of soil. The top of the hotbed is roofed with one or more sashes, which usually measure about 3 x 6 feet each. At night a straw or other mat is laid over the glass to keep out the cold.

Hotbeds are usually made of inch boards. If the boards on the back of the frames are twelve inches above ground, those in front should be several inches lower; thus giving a slant to the sashes, enabling water to run off quickly.

Throw the manure into the hotbed in successive layers, continuously tramping. Fill the pit to within four or five inches of the top of the frame on the front side. The manure will settle several inches before time for sowing the seed. Place sash on the frame immediately after filling.

The heat in a newly-made hotbed will rise rapidly until it reaches a temperature of at least 120 degrees. A high temperature may be maintained for a week or more, but it will not do to sow seed over such hot material. Wait until the temperature drops below 90 degrees, then place two or three inches of good soil over the manure if flats are to be used, or about four inches if the seed is to be sown directly in the soil.

PLANTING TABLE FOR VEGETABLES AND BERRIES

VARIETY	For Horse Cultivation Have Rows	For Hoe or Wheel-Hoe Cultivation Have Rows	Distance Apart in the Row	Depth to Cover	Time to Plant in the North, Outdoors (See Foot-note)
ASPARAGUS, Seed	2½ ft. apart	1 ft. apart	3 in. transplant in 1 year	1 in.	March-April
ASPARAGUS, Plants	4 ft. apart	3 ft apart	2 ft.	5 or 6 in.	March-April
BEAN, String	2½ ft. apart	2 ft. apart	Thin to 4 in.	2 in.	May 10-15
BEAN, Lima	Pole, 4 x 4 ft. apart	4 x 3 ft. apart	Thin to 3 plants to a pole	1 in.	May 20-25
	Bush, 2½ x 1½ ft. apart	2 x 1½ ft. apart			
BEET	2½ ft. apart	1 ft. apart	Thin to 5 in.	1 in.	March-April
BLACKBERRY, Plants ...	8 ft. apart.	6 ft. apart	2 ft.		April. Or in the fall
CABBAGE and CAULI-FLOWER, Plants	2½ ft. apart	2 ft. apart	16-24 in.		Early kinds, April; late kinds, June
CARROT	2½ ft. apart	1 ft. apart	Thin to 5 in.	½ in.	March-April
CELERY, Plants	3-4 ft. apart	2-3 ft. apart	6 in.		Early crop, May; late crop, early July
CORN, Sweet	4 ft. apart	Same	8-12 in.	2 in.	First sowing, early May
CUCUMBER	5 x 5 or 6 x 4 ft. apart	Same	Scatter 15 seeds in hill; thin out later	½ in.	May 15
CURRANT and GOOSE-BERRY, Plants	5 x 5 ft. apart	5 x 4 ft. apart			April. Or in the fall
EGGPLANT, Plants	2½ x 2½ ft. apart	2 x 2 ft. apart			June 1
LETTUCE	2½ ft. apart	1½-2 ft. apart	Thin to 6-10 in.	½ in.	March-April
MELON, Musk	6 x 4 ft. apart	Same	Scatter 15 seeds in hill	½ in.	May 15
MELON, Water	8 x 8 ft apart	Same	thin out later	½ in.	May 15-20

PLANTING TABLE FOR VEGETABLES AND BERRIES--Continued

VARIETY	For Horse Cultivation Have Rows	For Hoe or Wheel-Hoe Cultivation Have Rows	Distance Apart in the Row	Depth to Cover	Time to Plant in the North, Outdoors (See Foot-note)
ONION, Seed	2½ ft. apart	12-15 in. apart	Thin to 4 in.	½ in.	March-April
PARSLEY	2½ ft. apart	1 ft. apart	Thin to 6 in.	½ in.	Early April
PARSNIP	2½ ft. apart	1 ft. apart	Thin to 5 in.	½ in.	March-April
PEPPER, Plants	2½ ft. apart	2 ft. apart	20 in.		June 1
PEAS	3-4 ft. apart	2½-3 ft. apart	Continuous row	3-5 in.	March-April
POTATO	3 ft. apart	2-2½ ft. apart	12-18 in.	4 in.	Early, March-April; late, May-June
RADISH	2½ ft. apart	1 ft. apart	Thin to 3 in.	½ in.	March-April
RHUBARB, Plants	4 ft. apart	3 ft. apart	3 ft.	2 or 3 in	March-April
RASPBERRY, Plants ...	6 ft. apart	5 ft. apart	Red, 2 ft. Black, 2½ ft.		Early Spring
SPINACH	2½ ft. apart	1 ft. apart	Thin to 5 in.	1 in.	March-April (or fall)
SQUASH-PUMPKIN ...	8 x 8 ft. (Bush Squash 4 x 4)	Same		½ in.	May 15-20
STRAWBERRY, Plants	4 ft. apart	3 ft. apart	15-20 in.	Have crown level with ground	April. (Pot-grown plants in August)
TOMATO, Plants	4 x 4 ft. apart	4 x 3 ft. apart			May 25-June 1

NOTE.—Planting time varies according to season and locality; dates given above are only approximate, and are based on latitude of Pennsylvania; allow about five days difference for each 100 miles north or south of this State. Do not work soil in spring while it is very wet and soggy; wait. Plants set in autumn must be well mulched with strawy manure, leaves, etc., during first winter. Successional sowings of corn, peas, etc., may be made later than the dates given.

How to Trap the Raccoon and Opossum

The raccoon and opossum are quite hard to capture, especially the raccoon. They are found in wooded country, usually not far from water. The raccoon is distributed all over America, but the opossum does not range very far north. Generally speaking, the animals are most numerous in swamp and marsh land. The flesh is good to eat, and can be sold in many cities. When the pelt hunter can market both the skins and the carcasses he makes more money trapping the animals.

The beginner must pay attention to the traps that are used for taking the raccoon. The animal is large and strong and often pulls out of holds which would prove effective for other fur-bearers. Unless fastenings can be put in deep water it is best not to employ stakes. Wire the chains to rocks or logs instead. The rocks or logs should weigh at least twenty-five pounds. A captured raccoon can drag these but a short distance, and does not get a straight pull nor much of an opportunity to work its leg from between the jaws.

Nothing smaller than a No. 1½ trap should be put out. Catches can be made with No. 1 traps, but it is best not to take any chances of losing the skins. If stakes are utilized, do not use soft wood. The raccoon has strong, sharp teeth and uses them to

advantage. It is not unusual to have them gnaw the stakes and escape with the trap. This causes needless suffering—a thing the trapper should prevent at all times.

The raccoon, like its larger brother, the bear, has a keen scent and always seems hungry. A good bait is almost a necessity in trapping the animal. Among the baits used by professionals a r e h o n e y, fis h, fresh, smoked and canned—corn, clams, etc.

When making sets, the trapper should be careful to hide all boot marks or other evidence of his presence. Unless this is done but few skins will reward him.

Sets for the raccoon give best results when placed in water. A good method is to build V-shaped pens of rocks near shore, place a bait in the back part and guard it with one or more traps. If honey or canned fish is used for bait, it must be above the water. If a small perch or sucker is used, stake it just below the surface so it can easily be seen. The idea is to give a natural appearance to the fish.

Do not overlook partly submerged

hollow logs anchored along a stream. Place traps at each entrance. If the water is too deep, build a base of mud, stones, or other material for each set; when the water is shallow make an excavation for the trap. No decoy is needed for a set of this kind, for the first animal in passing will try to enter the log.

Search for raccoon tracks where small streams enter larger ones. The tracks resemble the imprints of a small baby's foot. Once seen they are not easily forgotten. As a rule, in places like those just mentioned, concealed traps will prove very effective. In case there are no distinct paths, use a bait and arrange the set so that the animal in investigating is bound to be taken.

In shallow water, not more then ten feet from shore, open clams and arrange them in a pile so that the top is slightly above the surface. Surround the pile with traps. For this plan the water must be reasonably clear and the set must be hidden by grass, moss or similar material from the bottom of the stream or pond. Make a trail of canned salmon leading to the spot.

The raccoon often travels under shelving banks, and it is there that some of the largest catches can be made. When such a location has been found, conceal the traps where the tracks enter the water.

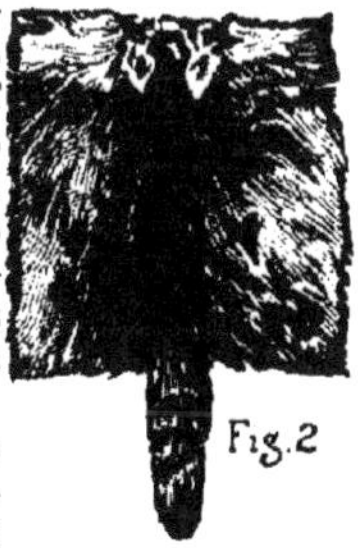

Land sets like those used in taking skunks and civets may fre-

Fig. 2

quently be of value if care is taken in the arrangement. Traps can be concealed in trails leading into cornfields.

Raccoons never fail to investigate any bright object in the water. Knowing this, pelt hunters often place a piece of bright tin near their traps, so that when a raccoon approaches to investigate he is taken.

In steep banks, dig pockets about eighteen inches deep and four or five inches in diameter. Have these so that the entrances are in water but the back parts above it. Put bait in these excavations. Smoked fish is good bait because it gives off an odor which the animals can detect and locate. Conceal these sets with moss or water soaked leaves.

To skin the raccoon cut from the tip of the jaw to the root of the tail. Remove the fur in the same way that the hide of a steer is removed. The raccoon is the only one of the smaller animals that should be skinned open.

Fig. 1

Raccoon trapped in northern and central sections should be handled as shown in Fig 1. Do not overstretch the skins. The heavier and more dense the fur, the greater the value; overstretching injures the fur and depreciates the value. Raccoon trapped in the southern sections should be stretched square, as shown in Fig. 2, as most southern raccoons are used for coat purposes and will bring considerably more if stretched as nearly square as possible.

Raccoon and Opossum   Trapping

The opossum is not so hard to trap as the raccoon. One of the best ways is to hang sardines about a foot from

Raccoon Tracks

the ground in bushes and conceal traps under them, covering with something natural to the place. The opossum can smell the sardines for long distances.

Good places to trap opossums can be found along streams. They travel in such places, generally following a trail in which sets may be made. If there are no distinct paths, use rabbit, fowl or muskrat for bait, suspending the bait above the trap on a stake. If fowl is used, pluck and leave the feathers near. Scatter particles of canned fish where the opossum travels, in trails leading to the traps. Sets should not be too close together. Several trails 100 feet long will accommodate about two or three traps each. If opossums are numerous it is possible to capture a half dozen during a night with the method just explained.

Opossum Tracks

In preparing opossum skins for market do not split the pelts open, but case them pelt side out. Scrape off all superfluous meat and fat and chop off the tail, as this is worthless and only adds to the weight of the skin. See Fig. 3 for correct way to stretch. The pelts need to be dried only long enough to hold their shape.

Skinning, Stretching and Drying Pelts

Instructions That Will Insure Results

Next to catching fur-bearing animals, too much stress cannot be placed on the importance of properly skinning and stretching, so that the best prices may be obtained. Skins improperly removed and stretched will often bring from ten to thirty per cent. less than they would if properly handled. If well stretched, you can tell better what size they are, and therefore grade them properly, enabling you to know whether you are receiving full market value.

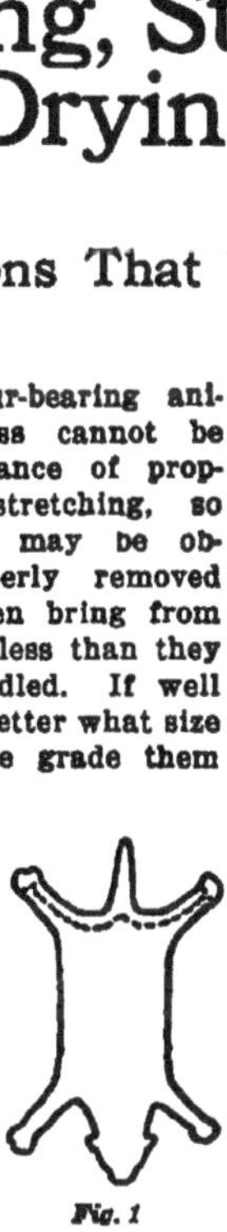

Skin animals as soon as possible after taking them from the traps, but not until they are dry. If you have a water-soaked muskrat, mink, or other animal, rinse it carefully, then take it by the hind legs and tail and crack as you would a whip. Next take the head and crack in the same way. The animal will soon be dry.

Fig. 1

To skin an animal cased, make cuts as shown in dotted lines in Fig. 1, cutting from the toes of one hind leg. Make no other cuts. Be careful not to cut the skin or leave any more fat or meat on the skin than can be helped. Skin down over the head and eyes, skinning even the nose. Skin very carefully over the ears and eyes. Cut tails from muskrats and opossums only. Skin the feet out of valuable animals.

To skin open, make additional cuts from one front foot to the other, and from the base of the tail to the under lip. Stretch the skin in natural shape on a board, tacking it fast at the edges.

After skinning cased, stretch as soon as possible in natural position on the board. Stretch fully, but not too much. Keep the pelt side out. It is not necessary to turn skins, but sometimes mink and fox skins are turned. Turn in from twelve to twenty-four hours. Pelts that are skinned cased should be tacked at the large end of the board. When the skins are dry enough to hold their shape well, remove them from the board.

The boards for cased skins should be planed smooth, with beveled edges, and taper gradually. The thickness should be from one-quarter to three-quarters of an inch, according to the skin. Shingles are fine for making muskrat boards. For convenience in handling the boards should be a few inches longer than the pelts. The

Drying Pelts   **Trapping**

sizes vary in different parts of the country, but the following measurements for large, medium and small animals will do for making boards in most cases. The sizes are in inches, and do not include the tails:

ANIMAL	Length Inches	Width Shoulder	Width Base		ANIMAL	Length Inches	Width Shoulder	Width Base	
Martin	28	4	4¾	L	Oppossum	26	5⅝	7¼	L
	25	3⅝	4⅜	M		21	5	6¾	M
	22	3⅜	4	S		16	4¾	6⅝	S
Raccoon	32	8⅛	10¼	L	Badger	32	8¼	10¼	L
	28	7⅞	9⅝	M		28	7⅝	9⅜	M
	24	6⅝	8¾	S		24	5⅝	8¾	S
Fox	45	6	7⅜	L	Skunk	30	6¼	8	L
	38	5⅝	7¼	M		25	5⅝	7½	M
	32	5¼	6¾	S		20	5	6¾	S
Otter	64	6¼	8¼	L	Mink	21	3½	4	L
	56	5⅝	7⅞	M		18	3⅛	3¾	M
	48	5¼	7¼	S		15	2⅞	3½	S
Weasel	16	2	2½	L	Muskrat	14¾	6	7¼	L
	14	1⅜	2	M		13¼	4½	6⅝	M
	12	1⅝	2¼	S		11¾	4	5½	S

After the skins are stretched, they should be placed in a cool, shady place (never in the sun or near a fire) to dry. When partly dry, scrape with a dull knife to get the fat and meat off, but do not go too deep. Freezing does not hurt skins. When shipping, pack flat, fur side to fur side, or pelt side to pelt side. Always put a tag on the inside, and insure your furs, if shipping by parcel post.

1. Muskrat board.   2. Plain mink board and key.   3. Three-piece board.

The Hessian Fly

Frequent warnings to speed the plow in order that the food supply shall not suffer, may induce some to begin seeding the winter wheat shortly after small grain is harvested, or immediately after the corn is removed and placed in the silo. And that is just why the bug editor is inclined to say, "wait a while."

It is only by waiting that the crop can be made secure against the ravages of the Hessian fly, an insect which causes more damage to the wheat crop in the United States than any other insect pest. During seasons when the fly is especially abundant, hundreds of thousands of acres of wheat are totally destroyed or so badly injured that the yield is reduced fifty to seventy-five per cent. Money losses run far up into the millions.

Disking stubble ground, or burning stubble immediately after harvesting the grain, thorough preparation of the seedbed, late seeding and the use of good seed, are effective measures for controlling the pest in winter-wheat growing regions. A trap crop of wheat may be sown immediately after harvest and disked under later in the fall before seeding the main crop. In spring-wheat growing sections late seeding will not apply; on the contrary, the earlier it is sown in the spring, the less it seems to suffer from this pest.

The general rule for winter-wheat seeding is that there should be a difference of one day for each ten miles of difference in latitude, and seeding should be approximately one day earlier for each 100 feet of increase in elevation. There is usually, however, a period of several weeks in all the winter-wheat area where sowing may take place with about equal results. This period is longer as one proceeds to the southward.

The Grasshopper

After seventeen years of study the Kansas grasshopper has been reduced to a harmless quantity. The grasshoppers that do the damage are native. That is, they develop and perpetuate themselves on one farm; they do not move about. Of course, no one should confuse these native grasshoppers with the hordes of small red ones that used to sweep down from the North in armies. The latter, raised in arid land, are forced to migrate to obtain food.

In the counties that provide the materials, poison is spread on the farms. The formula used is the following, obtained after years of experimenting: No. 1. Two and a half pounds Paris green or white arsenic; fifty pounds bran (mix these dry). No. 2.

Six oranges or lemons, chopped up fine, rind and all; four quarts syrup; five gallons water (mix these three together thoroughly). Mix Nos. 1 and 2, then add sufficient water to make a wet mash.

The lemon and orange in the mixture attract the grasshoppers, who find it irresistible and deadly. A scientific count showed that from two-thirds to three-quarters had been killed.

Alfalfa should be disked and cross-harrowed early in the spring as soon as the frost leaves the ground. This throws out the eggs of the grasshoppers, to be destroyed by the weather and eaten by the birds. This method of culture, first advocated by the University of Kansas, not only lessens the number of grasshoppers, but also has been proved to increase the yield of the alfalfa fully one-third. Scatter the poison, disk the fields and say "Good-bye" to the Kansas grasshopper.

Three Autumn Pests

Most of the warfare against enemies is concentrated in big "spring drives." In the war against insects there are quite a few enemies which can be controlled almost as effectively by "fall drives." Three such enemies are the grape cane-borer, the fall canker-worm and the white-marked tussock-moth.

The first of these, the grape cane-borer, is a serious pest in the grape-growing sections of the Central States. It has caused the destruction of many large vineyards and necessitated extreme measures of control in other plantations.

This pest is called the grape cane-borer because of the manner in which it bores into the canes and hollows them out, as shown in the illustration. It breeds in diseased and dying wood of the grape, fruit trees and shade trees. As a result of its work, the canes break off when pruned, or under a heavy load of fruit. Control methods are simple if started in time. The dead, diseased and dying wood of the vines should be removed and burned. All trash and rubbish allowed to remain in the vineyard serve as first lines of defense for this enemy.

Fall Canker Worm

The second pest that effectively attacks in the fall, the fall canker-worm, is destructive to the buds and foliage of apples. A "fall drive" on the moths consists in placing bands of tanglefoot or other sticky substance on the trees. Moths will be caught by these bands when they ascend the trees to lay their eggs. Printer's ink, bod lime and raupenleim may be used for the bands if a band of cotton-batting or heavy building paper is first placed on the tree to prevent injury to the bark. Tanglefoot may be applied directly to the bark.

Tussock-Moth

A third enemy is the white-marked tussock-moth, which does great injury to apples, the caterpillars gnawing into the fruit and disfiguring it. However, their appetite for fruit does not keep them from doing a great deal of damage to shade trees, particularly on the foliage of elms, poplars and chestnuts.

The most effective measure is that of gathering and burning the egg masses which are deposited early in the fall, and which can be found on the bark and dead leaves of trees in October and November. These egg masses are gray in color and somewhat fuzzy in appearance. Often school children are offered prizes for the largest collection of eggs.

Grain Weevils

Grain infested with weevils loses in weight, is undesirable for seed, and is unfit for human consumption. Nor is such a grain good feed for livestock. Millions of dollars are lost each year simply because many farmers do not understand how to deal with the weevils.

The mature Granary Weevil is only about one-sixth of an inch in length and the color is a shining chestnut brown. This species is unable to fly; but it doesn't worry on that account. No, indeed! For it easily makes up in grain-puncturing and egg-laying power all that it lacks in wing power.

The female Granary Weevil attacks all kinds of grain, but prefers that which is husked. After puncturing the grain she inserts an egg; this hatches into a larva that devours the mealy interior. This egg-laying process is continued for an extended period, and in a single season one pair of weevils will, it is estimated, produce 6000 descendants!

The Angoumois Grain Moth came to this country from France nearly two hundred years ago. The color is light grayish brown, lined and spotted with black. This insect is very apt to deposit its eggs in unthreshed grain in sack or mow. Where the moths appear in force it is wise to thresh the grain quickly and hurry it to the mill, rather than attempt to store it.

Preventive Measures

Now for general remedies. Careful attention to the following preventive measures may bring relief:

First: Never store fresh grain in bins or granaries (or even under the same roof) where there is, or has been, weeviled grain. Before using such storage places remove all grain and thoroughly scrub, clean and fumigate the bins, using bisulphide of carbon.

Second: Remember that damp, warm bins foster the rapid increase of insect life. Endeavor to have the granary cool and dry.

Third: Build the granary as nearly vermin-proof as is possible. Cover windows with fine wire gauze. See that doors, floors, walls and ceilings are tight.

Remedies

As regards aggressive remedies, there is one which is a grand success—carbon bisulphide. This is a colorless liquid which rapidly vaporizes into a heavy gas which works downward. Rightly applied to infested grain or seed the cost is slight and no injury results to edible or germinative qualities.

How to use bisulphide: See that the grain receptacle can be tightly closed. Figure out the cubical contents of the receptacle or bin, and apply the bisulphide at the rate of about one pound for each 1,000 cubic feet of interior space. Place the liquid on top of the grain in shallow pans; about a teacupful in each. Then quickly close the bin for twenty-four hours.

Cautions: The vapor is highly inflammable and poisonous. Do not breathe it nor allow any light near. Thoroughly air the bin or building after fumigation.

The Cabbage Worm

This pest is the larva or caterpillar of a white butterfly which appears early in the season and which can be seen flying about cabbage fields until late in the fall.

Remedies: The main secret of success is regular, persistent treatment nearly every week. One treatment alone does little good, owing to the fact that new egg supplies are being placed on the cabbages by the butterflies all summer. There are many remedies, and below we give some of the safest and best known:

Pyrethrum (also called California buhach and Persian insect powder). This may be diluted with five or six times its bulk of flour and dusted on the plants in the evening or early morning when wet with dew; or it may be mixed with water—one ounce to four gallons—and sprayed on at any time.

Hot water: Water at a temperature of 130 degrees will kill every worm it touches without injuring the plants.

Kerosene emulsion. An excellent remedy while the plants are young, but may give the heads a bad taste if used too late in the season.

Air-slaked lime: Some growers say that this (or, in fact, fine dry road dust, or any powdery substance) will kill every worm it covers.

Hand-picking: In small gardens the worms can easily be controlled by picking them off and killing them at regular intervals.

Preventive measures: The practice of leaving cabbage stalks in the field after the main crop is off is a reprehensible one. All remnants should be gathered and destroyed, with the exception of a few left at regular intervals through a field as lures for the females to deposit their eggs. Such stalks, being useless, should be burned later on.

The Ox Warble

The losses caused by this insect each year aggregate millions of dollars. Not only is the hide of the infested animal punctured by the emerging larvæ, causing a reduction in value of one-third, but the presence of the grubs in the animal's back is a source of loss which cannot be neglected. The latter is very often overlooked.

The poor condition of the animal, its inability to take on flesh, or poor showing at the milk-pail, are factors attributed to other causes. Most uninformed cattlemen attribute these conditions to poor care, lack of proper nourishment, or physiological troubles. The presence of fifty or sixty burning, running ulcers on the back of the animal seems a matter of small importance, and is considered lightly by most people. But, Mr. Farmer, how much would you accomplish if you had fifty or sixty boils on your back? The two conditions are analogous and conducive to the same results.

The Fly That Makes the Trouble

The insect causing all of this trouble is one resembling, in the adult stage, the horse bot-fly, or as it is sometimes called, the "nit" fly, but somewhat larger. The adult is seldom seen about the cattle. It is timid and appears only when everything is quiet. The eggs are deposited upon the hairs during the spring or summer, and the animal, upon licking them, carries the egg or larvæ into its mouth. The young maggot passes into the gullet. From the gullet it migrates slowly through the tissues toward the back. It arrives beneath the skin, and a lump or excrescence begins to appear about midwinter. This lump gradually grows larger until the middle or latter part of April, when the full-grown grub, which caused the lump, emerges through a hole cut in the hide some time previous, and falls to the ground. It then burrows into the ground and transforms into a pupa, and the adult two-winged fly appears from three to six weeks later. This completes the life cycle, which occupies about one year.

Easy to Control

This is one of the easiest insect pests to control that we have. In the winter and early spring all the insects are in the larval stage in the backs of the cattle. If every one owning cattle would squeeze the grubs from the backs of his animals and destroy them, there would be no nucleus for a new generation, and consequently no ox warbles the next year.

When the grub "ripens," *i.e.*, when it matures, a large hole appears in the lump, bordered with pus. This condition usually appears in April and May. When it comes, wrinkle up the hide containing the grub, get the two thumbs and first two fingers on each hand beneath the lump, and squeeze. The grub usually flies to the ceiling like a wad from a pop-gun; now tramp on it, and the job is completed.

The Woolly Aphis

In the fall of the year the woolly aphis can be seen in clusters around wounds in trees, or even around twigs that are not damaged, and at the bases of leaf-stalks and other places where they can get a start. The females deposit their eggs beneath the edges of bark or in cracks, and then proceed to go to the roots of the trees. It is fortunate that the horticulturist can see the pests before they reach the roots, and can thus tell when the trees are liable to become seriously infested.

It is very important for every apple grower to look over his orchard in the fall of the year, and see whether the woolly aphis is present. If so, he will be able to observe small tufts, like cotton, which upon crushing, are found to yield a brown liquid. By careful examining without crushing, he will find the dark colored bodies of the pests beneath the cottony protection.

Aphis is Rain- and Water-Proof

This cottony substance is a very effective protection against rain, and against most spray solutions made in water only.

There are two or three ways of combating such pests while yet on the branches. One is to rub the clusters where they occur and crush them. If the operator is careful to do this thoroughly he may kill most of them. With this he can combine cutting off the worst infested branches and burning them. They should not be dropped on the ground because of the danger of the pests reaching the branches or roots.

A second method is to paint the attacked spots with brushes dipped in a very strong solution of soap or nicotine extract, or a combination of both. The best preparation is made by mixing one ounce of strong commercial nicotine extract in about four gallons of water, containing at least one pound of brown soap shaved fine and dissolved in water (or soft soap, or naphtha soap, or fish-oil soap may be used).

The third method is to spray with a ten per cent. kerosene emulsion, or one pound of fish-oil soap in five gallons of water, or one pound of ordinary soap in three gallons of water, or perhaps, best of all, with the combined nicotine extract and soap solution mentioned above.

The spray liquid should also be directed to the base of the trunk of the tree, so it will reach the collar of the tree where the pests are liable to crawl down the trunk and get into the ground. Mound the trees with earth, and spray the top of the mound.

Remove the earth over the roots, and see if the woolly aphis is at work on them. If so, cover the exposed roots with fine tobacco dust, or pour over them one of the spray liquids mentioned above, and replace the earth.

The Nasty Roach

The ordinary roach or cockroach comes of an ancient (although not distinguished!) lineage which dates back farther than the "family tree" of the proudest human family. Fossil remains prove that this insect existed away back in the Carboniferous age.

About 5,000 different species of the roach family are believed to exist in different parts of the world, most of which, however, live out-of-doors and subsist on vegetation. Only a very few species are engaged in making trouble for good housewives.

Habits

Roaches are lovers of the dark; it is then they roam around pantry and rooms in search of mischief and food. At the approach of a light they scurry away, like a pack of cowardly thieves. Any kind of food tastes good to them, whether it be shoe leather, apple dumpling or book covers. Toothache, loss of appetite or dyspepsia are, we believe, unknown to the roach family.

But the roach has one good point. Alas, only one. It is this: He has a fondness for bed bugs, and, by eating them, he often does the housewife a real favor.

Remedies

Now we come to "remedies," and here the trouble begins. Nothing short of eternal vigilance will rid a house of these pests when once they gain a foothold.

Unfortunately (for the housewife) the roach seems to be endowed with remarkable intelligence as regards poisoned foods. Arsenic, no matter how disguised, he refuses (with thanks) nearly every time. However, it is said that a preparation of sweetened flour paste containing phosphorus will often fool him.

Another remedy often used is fresh pyrethrum powder or buhach. This, when liberally dusted on shelves, etc., usually affords temporary or partial relief. A better use of this powder, however, is to burn a quantity of it in an infested room and then tightly close the apartment for ten hours. Bisulphide of carbon is sometimes used in this way also, but its vapor is more dangerous to have in a dwelling house.

Trapping the insects is another remedy. Roach traps may easily be made at home as follows. Take any deep vessel or jar and place it where the roaches congregate. Fill it partly full of sweetened liquid paste. Then take several thin, narrow pieces of wood, bend each one into an inverted V, and hang them on the jar—one end almost in the liquid, the other on the shelf or floor. The idea is to make several "gang planks" up which the roaches can crawl, with a steeper gangway inside, down which they will slide into the liquid—never to return.

Some of the prepared "roach powders" that are on the market are also effective when perfectly fresh.

The Buffalo Moth

The carpet beetle (often called the "Buffalo Moth") has proved to be a very annoying and destructive pest throughout the northern part of the United States. It was imported into this country from Europe about the year 1874, and has spread from the East to the West.

All the year, but more often in summer and fall, an active brown larva about a quarter of an inch in length feeds upon carpets and woolen goods. This larva is decorated with stiff brown hairs, which are longer around the sides and still longer at the ends than on the back. It works in a hidden manner from the under surface of a carpet; sometimes making irregular holes, but more frequently following the line of a floor crack and thus cutting long slits in the carpet.

Description

The adult insect is a minute, broad-oval beetle, about three-sixteenths of an inch long, black in color, but is covered with exceedingly minute scales, which give it a marbled black-and-white appearance. It also has a red stripe down the middle of the back, widening into projections at three intervals. When disturbed it "plays 'possum," folding up its legs and antennæ and feigning death.

Prof. J. B. Smith says: "The Buffalo Moth lives during the winter under scales of bark, in crevices and wherever else it can find shelter. It is the beetle that lives over, of course, and in the spring it congregates sometimes in great numbers on blossoms, favoring those in gardens, and from them it finds its way into houses nearby. I do not think that I have ever found larvæ in houses under ordinary circumstances in winter; but I am quite ready to believe that in places kept uniformly warm at all times, breeding may go on in winter as well as in summer."

We believe that only where carpets are extensively used are the conditions favorable for the great increase of the insect. Carpets when once put down are seldom taken up for a year, and in the meantime the insect develops uninterruptedly. Where polished floors and rugs are used the pest ceases to be a serious one.

The beetles are day-fliers, and when not engaged in egg-laying are attracted to the light. They fly to the windows, and may often be found upon the sills or panes. Where they can fly out through an open window they do so, and are strongly attracted to the flowers of certain plants, particularly the spirea.

Remedies

There is no easy way to keep the carpet beetle in check. When it has once taken possession of a house nothing but the most thorough and long-continued measures will eradicate it. The practice of annual carpet-cleaning, so often carelessly and hurriedly performed, is, as we have shown above, peculiarly favorable to the development of the insect. Two carpet-cleanings would be better than one, and if but one, it would be better to undertake it in midsummer than at any other time of the year.

Where convenience or conservatism demands an adherence to the old house-cleaning custom, however, insist upon extreme thoroughness and a slight variation in the customary methods. The rooms should be attended to one or two at a time. The carpets should be taken up, thoroughly beaten and sprayed out-of-doors with benzine and allowed to air for several hours. The rooms themselves should be thoroughly swept and dusted, the floors washed down with hot water, the cracks carefully cleaned out, and gasoline or benzine poured into the cracks and sprayed under the baseboards. The extreme inflammability of gasoline, and of its vapor when confined, should be remembered and fire carefully guarded against.

Treatment of Floors

Where the floors are poorly constructed and the cracks are wide, it will be a good idea to fill the cracks with plaster of Paris in a liquid state; this will afterwards set and lessen the number of harboring places for the insect. Before relaying the carpet tarred roofing paper should be laid upon the floor, at least around the edges, but preferably over the entire surface; and when the carpet is relaid it will be well to tack it down rather lightly, so that it can be occasionally lifted at the edges and examined for the presence of the insect. Later in the season, if such an examination shows the insect to have made its appearance, a good though somewhat laborious remedy consists in laying a damp cloth smoothly over the suspected spot of the carpet and ironing with a hot iron. The steam thus generated will pass through the carpet and kill the insects immediately beneath it.

The Arch Pest—Rats

At one time our premises were so overrun with rats that we sustained quite a loss from their devastation. A plan for their destruction was devised, as follows: Filling an iron kettle three-fourths full of barn sweepings, corncobs, and a little mixed grain, we set it in an empty stall in the horse stable where the rats seemed to predominate most, and left it this way for some time, keeping plenty of grain in with the rubbish as an enticement for the rats. We laid several boards sloping from the kettle to the floor, so that the rat could easily run up and down and into the kettle.

The Kettle Trap

At the end of about two weeks, or when we thought a great number of rats had become accustomed to frequent the kettle, we emptied the kettle of its rubbish contents and filled it three-fourths full of water and covered the water about an inch or more with light chaff, leaving no water exposed. (If water remains entirely undisturbed the chaff will not sink over night.) On the chaff we scattered a little light grain. There was something going on that night! The rats had a party or something; at any rate, the next morning when we went to fishing we scooped out about a half bushel of rats, big and little. The next morning our haul was not quite so large, but we got quite a number; and so on until the rats either got wise or there were no more rats. If we did not get all, we at least got a large majority of them.

The Tar Cure

At another time when rats were getting altogether too plentifully, we caught a rat in a box trap. This rat we let run into a grain bag and there we caught it by the nape of its neck, guarding carefully against being bitten; then we let all but the head and neck come out of the bag and painted all over the exposed parts of the rat thoroughly with tar and let the rodent go. We had heard that doing this to one rat and letting it go would clean the premises of all other rats, as they object to the smell of tar, or are frightened at the strange appearance of one of their party. It seemed to work in our case and work well. We had no trouble with rats for several years after that. Lonesome, heartbroken, or what, I don't know; but one morning shortly after we had tarred this rat we caught the same fellow again in the same trap we had caught it in before. However, this time we did not let it go.

Steel Traps

It seems that in no other place are rats so hard to catch as in the cellar. Located there they seem to be able to evade all traps and trapping. But I found a way to get Mr. Rat in the cellar. I set a steel trap and put it in a shallow, discarded bread pan and covered the trap completely with wheat bran; the bran being light, did not spring the trap nor hinder the working of it. Over and about the trap on the bran I scattered a few bread crumbs or meat scraps. This method has never failed me in getting rats in the cellar; although it has when tried in other places. The bran and foregoing baits differed so much from the edibles the rats in the cellar were accustomed to diet on that they jumped for the chance of a change, and consequently were easily caught in this manner.

I have found that rats often gain entrance to a cellar through the cellar drain, and for this reason the out-let to the drain should be screened so that no rats can enter.

Chloride of lime, if generously sprinkled over the runways of rats will also clear the premises of the pests. It gets into their nostrils and burns their feet. Rather than brave many repetitions of it they leave the premises.

Prevention is sometimes better than cure. Where possible to do it, use concrete for floors, foundations, etc. The additional cost of thus making buildings ratproof is slight as com-pared to the advantages. With ce-ment even an old cellar may be made proof against these pests.

Rats are expensive, they are de-stroyers of property. They are a menace to health, carrying in their fur disease germs; they are trans-mitters of plagues, a general nui-sance—biff the rat!

Grow Your Own Posts

The many disappointments in growing catalpa trees are attributable largely to unfavorable site and stock of an inferior kind of catalpa. Crooked, limby trees often result when the trees are not cut back; and where the limbs, after attaining some size, are broken off, decay enters, and the heart rot, so injurious to the tree, begins.

In the first place, the seedlings must all be of the particular species known as "speciosa," or hardy catalpa; the others, for production of wood, are not worth the planting. The soil must be rich and rather moist, must not contain too much alkali, and the location should not be much farther north than the Central States. Whenever the young trees are making a low branching growth so that they will be unsuited for posts or poles, they should be cut back—that is, cut off close to the ground when two or three years old, just as if they were large enough for market. Several sprouts immediately spring up from each stump and grow vigorously, and the competition for sunlight stimulates height growth and encourages natural pruning. This enables the tree to form a straighter stem with fewer branches. At the end of the season the best sprout is nearly as tall as the three-year-old tree would have been. Remove all inferior sprouts.

After ten years under best conditions, the first crop reaches the size at which it can be cut most profitably for posts.

Each tree should then produce one first-class post, one second-class post and two or three stays. From the

Baby trees that will soon be grown-up

small and crooked limbs, desirable firewood is secured. And note this: Another crop will grow from sprouts in eight years!—and then do it all over again!

Investment Per Acre

Cost of cheap land, say.........$22.00
Cost of home-grown seedlings...   .80
Cost of transplanting...........   3.20
Cost of cutting back (3d year).   2.50
Superintendence, implements, fire
    guards, etc..................   3.96
Cutting.......................   20.00
                        $52.46

Returns Per Acre

Value of posts, about.........$315.00
Value of firewood.............   12.00

Total gross returns.........$327.00

Raising Guinea Pigs

The guinea pig is a native of Brazil, and comes in three different colors— white, black and fawn. Some of the white ones have red eyes.

Before starting in the business of raising guinea pigs you should carefully consider several things. If you have hay, apples and similar feed on the home place, it is all right; if not, it may be a mistake to start in the business, as these feeds cost too much. Grain must be purchased, but that is a small expense compared with the other feed.

Then there must be a good place to keep the little animals. They won't thrive down in the cellar, nor out in the shed, nor up in the garret. They must have a place where a fire can be kept in cold weather.

They must be attended to as regularly as other farm animals. They must be watered once daily, fed two or three times and hutches cleaned out every day.

When you get 200 or 300 pigs, which would be necessary to have a steady income, you will find it work—not hard labor, but work you cannot shirk. A few pigs will show no profit. The more you have the easier it will be to sell them.

Harvesting the Ice Crop

The larger the amount of ice packed in a body the better it will keep. It will not keep without drainage and ventilation. The ice house should stand on sand or gravel or have ample artificial drainage put in to carry away the melted ice; if there is a drain it should be trapped to prevent air from entering the house. The ice must be packed on a bed of sawdust or marsh hay two feet deep and be packed in a solid layer of cakes that are sawed with square angles and as large as can be handled conveniently, say 22x30 inches. Place these on edge, all one way and twelve inches from the sides of the building until solid, and if spaces occur between

Fig. 1

them fill with pounded ice. After the upper surface has been leveled in the same manner, and the sides next to the wall filled with sawdust tramped hard, place the next layer. This is to be continued until the house is full, say to within three feet of the eaves. Over the top of the ice place eighteen inches of sawdust. Above must be plenty of ventilation.

As mild weather approaches, the ice should be inspected two or three times a week and the side packing kept tramped hard, to make sure that no ventilating tunnels occur in the sawdust, for these quickly waste a ton of ice. A stick may be needed to

Fig. 2

punch down sawdust into such spaces.

As the sawdust is taken from the top to fill around the sides, more must be put on, keeping it eighteen inches deep; but as ice is removed, see that sawdust does not accumulate much deeper than that, as it will generate heat and the ice will not be ventilated and will waste rapidly.

Many persons in a rural community upon learning that ice may be bought, will come and buy it, thus saving delivery. It should sell to these neighbors at thirty-five cents per hundred and be delivered at fifty cents and sometimes higher. Of course ice will wet a wagon if placed in it with no protection, but a sheet of galvanized iron turned up at the sides and front for an inch and allowed to project under the endboard, will keep the wagon dry.

To aid the farmer in securing his ice harvest there are some very helpful tools on the market, such as ice plows, saws, hooks, chisels, tongs, etc. We show a simple loading platform, Fig. 1; and a homemade lift, Fig. 2, for loading ice into sled or wagon.

Roberts' Portable Fence

A Boon to the Farmer

The accompanying drawings are not made to scale, but plainly illustrate the construction of the fence.

Note how the upright sticks lap at E. The projecting end at F is used as a handle; you will see that by lifting its raises the panels B D. Each corner is locked in the same manner

as the one illustrated, and not only locks but acts as a hinge so that each corner is a gate; F swings out to the right and will open wide enough to let a wheelbarrow in with ease; this can not be done unless notches as indicated at I are one and a half inches wide at bottom.

To have gates swing to right, the mode of construction must be reversed at each end of the same panel.

Note that the notches G are cut close to upright, while H is one inch from upright. C is six inches from ground, and the wire netting to within one inch of ground. From ground to F should be about three feet, and the height of posts above F as high as you like. I have found 1 x 3 inch stuff a good size to use.

The yards can be made as large as 20 x 20 feet, but should have a post in the middle of each panel when made that large.

Some advantages of such a fence will come to the mind of the reader at once, but to be appreciated must be tried out. It can be used to keep chickens out as well as in, and therefore comes in useful to fence a garden patch in and the chickens out. It can be moved anywhere at any time, as no posts are in the ground. It can be taken down and placed one panel over the other when not in use. It can be filled with broody hens and moved to a strange spot and the hens' broodiness soon broken up. New fowls can be tried out in it. Sick fowls can be placed away from others in it. Move your fence about and plant your garden where it stood, if you want a fine garden.

When your young stock get troublesome, cage them up. When you want to fasten cockerels for broilers, cage them in a small yard, and when they have gone to the butcher fold your fence away. Use the panels to protect young plants by letting them rest over the plants and against the sides of the building. There are many other ways in which this portable fence will be useful.

Home Water Works

The system of underground pressure for home water supply is practical and not expensive, and makes possible hot and cold water for the kitchen, stock water and fire protection to both farmer and villager.

There is an eight-foot wind wheel, a force pump over a large well in which is placed an automatic cut-off: About 900 feet away is the boiler, B, connected by an inch and a quarter galvanized iron pipe. Both boiler and pipe are underground below frost. From the boiler a pipe conveys water to the stock troughs, another to the hydrant, and a third to the house and lawn.

Primarily one needs only a force pump, pipe, cut-off and faucet to give sufficient pressure as long as the wind blows, but if you wish to store reserve for a rainy day, or rather, for a calm day, add to your equipment a boiler, old or new, but the larger the better. Any condensed steam boiler will do. Both inlet and outlet, you notice, are at the bottom of the tank. In the beginning the boiler is filled with air only. As soon as the mill starts the vents are closed, then, if it is airtight, and the mill continues it will soon compress the air into one-half its volume or less, forming an air cushion which gives a pressure of twenty, thirty or fifty pounds to the square inch, equal to any Holly or reservoir system.

Any plumber or windmill man can furnish everything necessary. A few feet of hose will then enable one to throw water over the farmhouse, sprinkle the lawn or strawberry bed, provide for bath room, hot and cold water, etc.

Hitch Your Telephone to the Fence

You want a telephone, of course; but the poles and wires cost real money, to say nothing of the time and labor that it takes to put up the line. If you have a barbed-wire fence that runs in the right direction, hitch your telephone to it.

Let's suppose you want to connect with a neighbor's farmhouse and that a barbed-wire fence, on wooden posts, runs across. You buy two telephones with magneto calls, four dry batteries and a few feet of insulated copper telephone-wire for use inside the houses. All these can be bought for $20 at the outside.

Set one instrument in your home and connect a pair of dry batteries to it according to the directions that come with the outfit, or consult a telephone lineman in the nearest town, and get him to show you which wires fasten to which terminals. Run another wire from the telephone to the top wire of the fence; see that this strand runs unbroken all the way to the other telephone. When you come to a gate, replace the gate-posts with

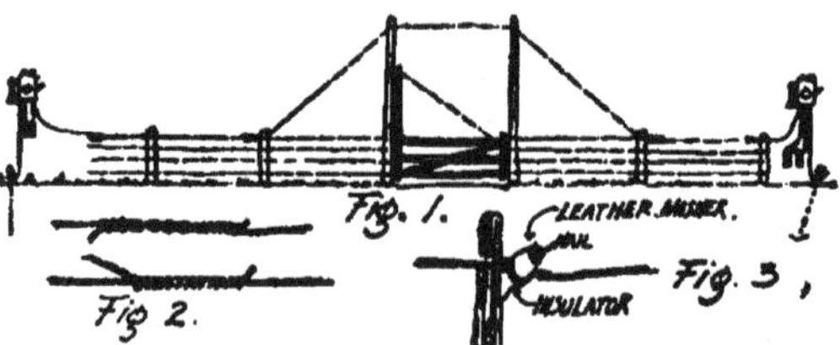

tall poles and carry a wire across as shown in Fig. 1.

Be sure that all joints in the wires are properly made, as in Fig. 2; the upper sketch shows the right way to splice two ends, while the lower one shows how to join a wire to a main line. Clean galvanized wire will make a good electrical joint, but if you have a rusty wire you must scrape it bright, twist it together, and solder the joint so that rust will not form between and cut off the current.

For short distances, up to five miles, you need only one wire. The ground can be used to complete the circuit by running a wire from each telephone to an iron rod driven into the ground until the lower end touches earth that is always damp. A still better way is to hitch your wire to an iron pump or water pipe. Naturally you must be sure that your fence wire does not touch the ground or any grounded metal. Therefore, metal fence posts cannot be used on the line. Neither will woven-wire do unless it is raised clear off the ground. Green, sappy posts will let some of the current leak away the same as metal posts.

When you run across from the fence to the house, carry the wires on glass insulators fastened to trees, poles or buildings. A bottle neck, Fig. 3, held by a large nail driven through a leather washer will do just as well as a regular glass or porcelain fixture. Use insulated wire for inside work; connect it on just before the wire enters the house.

Getting the Saws Sharp

When a saw is to be filed it should be placed in a vise or saw-clamp. When a vise is used, it is well to place a thin board on each side of the saw to keep the vise from damaging the saw.

In filing hold the file as nearly horizontal as possible. If it is held horizontally and at right angles to the blade of the tooth which is being filed, and the tooth is filed straight across, a perfect chisel edge will be produced. In filing teeth to a chisel edge, if the teeth are all filed from one side, a wire edge will be produced on the opposite side, which will make the saw run crooked. Hence it is best to file the teeth from alternate sides, filing half from one side first.

It is also necessary to joint a saw before setting or filing it. This is done by putting a flat file in a clamp so that it will be held perfectly level, and running this over the points of the saw-teeth. Continue this until all the saw-teeth are dressed down to the same length. Circular saws are jointed by holding a file stationary and turning the saw with its mandrel. A clamp for holding files, or a "jointer," as it is called, may be purchased.

Ripsaw-Teeth

Ripsaw-teeth are chisel shaped, as these teeth cut a log or board lengthwise with the fiber. They should be filed across the saw at an angle of 90 degrees with the edge. The front of a ripsaw-tooth is perpendicular and the back is inclined at an angle of 60 degrees from the perpendicular front of the tooth. Ripsaw-teeth should be set slightly to insure easy and smooth work. Soft woods require larger teeth and more set to prevent binding.

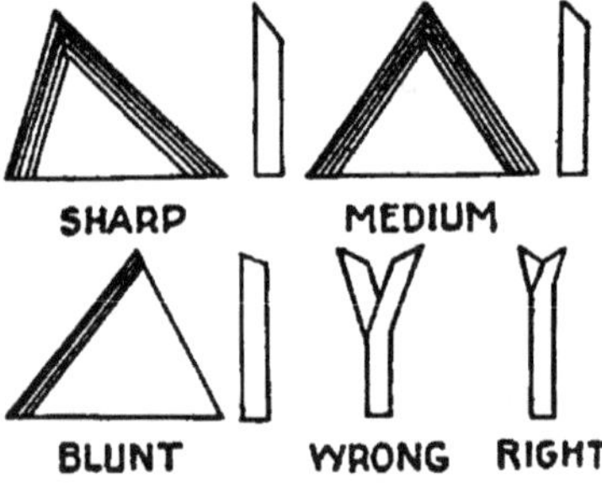

Soft Wood or Hard

A saw that is to cut soft wood, such as basswood, redwood, butternut or pine, should be very sharp—similar to the tooth marked "sharp" in the drawing. The bevel or fleam on both the front and back of this tooth is very wide and therefore gives a sharp, weak tooth. It also has a greater angle or "rake" in front than in the back of the tooth, the front being almost perpendicular. This will insure fast cutting in soft wood, but in hard wood will cause saw to buckle or hang up. A tooth filed similar to the one marked "medium" is suitable for moderately hard woods. It has a moderate bevel at front and back with an equal front and back rake of 60 degrees.

A tooth similar to the one marked "blunt" is suitable for very hard woods. It has a narrow bevel in front and none at the back, and the rake is a trifle greater on the back than on the front of the tooth. A saw filed in this way will stay sharp even when used in very hard woods.

Setting the Teeth

To set a saw the teeth are turned alternately to the right and left. This causes the saw to cut a slot wider than the blade of the saw and thus prevents the wood from binding. A considerably greater set is required for saws to be used in soft, green wood than for those to be used in dry, hard wood. Do not try to set the entire tooth, but only the point, as indicated by the two drawings in the lower right-hand corner of the accompanying sketch. This setting should be done with one of the many tools made for this purpose, and every tooth should be set at the same angle. Irregularity of set is bad.

When a saw has been filed, it should be side dressed by laying it flat upon a bench and rubbing once or twice over the teeth with an oilstone. If you are to saw soft or green wood, side dressing is unnecessary; but if you are to saw hard wood considerable side dressing will be of advantage, as it will insure a very smooth cut; in fact, smoother than one can plane the end grain with a block plane.

Fighting Poison-Ivy

Poison ivy should not be allowed to go to seed to contaminate the farm for years to come. It is frequently found along fence rows and roadsides, where it is too often neglected. It may be distinguished from Virginia creeper by the three leaflets per group (see cut) as compared with the five leaflets of the Virginia creep-er. (The common Boston ivy has a simple three-lobed leaf like the maple.)

Concentrated sulphuric acid will kill poison ivy. Dose each plant with a half-teaspoonful to each stem, making the application during the growing season, every three weeks. If a large area is covered by the plants, spraying with arsenic of soda (one pound of the poison to twenty gallons of water) will kill all vegetation. One application, if the growth is young and tender, will do this.

Destroying by Fire

Here's another way: A friend of ours puts straw along the stone fences, etc., infested with poison-ivy, and then sets fire to the straw, repeating the operation at intervals until the plants give up trying to grow.

Poison-ivy may be grubbed out by one who is immune to the poison. All parts of the plants should then be gathered into a pile and burned. The resulting smoke should not be inhaled nor allowed to get into the eyes.

Fire Protection That Protects

"Well, you saved the barn, anyway," I said, consolingly.

"Yes—by sheer good luck," grunted the owner of Chesapeake Farms, picking a dented fire pail from the cinders. "The wind happened to be blowing the other way; that was all."

"Couldn't you get a fire stream on it? I thought you had a good water supply!"

"I thought so, too. I had a pressure-pit under my shop and a gravity tank over it, on a high iron tower. But the fire started in the shop and burst through the roof before we dis-

Fire Protection

About the Farm

covered it. In two minutes the iron supports of the tower were red hot and crumpled up—there the thing lies." He pointed to what looked like the blackened, tangled framework of a wrecked Zeppelin. "Of course, when the tower-tank fell, it landed on the pressure-tank, smashing the valves off that; my gasoline engine and pump were in the shop, too; the fire-buckets had been carried off to slop the hogs —and there you are!"

Simple Things to Remember

Now, all this isn't an argument against fire protection; it's precisely the opposite. My friend did *not* have a good fire system; and so he lost several thousand dollars' worth of farm buildings, with all their contents. Iron is far less fireproof than stout timbers; it bends like wax, when hot, and should never be used for a tank-tower, unless set away off by itself. The pressure-tank should have been buried in the ground. The pumping engine ought to have been in a small, isolated building. And so on.

In these busy days a farm fire is as much a national calamity as the destruction of a steel mill or a shipbuilding plant; and it's a patriotic duty for all of us to protect our farm buildings more carefully than we have been doing.

Common whitewash, with a little salt added, makes the best possible fireproof paint. Did you know that? And Pacific coast redwood scarcely burns at all. In the San Francisco fire some redwood houses actually stopped the progress of the flames.

In a large, connected mass of farm buildings, fire partitions can be run up, so that a fire can be kept from spreading. These partitions should of course cut right through the roofs and frame walls, and can be made of brick, cement block, hollow tile or metal lath plastered with cement. All doors through such partitions must be tightly covered with tin on both sides.

Fire-extinguishers are good things to have handy.

Your Water System

If you have a water-system it should keep head enough to throw a good stream against the highest point of any building. A pressure at the ground of thirty pounds will shoot the water about forty feet in the air, using two-and-a-half-inch fire hose.

If you have only the ordinary garden hose, a very much greater pressure is necessary; the concern you buy your water-tank from will figure it all out for you.

But the best possible fire protection is a "sprinkler system;" there are dozens of good sorts on the market, and practically every factory, large or small, is equipped with one.

Then, there are all sorts of things you can do to keep fires from starting. When I visit an old farmhouse I always examine the chimneys very carefully; nine times out of ten I find gaping holes right through the brickwork, just under the roof! And then there's the danger of spontaneous combustion from greasy rags; the danger from lightning, etc.

Ten Rules for Thinning Woodlots

1. Prepare in advance a list of all the different kinds of trees in the woodlot and arrange the names in order of their desirability. This list may also include facts about the size and kind of products that can be used or sold.

2. Mark on the same side all trees that are to be removed, using the side from which the chopping will naturally progress. If the trees are not to be cut by the owner, he should blaze beforehand all that are to be taken.

3. Cut for firewood only those trees that can not be utilized for timber or other products of a higher grade than fuel wood. Spare young, thrifty-growing trees that can later be put to the better uses. Examine each tree carefully as to straightness, soundness, salability, and relation to neighboring trees.

4. In a fuel-wood cutting, remove first all dead or badly decayed trees.

5. Remove also defective and inferior trees to insure better growth for the good trees that are left. Even if nothing but poor fuel comes from the first thinning, it is likely that the work will pay in the improved growth of the good trees that are left.

6. Have a definite reason in mind whenever a tree is selected for cutting; and do not mark two adjoining trees except for a very good reason, such as great overcrowding.

7. Frequent moderate thinnings give better results than infrequent heavy ones; never thin a stand of young timber heavily.

8. Leave a dense wind-mantle along the edge of the woodlot; nature put it there for a good purpose. So never thin this outer strip.

9. Be conservative; it is better to leave some poor trees than to sacrifice one of great promise.

10. Grade the product, pile the different grades separately, and be sure to know the range of local prices.

What to Do With Sewage

Sewage disposal, in the last fifteen years, has grown to be a science. The labors of eminent engineers, chemists and bacteriologists have resulted in various methods and contrivances.

For handling the waste of big cities, large beds of sand, properly under-drained, are used with the greatest success. Several of these beds, generally about one acre each in size, are constructed, and are used in rotation. This method has come, therefore, to be known by the name of "intermittent filtration."

Sewage Disposal   **About the Farm**

A method which is in use in England, in many places on the continent, and in a few places in this country (notably at Vassar College) is termed "broad irrigation." It consists in preparing a field in the same way as is done for irrigation in the West and allowing the sewage to flow through the ditches and distribute itself as evenly as possible over the surface. Corn is the principal crop grown, and for silage a very rank growth is obtained. The success of this method, which in some instances might be recommended for the farm, depends upon the proper understanding of the problems involved, in order that the field may be maintained under the proper sanitary conditions. On a clay, or naturally wet land, this method is out of the question.

For small quantities of sewage, such as the product of a single house, the most practical method is the one which combines a moderate cost of installation, a minimum of care and attention and a maximum of efficiency in purification. This combination will be found in the septic tank. Septic action denotes decay and putrefaction. By this means much of the solid matter is reduced to water and gas. Since this action is produced by bacteria which work to the best advantage away from the presence of air, the tank may be placed in the ground and covered. The ordinary cesspool is, roughly speaking, a septic tank, and its use is to be recommended on the farm. Where there is a porous soil, the excess of water will drain away and no attention will be needed other than a cleaning out of the solid matter once a year. It is taken for granted that the farmhouse is to be equipped with running water and a modern bathroom. The next step is to construct a drain from the house to the tank. Five-inch sewer tile should be used and the joints cemented for the first 100 feet (or farther, if at this distance it is within 100 feet of the well). After this the joints may be packed with clay. The tile should be laid to grade with a fall of not less than one foot in 100.

A cheap tank may consist of a hogshead sunk in the ground so that the top of the drain shall enter it six inches below the top, and a circle of inch holes may be bored two inches below the bottom of the tile. The filling about the top of the hogshead should be of cobblestones, in order that these holes may not become clogged. Should the soil about the tank be a heavy clay it will be necessary to drain the tank; and the purification can be increased if, at the point where the drain comes to the surface, a bed of small cobbles is so placed that the water may flow over them. The tank is then covered with plank, and may be further covered with several inches of soil. In this case a quarter-inch pipe should extend from the top of the tank to the surface of the ground to give the necessary vent. The distance at which the tank may be placed from the well varies with the character of the soil, slope of the strata, depth of the well, and other geological conditions, and should be given the most careful consideration. (The distance of the barnyard from the well is a matter of sanitation too often overlooked or negleced). Trap the kitchen sink drain and connect to the main tile with a V,

Making a Septic Tank

Here is the theory of sewage disposal: The germs of typhoid fever and other diseases—the bad little germs—live and thrive in darkness and dampness; but nature has provided certain good little germs whose business it is to eat up these bad little germs. The good germs swarm in most topsoils, but they must keep fairly dry and have plenty of air. If the ground stays wet and soggy they eventually die or move out.

Now, keeping these facts in mind, we'll look at Fig. 1. The sewage flows through a line of four-inch glazed pipe, with tightly cemented joints, fifty feet or more to the septic tank—a big concrete box buried in the ground. Passing through this tank the sewage flows onward into the porous drain-tile about four feet apart, just under the surface of the

germs' comfort; so we must remove these two difficulties by means of the modern septic tank.

Figs. 2 and 3 show a plan and a section of such a tank, suitable for the average family of not more than twelve persons. The walls and bottom are four-inch concrete, reinforced with chicken wire or metal lath; the dimensions are as marked. The raw sewage runs into the first section, or "setting tank;" the solid matter sinks to the bottom and slowly dissolves, letting the liquid flow over the top of the partition into the second section, or "syphon chamber." When this chamber gets full, say about once

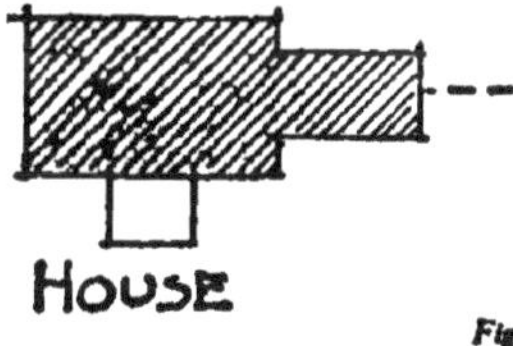

ground. Here the liquid seeps out into the soil, and the good little germs get busy, eating up the bad little ones. Very simple, isn't it? But wait a moment: if the raw sewage, just as it comes from the house, is shot into the drain-tile, the solid matter will soon choke things up. And further, a continuous dribble of liquid will keep the ground too wet for the good little

every twenty-four hours, the big cast-iron syphon at the bottom starts up, shooting the whole contents of the chamber at one gush out into the disposal field. Thus the field gets a chance to dry out between times.

Septic Tank   About the Farm

After a few weeks a thick green scum collects on the water surface of the settling tank. This makes a breeding ground for certain bacteria that help to break up the solid sewage, and it must not be disturbed.

Fig. 1 and Fig. 4—with joints about one-quarter inch apart, protected by a bit of tarpaper, broken brick or china. The bottom of the tile is not more than six or eight inches below ground, and the fall is extremely

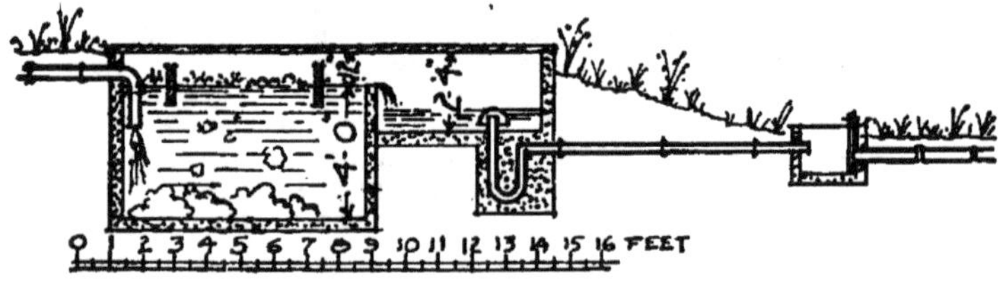

Fig. 2

The inlet-pipe, you'll notice, turns below the water-line, and two oak boards are set crosswise of the chamber to keep the scum quiet and pre-slight—one inch in sixteen feet or less. There are two fields, used alternately, week about; a simple concrete valve box with a wooden gate controls the

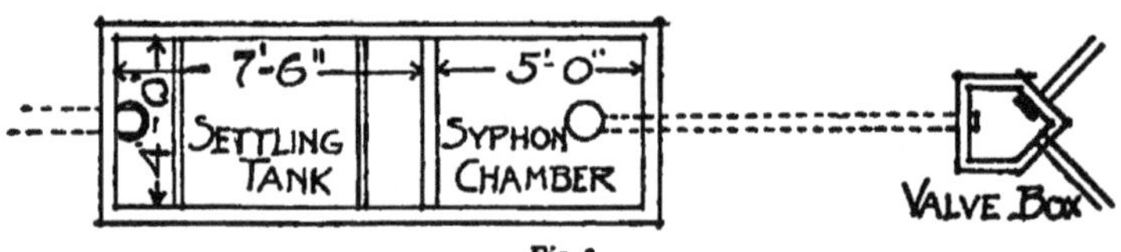

Fig. 3

Fig. 4

vent it floating away over the partition.

The three-inch porous drain-tiles in the disposal field are laid as shown in

flow. In ordinary soil 300 feet of tile in each field will do; but in stiff clay this must be nearly doubled.

Of course the septic tank should be covered, either with plank or a concrete slab; in the latter case, manholes of some kind must be put in. Iron manholes, and the syphon, can be bought from plumbing or hardware dealers.

How to Empty a Barrel

Those Who Usually Think It Hard Will Find This Way Easy

Wooden barrels containing fifty gallons, weighing 600 pounds and upward, are very hard to handle except by rolling. In order to withdraw the contents by means of a faucet it is necessary that the barrel be placed on a raised platform. By using the following method of drawing off part of the contents the barrel can be handled much easier:

With the barrel lying on its side on the ground, place a stick of wood under one end to cause that end to be slightly higher than the other, as in A. Bore a three-eighth-inch hole in the head at the uppermost point next to the chime. About six inches to the right bore another hole with a one-inch bit, which makes the right-sized hole for a piece of three-quarter-inch pipe. Then screw a piece of pipe four to six inches long in the second hole, and fit a plug into the smaller one. To withdraw some of the contents, place a pail under the pipe and roll the barrel to the right, as shown by dotted lines,

until the flow starts, then withdraw the small plug and admit the air so the flow will be increased.

To check flow, insert the plug in the air-hole and roll the barrel back to the left. If barrel is chinked in such a position that the pipe is at the top of head, it will not even need a plug in the pipe.

This idea can be readily used when creosote, kerosene, sheep dip, molasses, oil, spray material or anything of a like nature is received in barrels. When the barrel has been about half emptied, it may be fitted with a faucet and then rolled upon a raised platform.

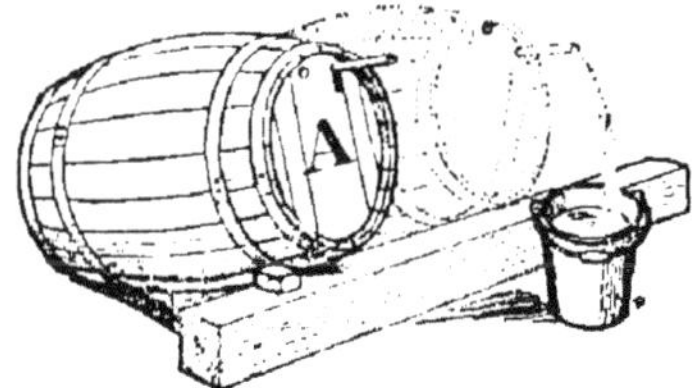

The Plan in Operation

A Standard Farmhouse

"Blueprints from The Farm Journal office," said the draftsman, laying a bulky bundle on my drawing-board. Blueprints? . . . that was queer; what did it mean? I opened the letter from the Editor, and read a few lines.

". . . . . So I'm sending you these standard farmhouse plans, just received from the University of Missouri; if you think best, you might let Our Folks see them."

Then I understood! And very interesting the floor-plans were, too;

simple, compact, inexpensive. But some few of the arrangements seemed as if they might be improved a trifle; the showers of letters that Our Folks send in show me just about what the farmer's wife really wants. Besides, the outside of the house was a bit too bare and barnlike.

So, as architects will, I have used my blue pencil on this design; cutting out something here, marking in something there, and so on. I think I have improved it; but quite possibly Professor Fenton (who first designed it) may think just the opposite!

It isn't necessary to do much explaining; you can understand what I've shown by looking at my drawings. On the first floor, the closet beneath the stairway gives a passage from kitchen to bedroom; any housekeeper will grasp the comfort of this.

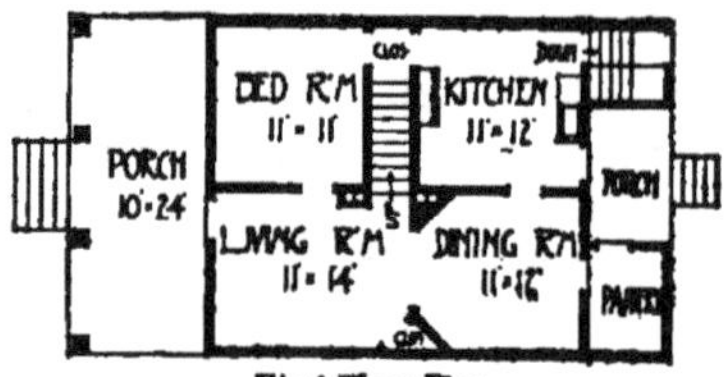

First Floor Plan

The cellarway goes down from the kitchen, and has an outside door at the ground level; you get the idea? In the second story the balcony is meant as a place for airing bedding, moving furniture while housecleaning, and so on; but it's also a fine place to sleep on hot summer nights. (I know, for I have one on my own farmhouse!)

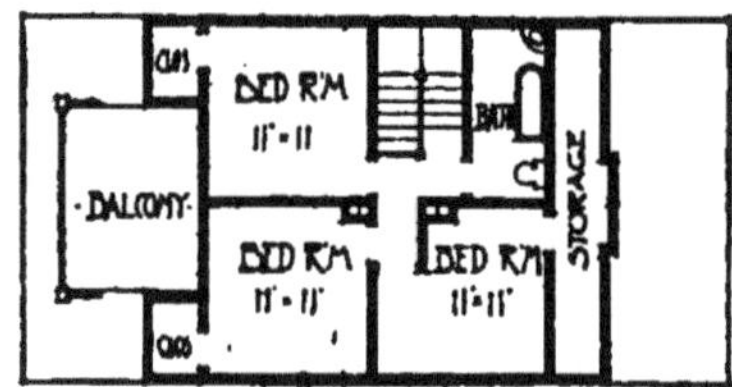

Second Floor Plan

The storage space at the rear is rather bigger than I've shown; it really runs out under the eaves and gives

House Plans | **Making Over**

a lot of room to pack away all sorts of useless stuff. (You know, women *will* keep the most miscellaneous lot of old junk, hoping that sometime it will come in handy!)

The chimneys run up separately in the first and second stories, but rack over and join before passing through the roof. If there is a cellar and a heating plant, one of these chimneys may be cut out; but I know from experience that there are some places where the water lies so near the surface that a cellar is simply out of the question.

I have imagined that the house is built of frame, covered outside with cement stucco on metal lath; but, of course, one might use almost any other material.

Just one final word: Let the front porch face south. That brings the cheery morning sunlight into the dining-room, shelters the living-room from the cold northwest winds, but lets the summer breezes blow across the porch and through the front bedrooms.

Made-Over Farmhouse

The farmer's wife rolled up her ball of olive-drab yarn, and laid it carefully aside. "Mary and the children are coming to live with us," she said.

ing that I've marked Fig. 1 is the house as it now stands; Fig. 2 is what it will look like if you follow

Fig. 1

Fig. 2

"Of course the house is entirely too small for such a large family; so my husband wrote you to come here and tell us what to do."

"H'm, yes; let's see." The architect sat down at a table and sketched rapidly for some minutes. "Here you are," he said, at length. "This draw-

my suggestions. Very simple; we just run the old roof-lines right on down to the first story, front and back. Fig. 3 is the first floor plan; I've drawn all the old parts in outline, and blacked in the new parts solidly. Here are the old kitchen and the old dining-room unchanged. But your

The Square House House Plans

living-room will never be big enough for your enlarged family, so I've added a den at the rear and cut a big archway to connect it with the living-room. Over to the **r i g h t** I've built on a **b e d**-room, as you see."

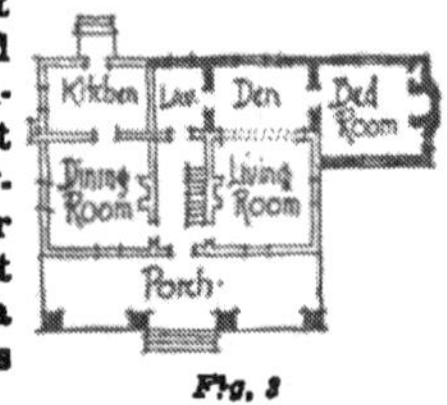

Fig. 3

"What's that little place marked 'lav'?" asked the farmer's wife.

"A lavatory; you'll find it mighty convenient. Fig. 4 is the second story. I'll fix you up a splendid pair of new closets under the slant of the new rear roof. The bathroom is also worked in under this new roof."

"But I wanted another bedroom up here."

"Exactly; see that big new sleeping-porch? In summertime it will make the best sort of bedroom for your son's children; in winter you'll enclose it with glass, and go on using it, and it won't

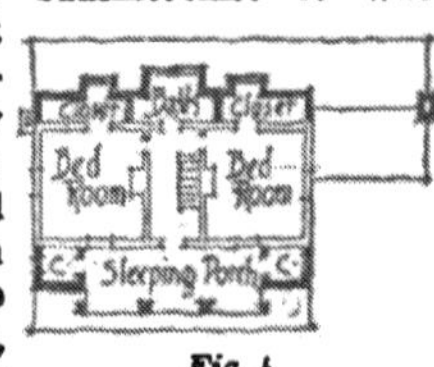

Fig. 4

be so very expensive, for I've left the old house practically untouched, and merely added things on to it. Changing things would cost money."

An Almost Square House

It was a letter from T. M., out in Rainier, Ore., and it read as follows:

"I am an old subscriber to your good and useful paper, The Farm Journal, and I'd like to get some information in regard to a house plan. I want to build a two-story square house, with basement, about twenty-four or twenty-six feet square, and if convenient, I'd like but one chimney in center of building."

Well, now, Mr. T. M., let's see what we can do for you. But an exactly square house is apt to look rather ill-proportioned; so we'll pull and push, and work out an almost square scheme. Here it is: twenty-nine feet eight inches long, and twenty-three feet eight inches wide; the second story is the same width, but a foot longer at either end. At

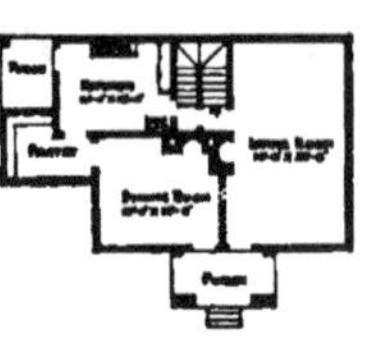

the front comes a porch twelve feet long and ten feet wide. Two doors open from the porch; one goes to the long living-room, and the other to the dining-room.

There really isn't any need to talk much more about the plan—it explains itself, in most cases. The central

House Plans   Turn It Around

chimney is a rather difficult thing to work out; still I have done it, by putting an archway between the flues in the second story. Of course these flues "rack over" and join before reaching the roof. The cellar stairs pass down under the main stairway; and, by the way, there is an outside cellar entrance, opening from the ground level to the lower landing. The third floor is a mere attic, unfinished, but capable of being made into two fairly comfortable rooms, if needed.

I've imagined that the house is to be weather-boarded with gray-stained redwood shingles; they are a Pacific Coast product, yet we use them to a tremendous extent in the East, and mighty pretty they are, too. There can be red brick foundations and chimney, red brick porch paving, white painted columns and cornices, moss-green shutters. There! Isn't that a fine color scheme against the soft silver-gray of the shingles?

Indoors there can be something picturesque, too—rather low ceilings, showing the heavy, brown-stained beams, and wall-boarded side walls paneled off with strips stained to match the ceiling beams. Brother T. M., how do you like it?

Turn-Around Farmhouse

What do I mean by calling it that? Why, I mean—but let's begin at the beginning:

When a man and his wife out on a farm decide to build a home, they usually collect a more or less complete assortment of plan-books, house-building pages, and so forth; then they choose this or that design, send for the plans, and start work. But these plans, nine times out of ten, were really designed for village or suburban houses; designed, therefore, to fit a small, narrow lot. So the porch is always at the front and the chief rooms always face that way, too.

Thus it comes about that you so often see a newly-built farmhouse resolutely turning its back on some wonderfully beautiful prospect of mountain, lake, or breezy prairie, and thrusting its porch and its living-room toward the bare, dusty country road.

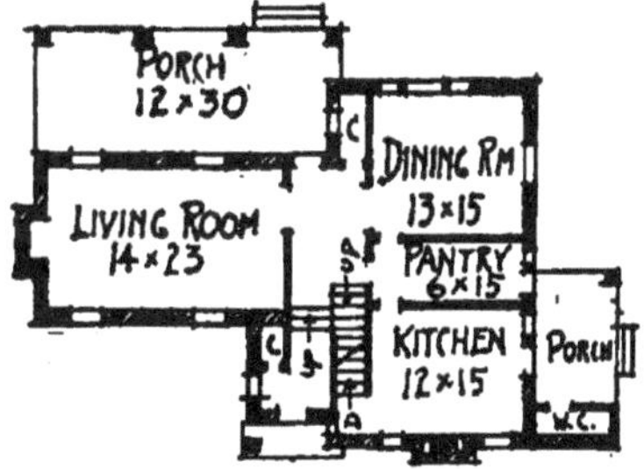

Quite likely the wife objected to this; but the builder told her, as builders will, "can't do anything else; the front door's got to face the road, of course."

Turn It Around · House Plans

So here is a farmhouse that may be *turned around* in almost any direction, regardless of the position of the road; and yet there is always a front doorway greeting you as you drive up!

As actually built, the house stands on the shore of a river, with the porch looking out over the water; the picture shows the landward side, with the little entrance-porch turned toward the driveway. But if, for example, the road had come winding up from the river valley to the hill where the house stood, why then we would have driven straight in to the larger porch and called that the front of the house.

In other words, the house is designed to look well from any direction. The hall runs right through, as you see, with doorways at each end; and of course either of these ends may serve as the main entrance. The big porch will probably face south to catch the summer breezes; I've put the vestibule at the other end to keep the bitter north winds from rushing through the house whenever the outside door is open for an instant. The living-room has windows on both sides; so you are sure to get a pretty good outlook, one way or another.

The general layout of the first and second floors is quite simple; it's hardly worth while for me to say much about it, for the plans explain themselves. The pantry, however, is worth discussing. A door leads to the kitchen porch; ice and all sorts of things may be brought in this way. There's another advantage, too: "I want some place where my husband can come in, wash up and leave his gum boots and overalls, before he goes into the rest of the house," wrote one of our folks to me the other day. Surely, she will approve this big pantry with its convenient back door, its sink and its coat closet!

Occasionally a housewife will want her kitchen directly adjoining the dining-room. Of course this plan can be changed to suit her by cutting out the pantry and moving the dining-room farther back; but that will mean some re-arrangement of the second story.

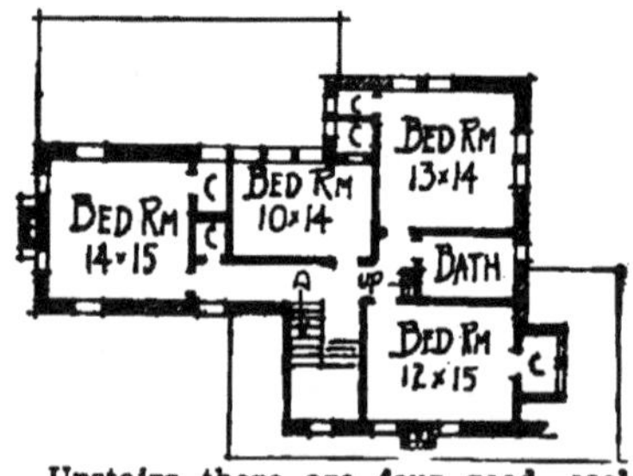

Upstairs there are four good, cool bedrooms, with plenty of closets.

Another Made-Over Farmhouse

Fig. 1 is the rough sketch, slightly smoothed out, of the front of the house (strictly speaking, it's a story-and-three quarters; but we'll let that pass); and Fig. 2 is the present floor plan. Mighty inconvenient and ill-arranged, that's a fact; the kitchen is so surrounded by cubby-holes that never a breath of air can get to it. And that little bedroom adjoining—whew! So we'll tear out a few walls and partitions, putting porch posts in their place; you'll see the result in Fig. 3. The old bedroom and wash-room are turned into a back porch, letting the breezes blow right through the kitchen. The old pantry becomes

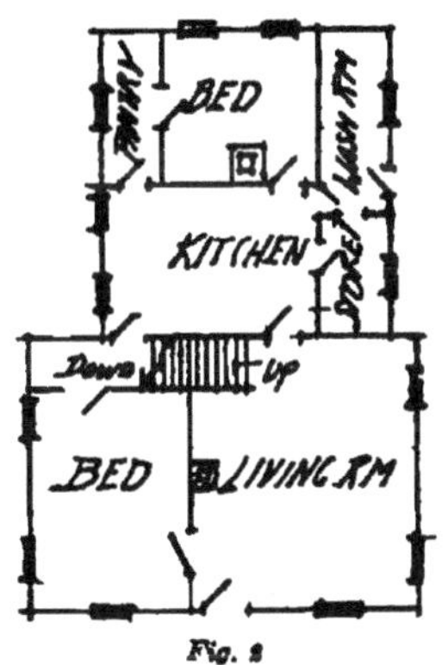

Fig. 2

a storeroom and toilet; the wash-room disappears, for the porch is the proper place for washing, except in midwinter. In cold weather this porch

can easily be enclosed with temporary panels. The two front rooms are thrown into one big living-room, with a dining alcove at one end; a very convenient and space-saving scheme, by the way. The old storage-room is changed to a pantry, forming a short and direct passageway from kitchen to dining alcove, as you see.

My friend didn't send the second-story plan; but I assume there were two rather inferior bedrooms up there. So we must improve these, and get at least one more bedroom; that's evident. Also we want to make the outside of the house a little less barnlike! So, we'll build a porch at one end, and carry an earthen terrace all along the front, sheltered by an overhanging hood, Fig. 4; this hood vastly improves the effect at very little cost. Then, over the porch we build a half-story room, with Dutch dormers to give head room and light. Any good insulating fabric, nailed to the rafters, will make this room cool and comfortable. A bathroom is worked in at the other end of the second story, Fig. 5;

Fig. 1

the two old bedrooms have ranges of windows added, in front. If still too hot, the ceilings can be lined with insulating fabric.

Making Over   House Plans

I don't know what material is used on the outside of the old house—the letter doesn't say; clapboards, I imagine. And probably there is no sheathing, so that the wind whistles through the cracks at a great rate.

Fig. 4

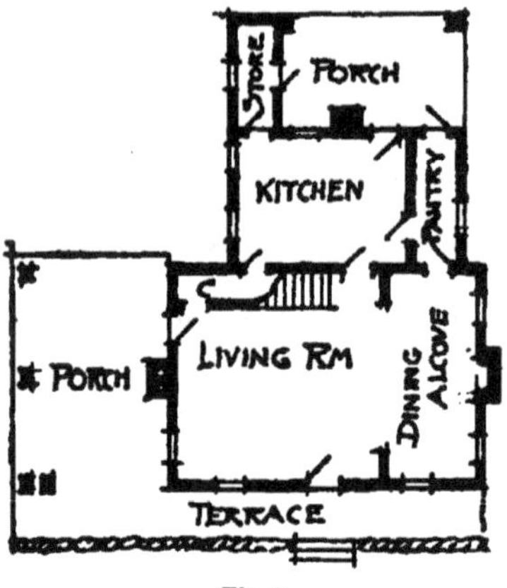

Fig. 3

Fuel is too precious to waste in heating such a building; therefore it's good economy to paper and shingle all outside walls right on the old clapboards. Dip these shingles in gray or white shingle stain; colors are costly and fade quickly. And, before I forget, let me warn you to pack the floor of the room over the porch with sand or sawdust, else you'll fairly freeze in such a winter as we've just had!

But, maybe three bedrooms aren't enough; in that case we'll try to work in a fourth over the kitchen, getting light and air by ranges of Dutch dormers on either side. If this isn't pos-

sible, a little one-story bedroom may be built off from the dining alcove. Another scheme would be to build a nice little bedroom at the right, to balance the porch at the other end; this, of course, would give us four second-story bedrooms instead of three. In that case the bathroom would have to be crowded off to the rear, above the kitchen, and lighted by one or two dormer windows.

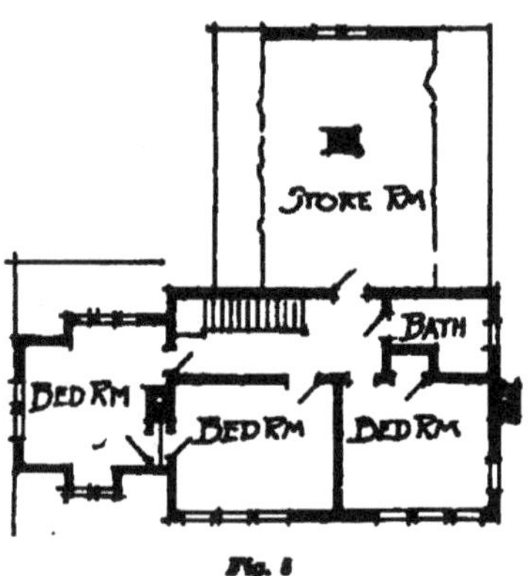

Fig. 5

A Western Farmhouse

The Middle-West farm with its broad, level acres, demands a house

that looks well from all sides; in other words, a house without a back; and this is such a house.

Straight up to the roadway front you drive; the big windows of the living-room open to a bricked or graveled terrace, and through these windows you may walk right into the house should it be summertime. Off to the left, the driveway swings around an ample porch (where you may come in, should it be raining); and so on to the rear entrance. In wintertime when storm-sash seal the big windows, this will be the main doorway; indeed, the family will most likely use it the year around, for it faces so conveniently toward the barns and other farm buildings.

The hallway does not cut wastefully straight through the center of the house as in many eastern homes; all compact, without an inch of lost space. The long pantry will be a joy to the house-

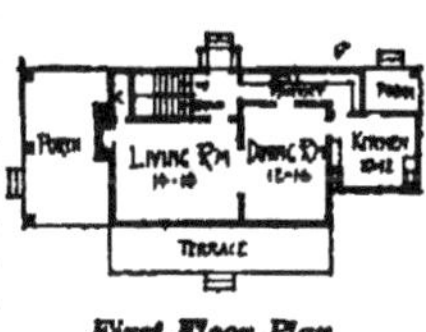

First Floor Plan

keeper; and the kitchen, you'll notice, is brought right up to the dining-room. (A lot of extra steps the housekeeper has to take when the pantry separates these two rooms, as in most house plans.)

The kitchen may be any size at all; I've shown it 10x12 feet, but in this particular design you may swell it or shrink it without affecting the rest of the house in any way. Grain farmers will want a big kitchen to take care of the harvest hands, but fruit farmers and truckers have different needs. And the larger the kitchen the more steps the mistress of the house must take.

In the second story are three ample bedrooms, besides the bath and plenty of closet room. The low loft above the kitchen is meant for storage space. The balcony may have a roof put over it if desired; then it will make a splendid sleeping

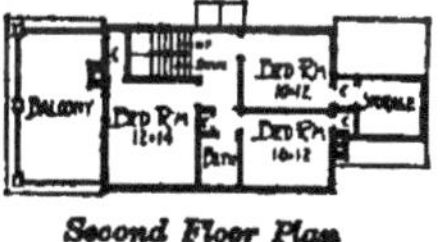

Second Floor Plan

porch. The third story has two more sleeping rooms, not the usual miserable little attic, but comfortable, breezy sleeping quarters.

A house of this sort may be built of almost any material, but I've imagined wide white clapboards for the walls, relieved by green blinds and topped off with a red, patent-shingle roof.

The cost? Don't ask me, in these times of leaping prices! But I do know that such a home as described here may be had at absolutely the lowest possible figure, for every detail spells war-time economy. Cut-up roofs, corners, bays and waste room all cost money; and this house has none of those things.

A Remodeled Tenant House

The doctor pointed it out with an ancient hoe handle. "There it is— just across that muddy hollow. Rather dilapidated, eh?"

The architect whistled thoughtfully and peered across. "And you want me to make a decent tenant-house out of that old wreck?"

The doctor laughed. "Oh, it isn't so bad as it seems; it has a good white oak skeleton."

"So you want me to operate? See if I can do a bit of constructive surgery? All right, doctor! but don't blame me if the operation isn't a success. Looks to me as if the patient is in a pretty bad way."

It was just a little story-and-a-half shack with one big room downstairs

one end knocked out and the chimneys and a smaller sleeping-room above; all torn away. But the architect tackled the job, nevertheless, and here is the result.

A dining-room was partitioned off one end of the big living-room; then at the right, a bedroom was built, with a little bathroom adjoining. Over to the left a kitchen wing was added with a concrete paved porch at the front. In the second story the old bedroom was fixed up, and another bedroom built in the new part.

There was no entrance porch; so the eaves were dropped down a foot or so over the doorway, making a little hood. A latticework shut in the new kitchen porch.

The old roof was utterly rotten and unsafe; so a new one of green shingles was put on. The clapboards were painted white, forming a most satisfactory background for the green of the roof, the shutters and latticework. The warm red bricks

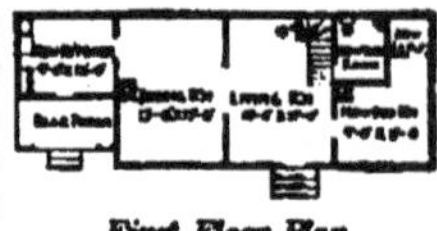

First Floor Plan

of the chimneys and front steps added just the needed note of color.

The bathroom was equipped with tub and toilet; rather unusual in a tenant-house, but the doctor insisted. "I don't want the usual horrible outdoor toilet, breeding germs by the million," he declared. "I'm going to

House Plans   **Adding Wings**

make the tenant-house just as sanitary as my own home—it's really a matter of protecting my family as well as my tenants." And he was right.

The new chimneys were carefully placed—two of them—so that every room could have its own fire; even the bathroom w a s so arranged that a little c o a l-stove could be kept going to keep the water-pipes from freezing. Of course it took a bit of planning to heat seven rooms, with only two chimneys; but it was done. Seven rooms? Yes—living-room, dining-room, kitchen, bath, three bedrooms. Indeed, we might almost count eight, for the old hallway in the second story was fitted up with linen shelves and so on, to serve as a sort of storeroom. Every inch of space

Second Floor Plan

was utilized; there wasn't any waste in the plan.

The doctor was delighted with the result, as he might very well be. "I wonder why it is that the farmers hereabout don't put up tenant-houses of this sort?" he asked.

"Too expensive," laughed the architect. "This house cost twice as much as the four-room shack that the average farmer provides for his married help. Still, I think the farmers would save money by building bungalows like this one."

"Save money?"

"Yes, in the long run. If a young man amounts to anything he wants to provide his wife with all possible comforts and conveniences. If he thinks he can't give her these if he works as a farm-hand, he will take her to the city."

"I recken you're right," mused the doctor.

The Elastic Tenant House

"You might tell me just how many rooms you want in the house," suggested the architect.

"Wait, that's just the trouble. You see, the sort of a hired man that I'm likely to get this year will probably be satisfied with just the ordinary five-room affair; but next year my brother-in-law expects to come and help. Naturally, I want something a little better for him; but a cog may slip somewhere between now and next winter, so I hardly feel like building a big house until I'm sure."

"Then you want an 'elastic' house," said the architect, pulling out the

tacks and taking down a sketch. "Here it is—I worked it up for a fellow who was in just about your di-

A Bungalow   House Plans

lemma." He jerked open a wide flat drawer and dragged out a set of tracings. "Look; I drew the original part in outline, then I marked the new parts in solid black. See?"

"I guess you'll have to explain."

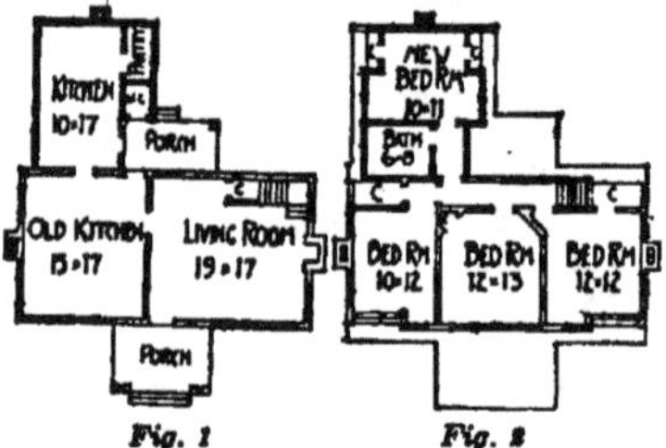

Fig. 1   *Fig. 2*

"All right; Fig. 1 is the first floor plan. As we originally built the house there were just two rooms down here —a large living-room and a big kitchen that served as a dining-room. The front porch was eight feet deep, and twelve feet wide, as you see; the back porch was the same width, but only six feet deep. Upstairs, Fig. 2, we had three bedrooms and that was all there was to the house. The owner meant to put a negro farmer and his family in the cottage, but wasn't sure about it. So I drew plans for a rear wing that could be added."

"Didn't you have to pull things down when you added it?"

"Not a bit of it. See, here's the new kitchen; it is slipped in right beside the old porch. Then, to square things out, I planned a pantry; also, a small wash-up place, opening off the back porch and provided with stationary washstand and frost-proof toilet."

"My wife surely would approve of that wash-closet!" said the man.

"No doubt," agreed the architect. "Now here's the second-story addition —a large bathroom, all complete, and a good-sized bedroom. We cut one doorway in the old wall, up here, and another one down-stairs; that was all the tearing out we did."

"Where's the cellar-way?"

"The cellar-way will go down beneath the main stair; or, if you prefer, the kitchen wing may be made about three feet longer and a back stairway worked in close against the old house. The cellar stairs can run down under this."

A Maryland Bungalow

Yes, a Maryland bungalow—but it might equally well be a Maine or Montana bungalow, save perhaps for a very few minor changes. Some types of houses are so strictly sectional that they never seem to grow very well outside their own particular State; but it isn't that way with the bungalow. True, it first took root in California; but like the San Jose scale and other things of that sort, it was soon heard of far beyond the borders where it started. So today we find bungalows spread over the whole country.

House Plans   A Bungalow

And small wonder! for to have a home where one doesn't perpetually trot up and down stairs is the dream of many a housewife.

Therefore, here's a bungalow, designed (as it happens) to be built in one of the country towns of the "Eastern Shore"; yet, with almost no change at all, it will suit wonderfully well as a farm home in any section of the United States. The plans explain themselves, I think.

Possibly s o m e people would prefer a little larger dining-room a n d kitchen, particularly if this home is built as a farmhouse; but it's very easy to make that change. The space now given to office and rear porch may be used for enlarging the kitchen;

First Floor Plan

a new back porch and office may then be planned in a wing stretching out to the rear.

The material of this bungalow is frame. There is a cemented cellar underneath and the floor of the porch is cemented. The roof is shingle, bleached to the good gray tint that all roof shingles take if left to themselves; but the walls, while also c o v e r e d with shingles, are colored white. Not the o r d i n a r y g l a r i n g white paint, but a newly-devised stain, giving a c l e a r transparent color something l i k e the effect of a very thin coat of

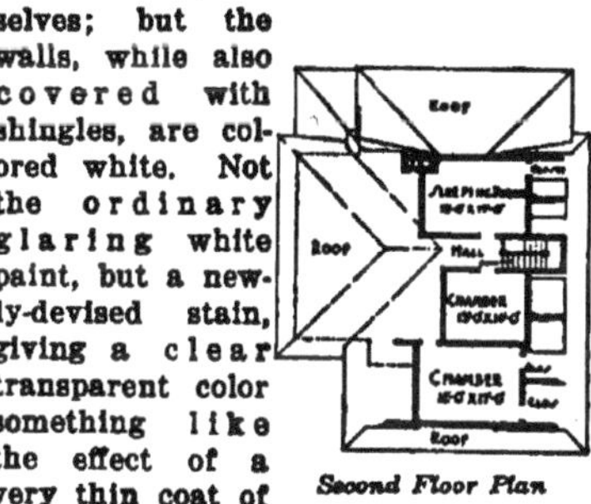

Second Floor Plan

whitewash, only far more beautiful and permanent. The shutters are a warm green. The bases of the heavy porch-posts are of good red brick. Could there be any better color scheme for a background of green fields and blue sky?

For a small family or a newly-married couple the bungalow type of house is often very appropriate. It saves work and it has that "cozy look" which is so appealing to many. It should be remembered, however, that some folks like an attic and object to sleeping-rooms on the ground floor; and to such folks the bungalow does not appeal.

Planning the Porch

" . . . Just exactly like *that!*" finished the farmer's wife, flipping the photograph down on the drawing-board.

"But, my dear lady . . ." protested the architect.

"No, that's the porch I w a n t, and you can't talk me out of it!"

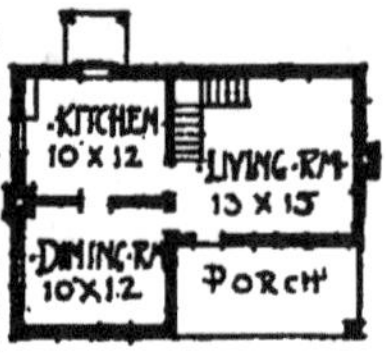

Silence, then a little laugh from t h e farmer's wife. "Oh, y o u k n o w w e l l enough that my curiosity can't stand that! Come now—*why* don't you like that porch?"

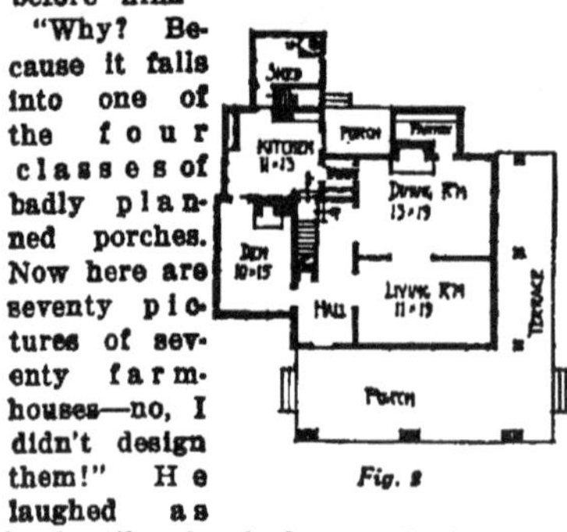

Fig. 1

The architect opened a drawer and lifted from it a huge handful of photographs; carefully he spread these out before him.

"Why? Because it falls into one of the f o u r c l a s s e s of badly p l a n ned porches. Now here are seventy p i c tures of seventy farm-houses—no, I didn't design them!" H e laughed a s he hastily sketched several plans on

Fig. 2

the back of a photograph. "Here you are, Fig. 1, only a portion of the porch can get the breeze; half of the space is pocketed by the angle of the house. Sixteen of my farm pictures show just that defect, and twenty-five are like Fig. 2."

"But I *like* a porch that turns around the side of . the house!" The farmer's wife smiled a bit defiantly. "You can always get a good breeze, and you can keep out of the sun, too !"

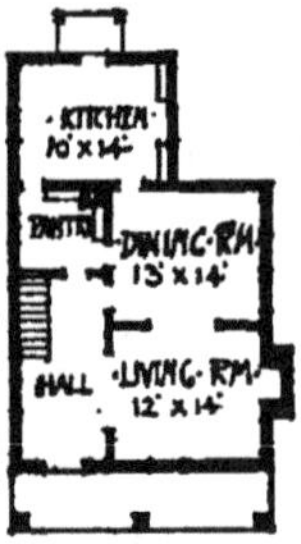

"Quite true, but I'll explain presently how you can get both these advantages in

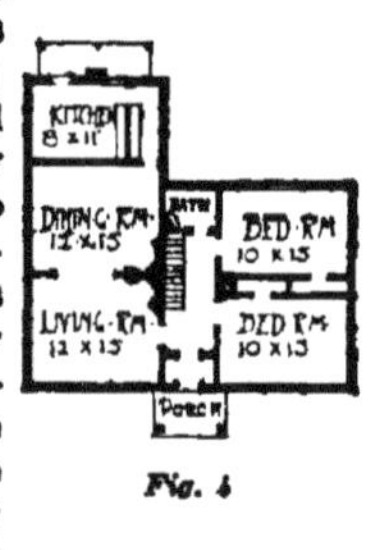

Fig. 3

another way, without the excess cost of so much porch. Besides, a round-the-corner porch usually covers all the windows of one or more rooms; very dark and gloomy, that! And here are twelve houses with porches all across the front—Fig. 3; that would be well enough if o n l y these weren't so narrow. Remember, a porch is really an outdoor living-room; especially so on the farm. Now, the *very narrowest* room is at least ten feet wide; yet we often see

Fig. 4

House Plans   **Home-Made Greenhouse**

porches of seven feet, or even less. You can't form a family group; you

Fig 5

string yourselves out in a row, minstrel-fashion, and talk back and forth as best you can. Then, there are ten more with merely a cramped little entrance-porch, like Fig. 4; and finally seven porches that really are porches. Seventy in all; which means that only ten per cent. are really good!"

The farmer's wife gasped. "But—I don't quite understand; what makes those seven so good?"

"Look at Fig. 5 a minute; twelve feet deep and twenty feet long. That makes a very cool, comfortable outdoor room, indeed."

"But supposing the breeze should change?" was her prompt question.

"Well, let's suppose the house faces south; then you get all but the north wind, and you don't want that!"

"But if it doesn't face south?" insisted the farmer's wife.

"Then we will put just a little entrance-porch

Fig. 6

at the front like Fig. 6, and tack our main living-porch to the south end of the house. See?"

"Yes, I see!" admitted the farmer's wife.

Home-Made Greenhouse

Examine the accompanying plans carefully. Elevation shows walls and roof supports. Paint all the woodwork before erecting.

The posts are of five-inch round cedar, set below grade about three feet and tamped with concrete. These posts should be set between five and six feet apart, according to the size of the house wanted and the length of lumber you intend to use for siding.

Now put on the plate, and measure down the depth of the side glass, and put the glass sill on. The siding should be double, with good waterproof paper between.

Next put on the ridge and rafters. Be sure to allow one-eighth inch for the glass to play—in the glass groove or rebate; then put on the vent, sash and headers (see cross-section showing post in concrete, siding, glass sill, plate, header, vent and sash).

Separate hotbed sash should be on the south side of the building, if possible; they are of great assistance to the greenhouse.

The ground plan shows the benches

Home-Made Greenhouse   House Plans

and walk arrangement; also heater, potting and workroom.

Sectional view of bench shows posts set in ground eight inches, with a 2 x 4 across the top, and 1 x 6's on the sides and bottom of bench.

The heating plant should be a sectional hot-water boiler; you can then increase the capacity of the boiler by additional sections at any time you may wish to enlarge the greenhouse.

With the aid of the plans here given, any local carpenter, or any man handy with tools, should be able to erect a satisfactory greenhouse. Probably your local lumberman or planing mill can furnish the desired materials, and your hardware man or plumber can furnish the boiler and piping. The necessary glass, glass points, putty or mastica, etc., are easily obtainable in almost any town.

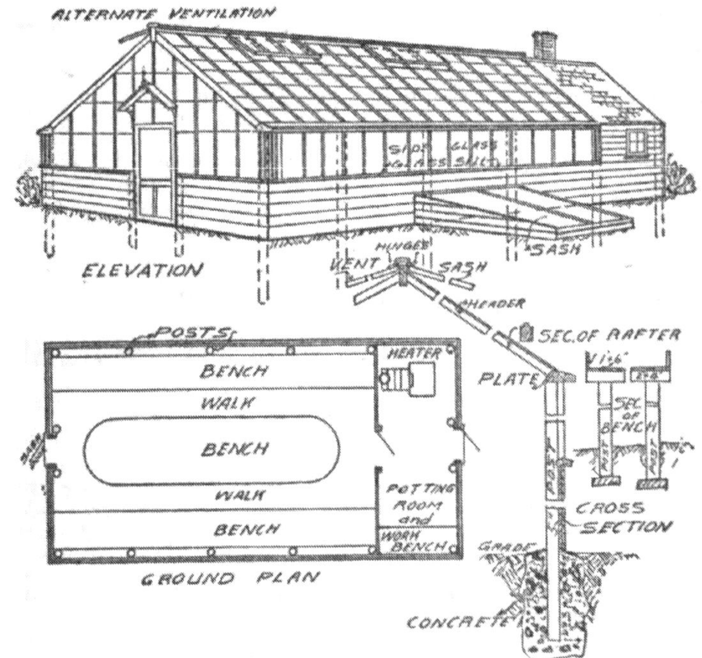

Plans for a Thoroughly Satisfactory Greenhouse—One that Has Been Tried and Not Found Wanting

A Storage Cellar for Roots and Vegetables

Concrete is one of the most satisfactory materials for storage cellars, not only because of its permanence, but because its use makes it possible to maintain definite temperatures. When the frame, roof and walls are made of wood, brush or wire, the storage cellar has to be rebuilt every few years, because of the changes in the moisture content of the soil coming in contact with the wood or wire. The moisture causes the wood to decay and the wire to rust. The best wood cellar is only temporary.

A root or vegetable storage cellar of concrete is built to last. Like a concrete silo, a concrete storage cellar of proper size should pay for itself in a year. Marketing can be controlled in strict accordance with supply and demand and the most favorable conditions, and waste by rot entirely prevented, if crops are stored carefully and at the right time.

The storage cellar described here is suitable for either fruit or root crops. Its size can be varied to suit different sized orchards or farms, simply by adding to or taking from the length, leaving the other dimensions unchanged. The plans show a cellar twelve feet wide, fourteen feet long and nine or ten feet deep inside. Floors, roof and walls are of concrete, with tar joints where the walls join the floor. The cellar is moisture proof and secure against rats and mice.

End walls are ten inches thick. Side walls are sixteen inches at the base and taper to six inches at the crown. The floor is five or six inches thick. The materials required for a cellar of the size shown in the plans, are forty-five barrels of cement, fourteen and a half cubic yards of sand and twenty-two and a half cubic yards of pebbles. The concrete for the walls is mixed in the proportion of one part of cement, two and a half parts

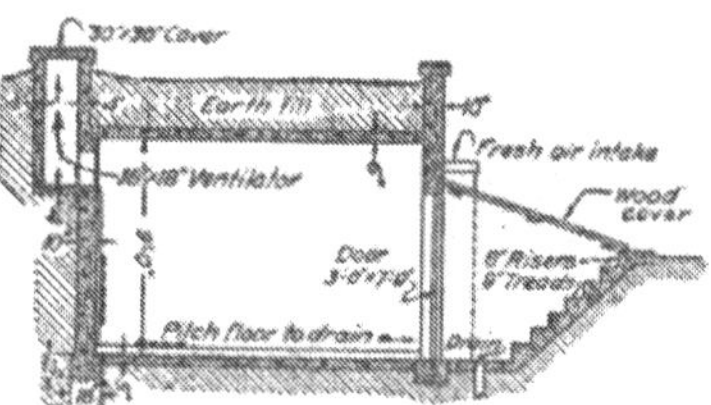

A section through the cellar, lengthwise, showing ventilator and fresh air intake, together with dimensions.

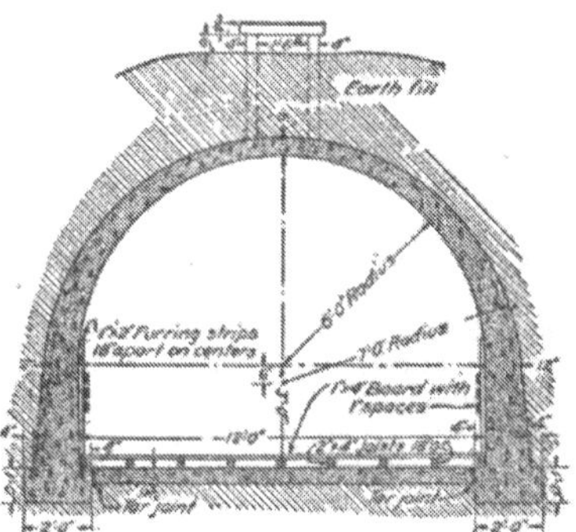

A section through the cellar, crosswise, showing how the false floor and walls of boards are put in. Notice the tar joints to allow for expansion.

of sand and four parts of pebbles or stone. The floor and arch of the roof are made of a 1:2:3 mixture. The floor is sloped toward the entryway where the drain is located.

This design has been prepared with special reference to ventilation. During cool evenings man-hole and cold air intake covers are removed and the cold air permitted to pass down the intakes, circulating through the passage between the concrete floor and the false floor of the bins. The false floor is made of 2 x 4 joists, covered with 1 x 4 boards nailed one inch apart. Openings in the floor allow the air to pass up through the stored contents, thus cooling them.

Outside walls are built so that cool air can circulate up along them. The warm air passes out through the manholes. In the course of one night the entire air in the storage cellar is in this way changed many times, thus thoroughly cooling the cellar before morning.

The cellar holds about 600 bushels of potatoes. On the average farm, where the whole cellar is not needed for vegetables, it can be divided by a partition and part of it used for other purposes. During the summer months it can be used for a milk cellar or a storm cave.

Where Shall I Place the Farm Buildings?

In the corn belt many of our folks are using just such a farmstead layout as this; very practical and convenient they find it, too. But unless we explain it a bit, you may not quite understand it. Well, then:

On the front stands the farmhouse, F. A group of sturdy shade trees make a wonderfully fine background; nothing is so dreary as the bare bald homes that so often stand exposed on otherwise fine farms. The windbreak of evergreens, bordering the whole thing on the north and west, is very necessary in prairie country, and adds still more to the beauty of the whole scheme. Some of our folks who live in other sections may not need this feature.

So, let's drive on past the house. To our right stands the machine shed, M, with the windmill and well, W, at one end. Over to the left, the dairy, D; the gasoline engine is here, too. Some folks have the well here, so as to get direct water for washing the milk cans, etc.; but we should consider that entirely too near the poultry house, P, with its runs. We once found a well badly polluted and full of dangerous disease germs, merely because it had been driven too near the poultry yard; so don't take any chances. It would be safer to put in pipes and force the water across from the well. Keeping on, we next come to the corn-cribs, C, with, possibly, grain-

Farm Buildings   Model Barn

bins overhead. Beyond that, you'll see the hog house, H, with runs to the east; over on the other side of the driveway is the big barn, B. Cattle can be kept on the south side next the cow pound, while the horses are housed on the north of the center gangway.

As to the distances apart of these various buildings—well, that depends on circumstances. We suggest from twenty-five to thirty feet as a good space between the hog house and the barn; and there ought to be at least 100 feet between the farmhouse and the cow pound fence.

Now, there is a definite reason for the layout—several definite reasons, in fact. The quarters of the large animals—with flies and foul odors—should be as far as possible from the house; so we have set the barn and hog-pen at the very back of the group. The poultry yard is nearer the house, to protect against marauders—human and non-human; besides, the farmer's wife should not have too far to walk as she goes egg-gathering

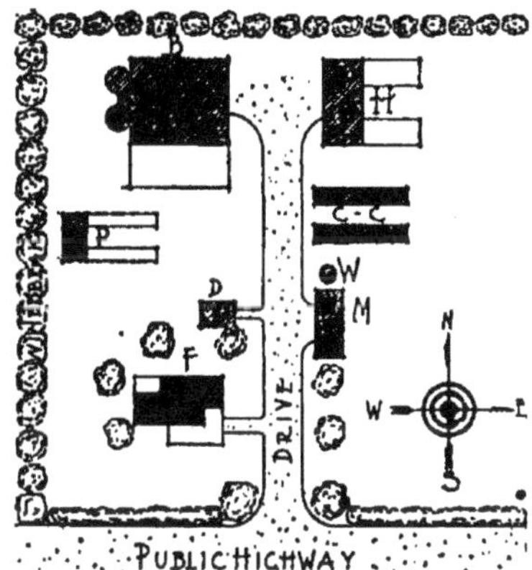

and fowl-feeding. The dairy, too, had best be rather near the house for various reasons; and the machine shed, set just across the driveway, balances the group very well.

Naturally, the corn-crobs must be centrally located, convenient to barn, hog house and poultry yard.

The Barn I'd Like to Have

I have a small barn on my own place, but I'm not going to show the plans. Why? Because the former owner built it, and he didn't know much about planning a barn; that was plain. So here's the barn I'd *like* to have—not the one I have, unfortunately.

The first story, as you will see by the plan, is exactly 23 x 44 feet; the broad front should face south, to give the cows the greatest possible amount of warmth and shelter in winter. Each cow stall is three and a half feet wide, in the clear; the horse stalls are four and a half feet, or possibly an inch or more for large draft horses. Nine feet, including the manger, is a good length.

The box stall about ten feet square is absolutely necessary for a sick horse or a cow with a calf. The shop is nearly the same size. Harness, saddles, etc., can be kept here. The wagon house, 16 x 22 feet, will hold two wagons, two carriages, and various other stuff, very easily. Your windows in the wagon house admit plenty of light.

The cross-section shows the construction clearly enough. The wall-studs are 2 x 4, set about two feet on centers, with 4 x 4 sills and 2 x 4 plates. The center posts are 6 x 6, carrying a girder of three 2 x 12 joists.

The second-story joists are 2 x 10, set two feet on centers. In most cases they are 12 feet long, but above the stalls half of them must be fourteen feet long, because there the girder isn't in the exact center of the stable. The rafters are all 2 x 4, twelve feet long; the diagonal braces are 1 x 6, about twelve feet long; while the little short pieces are 1 x 4.

I should advise using six-inch lapped siding for walls and the gables; the roof may be shingled, or covered with some sort of patent roofing. Ventilators may be put on the top, though a simpler scheme is to make slatted openings high up in either gable. The foundations had best be of concrete. Cork bricks are good for stall bottoms.

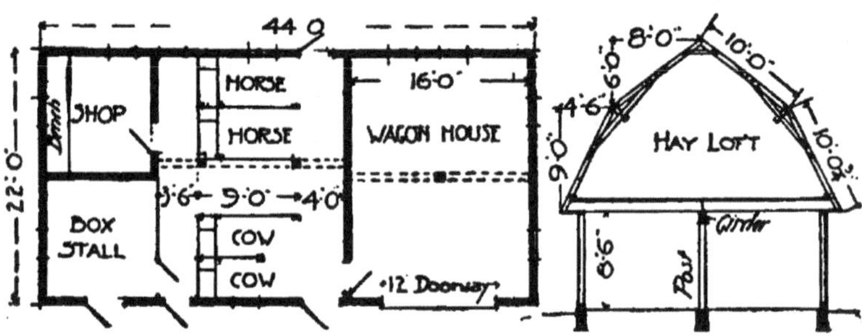

Expert on Lightning Rods

On July 4, 1918, I had been forty years in the lightning-rod business. I began at the bottom, learning from the best practical and scientific men of the day. When I first started out I had partners, but for twenty-five years I have been at the top of the business for myself, and have stood the test of insurance companies, architects, builders, business firms and business

Farm Buildings   Lightning Rods

men, and have erected, on an average, 32,000 feet per year.

Insure Against Damage

Why erect lightning-rods? To protect buildings, trees, stacks, etc., from lightning. A building that is properly rodded with the right kind of rods is

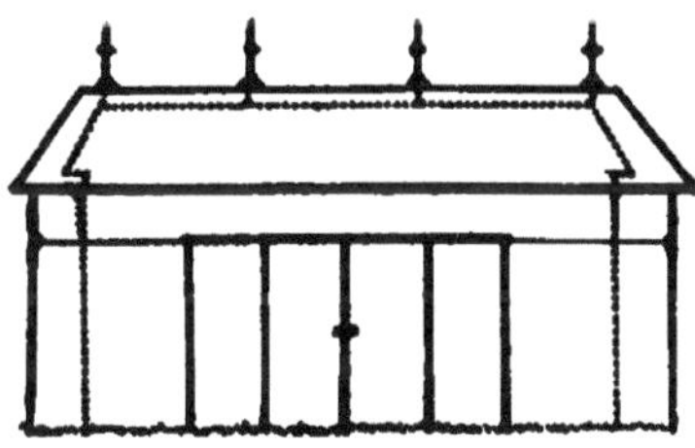

All points must be connected, and there must be at least two ground rods in the circuit system of rodding

insured against damage. As far as I know, I have lost no buildings. I have had no complaints from damage done by lightning, except for points melted, rods ripped up and holes burnt in elevations, rods, being able to take care of all currents, when in circuit form. By the circuit form I mean when all the points are connected and where there are at least two ground rods, as shown in the illustration.

Points Must Be Sharp

Oftentimes, after storms, people whose buildings I have rodded, tell me that they are sure the lightning struck near their homes, because of the violence of a discharge. Sometimes I find that the points of the rods on their barns or houses have been melted by the heat of the electricity, so that the points are blunt instead of sharp. They have to be renewed, for a blunt point will not protect the building. Once in a while the electricity will burn holes through the rods, and rods may be torn loose from the buildings if the current is too heavy.

There is no truth in the statement that lightning never strikes twice in the same place. Some places seem to draw lightning more than others.

Well Grounded

One of the most important things about lightning-rods is to have them well grounded. It is not enough to have the rods stuck in the ground for a few feet; and in some cases, where the soil is dry, a greater depth is not enough, for dry soil is a poor conductor. Groundings should always go to permanent moisture, at least eight feet. There should always be plenty of points. The distance between points should be four times the height. That is, if the points are four feet above the ridge of the roof, they should be sixteen feet apart. Insulators are not desirable. It is best to connect metal water-troughs to the system, so they can carry off the electricity on the buildings. Wire fences ought to be grounded, too, to protect stock. A suitable connection between the fence and the ground can be made of a piece of heavy galvanized-iron wire. One rod every 150 feet is enough.

How to Ventilate a Barn

Since air makes up the greater part of stock rations, the air animals breathe should be as pure as the food they eat. To keep the air in the barns fresh all the time requires a set of openings or flues to admit fresh air, and another set to carry off the impure air and gases from the animals and the litter. These are not necessary in barns where the cracks admit plenty of air; such barns need battening rather than ventilating.

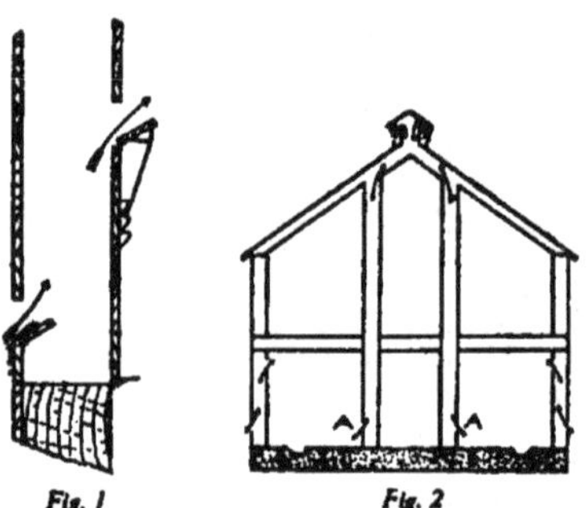

Fresh air should enter at a point above the level of the animals' backs. When barn walls are double thickness, a hole may be cut between the studding on the inside, about six or seven feet from the floor, and the piece hinged to open as shown in the cross-section of wall in Figure 1. Outside, a foot from the ground, another opening is made which may be covered with screen. Where the walls are single, a trough open at the lower end and closed at the top may be nailed on the outside of the barn, leading up to the hole. The entrance and opening into the barn should be several feet apart. Often windows may admit fresh air, if they are high enough, hinged at the bottom, and open in. The air will be deflected toward the ceiling. The opening may be covered with burlap or muslin to check the force of the air.

Flues for carrying off impure air should reach from near the floor to the roof. They can be made of six-inch or eight-inch boards nailed together. They must be tight; leaky flues will not do the work. The impure air, because it is warmer and lighter, will be drawn up the flues, entering near the floor as shown in Figure 2. It will not cost a great deal to put in several of these flues, and the system will pay for itself in a short time. Last winter a hog feeder in the Middle West was losing hogs every day, apparently from cholera. Vaccination did no good, and the disease was not checked until the herdsman cut several holes in the roof for ventilation. The building had no means of ventilation before. In half-monitor types of hog and sheep houses, if the doors are upper and lower, the upper door can be left open for admitting air and sunshine. Impure air will escape through the windows at the top.

No Successful Dairy Without a Silo

If there is anything a farmer is justified in going into debt for it is a silo. No implement will pay for itself so quickly as the silo. The question, Will it pay? is no longer debatable. With a dairy of twenty cows it will pay for itself in two winters' feeding. The only questions nowadays are, How can I get one? What kind shall I get? Where shall I put it, and how shall I pay for it?

An acre of corn made into good silage has about forty per cent. greater feeding value than when fed as crib corn and dry fodder.

Cows need succulent feed during the winter months and silage furnishes it in convenient, economical form. Steers and lambs make faster and cheaper gains when silage is part of the ration.

The Bank Will Lend the Price

Wideawake bankers now recognize the silo as a wise investment, and will usually gladly lend the necessary funds when needed.

The absolutely essential features about a good silo are: The walls must be air and moisture proof, the inner surface must be smooth and perpendicular, the walls must be strongly reinforced, and the doors must be air-tight.

There are five kinds of silos on the market, those made of wood, solid cement, cement blocks, hollow tiles and iron.

The wood silo was the first commercially introduced and has been greatly improved. In selecting a wood silo, the writer would go to the expense of having the staves made in one piece. This costs a little more. For instance, a ten by thirty two-piece stave silo is listed at $190, and $210 if each stave is thirty feet long.

The first silos built were of the pit type, dug in the ground. It was soon found that the ensilage in these pit silos rapidly became moldy and unsuitable for feed. The silos first constructed above ground were of wood, being square in shape or eight-sided; but they were not a success on account of the air pockets in the corners, which caused the silage to spoil. It was not until the round stave silos with hoops that could be tightened or loosened as the silo swelled or shrunk, were put on the market, that silos became a practical success.

The Wooden Silo

The expense of wood silos depends on whether they are made of long-leaf yellow pine, Oregon fir, spruce or redwood. The latter are more expensive.

A silo ten feet in diameter is the best size for from twelve to fourteen head of cattle, as enough silage is

The Silo   Farm Buildings

taken out each day to keep it fresh. This is a very important point. A silo twenty feet in diameter would require a herd of from thirty-five to forty head of cattle to eat enough off each day to keep the silage fresh. A silo ten feet in diameter and thirty feet high is the best size for the small-sized herd. This will hold forty-six tons of ensilage, enough for eleven cows, each getting forty pounds a day for 200 days, and allowing some for loss.

The Cement Silo

Cement silos, and silos made of hollow cement blocks, have been used, but do not seem to be growing in favor. Every silo, however, no matter of what material it is constructed, should have a solid foundation of masonry or cement, preferably cement.

Hollow Tile

Seven or eight years ago silos began to be made of hollow tile, and are steadily growing in popularity. The manufacturers claim that they will last for generations, are wind and moisture proof and fireproof.

Metal

Metal silos are being built and extensively exploited, being put up in sections and bolted together, the joints being filled up with suitable paste. The manufacturers claim the work of erecting these silos is so simple that a farmer with ordinary judgment can put them up, using the help on the farm. When the silo is complete and erected, it resists the action of heat and cold, the walls being absolutely air-tight, and the silage will keep well; but there is no doubt that the silo of this type should be kept painted, particularly inside, where it might be eaten with rust.

Get Catalogs

All varieties of silos have some good points which some of the others do not have. The Farm Journal knows no better way than to advise getting catalogs of the different manufacturers whose announcements are in our advertising columns, and picking out the silo best adapted to your requirements and your neighborhood. These catalogs are filled with figures, facts, and much helpful information, the result of many years' experience. Consult your neighbors and see what they have to say about the silos they have.

No matter what silo you buy, you can get it with the assurance that it will have some advantages, either in cost, looks, endurance, upkeep, or simplicity of construction, which some of the other silos do not have.

Terms Liberal

Almost any manufacturers will make liberal terms to farmers who have not the ready cash to buy a silo and who may need the summer's crops to help pay for it; but there is no one way in which The Farm Journal can be of more service to its readers than to urge every man who has a herd of ten cows or more, who is selling milk, making butter or feeding stock, to take immediate steps to get a silo.

Easy-to-Build Ice House

Some farmers are thoughtful enough in winter to look ahead and store up ice for the summer, and they truly enjoy the fruits of their labors. On the farm ice is a necessity and also a luxury.

Ice is used for a large number of purposes. It deserves a place on every farm. Ice-houses can be built with only a small outlay of money, as skilled labor is not necessary for the construction if care is used, and in almost every locality there will be found running water that can be dammed up into a small ice pond.

The ice-house shown here is made of twelve-inch-thick hollow building tile, with a one-inch cement coating inside. It keeps ice with but little cost during the summer months, and is permanent. In size it is 14x20 feet, and the capacity is fifty tons. In many places the materials needed for this structure can be had for approximately $175. The floor is dirt; t h e footings for the tile walls are conc r e t e, set down deep so as to be under the frost line, and proper drainage is provided. Mix the concrete 1:3:5. Lay the tile

Fig. 1

blocks in the wall with a lime and cement mortar, using only enough lime to make the mortar sticky. Build the walls true and in perfect line to prevent any possibility of cracking or bulging. The walls go up ten feet.

At the top bolt the wood sill every five feet to the wall, so that the rafters can be securely spiked in place. This will make a solid connection between the tile and the frame roof, and prevent the danger of wind blowing the roof off. The 2 x 4 inch rafters are set two feet centers and at one-third pitch, with a cross-tie on every set of rafters so as to form a support for the ice carrier track. The gable ends of the ice-house are of frame and are fitted with ventilators. The rafters are covered with tight boards, preferably sheathing lumber, laid with tight, smooth joints to make a trim and solid foundation for the roofing felt.

At one end of the house the door extends from the ground up to the plate. This is built in sections for convenience, and is best made double-thick with building paper between the layers of boards. Cut some short lengths of planks for loose leaves that can be set in across the doorway so that the ice and the sawdust floor will not crowd against the door.

Materials Needed

Materials needed for the ice-house shown:

  10 bbls. cement for wall footings, etc.
650 hollow tile blocks for walls
   2 pieces 2x12, 14 feet, for plates
   2   "   2x12, 20 "   " plates
 22   "   2x 4, 10 "   " roof rafters
 10   "   2x 4, 8 "   " cross-ties
  2   "   2x12, 10 "   " door frames
530 feet of roof sheathing
100 feet of six-inch flooring lumber for
    doors and gables
   4½ squares three-ply roofing felt.

The location of an ice-house is important. Often it is placed near the pond or stream where the ice is secured, thus doing away with a long haul for the ice. This is handy at filling time, but not very convenient in the summer when ice is needed and all the help is busy in the fields. Have the ice-house near the dairy house and not far from the house. Then the farm-hands will have time to fill the refrigerators and ice-boxes during the noon hour, without hitching up a team. So much the better if the building is shaded and open to the north.

A House for Growing Sprouted Oats

It seems that each year sprouted oats are becoming a more important ingredient in the bill of fare. They certainly do solve the winter green food question to a great extent.

The illustration shows a cellar constructed and used for the sole purpose of sprouting oats. As will be seen, the "house" proper is underground; the entrance is made by cellar steps. Heat is furnished by pipes connected with a regular brooding-house stove.

The arrangement is complete and can be built at a small cost.

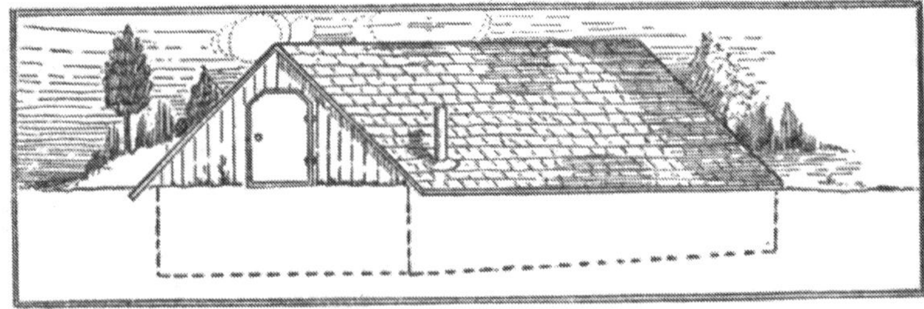

A Good Cellar for Sprouting Oats

Model Poultry House

Elevations and floor plan: 12 x 18, eight feet high in front, five feet in rear. Ends and rear covered with No. 2 matched spruce. Windows must be removed in warm weather, as they

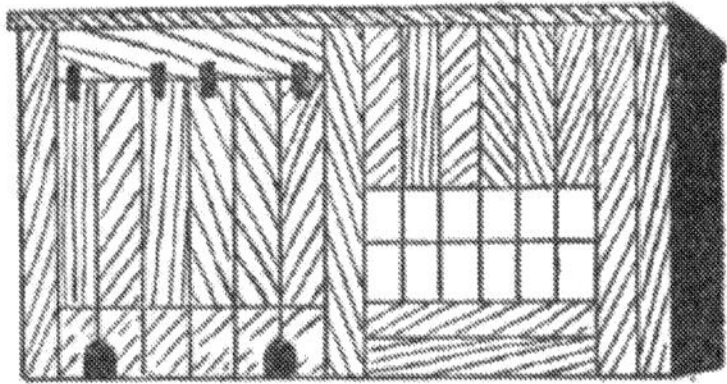

Fig. 1

are designed only for the coldest winter nights. At any other time of year open air is best. Wire netting is tacked inside the window sash.

Drop doors are each three feet wide, five feet four inches long. Board above one foot wide; below, twenty inches. A 1 x 4 strip, five or six feet long, is hinged to lower cleat. This serves to push the door outward and upward as shown in Fig. 3. A slot in the strip fits over an iron pin driven into studding below door and holds it in position.

Dropping board, two and one-half feet high in front, three feet at rear. Roosts, 2 x 4 (planed and corners rounded) six inches higher. Nest boxes extend along the front and end

as shown in Fig. 3, and are entered from the rear. Hinged boards in front are easily raised to gather eggs.

Roof is covered with paper roofing.

A small room, 4 x 6, can be partitioned off, if desired, for storing supplies and dressing fowls.

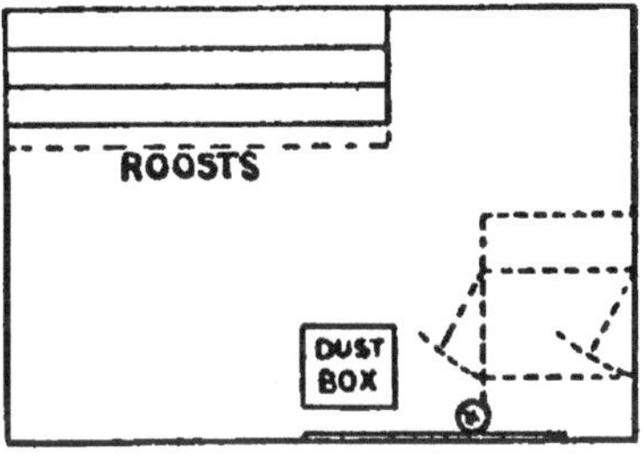

Fig. 2

Following is the way the Editor has his poultry houses arranged: The roosts are two feet from the ground. Six inches beneath the roosts is a solid board platform to catch the manure of the night. Under this platform are placed the nests.

A corner shelf is made about six to eight inches from the floor, on which is put the drinking fountain. This keeps the water out of the sun in summer and the severe cold in the winter. Hopper feeder is hung on the back wall, to be ready for use during bad weather

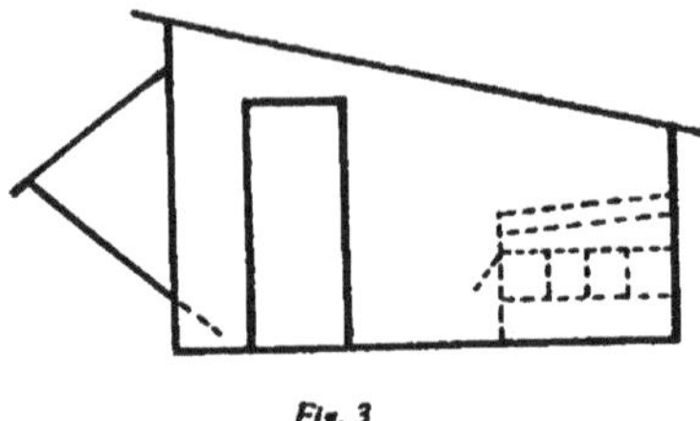

Fig. 3

when the fowls are necessarily confined. The north walls of the house are double-walled with heavy lining paper between. A dust bath is placed under the window.

If possible, each poultry house should be arranged for both a scratching shed and a roosting pen—two-thirds the amount of space being used for the former and one-third for the latter. Earth floors in the hen house are the best.

The Farm Journal Poultry House

For about twenty years I have been experimenting in the construction of poultry-houses, carefully studying every matter of convenience, comfort and economy. As I did not have very much working capital when I started, economy was an important factor with me at first. Throughout these years of experimenting I made many changes, changing my ideas completely in some respects.

In my boyhood days the poultry-houses on my father's farm were built so snug that hardly a breath of air could enter them. "Keep out the cold" seemed to be the idea; and yet these houses were cold and damp. The heat from the fowls at night would cause moisture to collect on the ceiling and walls, and roup and colds were constantly present. What convinced me of the importance of fresh air was that those fowls which roosted out-doors or in the open wagon shed, never seemed to suffer with colds. This led me to plan a house that would give all the outdoor conditions, with the added protection from cold

winds, rain and snow. If given proper protection a fowl can stand more cold weather than can a human being. I never knew a hen to freeze to death if she had a dry pen to scratch in.

The scratching shed idea appealed to me very strongly, too, and as time went on my faith in this "workshop" idea strengthened.

So my first house, built on the scratch-shed style, closely resembled the one illustrated, except that there was a window-sash in the roosting pen, and a muslin curtain dropped down in the scratching shed. This curtain was tacked to a frame and hinged to the roof. During the winter it was hoisted in the morning and lowered at night. Each pen was 7 x 16 feet, of which 7 x 10 feet represented the scratching shed and 6 x 7 the roosting pen. Later, the width was increased to twelve feet.

I used that style of house for several years before I made a change. It was quite satisfactory, except that every spring and fall there were numerous cases of colds among my fowls, and it was some time before I discovered the cause. During the daytime I would leave the door open between the scratching shed and the roosting pen, and close it at night. After investigating I found that when I opened the door in the morning there was moisture on the walls and ceiling over the roost, and at night

colds, and seldom have frozen combs and wattles.

I prefer to have the roosting pen separated from the scratching shed, as it gives the fowls more protection at night. The north wall is made double with heavy lining paper between, and the roof is constructed of matched boards covered with three-ply roofing-paper. In the roosting

Interior of Roosting Pen, 6 x 12 feet

when I went to close the door the moisture was gone. I took out the window-sash and covered the window with one-inch netting. I never had any trouble with colds after that.

Then I had trouble with the curtains in the scratching pen. The rain would beat against them and the dust raised by the hens scratching in the shed would settle upon the curtains and they became so encrusted that, do what I would, I could not get them clean so that the pure air might sift through. Consequently, I removed the curtains. Now, even in zero weather, my fowls never suffer from

pen there is room for the drinking vessel, grit and oyster-shell boxes and a row of nests. The floors are made of cement and covered with several inches of road dust. Hay, straw or leaves are thrown upon this. Such a floor keeps out rats.

After the morning meal is over, the feed-trough is fastened to the wall of the scratching shed with a hook and eye. The front of the scratching shed is covered with one-inch wire netting. At first I used two-inch netting, but one season's experience changed that; it was too easy for the sparrows to get in, and I could not well afford to buy grain for them.

So here, as the result of my experiments, is The Farm Journal poultry-house, built upon sanitary principles. It is about the least expensive to erect, has all the conveniences, and freely admits sunlight and fresh air. In this style of house I get eggs throughout the entire winter and keep my fowls in the best of health. These ing pen instead of leaving it open. Some poultrymen prefer to have a drop curtain about a foot from the roost in addition to the muslin. This is fastened to the ceiling, dropped down at night after the fowls are on the roost, and pulled up in the morning. As this curtain, which should not be longer than a foot below the

Right section of house is scratching shed, 10 x 12 feet.
The open front admits air and sunlight.

outdoor conditions give my birds a ruggedness that is of untold value in producing eggs.

The houses can be built in a row— that is, a continuous house, with all the doors on the inside, so the attendant can go from one pen to another without going outdoors. In extremely cold climates it may be well to tack muslin over the windows in the roost- level of the roosting platform, drops down a foot away from the roost, there is sufficient space for the fowls to get off their perches in the morning before the curtain is rolled up. But only in extremely cold climates is the drop curtain really necessary, as the roosting pen, as designed, will afford ample protection.

Chicken Coops

"I want some new chicken-coops," said my wife, decidedly. "But there are at least a dozen down at the barn; they were thrown in with the other things when we bought the farm," I objected.

"Those miserable, dark, drafty little dungeons? No, indeed! I want something nice and sunny and sanitary; so run along, like a dear boy, and design it."

So I sat down at my drawing-board and worked out something better than the old-fashioned open wooden cage or closed box, chill and musty. Then I took up my tools and built the coop I had planned. Very satisfactory it is, too; four years of use haven't sug-

Poultry Houses   **Chicken Coops**

gested any improvements. so I venture to show it to Our Folks.

The material is mostly ten-inch rabbeted barn-siding, such as every village lumberman carries in stock; but nine-inch rough boards will do in a pinch. Here is the list:

```
2 pcs. 1 ft. 10 in. long, for ends      (Fig. 1)
1 pc.  2  "   1  "    "    "  one end (Fig. 2)
1 pc.  1  "   4  "    "    "          "  (Fig. 3)
1 pc.  1  "   1  "    "    "  door       (Fig. 4)
1 pc.  2  "   1  "    "    "  top of front, above sash
2 pcs. 2  "   5  "    "    "  roof
3 pcs. 2  "   1  "    "    "  back
```

Also the following rough lumber:

```
1 pc.  1 x 2 ins., 2 ft. 1 in. long for bottom brace
3 pcs. 1 x 3  "    2 "  3  "    "    "  floor battens
2 pcs. 1 x 6  "    1 "  4  "    "    "  end battens
3 pcs. 1 x 9  "    2 "  1  "    "    "  floor
2 pcs. 1 x 2  "    2 "  4  "    "    "  floor battens
```

Then, we need a four-light sash, glazed with 8 x 10 glass (it is about 2 feet 1 inch x 1 foot 8 inches); a

Fig. 1   Fig. 2   Fig. 3   Fig. 4

piece of heavy galvanized netting, 2 feet x 1 foot 10 inches; 2 loose joint hinges, a hook-and-eye, and the necessary nails, staples, screws, paint, etc.

Fig. 1 is sawed from corner to corner to make the wedge-shaped pieces for the ends; but both of these pieces will be left-hand, so a pair of right-hand ones must also be marked out on the other piece, with the line running the opposite way. Fig. 2 is the middle piece of the blind end; Fig. 3 for the

Fig. 5

other end is shorter, to allow for Mrs. Henrietta Rooster's doorway. Put the ends together with battens, like Fig. 5; nail on the back, and notch in the brace across the bottom of the open front. Now, staple on the wire netting and put the bar across

the upper edge of the front. The sash is hung at the top with two loose-joint hinges so that the whole thing can be slipped off and laid aside in fine summer weather.

The guides for the sliding door are rabbeted edges that were ripped off in fitting the upper bars at back and front; the door itself, Fig. 4, is made as I've shown, by nailing a reversed piece of rabbeting to one side. I haven't shown the guides in Fig. 5, but the picture of the finished coop

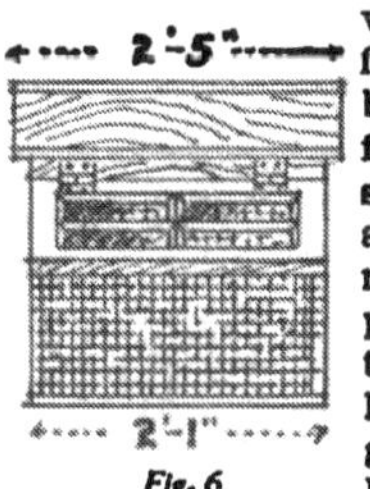

Fig. 6

shows you how they go. The roof boards are put on as you see in Figs. 5 and 6; the ridge-comb is a piece of rabbeting, sawed off from the edge of the roofing. The four holes in the front gable are for ventilators, although the top one takes a peg to hold up the sliding door.

The flooring is made of rough boards,

The finished coop

backed with battens; the coop merely stands on this, and isn't fastened down in any way. This permits thorough and rapid cleaning or whitewashing of the whole interior. Remember, though, to make the flooring a neat fit; projecting edges will catch rainwater and run it inside.

Water Systems Save Labor

The saving of steps and labor are important factors in obtaining profits in poultry farming. The wide-awake poulterer has everything convenient in order that he may lose no time nor do unnecessary work in the care of his fowls.

Clean, fresh water is as necessary for making eggs and maintaining health as is good, sound grain; but the average man who has to carry the water from a well or pump finds it too great a task to perform more than once a day.

During the hot weather, water given in the morning will be stagnant by night, and during cold weather ice will form on standing water before noon. All this can be avoided by adopting a water system whereby water is kept moving.

The first illustration shows the pipe running under ground from the house to the outside. The water is piped from a water-main. When the fowls are outdoors the faucet in the house is turned off, and when indoors, the one outside is turned off. The water is constantly dripping, and when the vessel is filled to a certain point the surplus runs off. Hence, the water is always fresh, and the hens in consequence drink plenty of it. The second illustration shows a method used in brooding houses, where either broilers or stock birds are reared.

Still another method is a long trough running the entire length inside the house, and a hydrant placed at the head of it. In the morning when the attendant first visits the pens he turns on the water until the entire trough is filled. Several times a day he empties it by pulling out a plug at the extreme end of the trough, and again fills the trough with fresh water.

On some farms the water comes from the town water-supply. On others it is furnished by a windmill tank. The long trough plan would be suited to places where the water must be pumped and carried. The trough can be quickly filled, and while it would not be so easy a job as when the water comes from a hydrant, it would be a big saving over

the old style of filling individual drinking vessels, and it would be cutting down the labor by fully one-half.

Poultry Fixtures of Value

Any one handy with tools can construct nearly all the fixtures needed in a poultry house, and much of the waste lumber lying about the place can thus be utilized. On rainy or stormy days when there is very little that can be done outdoors, many of these contrivances can be made, which will be putting spare time to good use.

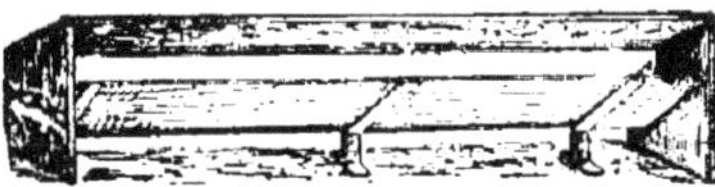

Covered troughs protect feed from rain and sun

Feed-troughs are in the greatest demand. Generally these are built V-shape, but some poultrymen prefer them flat—that is, using an eight-inch board for the bottom, and adding two-inch sides and ends to it. There is no "best" style, however. But unless troughs are protected, the stock will soon befoul them.

In order to overcome this, The Farm Journal artist has designed a trough that will protect the food that is in it, as the fowls cannot get into it with their feet, and the cover shields the food from rain and sun. Cleanliness is probably more important in the matter of food than anything else. This style trough, to a great extent, prevents crowding.

There are no special instructions for constructing it. The illustration is so plain that any one handy with tools can build this trough. It might be added, however, that good, heavy lumber should be used, especially if the troughs are to remain outdoors, exposed to all sorts of weather.

The other illustration shows roosts improperly laid. Experience has proved that the most satisfactory roosts are 2 x 3 scantling, planed smooth and the edges rounded. But in placing them the two-inch side should be upended for the fowls to perch upon. When the three-inch side is placed upward, as shown in sketch, the roosts will in time sag, due to the combined weight of the fowls.

At one time it was thought that the model roost was a round perch, so the fowl could take a good hold and not tumble down when asleep. This idea, no doubt, originated from the fact that fowls love to perch on the limbs of trees. But it has since been proved that a more comfortable way is to permit the feet to spread out. There is no danger of the bird falling off when asleep for the reason that on going to roost she balances herself before sitting down, and when once rightly fixed she tucks her head under her wing and goes to sleep.

To get the strength of roosts, they should be placed narrow side up. Sketch shows the wrong way

Every roost should have a solid platform built beneath so that the droppings may be caught. If sifted coal ashes or road dirt is lightly sprinkled over the platform, it will prevent the manure from sticking, and it can be readily scraped off.

A strongly built table, as indicated in the sketch, answers the purpose of a platform. The roosts can be placed on this, and held in position by driving nails alongside the crosspieces. This will prevent the roosts from shifting. If the pen is of good size, this table can be placed in the center of the room.

Roofs, Windows and Walls for Poultry Houses

The types of roofs for poultry houses are plainly shown in Fig. 1. The shed roof in the upper left-hand corner is one that is most commonly used, the highest point at the front of the building. This style carries all of the water to the rear. If the shed faces the south, it leaves the south front dry and permits more sun-

the half-monitor style at the right, the front or low part is sometimes used for a scratching shed. This style, however, is not practicable for long houses.

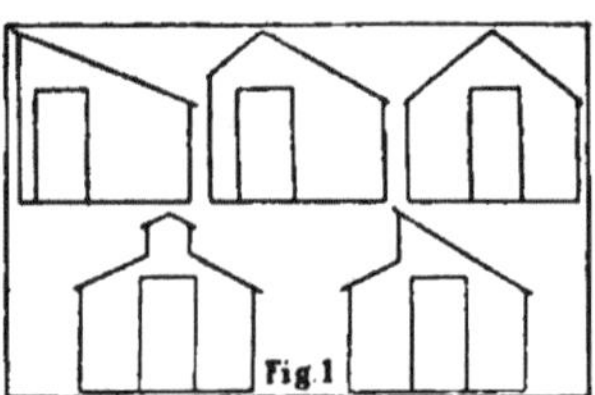

shine to enter the house directly. Next is the two-thirds span roof; it is a modification of the shed roof style, and is generally used on houses that are more than fourteen feet in depth. Then comes the equal-span roof, also known as a double-pitch room, which is largely used for colony houses. The monitor style at the left, below, has never been in great favor with poultrymen, as it is too complicated in construction. It is especially objectionable in cold climates, as it permits too much heat to escape. In

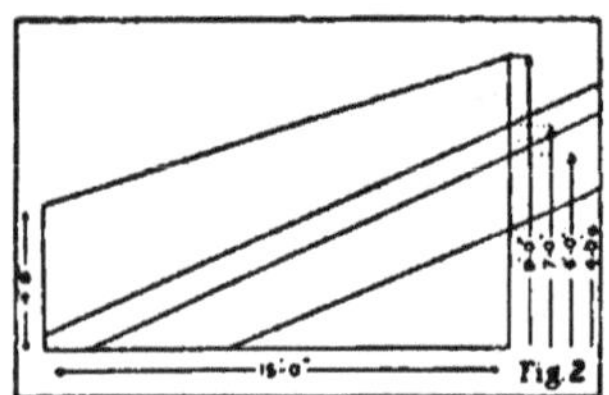

Fig. 2 shows the average direction of the sun's rays. Sunlight reaches farther back into the house when the windows are placed high. The top slanting line indicates the roof of a building nine feet high on the front side. Height is denoted by vertical lines at the right. The next slanting line indicates the direction of the sun's rays when the window in the front of the house is placed seven feet from the ground; the next slanting line below that indicates the direction when the window is six feet high; and the lowest slanting line when the window is only four feet from the ground. A building with a nine foot front should slant to four feet and six inches in the rear.

Windows can be hinged on side, top or bottom, and made to swing inward or outward as desired

Windows may be swung in different ways as shown in the above illustration. Beginning with the top row, left to right, the first window swings from the side on heavy hinges; the second has upper and lower sash; the third slides open; the fourth has hinges on top and opens outward at the bottom. In the bottom row, left to right, the first window also has hinges at top, and opens inward; the second has hinges below and opens at the top; the third works on pivots in the center; the fourth has pivots above and below, and swings around; and the fifth has lower sash firm, the upper sash swinging outward on hinges fastened at the top.

Concrete blocks are coming into favor in some sections of the country, and while they make a durable house, warm in winter, cool in summer, and are practically vermin-proof, the walls should be either plastered or covered with wall-board inside to prevent dampness, as concrete blocks will show more or less dampness during rain storms.

A Trap Nest

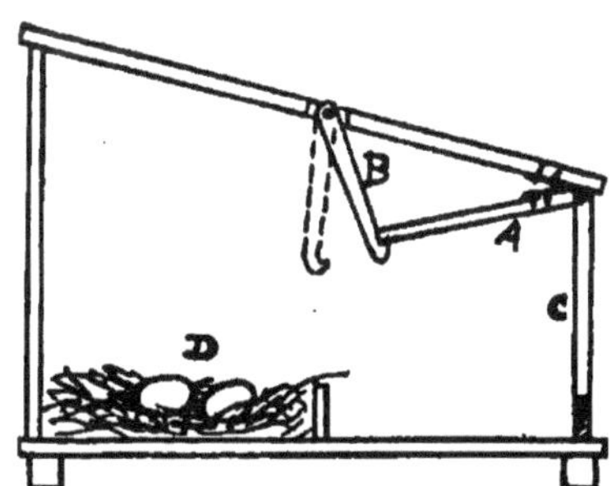

A trap nest is used for discovering the fowls that are laying and those that are not. In that way you are able to keep the good layers, and hence the good payers, and send the loafers to the market.

Get a dry-goods box in town, tear it apart, and make a box about two feet long and a foot and a half wide. The back should be higher than the front. In the front, put a door hinged from the roof. Cut a slot in the roof and hang a wooden hook through it, so that when the door is raised it can be hooked loosely on that wooden hook. If the whole door it set low enough so that when the chicken enters it she will brush up against the door and so knock it loose from the hook, the door will drop down and imprison her. "A" is the door, "B" the hook, "C" the doorway, and "D" the nest.

A Safety Device

B-R-R-R-R-R!! The electric bell interrupted my delightful dream. I sprang out of bed, raised the south window, and snatched up my shotgun (loaded with powder only).

"It's some one at the chicken house," whispered my wife. "Quick! before they break in and get some of our nice broilers!"

Bang!! Bang!!! Then, hastily pulling on some clothes, I ran out into the moonlight to investigate. The big padlock lay wrenched to bits beside the open door; the burglar had vanished, leaving a string of very hasty footprints in the mud. Something else, too, he had left—a big phosphate bag, holding eleven of neighbor Miller's barred Plymouth Rocks!

Now, you may not be so lucky as neighbor Miller, when the chicken thief, or any other thief, for that matter, comes around; so a bit of practical burglar insurance in the shape of a burglar alarm, may be a very good thing. Fig. 1 shows the general arrangement—an electric bell in your bedroom, a wire from this to a little switch near at hand (so you can cut off the current in the daytime), a wire over to the circuit-breaker on the chicken house door; then another wire back to the house, thence to the battery, and so on to the bell again.

Two dry batteries will do the work; the discarded cells from a gasoline-engine have enough current to run an electric bell for a month or two. The bell is the ordinary electric door-bell; the switch can be made as shown in Fig. 2, using a bit of spring brass or copper, on a wooden block. Iron telephone wire or barbed wire will do for the out-of-doors line, if it is wrapped around insulators made from bottle necks and mounted on trees, poles or fence-posts. Only one wire need be thus insulated; the chicken-yard fence will answer for the other. Inside the buildings, insulated electric-bell wire had best be used, for one side of the circuit, at any rate.

At the door you can put in a regular ready-made burglar-alarm connection; but I made something just as good myself (Fig. 3, open; and Fig. 4, closed). I first took off the hinge and reamed

Poultry Houses   Electric Alarm

out one of the screw holes a little larger than a lead-pencil; then, in line with this, I bored a three-quarter inch hole in the door-frame, about one inch deep; changing to a three-eighth-inch bit, I bored an inch or so deeper. Then I made my fixture as you see, with a hardwood spindle, a coiled spring, two washers, and a brad driven through the spindle. One wire is fastened to the coiled spring; the other is underneath the hinge-plate. Now, when the door is opened, the spring forces the spindle out until the metal washer touches the metal hinge; this completes the circuit, and the bell rings. Brass or copper washers and hinges are best;

rusty iron will not make a good connection.

Other doors can easily be attached by running two wires and wrapping the ends of these firmly around the outgoing and incoming main wires, respectively.

A double-hung window usually has a special fixture, rather too difficult for you to make, although you can get around this by setting a door fixture in the sill, and letting the sash shut down on it. A small movable chicken coop can be protected in the same way, using the loose floor-platform as the window-sill.

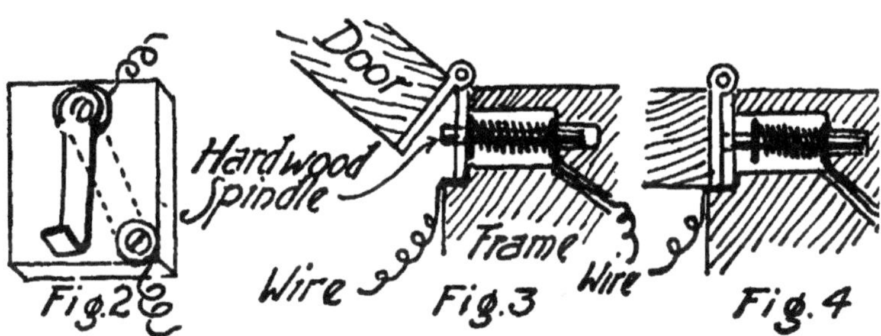

Fixing Loose Machine Bearings

If you examine any engine you will notice two half-circles of stuff that looks like lead surrounding the crank-shaft bearings. This is babbitt metal —a mixture of copper, zinc and tin. It's softer than iron or steel, so it takes all the wear and can easily be renewed. Fig. 2 shows one of these half-circles taken out; you'll notice the crossed oil-grooves (about a sixteenth of an inch deep) that are cut with a gauge on the inner face. These grooves run to the oil-hole to carry the grease into the bearing; and you'll also notice a little projecting bit of babbitt metal (about an eighth of an inch long, and the thickness of a lead-pencil) on the back. This fits into a hole in the iron casting, Fig. 1, and keeps the babbitt from turning around.

Fitting New Bushings

New babbitts for any particular bearings may be ordered by parcel post from the factory; but before you put them in, gouge the oil-grooves and also file one edge of each a little, like A, B, Fig 1. This is to let the oil slip underneath; and of course I'm supposing the shaft is turning over in the direction the arrow, C, points. If it were going the other way, you would file the other edges; that's plain, isn't it? Then put in thin cardboard or tin liners ("shims") between the castings, and tighten the bolts; experiment until the bearing is tight, but not too stiff to run freely.

Pouring Your Own Bushings

But suppose you have a worn mower bearing, Fig. 3, that has never been babbitted? Exactly; then you must bore it out with a machinist's drill until there is a quarter-inch clearance all around. File a groove, A, in the lower part, wrap the shaft with thin oiled paper, and clean everything well.

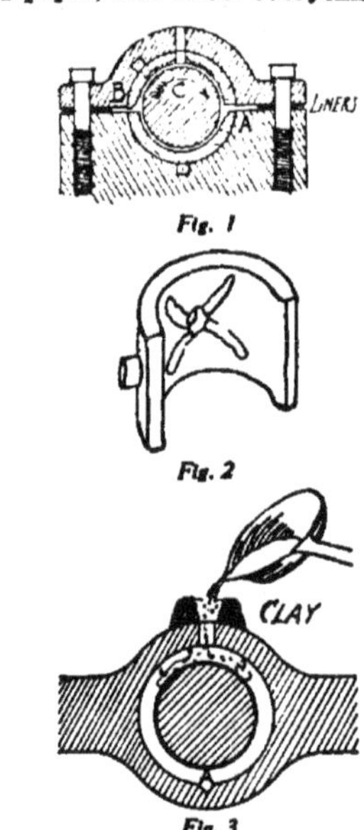

Then brace the shaft securely so it will be in the center of your bearing, and dam up either end with clay; also build a little funnel of clay around the oil hole. This clay should be damp; if too wet, steam will generate and blow it away.

Now, with a blow-torch (or any hot fire), melt some babbitt in a clean iron

Engines and Machinery   **Hydraulic Ram**

ladle, such as plumbers use; put plenty in, for you must pour the whole bearing at one clip. You can buy new bars of babbitt, or melt up old, broken pieces; but in either case scrape off the scum with a bit of wood or iron until nothing is left in the ladle but the clean, bright metal. When this is hot enough to scorch a pine stick, pour it into the bearing. After it cools, slide out the shaft, rebore the oil-hole and gouge your oil-grooves the same as in the half-bearing, slip the shaft back and the work is done.

Naturally the half-bearings in Fig 1 can be made in much the same way. Take off the top casting and pour the lower half, then remove the shaft and file the edges of the babbitt level. Then lay a bit of oiled paper over those edges, put back the shaft and casting, and pour the upper half similarly.

A Water Supply With a Hydraulic Ram

Many farms have streams of water running through them, and it is a very easy matter to lift this water by means of a hydraulic ram to such a height that there will always be a supply on tap in the home and barn.

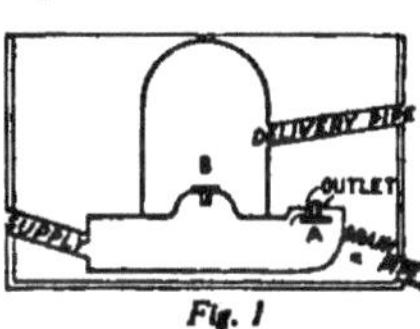

Fig. 1

To install a ram it is only necessary to construct a dam that will provide a fall of water. A ram is driven by the force of water behind it, and once

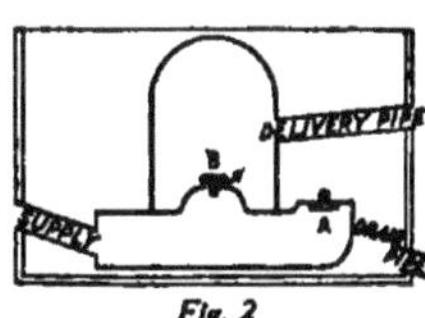

Fig. 2

it is installed, will require no attention other than an occasional oiling. The height to which a ram will lift water depends on the fall the water has. Where there is a fall of five feet the ram will lift one gallon out of seven to a height of twenty-five feet. Or, if the fall is ten feet, the ram will lift one gallon out of every fourteen to a height of 100 feet. The ram will lift water higher than this, too, but the amount of water lifted will be smaller as the height and distance from the fall increase. A spring or fall that furnishes seven or eight gallons of water a minute will run a ram. A ram receiving seven gallons of water a minute from a four-foot fall will lift nearly 500 gallons to a height of fifty feet or more, in a day. The ram works continuously, and even with a small supply of water it is really astonishing how much water it will lift.

A great fall of water is not necessary provided plenty of water is at hand, but a fall of four or five feet is best.

If you have a fall of four feet, you should place your ram about twenty-five feet away, in a well-drained pit, so the waste water will not flood the ram and stop it. All the pipes for conveying water to and from the ram should be below the frost-line.

The cost of putting in a ram, in most cases, would be well within $100, sometimes much less. The upkeep is a mere trifle. It will run for years before new valves are required.

The improved rams are supplied with devices for adjusting for different falls. All the valves are made of molded rubber with brass coil spring attachment providing effective and efficient closing of valves. An adjuster attachment controls the quantity of water supplied to the ram in cases where the supply is irregular.

The illustration in Fig. 1 shows the valve A open and water escaping

through the outlet. The rush of water through this valve soon becomes so fast that it closes the valve, and the shock raises the valve B, as shown in Fig. 2, forcing the water through and into the delivery pipe, which carries it to the supply tank on top of

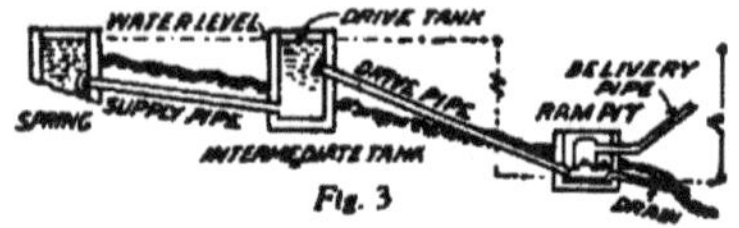

Fig. 3

the hill. The water which flows through valve A is wasted and flows on down the stream. Fig. 3 shows an intermediate tank which shortens the drive pipe and increases the fall.

Engine Troubles and How to Avoid Them

When for any reason a charge in an engine cylinder is fired before the proper time, we say it "backfires," or causes the crankshaft to turn the wrong way. This backfiring often occurs when an engine is being cranked, and is rather dangerous, as it may result in a broken arm. Backfiring may be caused by one of the following things:

1. Spark advanced too far.
2. Glowing carbon deposits in cylinder.
,3. Spark plugs rusty or dirty, causing points to become red hot.
4. Short circuit in timer.

On all high-speed motors there is a spark lever or control. This is placed there because it is necessary to advance the spark as the engine gains speed in order to secure efficient operation. But when the engine is to be started the spark lever should be in full retard, and "kicks" occur most often because the operator has carelessly left the spark advanced.

Backfiring

Glowing carbon deposits are another cause of backfiring, but this does not occur until an engine has been running for some time and become heated. Red-hot spark plug points cause the same trouble as the glowing carbon, for after an engine has been run a while the points may retain heat enough to fire the charge of gas as soon as it is taken into the cylinder. The spark plugs should be removed and cleaned when erratic firing occurs, and if this gives no relief you may be sure there is carbon to be removed from the cylinders.

The last cause of backfiring is a short circuit in the timer—that is, the commutator or distributor, or the mechanism that divides the current among the different cylinders at the proper time. This commutator consists of a hollow metal drum in the rim of which are imbedded as many contact points as there are cylinders. These points are insulated from each other, and a cam turning inside the drum makes contact with them at the proper time. If these contacts become uninsulated, cylinders will fire with no regularity. The only thing to do in this case is to buy a new part.

Of course, backfiring may be caused by having gears that operate the valves and timer set wrong, but I have assumed that you have not torn down your engine to misplace them.

Grinding Valves

One of the greatest troubles in internal combustion engines is leakage around the exhaust valve. The hot, burnt gasses pass out through this valve each time an explosion occurs, and it naturally becomes rough and pitted sooner than the intake valve.

The first step in valve grinding is to remove the valve spring, assuming that the cylinder head has been removed. This is done by the use of a valve lifter, or by the home-made device shown in the illustration, Fig. 1.

After the spring and foot are removed, lift the valve from its seat and clean the valve and seat thoroughly.

Fig. I

Plug the passage to the cylinder with rags or cotton waste. Now spread a thin layer of emery and cylinder oil on the surface of the valve and place it in its seat. With a screw driver inserted in the slot on the top of the valve, press the valve firmly down, and while exerting the pressure, turn the valve back and forth in the seat. Lift the valve occasionally, turn it to a new position and continue rotating. After a few minutes the gritty feel will have disappeared.

The valve should now be removed and the emery and oil wiped from both valve and seat. If a clean bright surface shows all around, the valve is sufficiently ground; but, if not, the operation should be repeated. Care should be taken not to press on the valve too hard, as this will do more harm than good.

Fig. 2

After the grinding is completed, be sure to wipe all emery from the valve, seat and other parts of the engine, for if the emery gets to working parts it will do much damage.

Some valves do not have a slot for the use of a screw driver in grinding, but have two small holes instead. A tool may be made for use in grinding these valves, as is shown in the illustration, Fig. 2, or may be purchased.

Piston Rings

Sometimes a set of new piston rings is all that is needed to change a weak, wasteful engine into a strong, economical power-producer.

When you find that your engine is dropping behind its rated horse-power —failing to do the work it should perform—see if it isn't due to worn piston rings in the cylinder. By replacing them you'll get the power back.

The Rings Hold the Power

Power in a gas-engine is created by first compressing and then exploding gas in the cylinders. Piston rings are depended upon to get this compression, and they can only do it properly when they fit tightly. Worn piston rings lose fit and let gas escape. Get a new set if you ever find that your engine is losing power.

The best cure for carbon trouble in a gas-engine is to use perfectly oil-tight piston rings. Carbon is caused by oil being burned in the cylinders. Keep oil out and you will stop carbon forming.

It pays to take proper care of your engine. Make a point to overhaul it thoroughly at least once a year, and see that the piston rings come in for some attention when you do it. They have a lot to do with power production—and power is what you use an engine for.

Don't say that your engine is worn out and plan to buy a new one, until you first make sure that it isn't only the piston rings that are worn. They can be replaced; and it's cheaper to buy new piston rings than a new engine.

If you don't get good compression in your engine, automobile, stationary or portable, it's the fault of the piston rings nine times out of ten. They're worn, have lost fit, and the gas naturally blows down past them. Try a new set of rings. Buy them anywhere. Put them on yourself. But be sure you get good ones, so that the trouble won't be repeated.

Bad Rings Waste Oil

You're not supposed to burn lubricating oil in a gas-engine. It's wasteful; and it's also bad for the engine, for it fouls the cylinders with carbon and causes backfiring. The best way to cure it is to put in a new set of piston rings.

Keep a record of what it costs to run your gas-engine; your auto-engine, too. When costs begin to climb, find out what is causing the waste. It shows that some part is beginning to fail—the piston rings, for instance. Worn rings can waste a tremendous amount of fuel and oil. Put in new ones.

If a horse isn't harnessed properly he only wastes energy instead of pulling his load—and he may break the wagon. It's the same in the case of the gas-engine, where the piston rings may be said to harness the power to the shaft. If they don't fit as they should, they simply waste power; and they may do a lot of damage, too. Don't forget them if you ever have power troubles.

Gas for the Farmhouse

Thousands of farm families are making and using gas today; very easy and inexpensive they find it, too. There are three good, practical systems. I'll briefly describe each, and you can take your choice.

The simplest of all uses compressed gas; exactly the same principle as the gaslights on an automobile. An iron tank about the size of a small steam-boiler is laid in the ground, just outside the house, and connected to the interior piping that serves the burners. This tank is leased (not sold) by the company, who send you steel tubes of compressed gas as you need it, to be emptied into the tank. Aside from the cost of piping, gas fixtures and gas stoves, the first expense is merely nominal; and the thing can not get out of order. The most serious difficulty is in getting shipment of the big tubes of gas in these days of freight embargoes and blockades.

Acetylene gas takes more equipment. A small frost-proof outbuilding, frost-proofed with sawdust, must be built, and the generator set up inside; Fig. 1. Then the machine is charged with calcium carbide and water, in separate compartments; and the water trickles through on to the carbide, generating a strong, rich gas—acetylene—that burns with a beautiful white light and is excellent to heat or cook with. As soon as enough gas is generated, the pressure forces back the water from the carbide. Anyone who has used the acetylene lights that automobiles carried several years ago will understand all about this. Calcium carbide can be readily bought and a supply can be safely carried on hand, provided it is

kept absolutely dry. The plant is simple, takes very little attention, and gives out a most satisfactory gas.

The third system is a much older one; it makes what is known as "gasoline gas." In the cellar of the house an airpump is set up, Fig. 2, worked by a clockwork device, to force air out to the carbureter in the underground gasoline tank. It is very much on the order of the carbureter in an automobile; the air passes over or through the gasoline, vaporizes it, and returns to the house heavily charged with the gasoline vapor. Pipes carry this to the various gaslamps and stoves. Of course, gasoline gas when heavily compressed in the cylinders of an engine is very explosive; but the pressure in a gas plant never exceeds two ounces at the most. The light from this system is not quite so brilliant as from acetylene, but nevertheless lots of folks use gasoline gas and like it.

Saving Gasoline

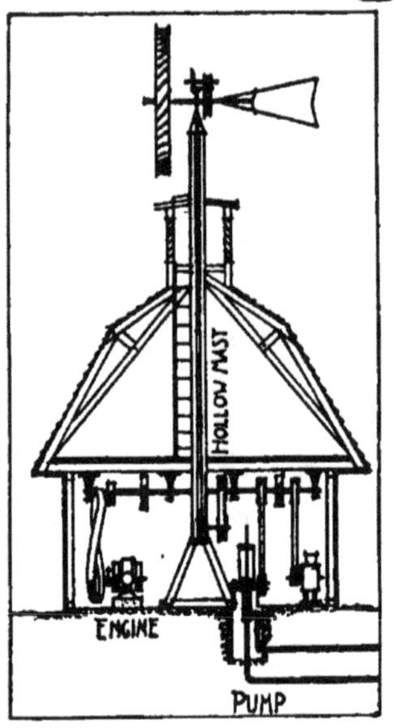

Many farmers who use a stationary gasoline engine will find a power windmill a mighty economical thing, and in calm weather there is the old faithful engine to fall back on. A sixteen-foot mill in a stiff breeze will develop about two and one-half horsepower; and a twelve-foot mill, one and one-half. The best place to mount a power windmill is on a cupola of the barn, as I've shown. A hollow mast, built up from ordinary lengths of 6x8 timbers, blocked six inches apart, is run up, as you see in the sketch; the base rests on a braced triangle of wood. Then a power windmill is set so that its center is at least fifteen feet above the ridge of the main roof, and fifteen feet above nearby tree tops.

Instead of being back-geared this mill is "geared up" to increase the speed; and by a pair of bevel-wheels, on the principle of the pinion and driving gear in the rear axle of an automobile, the power is transmitted to a long steel shaft that runs down the center of the hollow mast. At the foot another set of bevels puts the power into a pulley wheel; and it's an easy matter to belt this pulley to your line shaft. When the wind blows, put the windmill belt on and slip the engine belt off; reverse this, of course, in calm weather. The driving pulley on most power mills, in ordinary breezes, runs at about the same number of revolutions per minute as the average gasoline engine; but the engine has a smaller pulley, so you must have different-sized wheels on the line shaft.

It is sometimes possible to change over an old pumping windmill by taking out all the crank mechanism, replacing this with high-speed gears and bevel pinions, and putting in hollow wooden mast with its steel shaft; but this depends to a large extent on the make of the mill. Generally, it is more economical to sell or trade in the old affair and buy a complete new outfit. If you already have a good strong tower, mount the new mill on that and inclose the bottom for a machinery room. Keep on using the old pump— only you will work it now by means of a pump jack belted to the line shaft.

Wind power will give you satisfactory service for many kinds of work —pumping, sawing, grinding, running churns, separators, corn shellers, small shredders and so on. Threshing, however, is too heavy; and generating electric light is not always satisfactory.

The New Farm Helper— Electricity

A man who was called upon recently to speed up the production in a big factory, tells that in one instance, by the introduction of a little inexpensive machinery and the rearrangement of the working force, three men were able to do the same work that eight men had been doing before and do it better.

The same problem in a measure confronts the American farmer today—to be able to do the same work or more work with a smaller working force.

On a great many farms a force—new as a factor in farming but wonderfully practical and efficient—is being introduced to help solve the labor problem and to help speed up production. This force is electricity, and the introduction of small individual electric power and light plants makes it possible to use electric current anywhere.

This is an important matter for the farmer. He knows better than any one else about the difficulties that go with an attempt to secure man power for farm labor, and to keep it, once it is hired.

Electric power on the farm helps out in more than one way. It does many tasks that otherwise must be done by hand, freeing the hands for more profitable and often more congenial labor, and it furnishes improved conditions of living that appeal to the owner's family, to the boys and girls and to the hired help, and make it easier for them to be satisfied with conditions on the farm.

Drudgery Taken Out of Chores

It is influencing the farmer himself in a vital way. Life on the farm with modern conveniences in the house, and with the drudgery taken out of the chores, does not become distasteful as he grows older. He is content to stay out of the "retired farmer" class, and to live out his life in the farm home where he can keep his mind and body healthfully occupied and where the benefit of his advice and experience can still be had by the younger workers.

The benefits that electricity bring to the farm home are tremendous. Electric light about the barn is a time-saver. The chores can be done after dark with ease, and dispatch when securing the light needed is just a matter of snapping switches off at convenient points instead of carrying a lantern.

With electric light in the farm buildings it is easy to use all the daylight hours in the field and do the chores after dark.

Save the Housewife's Strength

Then electric power can be used in various ways to lessen labor of many jobs, as in the matter of running a churn or cream separator, a grinder, feed chopper, grindstone, fanning-mill and the like. It is proving quite a boon in the operation of pressure pumps for pumping water for the stock, for sprinkling the yard or gar-

Electric Power | Engines and Machinery

den, and for household use, doing away with a lot of hand pumping. It is running milking machines, cutting the time of milking in half and doing away with the mighty unpleasant task of milking by hand.

It is running washing machines and vacuum cleaners, saving time for the housewives and lightening their work; likewise electric fans—destroying the ill-effect of sultry weather and keeping everybody fit for more and better work.

An electric power and light plant will run one of the small motors which will operate ordinary light machinery, for two cents an hour if the fuel is gasoline; if kerosene is used the hourly cost will be about half that. In many farm homes where electricity is used, a time saving of from twenty to twenty-five hours a week is frequently reported. This is a considerable item, especially in seasons when labor is hard to get and when time is precious.

"Merely By Turning a Switch"

It is easy to operate your own electric power plant. There are plants on the market for which the only attention required under ordinary conditions is to start the engine once or twice a week, see that fuel is provided for the engine, and add a little distilled water to the storage battery occasionally.

There is a strong appeal in the possibility of electricity to the farmer and to those who labor with him. They have been accustomed to accomplish things by the outlay of actual strength, by toil that tries the muscles and oftentimes oppresses the spirit. To be able to accomplish the tasks merely by turning a switch and then watching the mysterious electrical force go ahead and do that task silently, tirelessly, efficiently and well, has a fascination for the farmer just the same as it has for the man in any other line of business who realizes that there is a benefit to him in letting a machine do all the work that does not have to be done by hand.

So we are going to witness—are witnessing—the introduction of this modern force on the farm at a rate that foretells great and helpful changes in labor conditions there, with a corresponding benefit to the producer.

Winter Care of Tractors and Automobiles

Cold weather brings several tractor operating problems—danger of water freezing in the cooling system, poor lubrication because of sluggish action of cold oils, difficulty in starting the motor, and the matter of proper storage.

Every year thousands of dollars' worth of damage is done by allowing water to freeze in the cooling systems of tractors. Iron throws off heat very rapidly, and consequently it takes an engine cylinder only a little while to reach the temperature of the surrounding air. The sheet of water around the cylinder is very thin and it quickly becomes the temperature of the jacket. Therefore, water will freeze in a tractor cooling system when there is no ice film on the surface of a pailful of water nearby.

The Only Way 100% Safe

Pure water is the best cooling medium. The only safe way to prevent freezing is to drain the system each time after a tractor is used in winter. Start early enough in the fall to be safe, too. Drain the cooling system while the motor is hot and every drop of water will be dried out.

Grease and Oil Thicken

There is also danger of a tractor not getting enough oil and grease in winter. Every kind of lubricant used becomes sluggish when subjected to cold. The custom which old steam engineers used of warming lubricants on frosty mornings, may be followed by the tractioneer to good advantage. A little warm oil added to that already in the lubricator will make the whole supply more easily handled. It goes without saying that every tractor owner should find out what kind of oil is recommended by the manufacturer of his machine for use in winter. Be careful to see that all grease-cups are not only filled with a good, light grease, but that they also are screwed down several times more before starting the tractor on a cold morning.

Starting a Cold Motor

Getting a tractor motor started in cold weather is often difficult. One of the most satisfactory methods is to fill the entire cooling system with hot water. At least, pour a bucketful of hot water into the system to warm it up, start the motor, and later finish filling the system when the engine is hot.

Where only a small amount of hot water is available, it probably may be used to best advantage by warming the carburetor. To do this wrap a cloth about the carburetor and engine intake manifold and pour the water on the cloth. This cloth will hold the hot water in contact with the cold metal for a greater length of time than if it were merely poured on the carburetor. Another help is to get some very high test gasoline for use in priming, or make a mixture containing one-half ether and one-half gasoline.

Before cold, bad weather sets in clean up the tractor and give it a coat of paint. If a tractor is to be idle any length of time, all bearings should be liberally oiled, and the lubricator filled to its limit. This protects the bearings.

It Isn't the Spark Plug

"Car misses? Doesn't pick up, eh?" asked the trouble man, thoughtfully, laying a screw-driver across the spark plug head.

"Oh, it isn't the spark plugs!" laughed the owner, scornfully. "I scraped the points bright and opened them a shade so as to get a good fat spark; and it didn't help things a bit. Besides, you can see for yourself what strong sparks jump off the end of that screw-driver when you hold it almost touching the cylinder head. No; it must be the carburetor; and you fellows have got to give me a new one."

"Certainly, certainly!" agreed the trouble man, with a cheerful smile. "But suppose we try the car on the road."

"All right; but wait till I get my overcoat."

Half an hour later the owner jerked out his gears, snapped off the switch and faced the trouble man. "Say, now; what *did* you do to this car while I was getting my overcoat? I wasn't three minutes, but it runs better than it ever did."

"Why, I cleaned the spark plugs; that's all."

Now this is a true story, or, rather, a lot of true stories; for, as every repair man knows, dirty or defective spark plugs are the cause of at least nine-tenths of all gasoline engine trouble.

How the Plug is Built

Look at the sketch a moment; it's a section through a jump-spark plug. The dotted parts are made of non-conducting material through which electricity can not pass—porcelain, stone or mica; the black or shaded parts are metal and let the electric spark pass freely. This high tension spark (at a pressure of several thousand volts in the secondary circuit) comes along the wire, passes down the little metal rod in the center of the plug, and jumps across the metal point that is sticking out to meet it. Thence, by way of the cylinder and other heavy metal parts, it travels easily and comfortably back to the

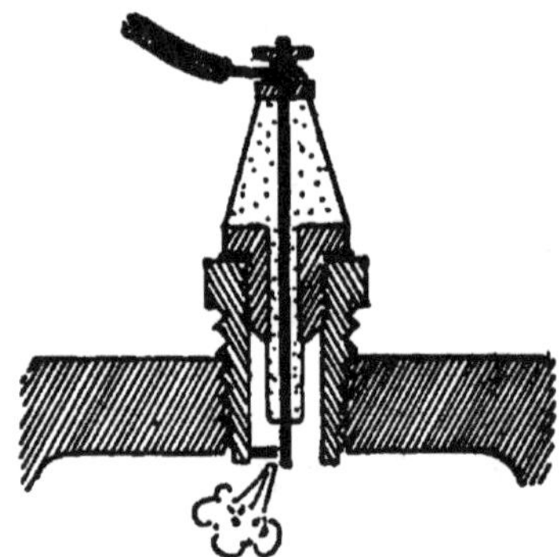

Cross-section through a jump-spark plug

magneto. But this spark current is at tremendously high tension, and is always trying to "break out" at the weakest point, like steam in a boiler. The weakest point is the air space between the ignition points; but if these are put too far apart, the current will try to hunt up some other, easier way. One-thirty-second of an inch, a little less than the thickness of a new dime, is about the right separation.

Carbon and water are good conductors of electricity; oil is a non-conductor. Therefore, the whole interior of the plug may be dripping with oil without any bad effect, but if there is a coat of carbon under the oil, the electricity finds it much easier to leak quietly away on this carbon coat, across to the metal, rather than to blaze violently over between the points. On a cold, damp morning a thin film of moisture often gathers on the non-conducting ma-terial, and short-circuits the electricity to the metal.

Four Things to Do

There are four things to do in case of trouble: 1. Take out the spark plugs, one at a time; connect the wire to the top, hold the side of the iron bottom of the plug against the cylinder, and crank the engine. If no spark, or only a weak or occasional spark, jumps the points, unscrew the plug in the middle; wipe and scrape clean the insulating parts.

2. See that the wire connections at the top are tight and clean.

3. If you still do not get a good spark, the insulating material is probably cracked or broken; you must put on a new plug.

4. If you still do not get results, the trouble is somewhere else—in the wiring or the magneto; but that won't happen once in a hundred times.

The Dangerous Backfire

When for any reason a charge in an engine cylinder is fired before the proper time, we say it "backfires," or causes the crank-shaft to turn the wrong way. This backfiring often occurs when the engine is being cranked, and is rather dangerous, as it may result in a broken arm. Backfiring may be caused by one of the following things:

1. Spark advanced too far.

2. Glowing carbon deposits in cylinder.

3. Spark plugs rusty or dirty, causing points to become red hot.

4. Short circuit in timer.

On all high-speed motors there is a spark lever or control. This is placed there because it is necessary to advance the spark as the engine gains speed in order to secure efficient operation. But when the engine is to be started the spark lever should be in full retard, and "kicks" occur most often because the operator has carelessly left the spark advanced.

Carbon May Do It

Glowing carbon deposits are another cause of backfiring, but this does not occur until an engine has been running for some time and become heated. Redhot spark plug points cause the same trouble as the glowing carbon, for after an engine has been run a while the points may retain heat enough to fire the charge of gas as soon as it is taken into the cylinder. The spark plugs should be removed and cleaned when erratic firing occurs, and if this gives no relief you may be sure there is carbon to be removed from the cylinders.

Or a "Short"

The last cause of backfiring is a short circuit in the timer—that is, the commutator or distributor, or the mechanism that divides the current among the different cylinders at the proper time. This commutator consists of a hollow metal drum in the rim of which are imbedded as many contact points as there are cylinders. These points are insulated from each other, and a cam turning inside the drum makes contact with them at the proper time. If these contacts become uninsulated, cylinders will fire with no regularity. The only thing to do in this case is to buy a new part.

Of course, backfiring may be caused by having gears that operate the valves and timer set wrong, but I have assumed that you have not torn down your engine to misplace them.

Carry an Extinguisher

A few weeks ago Jim Caldwell met with an accident that was both lucky and unlucky. Jim's luck was the indisputable fact that he happened to be in town at the time instead of and suddenly it went off like a cannon and flames shot up clear through the hood. Some gasoline had collected in the underpan.

A Handy Extinguisher Saved this Auto

be in town at the time instead of four or five miles from nowhere.

Jim had been tinkering with the carbureter of his auto, but its disposition kept getting worse and worse. It spit and it popped and it missed, Fires don't appeal to Jim and he didn't fancy sitting over the gasoline tank, so he jumped out and made tracks down the street. Somebody turned in a fire alarm. Then Jim saw a garage man in greasy overalls with a fire extinguisher under his arm running toward the blazing auto. Jim went back to help.

Well, the hose and ladder wagons came, but the man had the fire out long before that. The paint on the hood was scorched and the chemical had made a lot of sediment on the motor, but Jim's auto had been saved from going up in smoke. Jim peeled a greenback from his roll and handed it to the hero of the occasion.

The auto was badly damaged. Before Jim drove home that night he went to the hardware store and bought a fire extinguisher that's been hanging on the dash ever since.

Motor Troubles

What Causes an Overheated Motor

1. Low supply of water (add more).
2. Too rich a mixture (use less gas).
3. Carbonized cylinders (clean 'em).
4. Lack of lubricating oil.
5. Late ignition.
6. Broken water or oil pump (fix it).
7. Radiator stopped with mud or other matter.
8. Loose broken fan belt.
9. License tag obstructing front of radiator.

Why a Motor Knocks

1. Spark too far advanced (retard it more).
2. Too rich a mixture (change it).
3. Motor speed too slow on hills or bad roads for direct drive (shift to lower gear).
4. Loose connecting rod bearings (light knock at high speed).
5. Crank-shaft bearings loose (heavy pounding at slow motor speeds or hard pulls).
6. Worn valve tappets (light tapping sound).
7. Improperly adjusted tappets (light tapping at all speeds; adjust them).
8. Carbon in cylinders (peculiar ringing knock at slow speed). Use a better grade of oil and wash out crank-case more frequently.
9. Engine loose in car frame (tighten it).

Motor Won't Start Because—

1. Switch not on.
2. Out of gasoline.
3. Poor grade of gasoline or mixed with water.
4. Poor ignition.
5. Ignition contact points out of adjustment.
6. Ignition unit short-circuited.
7. Water on coils or terminals.
8. Overrich mixture by continued use of choke.
9. Motor cold (warm carbureter by pouring on hot water).

If the engine is not too cold, and has good clean gasoline in the carbureter and a good spark at the plugs, your motor will start if properly handled.

If a Motor Misses, Look For—

1. Short-circuited spark plug.
2. Partly short-circuited or broken terminals.
3. Poor contact between the various ends of wiring.
4. Loss of compression in one or more cylinders: a, valves may be stuck; b, valves may need grinding; c, valve springs may be weak or broken.
5. Water in gasoline, making motor run spasmodically (this is rare and difficult to distinguish from other causes. Look for it last).
6. Air leak between carbureter and intake manifold, or where the manifold is bolted to the cylinder.

When the motor misses you may locate the missing cylinder by opening the priming cocks on top of the cylinders one at a time. When the missing cylinder is located, replace the spark plug with a new one in good working order, and if it still misses, examine the wiring. If the trouble still continues, turn over the motor slowly by hand in an endeavor to detect a defect in the compression of the various cylinders.

When a spark plug has to be taken apart for cleaning, care should be taken to keep the porcelain from being cracked, and when reassembling it all the joints must be made gas tight. When the porcelain of a plug is cracked, throw it away, for the carbon will soon get in and form a short circuit. In buying new plugs, be sure to get the type for which your engine is fitted. Always keep an extra set of spark plugs with you so changes can be made on the road.

An Emergency Auto Repair Light

In traveling at night the motorist is often forced to change tires or repair a puncture when he has failed to take along a light. If a repair must be made that takes only a short time, a good emergency light may be had by heaping up a small mound of sand

or dust, moistening it with gasoline, and setting it on fire. The mound should be made at a safe distance from the car. In order to avoid having a widespread fire, the mound should be depressed in the center, like the center of a volcano, with the gasoline poured in the depression. This will insure the best possible light under the circumstances. (See cut.) The motorist also owes it to property owners along the highways to wait until the light burns out, or to see that it is entirely extinguished before leaving it. Never try to replenish such a light with more gasoline, unless you are in a hurry to join the angels.

Laying Out Land for Fall Tractor Plowing

In laying out a field for fall plowing with a tractor, one should do away with all unnecessary traveling, as it is simply a waste of fuel and time and wears out the tractor without giving returns. A field should be marked off, therefore, and plowed systematically.

Curve Plowing

One of the best ways to plow a rectangular field is by use of the continuous furrow. If this plan is followed, the first thing to do is to set stakes along the center line of the field. This line of stakes should extend ten or fifteen feet closer to the ends of the field than it is to the sides, which will allow for narrowing the furrows in turning at the ends.

The first tractor furrow should follow the stakes which have been set up. When the tractor reaches the end of the stakes, the plows should be lifted, the outfit swung to the right, and then back to the left in a complete circle. At each turn the ends should be rounded as much as possible so that after a few rounds have been plowed the outfit may be swung around the ends without lifting the plows. Thereafter a continuous furrow may be plowed. If properly laid off and plowed, the whole field will be turned except a small spot in each corner.

Curve plowing, however, is rather hard on the tractor gears, and a different method is usually recommended. This method is the back plowing or dead-furrow plowing. By this method, nearly all the plowing is done on a straight line and the outfit is turned with the plows out of the ground.

Dead-Furrow Plowing

To plow by this latter method, leave about forty-five feet all around the field on which to turn the tractor. With one bottom only in the ground, plow around the field to mark this distance. Then at one side of the field set a line of stakes, sixty feet from the furrow marked off forty-five feet from the fence, and parallel with that furrow. Sixty feet from the first line of stakes, set up another line, and 120 feet from the second line of stakes, set up another. The operator now has three lines of stakes to plow by—the first, sixty feet from the single furrow plowed as a guide line; the second, sixty feet from the first, and the third, 120 feet from the second.

Begin plowing at the right of the first line of stakes and throw the dirt toward them. When the length of the field has been traversed and the marking furrow reached, lift the plows, turn to the left and go to the third line of stakes. Let the plows into the ground at the marking furrow and throw the dirt towards this third line of stakes. When the opposite end of the field is reached, the tractor is again turned to the left and with the plows out of the ground driven back to the beginning. This round of plowing is continued till the land between the first and second line

Tractor Plowing

Tractors and Automobiles

of stakes, and half the land between the second and third lines of stakes, is plowed.

When that has been plowed, the outfit is turned to the left and the dirt thrown toward the second line of stakes. As the opposite side of the field is reached, the tractor is turned to the right, idled to the first line of stakes and then the dirt thrown toward them. These two lands are plowed in the same way as the first two—and then the plowing of the whole field continued in a similar way. After the center of the field is plowed the forty-five-foot border should be plowed by the endless furrow method.

In Hilly Country

Plowing hilly lands requires rare judgment. Where there is a valley in which there is no stream, the plowing may be started in this valley as though it were a rectangular field, the dirt thrown down hill from both sides and the ends idled across if it is thought advisable. Or where there is just a round knoll, it may be plowed with the endless furrow method, throwing the dirt down hill and finishing the plowing on top of the hill. The hill shaped like a horse's hoof may be plowed in the shape of a horseshoe and the space between the "calks" idled across. That is, one part of the hill may be so you can curve around it, but the other has to be mounted. In all cases try not to plow up hill if you can help it; plow *with* the hill.

How to Have Good Roads

The construction and maintenance of earth roads is a vital topic in every rural community. The most practical and successful system is that which originated with Mr. D. Ward King, and which is now in general use all over the country.

The keynote, or basis, of Mr. King's system is a simply-made road drag, fashioned from a split log about eight feet long, with the two parts about two and one-half feet apart. Any farmer can make one of these drags for himself, at a cost of a dollar or so—or less.

Speaking of this system, the Iowa Highway Commission says in a bulletin issued by the engineering department of Iowa State College:

"Water is the foe to good earth roads, and the whole object of earth road construction and maintenance is to get rid of the water and its bad effects. Three systems of drainage are needed:

"First, Tile or Sub-drainage. Wherever the soil is naturally wet from ground water, a line of four-inch tile should be laid to a regular grade longitudinally along the uphill side of the road, under the side ditch, at a depth of three to four feet.

"Second, Side Ditches. A good, big side ditch, built to a continuous grade as determined by a road level, so that the water will not stand in it at any point, should be provided on each side of the road. The road level should be used to make sure that the ditch is built to a grade which will not leave ponds of water in the ditches after rains.

"Third, Surface Drainage. Proper surface drainage, to shed the water promptly into the side ditches, should be provided by properly crowning the road, and by then keeping it hard and smooth with a King road drag. This drag is the cheapest instrument we have found for this purpose. The annual cost per mile of road treated with the King road drag where all the time has been paid for by the hour, has not been found to exceed $2.50 to $3.00.

"We advise farmers to start using the drag without waiting for the road officers to take it up. They will be well repaid for their trouble by the saving of time and expense in using the roads, and the increase in value of their land, due to a good road in front of it.

"We also advise road officers to adopt the drag, and to provide farmers with free materials to make them, and to hire the roads dragged where the farmers do not themselves undertake the work. There is no possible use of the road funds known to us which will yield such great returns for so small an outlay. In fact, the outlay will be more than saved by the lessened need for the big road grader, with its great cost of operation.

"Gravel roads, when cut up an inch or two deep in continued wet weather, should be gone over at such times with a King drag, the same as an earth road."

The correct method of using the King drag is about as follows:

Begin operations at once, and do not entirely abandon the work except when ground is solidly frozen. A few minutes' or hours' work, now and then, is better than a week's work all at once.

After each rain or wet spell drive up one wheel track and back on the other at least once, with the drag in position to throw the earth to the center. Ride on the drag. Haul at an angle of 45 degrees. Lay boards on the drag to stand on. Gradually widen the strip dragged as the road improves. To round up the road better, plow a shallow furrow occasionally each side of the dragged strip, and spread the loose dirt toward the center.

Thus the road gradually becomes smooth, hard and almost impervious to water. Rains run off the rounded roadbed, like water from a duck's back. By using the drag when the road is muddy (as advised) the earth packs and cements itself into a hard and nearly waterproof surface. And that is the idea, in a nutshell. 'Tis plain to see that if water can find no place to stand, no chuck-holes or ruts can develop.

A, B, C of Road-Making

The essential requirements of good stone road construction may be condensed into the following rules:

1. Cut the high places down to a grade not exceeding 1 in 20; fill up flats or low places so as to have a minimum grade of 1 in 200.

2. Construct subdrains to carry away all seepage water; also make enough cross-drains to dispose of surface-water. Fig. 1 shows a subdrain of drain-tile covered with stone. Fig. 2 shows a subdrain made of logs, and Fig. 3 shows one made of field-stone.

3. Make the subgrade firm and solid and give it the same curvature as the surface of the finished road.

4. Spread the bottom course of stone evenly, then roll and add a little fine material for a binder, and continue the rolling until the stones cease to sink and creep in front of the roller.

5. Spread the second course and roll it with the addition of binder and water until the whole surface is hard

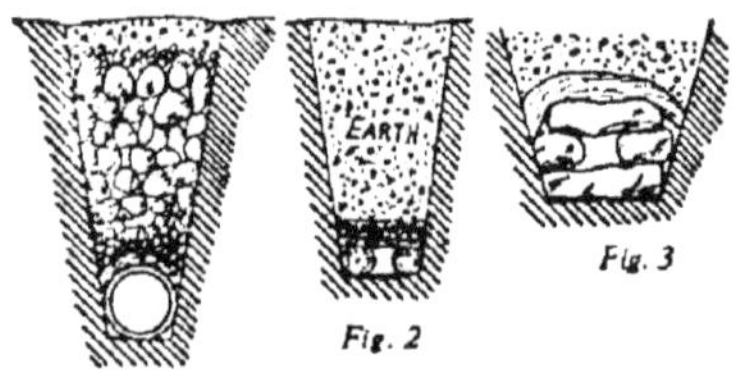

and smooth, carefully filling with stone any depressions that may appear; then finish the whole with a course of three-quarter-inch stone and

<table>
<tr><td>Good Roads</td><td style="text-align:right">The Snow Plow</td></tr>
</table>

screenings. This must be soaked with water and rolled until the surface is hard and unyielding. Always be careful to commence the rolling at the sides and gradually work toward the center; by so doing the crown of the road will be preserved. If this work is well and thoroughly done the result will be a road that is smooth, hard and convenient for travel at all seasons of the year. Fig. 4 shows a cross-section of a macadam road, with layers of stone compacted in place.

For a farming community the width of macadam need not be greater than ten or twelve feet. The width of stone surface should be sufficient to take care of all travel on the road; but on the other hand it should not be so great as to require unnecessary expense in the construction or maintenance of the road.

Fig. 4

When water has to be conveyed from one side of a road to the other it should be taken under the road by means of a culvert. A stone culvert is, of course, the best, but a vitrified tile-pipe or a corrugated metal culvert may be used.

Lastly, give the road a good coat of suitable road oil to prevent dust and retard damage by automobiles.

An Effective Snow Plow

A farmer who is a firm believer in good roads in winter as well as in summer, could not get any satisfaction from his town superintendent when snow blocked the six and one-half mile way to town. So instead of going to the town-father and bawling him out, the farmer knocked together a snow-plow and started out. The work done with the plow that day made it easy for a team to trot all the way to town and not get in trouble when it was forced to turn out for another rig. And the job was over before a gang of men with shovels could have started.

Directions for making the plow are as follows: Take a 2 x 8 inch plank, eight feet in length (A in the accompanying picture), and use it for the side which is to travel flush against the side of the bob runner. A ten-inch plank might work, but a twelve-inch one would not because it could not swing ahead of the rear bob when making turns. Take another eight-foot plank and saw off two feet for a brace (C) and spike this brace at right angles to the longer plank. It may be necessary to let out the reach so that the outer end of the brace will not be too far forward on the mold-board (D). The brace should be

raised an inch or two so that the snow which falls in the triangle will spill out readily. When spiked together the plow should have a comparatively small tread, say four feet. A wide spread means too heavy a drag and unnecessary labor. To hitch the plow

Farm Bookkeeping | **Farm Management**

to the bob use two clevices and a ring attached to the end of the left side of the plow (point marked B). In order to do effective work a man should ride the plow.

One distinctive feature of the plow's construction lies in the fact that the left side of the plow is two feet longer than the right side. The left side is made so it can be snugly attached to the right-hand front runner of an ordinary pair of bobs. When traveling the left (longer) side parallels the runner, and the right (shorter) side forms the whole plowing surface.

Directions for Farm Bookkeeping

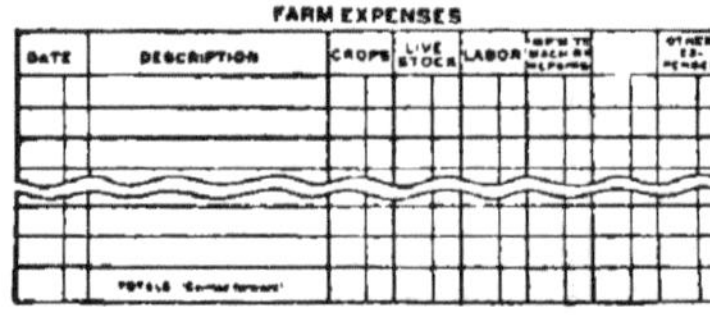

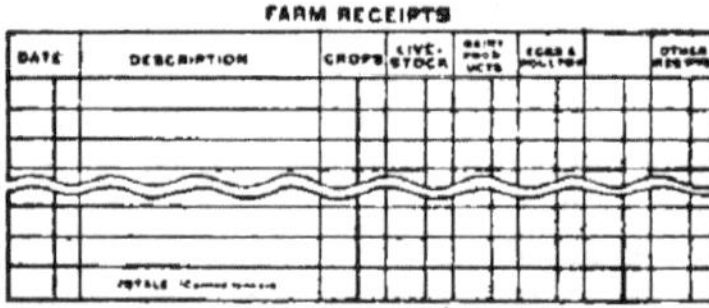

"It is too complex, and takes too much time." That is the way most people excuse themselves from keeping books on the farm. Contrary to that opinion, farm bookkeeping is easy, and need not take much time if a simple form of accounting is used.

The only things necessary to keep farm accounts are an account-book, determination, and a pen. If you have no account book, take an ordinary note-book and rule twelve pages (one for each month) for the farm receipts, and twelve for the farm expenses, as shown in the illustrations. The pages of the book should be at least 8 x 11 inches.

Under "Description" tell what the receipt or expense is for, and enter the account in the proper column at the right; if for oats or corn, enter the amount under crops; if for eggs, enter under eggs and poultry. The headings can be changed to suit the type of farming. One space is blank so that another heading can be inserted, if desired.

If a sale is made and the money is not collected, the amount should be entered on the receipt page at the time of the sale, or a memorandum of the transaction should be made on the memorandum page in the back of the book, and the amount entered on the receipt page when the payment is made. The same plan can be followed when buying a piece of machinery, or some pure-bred stock. Occasionally people buy a new mower or binder and they pay for it several months later when they sell hogs or cattle.

Living expenses, such as clothes, groceries and the like, should be kept apart from farm expenses. Generally the housewife will be glad to look after that part of the bookkeeping, and many housewives keep a separate book for the household expenses. Often the housewife looks after the farm account-book, too.

Farm Management   **Mailing Lists That Pay**

The expenses of the farm include labor, repairs, feed, machine work, hauling, fertilizer, seed, horseshoeing, breeding fees, veterinary services, insurance, taxes, new stock, equipment, etc. The loss of an animal should not be entered as an expense, for that loss will be expressed in the inventory at the end of the year. The cash receipts cover all sources of income from the farm itself.

At the end of each month the totals from each page are carried forward to the new page, and at the end of the year they are transferred to another page, which will show the total expenses and receipts for the year.

Besides the expenses and receipts accounts, it is necessary to take an inventory of all farm properties at the beginning and end of the year. These four records will show whether the farm has returned ten per cent. profit, besides the good living to which every farmer is entitled.

Mailing Lists That Make Money

Almost every farm can use mailing lists to advantage. In buying, they help to locate the cheapest and the most suitable article at once. In selling, they drum up trade, add new customers and help to obtain the highest market prices.

A farmer who had made a hobby of sweet corn, growing it for years until he had developed a superior strain, found that the local stores were glad to handle his seed, but offered a low price. He compiled a list of nineteen seedsmen operating in his territory, securing the names from farm paper and newspaper advertisements, and from personal knowledge. Some of them sold seeds by mail, some through retail stores, and others were city wholesalers.

Selling Seed Corn

Using a typewriter, he wrote a business-like letter to all of these, telling what he had to offer, and forwarding samples. His results were typical of mailing-list work. Eight concerns did not reply at all, and seven answered that they had adequate supplies arranged for. Four firms offered $1 a bushel more than the grower had been offered at home.

A good way to index customers

Selling Nuts

Another farmer with a big crop of hickory-nuts used a list of entirely different character. When in town he borrowed several city directories and wrote down the names of professional men, manufacturers and others that he believed had better than average buying power. These were classified in the directories and were easily copied. To every one of the several hundred families on his list he mailed a printed circular, one-cent postage, describing the superior sort of produce he had to offer and quoting prices in bulk lots. The prices were somewhat below what the city fruit stores were charging. He easily sold his entire crop in this way, and had a fine beginning toward a parcel-post trade in other farm products.

Selling Wood

A third farmer obtained a list of pulp-wood buyers and secured a price fifty cents a cord better than that which he was about to accept for his wood.

Purchasing agents of corporations send form letters to every concern manufacturing articles they are in the market for, giving specifications and asking for samples and prices. Farmers who are making purchases of considerable size can follow the same plan. From the farm papers a list of concerns is readily compiled. It is well to get prices from the local dealer, too. Whether the articles to be bought are fence-posts or farm implements, it pays to feel out the market thoroughly, and the mailing list is a cheap, effective way of going about it. The fellow who buys without comparison is often the disappointed one.

Mailing lists for most farm purposes can be compiled at home. There are concerns which make a specialty of furnishing lists, their charges running from about $2 a thousand up, with a guarantee of accuracy. If the list written to is a long one, it is oftentimes good business to use a printed form letter. With smaller lists, a typewriter will do. The typewriter lessens the labor in correspondence, and every farmer who does a great deal of writing should have one.

Mistakes in Buying Land

One of the first things to consider when planning to go to a new place, is to find out whether that particular section of the country is suited to the kind of farming you propose to do. Next, find out whether there is a good market for the things you will grow. Then there is the all-important question, the health of the community; next, the moral and social conditions of the people.

Real Estate Agent Not Always Reliable

Many people who change locations are induced to do so by some real estate agent. The individual decides that he wants to go to some other place to live, and writes to maybe a dozen real estate agents in as many different places; and the agent that claims the most impossible things for his special section of country is the one that gets the most consideration in far too many instances. As I have had some experience with this, I want to give some advice to people who contemplate moving to a new location.

What Kind of Land Do You Need?

Never buy until you are sure that you have the kind of land you need for your particular kind of farming; don't buy land with the idea of growing crops that you know nothing about. When you go to a new location to look at a piece of land with the idea of buying it, never be in too big a hurry; better pay a few days' hotel bill than to lose a few thousand dollars in the deal.

Now let me say that it is an easy matter to get fooled in regard to values in farming land. You may see a farm that looks as good as farms selling at $200 per acre near your old home, and in fact it may be just as fertile; but that is no proof that it is worth as much locally, or anything near it. While it may produce as much per acre, you may not have a market for your crops. There are too many things that influence the values of land for one to call attention to them all.

Talk to the Neighbors

The best way to find out the value of land is to go and talk privately to the people who own land around the piece you are thinking of buying. Ask them all about it, what it ought to sell for, etc. Next, go to the local banker and ask him what the land you want to buy is worth, and find out how much money he would loan you on it. Then go to the tax books and find. out what the taxes are on the property, and what per cent. of value property in that county is taxed. Better take this trouble than to pay two or three times the value of the property.

Don't Over-Extend Yourself

Another big mistake many people make in buying farms is to buy a too expensive farm for the amount of capital they have. You can take a very little money and buy a large or high-priced farm. You pay down all the money you have as first payment, then you find that you have to go in debt for your supplies. When your first note comes due you cannot meet it; hence you lose what you paid down on the place. If you had bought a smaller and cheaper farm you would have had no trouble in meeting the payments. Better buy a small place at first. And never pay out all your money—keep enough to run you until you make a crop. You can make more money if you have a little money to use as you go. Go slow at first; it is the safest way.

A Good Living and 10 Per Cent

We believe that every farmer who knows his business is entitled to "A Good Living and 10 Per Cent" and when we are asked to define exactly what we mean, here is how we do it:

Definitions

(a) "A Good Living": A farmer and his family should live comfortably and certainly not less well than a family in similar circumstances in a town or city.

(b) "And 10 Per Cent."—besides a good living. A farmer should earn at least 10 per cent. on the amount invested in the farm enterprise, which would be only 5 per cent. on his capital and only 5 per cent. actual profit besides.

A Good Living and 10%   Farm Management

(c) "Farmer": The term "farmer" as used in this connection means one who is engaged in any branch of agriculture as his chief means of livelihood.

Measuring the Income of a Farmer

The problem is different from that of a man engaged in most other enterprises, for in those cases a man usually receives a definite salary for his labor, out of which he pays his living expenses. A farmer's income, however, is composed partly of "living expenses" taken from the farm, and is only partially received in money.

(a) Living expenses obtained from farm include:

(1) A house to live in,

(2) usually fuel; and,

(3) to a large extent, food.

(b) The other income received by a farmer is the difference between receipts and expenses.

Receipts include:

(1) Income from sales.

(2) Increases in inventory other than mere increases in price of land, improvements and equipment.

Expenses include:

(1) Labor, except that of farmer, his wife and minor children.

(2) Materials purchased, such as feed, seed, fertilizer.

(3) Repairs to fences, buildings, machinery.

(4) Miscellaneous, such as taxes, depreciation on buildings, stock and equipment, rent (if tenant farmer), fire and farm insurance, interest on debts, losses by fire, diseases, hail, pests, etc.

In addition to these costs of production there must be an allowance for the living expenses not obtained from the farm. These include items of household and family upkeep, such as: Food not produced on the farm, clothing, recreation and travel, education, life insurance, benevolences, doctor, dentist, etc.

Where the 10 Per Cent. Comes In

From the "farm income," or balance left from sales after deducting cost, a farmer should have a balance left of 10 per cent. on his investment after paying all the expenses listed above. But this reservation must be made: The responsibility for spending the income of 10 per cent. necessarily rests upon the farmer himself; and, if he has chosen to spend this in luxuries, the choice was his—our part of the task is to make sure that farmers generally get this "Good Living and 10 Per Cent.," so that the nation's future food supply shall be assured.

It must be remembered that the amount earned (10 per cent.) will not necessarily be represented by a cash balance of that amount at the end of the year. The cash balance will be more or less than 10 per cent., according to whether the farmer has decreased or increased his investment, as shown by inventories taken at the beginning and at the close of the year.

Assumptions

The following assumptions have been made:

(a) That the farm is large enough to warrant the entire attention of at least one man.

(b) That the farmer possesses average ability and gumption, together with stick-to-it-iveness to carry through his undertakings.

(c) That the family is of average size.

(d) That only such buildings and equipment as are adequate and necessary shall be on each farm—in other words, plain, not fancy farming.

Butchering the Hog

Seventeen years ago the average American ate 180 pounds of meat a year—about a half pound a day. Three years ago the average amount was down to 156 pounds a year, or about two-fifths of a pound a day. Should that rate of decrease continue, meat will be gone from our tables by the year 2000. But it is not likely that the rate of decrease will continue so far. The average consumption will reach a stationary point, as it has in Germany and England, that point depending on the cost of production of meat animals and the economy observed in turning them into meat.

Of the several factors which will help to keep the average meat consumption high, home butchering will continue to be all important, because it is economical. Hogs are now slaughtered on only half of the farms in the country. If they were slaughtered on the remaining half, 11,000,000 fewer hogs would go to market annually, and the saving in profits and in freight, accruing to the farmer, would be enormous.

Do Not Feed Before Killing Time

Swine contribute most to the farmers' meat supply. To obtain the best results in killing hogs on the farm, they should have only good, solid food for thirty days before butchering. Keep the animals off feed for twenty-four to thirty-six hours before killing, for if the digestive tract is full at killing time, gases are generated and the meat may become tainted. Only healthy animals are to be selected. Excitement before killing is to be avoided, for meat often spoils during the curing process if the animals are excited. Hogs may be stunned and stuck with a six-inch knife. A short slit is made lengthwise in the throat, the knife inserted, pointing slightly backward, with cutting side up, then turned to the left and withdrawn. The arteries in the neck will be severed, and bleeding will occur freely, especially if the animal is rolled over on its side.

Scalding and Scraping

For scalding have the water 175 degrees, keeping the carcass in for about a half minute. If the temperature is as low as 155 degrees, a full minute is not too long. A small shovelful of ashes or a tablespoonful of lye in the water will loosen the scurf. A block and tackle make it easy to raise and lower the hog in the barrel. A hog hook will be needed, and the barrel is best slanted against a platform on which the carcass can be scraped quickly after scalding. The legs and head are scraped first.

Common Cuts of Dressed Hog

In the last few years gasoline is being used for singeing the hair from hogs, rubbing the gasoline on the carcass with a rag and igniting with a torch. This does away with scalding. The practice saves lots of time in cleaning hogs, and leaves a much better-looking carcass. There is danger of cutting the carcass when the hair is shaved off with a sharp knife. Packing-houses use a blow-flame which

Hog Scalding   **Butchering and Curing**

works satisfactorily. With this they can blow the flame into the depressions around the eyes, nose, feet, etc., where the hair is hard to scrape off.

Hanging the Carcass

When the carcass has been scraped, after scalding or singeing, the skin is split just below the hocks in both hind legs, and gambrels inserted. The carcass is easily swung to a movable derrick, if a small block and tackle is used. Otherwise, a pole may be used for a lever, attaching it to a tree or solid post. Wash the carcass thoroughly with warm water, and scrape to remove all dirt, following with cold water.

Cleaning

In opening the carcass split the hog between the hind legs and split the pelvic bone between the hams by cutting exactly in the center. Cut around the rectum and pull it down between the pelvic bones. Care must be used in cutting down the hog, not to cut the intestines or stomach. The skin may be held out with the fingers, cutting between the fingers. The entrails may be removed by cutting the cords by which they are attached to the backbone.

Pull up the gullet and cut it off. Split the breast bone, then remove heart, lungs, liver and esophagus. The head may be removed, the inside of the carcass washed, after which it is allowed to cool. After cooling, the backbone may be split or removed by cutting down on either side. The halves are then ready for cutting up and trimming.

Hog-Scalding Device

Here is a very handy hog-scalding pot which can be cheaply erected. I used a gasometer tank, made of galvanized iron, about as big as a barrel. This was placed on a tripod which was made of old wagon tires, riveting the legs to a circular ring, allowing about eighteen inches from ground surface to bottom of pot for firing room. The legs

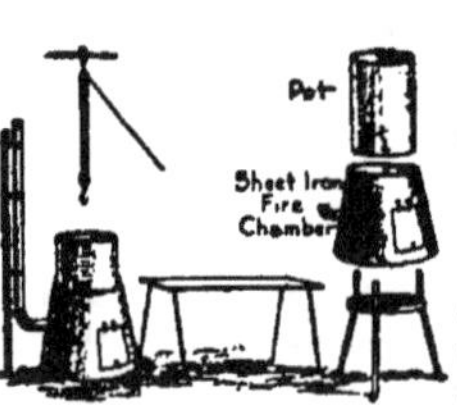

were allowed to extend above the supporting ring so the pot could not slip off tripod. A sheet of black sheet iron was bent around the tripod to form a fire-chamber. A door was cut in it, and fastened at the top to the sheet iron by split rings.

A few sections of smoke-pipe are erected and wired to a small pole for support. The water can always be kept hot by covering the pot with an old blanket to prevent evaporation. When the hog is not in it, this pot can also be used to boil lime-sulphur spray.

Usually butchering requires plenty of help, and everybody is running and fussing to keep the water hot and to handle the pigs. But with a rope and pulley one man can do it alone without trouble. This makes it possible to kill when a farmer has time, without waiting until neighbors find it convenient to come to help.

The accompanying sketch will serve further to explain construction and the manner of using this handy scalder.

Handy Butchering Helps

Devices for Saving Time and Labor

The cheapest meat a farmer can use is the product of his own farm. The custom of buying meat is increasing among farmers in spite of the fact that meat, especially pork, can be grown and killed at home for much less than the cost of purchased meat.

The equipment needed for slaughtering hogs is as follows: An eight-inch straight sticking knife, a cutting knife, a four-inch steel, a hog-hook, a bell-shaped stick scraper, a gambrel and a meat-saw.

Home-made water heater

A convenient arrangement for heating the water is made of a barrel and some pieces of old gaspipe; see the illustration that is shown above. The blacksmith will make the coiled part for you. The water for scalding should be at a temperature of 155 degrees, and the carcass should be left in for about a minute. Do as we say and use a thermometer for testing the temperature, so you can be sure it is just right. Otherwise the hair may set and you will have to shave the hog. Keep the carcass m o v i n g all the time it is in the water. A handful of wood ashes in the water helps to loosen the scurf.

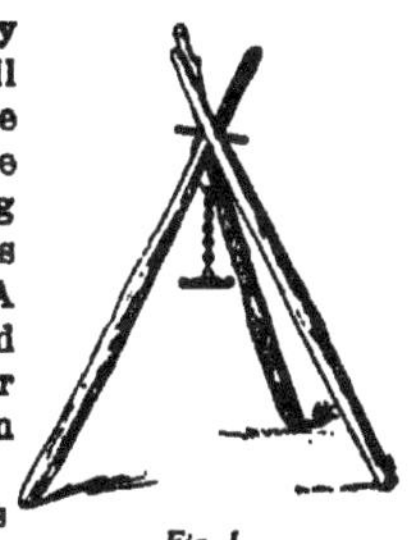

Fig. 1

A barrel makes a convenient receptacle in which to scald hogs. Place the barrel at an angle of about 45 degrees, at the end of a table or platform about two feet high, as shown below. The table and barrel must be securely fastened to prevent accidents to the workmen during the scalding.

Scalding Barrel

A small block and tackle will reduce the labor of handling the animal. If this is not to be had, use a strong three-legged derrick like the one shown in Fig. 1. Set it over the scraping table, and lower it. When the hog is placed securely on the gambrel, raise the derrick and move the table away.

It is getting to be a common practice to singe hogs instead of scalding them. Gasoline is rubbed on with a rag—not poured on. It takes about a pint of gasoline for an average-sized carcass. Light the gasoline with a torch fastened on the end of a stick several feet long. After singeing the carcass, scrape just the same as if it had been scalded.

Ways to Cure Meat

The two ways of curing pork are brine curing and dry curing. If brine is properly made it will keep for a reasonable length of time. If it becomes ropy, it must be poured off and boiled, or a new brine must be made.

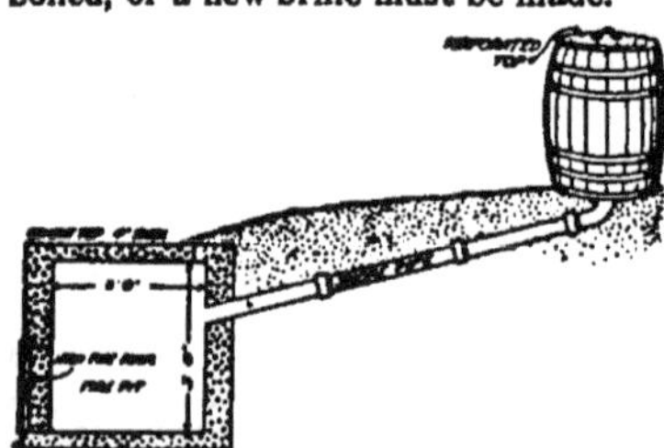

A cool cellar is the best place for both methods of curing. Rub the surface of the meat with fine salt and allow it to drain, flesh side down, for six to twelve hours before the meat is cured, either with brine cure or dry cure.

The Brine Cure

For each 100 pounds of meat use eight pounds of salt, two and one-half pounds of sugar or syrup, two ounces of saltpeter and four gallons of water. In warm weather nine or ten pounds of salt are preferable. All the ingredients are poured into the water and boiled until thoroughly mixed. Then let the brine cool. Place hams on the bottom of the container, shoulders next, bacon sides and smaller cuts on top. Pour in the brine and be sure it covers the meat thoroughly. In five days pour off the brine and change the meat, placing the top meat on the bottom and the bottom meat on top, after which pour back the brine. Do this again on the tenth and eighteenth days. If the brine becomes ropy take the meat out and wash it thoroughly, also the container. Boil the brine or make new brine, replace the meat in the barrel and cover with brine. Allow four days' cure for each pound in a ham or shoulder, and three days for each pound in bacon sides and small pieces. For example, a fifteen pound ham takes sixty days.

The Dry Cure

This requires more work than brine curing. For each 100 pounds of meat use seven pounds of salt, two and one-half pounds of sugar and two ounces of saltpeter. Mix all the ingredients thoroughly, rub one-third of the mixture over the meat and pack the meat away in a box or on a table. The third day rub on half of the remaining mixture and again pack the meat. The seventh day rub the remainder of the mixture over the meat and pack it to cure. Allow a day and a half

Butchering and Curing   **How to Prepare Meat**

cure for each pound in a piece of meat. A twenty-pound ham will take thirty days to cure.

Smoking

Smoking helps to preserve meat which has been cured. It also gives a desirable flavor if the meat is properly smoked with the right kind of fuel. A simple smoking arrangement which is cheap and convenient can be made according to the sketch above. The firebox and meat barrel should be at least six feet apart. Concrete and clay blocks are fine for smoke-houses. Wooden houses are not so good, for they are likely to catch fire—and down goes your meat house.

When meat is removed from the brine it should be soaked for about half an hour in water before being placed in the smoke-house. When removed from dry cure it must be washed in lukewarm water. The meat should warm up gradually in the smoke-house, and must not get too hot. Green hickory or maple wood is the best for smoking. Thirty-six to forty-eight hours is the time required to smoke meat. Slower and longer smoking will make the meat keep longer.

After meat is smoked it should be wrapped in heavy paper and put into muslin sacks to keep insects away. Tie the tops of the sacks tightly.

Preparing Home-Killed Meats for Family Use

Brine Cure

When the meat is cooled rub each piece with salt and allow it to drain over night. Then pack it in a barrel with the hams and shoulders in the bottom, using the strips of bacon to fill in between and put on top. Weigh out for each hundred pounds of meat eight pounds of salt, two pounds of brown sugar and two ounces of saltpeter. Dissolve these in four gallons of pure water and, after boiling and thoroughly cooling it, cover the meat with the brine. The bacon should remain in the brine from four to six weeks, the hams from six to eight.

Meat that is to be smoked should be removed from the brine two to three days before putting it in the smokehouse. Scrub it clean in tepid water and hang to drain for a day or two before smoking. When smoking hang it so that two pieces will not touch each other. Make a very slow fire of green hickory, apple or maple wood, smothered with sawdust of the same. Do not use pine or other resinous wood, as it will give an unpleasant taste. When smoked sufficiently let the meat air and dry, then wrap in paper and put in heavy muslin or canvas and cover with whitewash.

Westphalia Ham

Westphalia hams are imported and sell for a high price because of their piquant, appetizing taste, which is given them in the smoking. These foreign smokehouses are built with two, or sometimes three stories. The meat is hung in the upper one, the smoke ascending from the lowest story through holes in the flooring. The smoke is made by a fire of beechwood, on which twigs and berries of

How to Prepare Meat | Butchering and Curing

juniper are continually strewn. If the fire becomes too strong it is deadened with beechwood sawdust. It is the juniper berries which give the special flavor.

Plain Salt Pork

Rub each piece with fine common salt and pack in a barrel. Let this stand over night; the next day weigh out ten pounds of salt and two ounces of saltpetre to each one hundred pounds of meat, and dissolve the two in four gallons of boiling water. Pour this brine over the meat when cold, cover and weight it down. The meat will pack best if cut six inches square.

Sausage

For every fifty-five pounds of lean and fat pork chopped very fine mix together one pound of salt, six ounces of good black pepper, a teaspoonful of cayenne pepper and a handful of powdered dry sage. Mix these thoroughly through the meat. If you wish to stuff it in skins clean them thus: Empty the intestines of the pig, turn them inside out and wash well. Soak them in salt water a day or more. Wash again, cut into convenient lengths and scrape them on a board with a blunt knife, first on one side, then on the other, till they are clean and clear. Throw them in clear water and rinse. Tie up one end of each length, put a quill in the other end and blow them up. If whole and clear they are clean, but if there are thick spots they should be scraped off. Throw in clean, cold salt water until wanted. To use, put one end over the nozzle of the sausage stuffer and force the meat into them. This can be better done if the meat is first lightly sprinkled with cold water which is worked through it.

Pack the sausage to keep for winter use in stone crocks and run two inches of hot lard over it. That for summer use may be canned. Make into small cakes and cook about two-thirds enough for the table, or until all the water is out. Pack in the cans while still cooking, fill them full of hot lard and seal at once; or it may be stuffed tightly in muslin bags, then the bags rolled in melted paraffine, which should be heated in a large flat pan. When cooked next summer it will be more delicate if you pour off all the fat after it is fried, pour in a little cream, boil it and pour over the sausage.

Liverwurst

Boil the liver with about an equal weight of head meat, including the fat. After it is fairly well done, run it through a food chopper while yet warm, season with salt and pepper and pack it in a crock or into rolls. It should be sliced and fried for use.

Pigs' Feet

Pigs' feet should be thoroughly cleaned. First scrape and wash, then soak in cold water for several hours; scrub and wash well. Split feet and crack in several places. Cover with boiling water, add a little salt and let them simmer until thoroughly tender, about four hours. If you want them pickled add salt, pepper and a few whole cloves or allspice to half a pint of good cider vinegar, boil for a minute and pour this over the pigs' feet.

Canned Meat

Tenderloin and pork chops can be kept for some time by cutting into serving pieces, frying until rather more than half done, then packing into hot, sterilized jars or cans and running boiling lard over them. Do not put more than enough for one or two meals in one jar, as the meat will not keep after the lard is removed.

Corned Mutton

This is the way to corn mutton: To each hundredweight of meat take a half-peck of coarse salt, one-quarter pound of saleratus and two pounds of coarse brown sugar. Scatter some of the salt on the bottom of a tub, then put a layer of meat, then salt, then meat, until all is used. Let it remain one night. Dissolve the saleratus, saltpetre and sugar in a little warm water and pour it over the meat, adding enough cold water to cover it. Put a cover and weight on the meat to keep it under the brine, and it will be fit for use in ten days. For a less quantity use salt, etc., in proportion. One quart of molasses can be used in place of sugar.

Dried Beef

Get the tender side of the round out of a good fat beef. For every twenty pounds of beef take one pint of salt, a teaspoonful of saltpetre and a quarter of a pound of brown sugar. Mix these well, rolling out any lumps; divide into three equal parts and rub well into the beef for three consecutive days. Turn beef daily into the liquor it will make. It should not make much, but what there is rub into and pile on the beef. Rub a little extra salt into the hole cut for the string to hang it by. At the end of a week hang in a dry, rather warm place till it stops dripping, then in a cooler, dry place. Do not smoke it; it spoils the flavor. Before flies come in the spring wrap in paper and put it in a stout bag with a string out to hang by. If it molds some through the summer scrape and scrub the mold off, and always trim the outside before chipping.

Beef Tongues

Trim them and lay six or eight tongues into boiling water for five minutes. After they are cool rub them with a mixture of a quarter of an ounce of saltpetre, two handfuls of salt and either a quarter of a pound of sugar or a small cupful of molasses. Pack them in an earthen vessel, sprinkling each layer with the mixture and putting a weight on top. Turn them every other day, putting the top one at the bottom, and packing them closely. If there is not enough pickle to cover them, sprinkle lightly with salt. At the end of two weeks hang up to drain. When dry, wrap in paper, put into a bag, tie tightly and hang in a cool place. If you do not wish to use a whole tongue at once, it does not injure it to be cut in two; but it is best to dip the cut end into boiling water for a moment, in order to seal up the pores.

Corned Beef

Wipe every piece with a dry towel. To every fifty pounds allow an ounce and a half of saltpetre, a pound and a half of brown sugar, about nine or ten gallons of water and best salt enough to make a brine. Mix the sugar and saltpetre in the water and add enough of the salt to make a brine that will float an egg. Pack the meat in a clean, sweet barrel and pour the brine over, skimming off whatever floats. Cover with a thick cloth; watch the brine carefully for a week, skimming it every day. If not enough brine to cover, make more until there is. In about two months drain the brine off and make new in the same way. The meat will keep a year or will be ready for use in two weeks. Beef tongues may be cured in with it, but should be taken out at the end of a month, dried, wrapped in paper, bagged and hung in a cool, dark place.

Meat Supply from Rabbits and Hares

The Belgian hare is considered one of the best rabbits for table use.

In raising Belgian hares they may be kept entirely in hutches. If small runs or rabbit courts can be afforded to give the animals exercise in fine weather, so much the better. The hutches should be built of good lumber, and must have tight floors. There should be at least twelve square feet of floor space, and the hutch should be two feet high. Outdoor hutches should have sloping roofs and overhanging eaves to protect the hares from rain. The screened door should have a sliding wood cover to be fitted with a removable cloth cover. Small holes bored near the top of the hutch will afford all needed ventilation.

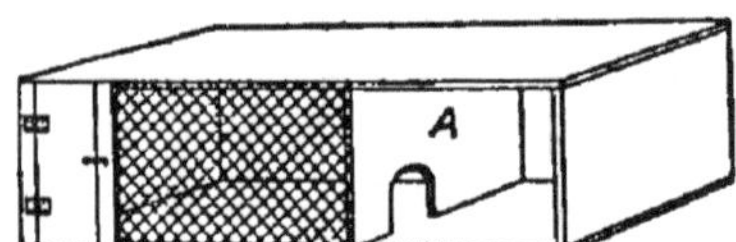
Division A must be movable so hutch can be well cleaned

Movable hutches have advantages; they may be carried outdoors in fine weather and taken back under shelter at night during storms. Long, narrow cleats projecting at both ends of the hutches are all that are needed to convert the ordinary hutch into a movable one.

Rabbits thrive on a diversity of vegetable food. Oats, whole or crushed (the latter preferred), is the best

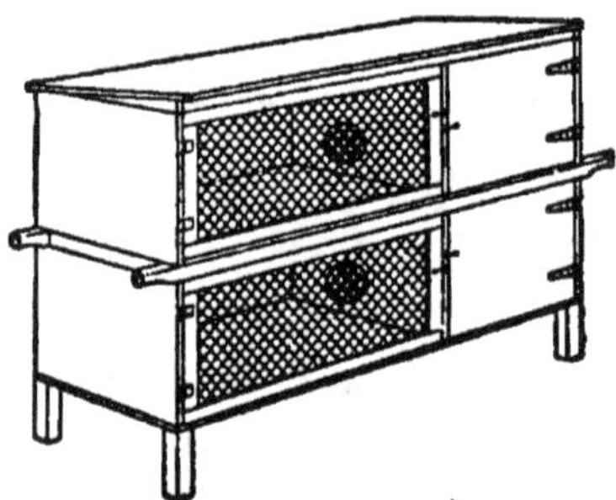
Handles make it easy to move them

grain that can be given. Hay, of the very best quality, free from moldiness and unsweated, is a necessary part of the diet. Cabbages, kale, spinach and rape leaves are also valuable. Rabbits should be fed twice a day, morning and evening. Suckling does should have a noon meal. Feed as much as the animals will consume. Hay can be constantly left in the hutch, as what they do not eat will do for litter. Care must be taken not to overfeed. A piece of rock salt should be kept in each hutch.

Rabbits under four months of age must be limited in the amount of green food, or they are likely to become pot-bellied. As soon as a young rabbit is seen to grow big about its belly, it is time to discontinue the feeding of green food and give the animal plenty of exercise.

Economy in Handling the Washer

While the washing-machine has come to be appreciated as one of woman's best friends, it deserves better care and attention than it receives on the average farm. On many farms a new washer is bought every few seasons, but with proper management, economy could be brought into practice along this line, as washers are very strongly made and with good care will last several seasons.

A good coat of paint would make the old washer look like a new one, and guard against rust and decay.

Keeping a bucket of water in the tub of the washer at all times will prevent bulging and warping of the bottom and shrinking of the staves.

A Light Load is Best

About a half-tub of water and a small amount of clothes require the minimum turning of the machine, and is a light strain on its working parts. Where the washer is jammed full of clothes it greatly increases the labor of the person operating the machine, while the clothes will not clean so well as a few.

Frequent adjustment of all the bolts, screws and other parts of the washing machine should be made, as it turns much harder with these things loose, and is a damaging strain on the whole machine.

Oil Persistently

Oiling the washing-machine every time one uses it makes wash-day less of a drudgery and prevents wear and tear of the machine, as well as adding to its length of life. Run kerosene through the gearing once a week to cut loose and remove refuse grease, oil, dirt, etc. Then give another thorough oiling with good separator oil after the cleansing. Use only first-grade oil.

Indoors, Please

Above all, keep the washing-machine in out of the weather. Either have a rain-proof building in which to do the washing, keeping the machine there at all times, or request "hubby" to help you store it in a dry place after each washing. Left outdoors, the action of the sun, wind and rain on the washer will soon deteriorate it in value till it is practically worthless. Good care in this respect, together with proper handling and oiling, as directed, will add much to its life, effectiveness and easy-running qualities.

Bob's Endless Clothes-Line

"I wonder if mother knows how tired I get lugging this old basket of clothes over the ground beneath this clothes-line," grumbled Bob.

"I wonder if you know how tired your mother gets every week walking along under the line hanging up clothes and taking them down," replied Uncle John. "Why don't you fix up an endless line for her so she won't have to

An Endless Clothes-Line   Household Management

walk back and forth, and so you won't have to lug that basket around?"

"An endless line? What do you mean, Uncle?"

"Didn't you ever see an endless line? Well, we can easily rig up one. Got any old pulleys? Got any extra wire for clothes-line? Got a couple of strong timbers for posts?"

"We've got about anything you can think of lying around the farm somewhere," replied Bob, interested at once.

Then they went to work. First they hunted up a big pulley and fastened it to a beam just inside the wash-room window. Then they found a piece of wire that suited, ran it through the pulley, fastened the end loosely, and stretched the wire along the ground to find out where to set the post. When the post had been set firmly, they found another pulley and attached it to the top of the post. Then they loosened the ends of the wire, ran one end through this pulley and then, after drawing up the slack, carefully spliced the two ends together.

"Now get a couple of boards, a hammer and some nails," said Uncle John; "and come into the wash-room. I want to show you something."

In the wash-room they set up a strong platform about two feet high, running from beneath the clothes-wringer to the window under the endless line. Uncle John picked up the clothes-basket and placed it on the platform beneath the wringer.

"Now suppose that basket is full of wet clothes," he said. "No one needs to lift it. Just slide it along the board to the window. Now we're ready to hang up the clothes. Don't have to stoop over for every piece when the basket is on this platform. Where are the pins, boy? Get a small box and nail it to the wall, put the pins into it, and they'll always be within your mother's reach. Well, we'll proceed to hang clothes. As we pin them on the line, we'll move the line along. Better oil those pulleys a bit. When your mother sees how fine this works, she'll want another, because two short lines are better than one long one."

"That's a fine idea!" exclaimed Bob. "Mother won't have to go out in the cold. And—and I won't——"

"You won't have to lug that basket around any more," Uncle John said.

"Don't ever be ashamed to admit that your head has saved your legs and arms. It's not the hard work itself that counts; it's the results."

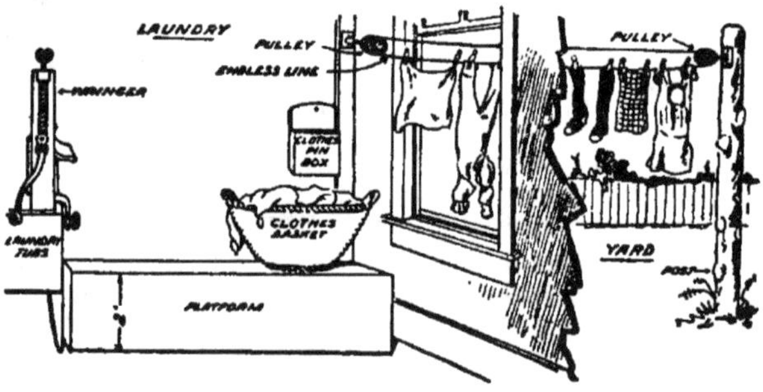

Dishwashing Made Easy

That same old hateful task of washing dishes—how women loathe the time and effort which would be so much better placed on something else!

But dishwashing is being touched by the hand of science; a few years hence dishwashing drudgery will have disappeared forever. Even today, in the most convenient kitchen, the task may be made easier by applying a few simple methods of work.

Analyzing the process of dishwashing from the time we remove the soiled dishes from the table to the time we put them away on their respective shelves, we see that dishwashing is not one single task but a group of small tasks. We see that it is not all "dishwashing," but that it is composed of these four parts: 1. Scraping and stacking dishes. 2. Washing them. 3. Drying them. 4. Putting them away.

The reason the work is hard may be due to fault or delay in any one of the four parts, and we have to find out in which part the delay occurs, and how it can be remedied, before we can improve the whole. Perhaps we do not thoroughly and carefully stack and scrape. Few women use the little scraper of wood and rubber, which looks like a paddle, Fig. 3 (which is better than a knife or fork), to scrape the dishes clean, so that the actual washing is not bothered by fragments, grease, etc. Or we can use

tissue-paper to remove grease, and thus make the work easier. Stacking is important, all similar sizes going together, all dishes stacked to the right and drained to the left.

Nothing does more to make dishwashing a drudgery than to have the sink, or table surface on which dishes are washed, too low. The height should be so arranged that the worker can stand at it with comfort, or, preferably, the worker can sit down to this, as well as any other household task of a similar kind. The writer always sits down to wash dishes, to iron, prepare pastry and peel vegetables.

The first secret of washing dishes is to have the water hot and soapy, the suds being made by a soap-shaker, Fig. 2, and not by floating a cake in the water. A good soap-powder is best for pots, kettles and pans. Use a string mop, Fig. 1, whenever possible; this makes the task easier on the hands, and is more sanitary than a dish-

Fig. 1   Fig. 2   Fig. 3

cloth. Two mops can be used, a small one for glasses and pitchers, and a large one for the regular dishes.

The Fireless Cooker | **Household Management**

Now, if we have a drainer, Fig. 4, we can cut down the time of the whole process more than a third, because the time consumed in wiping is greater than the time consumed in washing. Therefore, with a drainer, we reduce the drying time, because if dishes are arranged in a rack and have scalding water poured over them, they will dry themselves. It is neither necessary nor sanitary to dry each piece with a dish-towel of doubt-

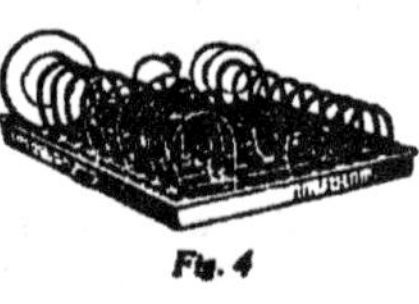

Fig. 4

ful cleanliness, and by the time the pots and small utensils are finished, the dishes will be completely dried and ready to be put away without further handling.

Much time is consumed in carrying the trays of washed and dried dishes to some distant place. Sometimes innumerable trips are required to carry all the dishes used at a meal to the pantry or shelves. Why not place the shelves for the pots and pans, and the kitchen dishes, adjacent to the sink, at the left? This has been done in several homes with the result that the dishes can be put away, when dried, without walking a step. Thus many minutes can be saved to be spent on something else, outside the kitchen.

A device which can be helpfully employed in the dishwashing process is a tray on wheels, which can be pushed or wheeled wherever desired. All the dishes from a meal can be stacked on these lower and upper trays, and wheeled at once to the right of the sink. If the dish-closet is some distance from the sink, the same tray can be used to receive the clean dishes and wheel them to wherever they are to be put away.

A new dishpan which can now be bought is rectangular in shape, more exactly to fit the sink. It is supported on legs, which are fitted with rubber tips to prevent marring the surface of the sink. In addition, it has a small mesh wire. Thus, instead of lifting up and tipping the pan to empty the water, the stopper can be removed, the drawer pulled out, and the water emptied and pan drained at the same time. This is a great improvement over the old round pan. With a convenient arrangement of sink, closet and shelves, proper tools and a good supply of hot water, time, strength, soap and towels can be conserved, all of which are worth saving.

How to Make and Use a Fireless Cooker

A saver of time, fuel and labor is the fireless cook stove, which can be made at home, absolutely without expense, and, though not adapted to all kinds of cooking, answers well for food that requires long, slow cooking to soften tissues, bring out flavors and conserve the juices, such as stews, pot roasts, soups, cereals, rice, tapioca, dried fruits, vegetables, etc. It consists of a kettle of agate or tin, inclosed in a box, with insulating mate-

Household Management   Making Fireless Cooker

rial between the two to prevent the heat of the kettle from escaping. Food brought to the boiling point over a fire, and inclosed, still boiling, continues to cook. This is the whole principle.

Choose a kettle with a tight-fitting lid and a box large enough to allow six or eight inches of insulating material. Line the box, bottom, sides and hinged-on lid with stout packing paper, or several thicknesses of newspaper. Make a firm, cylindrical shape to fit easily around the kettle and fasten a circular bottom to it. This might be of asbestos paper, or paper soaked in alum water and dried. Then no matter how hot the kettle there would be no danger of scorching. Fill the bottom of the box with packing, which can be of cotton, wool or ground cork (in which imported grapes are packed, and which grocers are usually willing to give away). Hay will answer, but does not pack so closely as these.

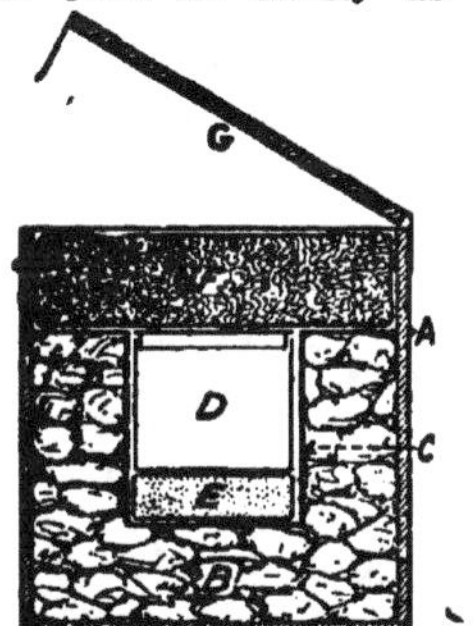

Longitudinal section through fireless cooker, showing details of construction: A, outside container. B, packing or insulating material. C, metal lining of nest. D, cooking kettle. E, soapstone plate, if used. F, pad for covering top. G, hinged cover of outside container

Pack hard to a depth of three inches, place the cylinder, containing the kettle in the middle, and pack tightly around it, even with the top.

The insulating material can be covered neatly with cloth, or a thin board with a round hole in the middle. A thick cushion will insulate the space between this and the lid, which must be fastened down tightly. If desired to cook several things at once it is best to have two or three such cookers, as the box should not be opened after the food is put in, except to reheat. Some persons prefer using a sort of double boiler, the inner kettle, containing the food, being placed in a larger one, partly filled with hot water. In this case the water in both kettles must be actually boiling. An additional vegetable can be put in the outside kettle, or water can be kept hot in it for dishwashing.

Ready-made cookers can be bought, but are rather expensive. Some of these will also bake and roast by means of thick disks of concrete which must be made very hot on the stove, then put under and over the kettle containing the food. The idea might be applied to the home-made cooker by heating soapstone griddles. These might be heated at the same time with a large iron pot. The meat or chicken, which should be seasoned, can be put in a kettle, a hot disk put in the bottom of the pot, the kettle set on this; the other disk put on top, then put the lid on the pot, and bury in the cooker. The pot, however, should be inclosed in asbestos paper to avoid possible ignition. It would be interesting for each housekeeper to experiment and invent improvements on the central idea. The time required for cooking vegetables varies according to their age and freshness, so only the approximate time necessary can be given. There is little danger of their being overdone, or at least injured by long cooking, and if underdone it is always possible to take out the kettle, reheat, and return to the cooker, or if needed quickly, to finish on the stove.

It is not worth while to use the cooker for food that takes but a short time to cook, such as corn, spinach, young peas, asparagus, etc., since the water for these must be brought to the boil anyhow, they can as well be cooked on the stove. Do not place the kettle next the flame, but always have a lid under it.

Potatoes

Five minutes over fire, an hour in the box. Potatoes must not be left overtime in box or they become watery.

Rice Pudding

Mix together in the kettle one-half a cupful of rice, a quart of milk, a tablespoonful of butter, one-half a cupful of sugar, a little salt and grated nutmeg. Boil on stove five minutes, in cooker six hours.

Bread Pudding

Soak one-half a pint of bread crumbs in a pint of milk, add a beaten egg, two tablespoonfuls of sugar and a pinch of salt. Beat with a spoon; heat on the stove till just short of boiling, stirring all the time. Put in the cooker an hour and serve with vanilla sauce.

Cricken Fricassee

Disjoint a chicken, roll in flour and brown in a little fat; as the pieces brown pack them in the kettle, and make some gravy in the skillet. Put this and a little water to cover the chicken. Boil twenty minutes then put in cooker over night.

Boiled Ham

If wanted for 6 o'clock dinner, put ham weighing six pounds in kettle at 9 a. m. Cover with cold water and bring to a boil; boil briskly fifteen minutes. Put the lid on the kettle when it begins to boil and don't take it off till it is taken out of the hay box, in which it should be put while still boiling. At 2 o'clock take out, boil up again, put in a few cloves and two or three peppercorns. At 5.30, take out, skin, put in a pan, fat side up, stick in a few cloves, sprinkle slightly with sugar and plentifully with bread crumbs and bake in oven till well done.

Onions

of moderate size, boiled ten minutes on the range, should be tender after four hours in cooker.

String Beans

Cut off the strings and slice down the middle; give five minutes over the fire, four hours in the cooker.

Cauliflower and Young Cabbage

Five minutes over fire, five hours in cooker.

Cereals started over the fire at supper time and placed in the box should be ready for breakfast with just reheating. Half a cupful of cereal poured into three cupfuls of boiling water, with a teaspoonful of salt is about the proportion.

A fireless cooker can be used for things to be kept cold as well as hot. Ice cream, if frozen, then packed in a kettle with ice and sunk in the box will not melt, and butter if put in it cool and hard will keep in the same condition, as the air is practically excluded.

Boston Baked Beans

Soak two cupfuls of beans in cold water a whole day. At supper time drain, cover with fresh water, put over the fire and simmer slowly for half an hour; pour off the water, scrape a quarter pound of salt pork, cut off a slice and push it down through the beans to the bottom of the pail; score the rest and put, rind side up, in middle of the beans. Mix a teaspoonful of salt, a tablespoonful

Household Management | **Iceless Refrigerator**

Pot Roast

Season the meat with salt and pepper, brown on all sides over a flame, and put in a stone jar, dry, no water whatever. Cover tightly. Put the jar in a kettle of hot water. Boil fifteen or twenty minutes. Place in a cooker for six hours. Even tough meat becomes tender and the juice at the bottom is very rich.

each of sugar and molasses, just a dust of mustard, a half teaspoonful of baking soda and a cupful of boiling water. Add enough more water to come to the top of the beans. Cover, and boil ten minutes; then put in cooker. In the morning reheat for ten minutes, return to the box and about half past five in the afternoon take out, sprinkle a tablespoonful of sugar over the top, leave off the cover, put in hot oven for half an hour.

The Iceless Refrigerator

An inexpensive refrigerator, or milk cooler, consists of a wooden frame covered with Canton flannel. Wicks made of the same material as the covering rest in a pan of water on top of the refrigerator, allowing the water to seep down the sides. When evaporation takes place the heat is taken from the inside, consequently lowering the temperature. On dry, hot days a temperature of 50 degrees can be obtained in this refrigerator. The following description will aid in the construction of this device:

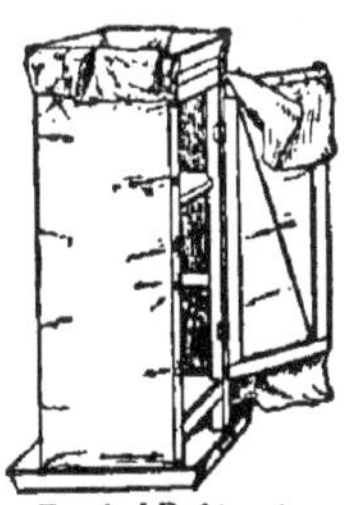

Finished Refrigerator

Make a screened case three and a half feet high, with the other dimensions twelve by fifteen inches. Place two movable shelves in the frame, twelve to fifteen inches apart. Use a pan twelve inches square on the top to hold the water, and where the refrigerator is to be used indoors have the whole thing standing in a large pan to catch any drip. The pans and case may be painted white, allowed to dry, and then enameled.

A covering of white Canton flannel should be made to fit in the frame. Have the smooth side out and fasten the covering on the frame with buttons or hooks, arranged so that the door may be opened without unfastening them. This can be done by putting one row of hooks on the edge of the door near the latch and the other just opposite the opening, with the hem on each side extending far enough to cover the crack at the edge of the door, keeping out the warm, outside air and retaining the cooled air. The covering will have to be hooked around the top edge also.

Two double strips, one-half the width of each side, should be sewed on the top of each side and allowed to extend over about three inches in the pan of water. The bottom of the covering should extend to the lower edge of the case. Place the refrigerator in a shady place where air will circulate around it freely. If buttons and buttonholes are used, the cost should not exceed eighty-five cents.

Canning Made Easy by the Cold Pack Method

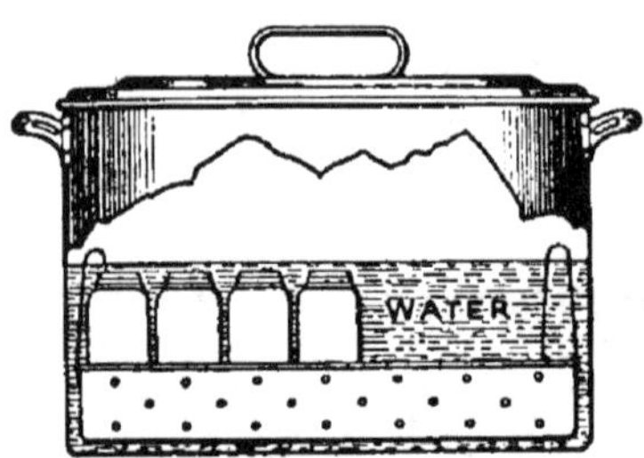

Here is the standard method for canning fruits and vegetables, as advised by the United States Department of Agriculture.

Wash the jars; wash the rubbers and test them for quality. Set the empty jars and rubbers into a pan of water to heat and keep hot.

Use only fresh, sound vegetables. Wash, then place in colander; blanch by setting in a tightly-covered vessel of boiling water or steam for from five to ten minutes for beans, one and one-half minutes for tomatoes, five minutes for sweet corn and beets.

Remove from the boiling water or steam and plunge quickly into cold, clean water for a moment only. Remove and pack immediately into hot jars; add hot water and seasoning, place rubbers and tops of jars in position, screw them down tight and and then give a quarter of a turn back. Place jars on a false bottom of lath or wire placed in the washboiler, submerge them two inches. Put the cover on the washboiler and let the water boil hard for two hours for beans, twenty-two minutes for tomatoes, three hours for sweet corn, and

one and a half hours for beets. Start counting time when the water begins to boil. Remove the jars, tighten the covers, invert to cool and examine them for leaks. If leaks are found, change rubbers and boil again for ten minutes.

Fruit that is to be canned by this method should be sound and fresh. Wash it carefully and remove any parts that are decayed. Place all fruit, except berries, in a square of cheesecloth or wire basket and dip into boiling water for one-half minute for peaches, and one and one-half minutes for apples and pears. Plunge for a moment into cold water. Skin the fruit if necessary and leave whole or cut as preferred.

Pack the fruit in hot jars, fill the jars with hot syrup or boiling water and proceed according to directions for vegetables, allowing the water to boil sixteen minutes for peaches and twenty minutes for apples and pears.

Make syrup according to formula 1, which requires three quarts of sugar and two quarts of water. Boil until the sugar is dissolved, skim off impurities and keep it hot; or use formula 2, which is not so sweet. This takes two quarts of sugar to three quarts of water.

Imperfect jars, caps or rubbers are responsible for the greatest percentage of spoilage. It is quite necessary to provide a good storage place which should be both cool and dark. If exposure to the light is unavoidable, wrap the jars in brown paper, or store them in the boxes in which they were bought.

Household Management   **The Kitchen Range**

Commercial canning outfits will in every case expedite the work, and should be found in every home having a large quantity of food to preserve, or in community kitchens. The get-together spirit is necessary for a movement of this sort. It has been demonstrated that community kitchens produce the maximum of food conservation and offer the most economical means for canning, five women in a community kitchen doing the work of from forty to fifty families.

Care of a Kitchen Range

The most important thing is to keep the range well cleaned out—top, sides and underneath—at the cleaning door. Familiarize yourself thoroughly with the uses and duties of the damper and different drafts, so you can obtain the heat and different temperatures necessary to cook and bake.

A chimney that is open to rain and snow permits the moisture to run down the flue, mix with the soot, and get into the back-flue of the range; this results in a mixture of soot and moisture, which forms an acid that eats out the range flue, thus doing away with half its life.

A Clean Chimney Should Draw

A range is very often condemned because it will not draw well. As a rule, upon investigation, you will find the flue clogged up with soot; this is a job for the men folks, although the cleaning up, a not overly pleasant piece of work, generally falls upon the housewife. In some cases the soot may be safely burned out. Sometimes a tinner can correct the faulty draft, but not often.

Not Too Much Fire, Please

Another common error in operating a range is the filling up of the firebox to the top. This means wasting fuel, overheating the range and burning the plates. This, in time, will warp the lids and make the top uneven. The best way is to use just enough fuel to keep the fire burning brightly—a fire-box about half full. The air is drawn up through the fire, and as it becomes heated it mixes with the fresh fuel-gas, and what would be a black smoke is immediately ignited to a bright flame. This flame is burned and utilized and gives the oven the best heat the fuel is capable of giving. Thus a little fuel given more frequently affords more heat, with a saving in many instances of at least one-third in the fuel bill. When you understand the relative value of the damper and draft, you can keep the temperature of your oven to any degree desired and for an indefinite period.

Shake from the Bottom

Try not to poke the fire from the top; shake it with the shaker. See also that the ashes are not permitted to accumulate until they entirely fill the ash chamber, as they choke off the free circulation of air to the bottom of the grate.

It is safe to say that two or three minutes of time, now and then, in looking to the proper condition of the range, will save many a baking, not to speak of the housewife's valuable time and disposition.

Care of Kerosene Lamps

How to Avoid Trouble

Much kerosene trouble lies in the lamp or wick rather than in the oil. If the flame is uneven the fault is usually in the trimming; or the burner may be coated with carbon scales which crumble down under the wick, and by their pressure cut off the flow of kerosene. Such a burner should be thoroughly cleaned and boiled, then dried and brushed clean of lint.

A lamp should never be allowed to burn dry. Besides destroying the wick, it chokes the burner with unconsumed carbon and renders a thorough cleaning necessary. With a low-test oil, too, there is considerable danger in permitting the reservoir to become low, as the heat evolved converts the oil into gas as rapidly as there is room for it to form, and the larger accumulation soon becomes a serious menace.

Trimming the Wick

A wick should be trimmed by scraping with a dull knife, rather than by cutting; then the carbonized fibers will all be removed.

If the wick is boiled in vinegar and then dried thoroughly, it is relieved of its tendency to smoke.

Never blow down a chimney to extinguish the flame. Blow across the top, or, if that is inefficient, blow against the hand held near the top. It may save a cracked chimney.

Watch the Temperature of the Chimney Bottom

The chimney of a lamp should never become hot at the bottom part, below the flame. If it does it indicates a hot burner and is a danger signal. Usually one can remove a chimney from a lighted lamp by grasping it at the bottom.

Lamps should be filled in the morning; then there is little space for gas to accumulate during the day. If this is neglected do not fail, before lighting the lamp in the evening, to remove the burner and agitate the wick enough to drive the accumulated gas out before lighting. This may save an explosion. For a similar reason lamps should be set away, during the day, in a moderately cool place.

Where good 150-test water-white oil is used, many of these precautions are unnecessary. They are recommended because of the variation in oil and its impurities. Some grades of oil, occasionally sold for the best, are much more highly charged with dangerous gases than others.

How to Carve Meat and Poultry

Volumes have been written about the "preparedness" for housekeeping, with which the bride should fortify herself; but one rarely sees it suggested that the prospective "head of the house" has some things to learn, and one of the most important of

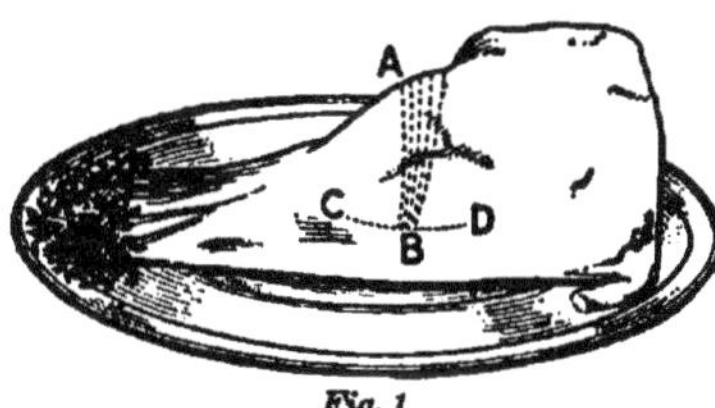

Fig. 1

these is skillful carving. Meats and poultry are now so high in price that perhaps they will be more carefully considered than they once were. Whatever the price, neat and intelligent carving adds not only to the attractiveness of the food, but prevents waste.

Without good tools, acceptable carving is an impossibility; so every housekeeper should possess one, if not two, carving sets of standard make. If there is but one set, let it be of medium size. A better plan is to have one fairly large set, and another of smaller size, known as game or steak carvers. The knives should be kept in the best condition, for a dull one is an abomination; therefore, it is no less important to be able to use the steel than it is to use the knife. Every once in a while the edges of the knives should be ground, for nothing dulls sharp edges so quickly as hot fat.

It helps greatly to have the meat put into shape before cooking by means of twine and skewers; a set of the latter, in steel, should be found in every kitchen. A large platter is also essential, for the carver must not be cramped for room.

While beef is probably served more than any other meat, we have prepared no diagrams of it, because the cutting varies in different localities, and what is known as a sirloin steak in one town, may, a hundred miles away, be called a porterhouse. Whatever the cut or the name, there should be one invariable rule: The meat should always be cut across the grain, not with it. With the right kind of knife the meat can be neatly cut away from the bones and should, in case of a roast, be cross-cut in thin slices; if a steak, it should be cut

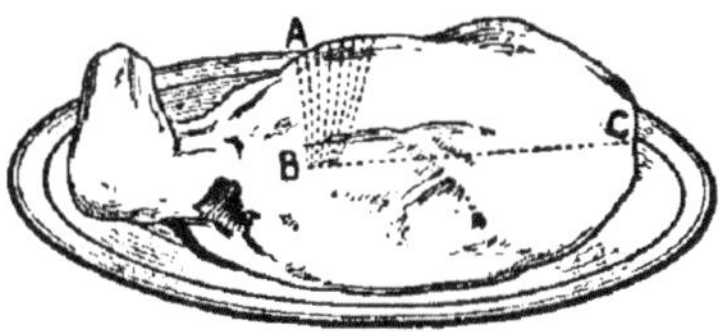

Fig. 2

in neat strips. Since people's tastes vary, be sure to ask their preference when serving.

A leg of mutton or lamb should be dished as it appears in Fig. 1, with the unsightly end of the bone covered with parsley or watercress. Stick the carving fork firmly into the small part of the leg and cut wedge-shaped pieces from A to B. Remove the first few

How to Carve   Household Management

slices and continue to cut in this direction until you reach the larger bone; then cut from C to D, loosening the slices afterward.

When a shoulder of lamb or mutton is roasted whole, carve as in Fig. 2, in wedge-shaped slices from A to B, then across from B to C.

The carving of poultry offers the most complications. Fig. 3 shows a turkey in the proper, therefore the most convenient position for the carver, and the directions which follow may be applied in a general way to the carving of all poultry. Insert the fork just above the spot marked G, placing it so that it holds firmly. Cut the leg from A to B, and the wing from C to D on the side farthest from you, first. If both sides are to be cut the same day, tip the bird

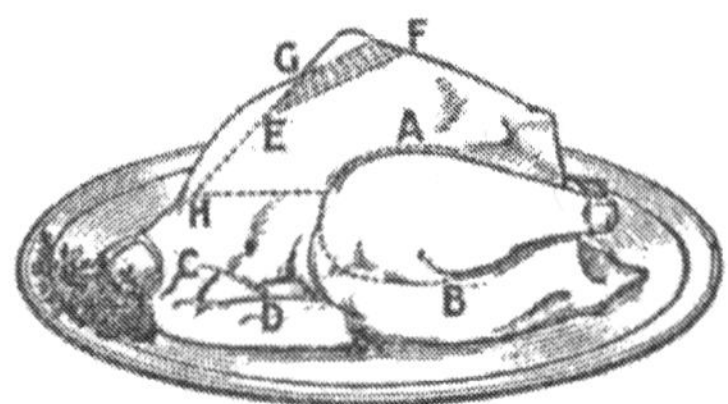

Fig. 3

over and cut the leg and wing from the side nearest to you. Cut thin slices from the breast, E to F, remove the wishbone by cutting from G to H, and separate the carcass by cutting across the line marked H. Cut the legs into second joints and drumsticks. Unless there is a preference, serve a portion of white and a portion of dark meat to each person. If the turkey is to be stuffed, instead of sewing up the opening, in-

sert a toothpick on each side and lace back and forth with a string. This is easily removed when the turkey is cooked. The opening should be large enough to insert the tablespoon with which the stuffing is served.

A slice of ham seems a homely thing to manage, but countless mistakes are made in serving it. It is often cut lengthwise, then cut into

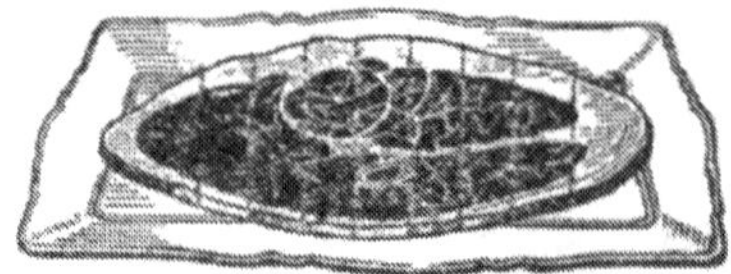

Fig. 4

smaller pieces and the choicer side served to those who like fat, and the other side to those who do not. The proper way is to cut the ham in vertical strips as shown in Fig. 4, so that each person receives his share of the part which is choice and that which is less so.

Bountiful serving has, of necessity, become a thing of the past. Along with skill in carving there should be a knowledge of the tastes and capacity of those who are being served. Small portions with a second helping are preferable to heaped-up plates, the contents of which are not consumed.

While the daughters of the home are being trained in housewifely arts, the sons should have their turns at the carving and serving. Their efforts in this direction should have the same recognition as the daughters' attempts at cooking and sewing, for they, too, have their place in the well-managed home.

One Solution of the Maid Problem

Through accident and illness the father's health failed, so there were two reasons for doing without a maid servant, viz., limited income and paucity of girls willing to serve in any household except that of their parents.

The mother had been taught in childhood many ways of economy, and had imparted this knowledge to her children, little by little, as occasions were presented.

Doing Without

There were two girls and five boys, but the older daughter taught school for a year, helping with fertile brain and deft hands during vacations; then she married. The mother and her children were very intimate in their daily life, and the daughter left at home assumed one duty and then another, finally including the laundry. Then the boys, who were accustomed to cutting, piling and bringing in the wood, mowing the lawn, etc., had to wash dishes. Sometimes the mother was ill; then the big sister gave orders, the boys (all younger than herself) obeying with more or less readiness.

What the Boys Did

It became the duty of one boy, of twelve, to wipe the stairs and banisters and oil the floor. The lad who, during his first years of voluntary locomotion, had been always at his mother's elbow to upset things when she was "mixing," made delicious cakes and doughnuts; and once, in a week's experience of "home without a mother" said: "I'm sick of father's bread. I can make bread!" (Ah! What stupendous power in the determination expressed in the four letters, "I can!") He made bread—good bread—for years, becoming an expert. It didn't spoil his muscles for throwing a ball, nor his brain for study.

By being much with mother and sister, they all soon knew how to make coffee, fry potatoes, cook meat and make pancakes, and one fourteen-year-old chap made cake-fillings.

They could all sew on buttons, sew up rips, and press their clothes, although mother and sister attended to the more difficult jobs of mending.

The boys did errands, of course, saved journeys upstairs and down cellar, whipped rugs, washed windows and floors, and even tinkered up things sometimes. It is surprising how well a boy can clean and arrange a book-shelf, mantel or cupboard. He likes a little responsibility; it helps his self-respect.

But conditions change in all families, and there came a time when the daughter who had filled all gaps so quietly, cheerfully and efficiently, chose her mate and went to grace a new home-nest. Then, indeed, "life was real, life was earnest" for the five boys. They washed, sometimes, and even ironed some things. A boy can iron flat pieces beautifully, and likes it better than preparing vegetables. But these boys did that work, too.

The Window Garden   **Household Management**

Making a Window Garden

It is late afternoon in December. The dark gray clouds, the cold drizzling rain and the wind sighing through the naked tree-tops oppress me. But as I step from gloom without into my living-room, sadness changes to joy—I have entered a garden, a small garden, to be sure, but a real garden, nevertheless. Before me is a large bay window framed by a gorgeous mass of color and foliage. Begonias, geraniums, coleus, heliotropes, fuschias, sweet alyssum, scarlet sage, impatiens and primroses shed their gladness and beauty about them, while the tender ferns, palms and climbing ivy cheer me.

Start Late in October

I started my winter garden about the end of October, before Jack Frost had become a constant caller. The first thing done was to accustom my plants to their new environment. Therefore I selected the choicest and most vigorous plants from the beds and borders about the house and put them into pots and boxes, according to their sizes, being careful to place some coal ashes in the bottom of each to prevent the soil from becoming sour.

The soil used for this work consisted of a mixture of one part well-rotted manure to three parts of the best garden soil I could find.

Selecting a sheltered but sunny corner of the front porch, I kept the plants there about ten days, during which time they received the benefit of full sunlight and careful watering. Here they made a good root growth and were better able to withstand the sudden change from the garden to the house.

The bay window of our living room opens to the south, southwest and southeast. I have noticed that the latter exposure suits the plants best, as they receive the brightest sunshine from ten in the morning to one in the afternoon. When I measured the width of the sills I found them a scant six inches. I widened them to eighteen inches, using chestnut lumber (which stands dampness well) planed and painted two coats of olive green. This extra width of sill permits the growing of several rows of plants, the small plants nearest the glass.

No Expense to Speak of

Instead of buying expensive saucers and pans to retain the excess water, I set the pots and boxes upon a three-inch layer of sphagnum (florist's) moss, packing the moss well about them until only the rims were visible. The moss not only absorbs all the drainage water, but prevents the pots from drying out quickly. Another thing in favor of the moss is its attractive appearance. A very natural effect is obtained by tacking strips of bark on the front and back edges of the sill (see cut) and allowing ivy or periwinkle to hang over.

In arranging my plants for the most artistic effect, I trained the climbing vines, such as smilax and morning-glory, along the sides of each window-sash, in order to hide the woodwork as much as possible. These

Household Management | **Saving Heat**

windows are opened only upon the mildest days. The room is aired by opening the windows and doors of adjacent rooms, while the plants are covered with muslin or newspapers to protect them from drafts.

The temperature of the living-room varies between 60 degrees and 75 degrees, although I aim to keep it as near 65 degrees as possible. A pan of water placed on top of the stove fills the air with enough moisture to prevent too rapid drying of the leaves.

There is a great danger of over-watering plants before the roots are able to withstand it. Only sufficient water should be given to prevent the foliage from wilting. Later on, in late winter, when the sun becomes brighter and the pots are filled with young roots, water may bo given freely.

How to Save Heat

Did you ever notice a buzzard sitting crouched on a chimney-top on a chilly morning? And did you ever think why he did that? To get warm, of course!

Exactly; that shows what a lot of heat goes to waste. Now, if we can catch some of this heat after it leaves the stove, and set it to warming an upstairs room, we are saving fuel; and fuel, you know, ranks next after food in importance. So here's a little scheme that I've worked out.

Go to the nearest plumber and ask him to sell you an old leaky kitchen tank. These tanks cannot well be mended and the junk-man won't buy them because they are galvanized iron; so, usually, it's possible to buy them very cheaply and in fair condition.

Cut a large hole the size of a stovepipe in one side of the top, and a similar hole in the opposite side of the bottom; this is something of a job, but your blacksmith can do it with a boilermaker's cold chisel. Set this boiler in a second-story room, directly over a stove on the first floor, run pipe through the floor (be careful to put in a burnt-clay pipe crock) and up into the boiler, nearly to the top. Then set a second pipe, as the drawing shows, starting near the bottom and going up to the chimney. There should be a damper just above the top of the tank to control the draft.

When you start a fire in the stove, the smoke has to double back on itself, exactly as it does in a steam boiler. In that way a lot of waste heat bumps against the iron and is radiated into the bedroom, instead of steaming through the chimney.

Make the joints very tight where the pipes pass in and out, else the room will be smoky and the air dangerous to breathe. A little bit of asbestos cement, which the plumber can furnish, will do for the joints. Go over the whole affair with stove polish, and your heating plant is ready to use. It is surprising how much heat is saved.

Paper-Hanging Made Easy

In certain sections of our country, "papering boards" were at one time considered as necessary a part of the household equipment as quilting frames. Both fell somewhat into disuse, but the pendulum swings both ways, and many householders are now finding it expedient to do their own papering.

Fig. 1

The work requires few tools and appliances, the first requisite being two boards seven or eight feet long and about ten or twelve inches wide. Place them side by side with the ends resting on a table about three feet high. This will make an excellent paste table. The other articles needed are a paste brush, paper-hanger's smoothing brush, a seam-roller and a pair of shears or a straight-edge and trimming knife.

Paper the ceiling first. Find the length of your room, unroll the ceiling paper, pattern up, cut and match

Fig. 2

enough for the whole room, being sure to leave the strips three or four inches longer than the length required, so that when hung two inches will come down on the walls at each end. Then turn the paper, pattern down, and it is ready to paste. Beginning at the left, apply the paste evenly. After you have pasted half the length fold it over as shown in Fig. 1, being careful to see that the sides are perfectly even. Then paste the other half and fold it toward the center, just as you did the first half. Now you have a full strip pasted and ready to trim.

Draw the paper toward you about three inches from the edge of the table and proceed to trim, commencing at the right. See Fig. 2. The great advantages of trimming after pasting are that the paste is more evenly distributed at the edges and it takes less trim.

Fig. 3

As a guide to hanging the first strip of ceiling paper properly, drive a nail in at each end of the ceiling sixteen inches from the side walls, chalk a piece of cord with charcoal or chalk, and tie to the nails, being sure that it is drawn tight, then take hold of the center of the cord, and pull it down and let go; the cord will strike the ceiling and leave a chalk line, as shown in Fig. 3. Take the first strip you have pasted and trimmed, unfold the end to your right and let the other end, which is still folded, hang over a roll of wall-paper which you hold in your left hand. See Fig. 3. Guide the paper along the chalk line, at the same time pressing it flat on the ceiling

Fig. 4

with your hand as you move along, and smoothing it out with a smoothing brush. When half the strip is on,

unfold the other half and continue to the end.

When the ceiling is finished, cut the side wall-paper, matching it and leaving about four or five inches to allow for any waste in matching. The upper ends of the strips will be uneven, but these will be covered by the border. The lower ends, which will stop at the baseboard, are either trimmed off with a base trimmer or a pair of shears. When using shears, paste the paper down in place close to the top of the baseboard. Use the back of the shears for marking by running them over the top of the paper where plaster and baseboard meet, lift the paper a little and trim along where you have marked and smooth down again with a smoothing brush.

Fig. 5

When hanging the side wall-paper, see Fig. 4, commence at any door, as corners are often irregular. This will assist you in hanging the first strip straight; then continue around the room until it is finished. The short pieces can be used over the doors and under the windows.

The border should be hung last, see Fig 5, and may be cut in five or six pieces, making it easier to hang. Paste, fold and trim the border just as you have done the paper used for side walls and ceiling.

A substitute for the usual wheat paste may be made one part by weight

Making Comforts   **Household Management**

of dried glue in ten parts by weight of water, melted in a glue pot surrounded by boiling water. To this should be slowly added four parts of laundry starch stirred up with ten parts of warm water. This produces a perfectly smooth paste, the consistency of which can be varied by changing the proportion of water used. If the paste is to be kept for any length of time, some preservative, such as oil of cloves, oil of winter green, or oil of sassafras can be added.

If wall-paper is marred or broken in places, it may be repaired by cutting from a remnant of the same paper figures or groups of figures corresponding to those that need repairing. The outline of the design should be carefully followed in cutting, and the whole should be matched and pasted exactly. If the new paper used to repair is hung in the sunlight and faded a bit, the mended places will escape notice. The colored crayons used by school children can, if they match, be rubbed over small breaks. The paints which come in the toy paint boxes will serve the same purpose.

Comforts from Stocking Legs

Warm comforts for the beds may be made out of old stocking legs. Cut five-inch squares out of unbleached

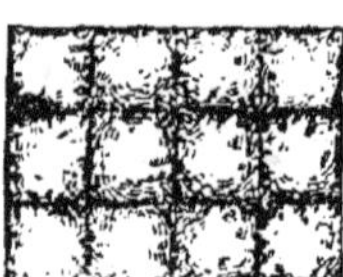

Fig. 1

muslin, calico or any pieces you happen to have; lay a tuft of raw cotton upon each square, then cover this with another square cut out of stocking legs. These should be cut slightly larger, and should be basted around the edges so as to form a slight puff, as shown in illustration No. 1. The squares may then be joined on the machine, and if several

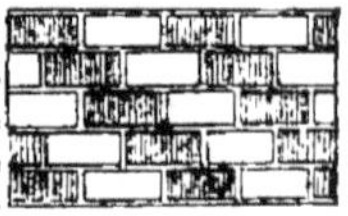

Fig. 2

people work together it goes very quickly. The comfort may be lined with figured chintz. Illustration No. 2 shows a brick pattern in which the pieces were oblong in shape, the light pieces are made of tan stockings, the dark ones of black.

The Typhoid Fly

The diagram herewith shows the number of infant deaths in Pennsylvania for one year, the greatest number occurring during the fly season. Evidence exists that flies aid in transmitting cholera, typhoid, diarrhea, dysentery and infantile paralysis. As flies breed chiefly in manure and garbage it follows that clean stables, clean houses and clean yards are the best preventives.

The Simple Art of Making Cheese

So much has been written about the use of skim-milk, buttermilk and whey for feeding calves and pigs, that it seems to have created an impression that those products of the dairy are not suitable for human consumption. It all depends on how they are used, to be sure. All those products, as well as whole milk, can be made into different kinds of cheese, all of which are palatable and have a high food value.

The Cheese in Congress

Making cheese from sour skimmed milk or buttermilk is not a difficult thing to do—at least not difficult enough to necessitate the expenditure of $64,000 to provide for government agents to go about through the country telling the people on the farms how to make cheese. The plan was gravely considered in Congress recently. One of the representatives, a practical farmer, remarked that there is not a ten-year-old farmer's girl in the country who does not know how to make cheese from sour milk by placing the clabber in a cloth bag and hanging the bag on a hook to drain over night, removing the curd from the bag next morning and seasoning with salt; after which the cheese is ready for use.

Not much needs to be added to the good common sense and plain directions of this western Congressman. He did not suggest, however, that the milk would be curdled sooner if heated to about 100 degrees, although he did prove that he knew all about the process even to eating the product, by saying that the quality of the cheese is improved by the addition of cream or butter.

In some parts of the country a kind of cheese is made by evaporating whey. The whey, not too sour, is boiled in a vessel, using a slow fire and being careful not to burn or scorch the whey. As the albuminous matters are coagulated, they are removed from the kettle and the boiling continued until the whole mass becomes syrupy. The albuminous matter is then returned to the kettle, the vessel removed from the fire and the contents mixed rapidly to form a thick mush. Some cream is added and the material pressed into molds, after which it is ready to be eaten or sold.

Cream Cheese

Cream cheese can be made from whole milk and the work of making it is often a welcome change from buttermaking. To make a cheese weighing three to four pounds, heat five gallons of milk to 100 degrees, stir into it a tablespoonful of salt; after dissolving one-half of a rennet tablet in a little cold water, add this to the milk. Remove it from the fire and allow it to stand five to ten minutes, when it will become curdled. The curd is carefully cut lengthwise and crosswise with a sharp, flat stick and kept standing a little while longer to get the whey to separate naturally.

Having dipped off most of the whey, the remainder is poured slowly into a strainer to drain out still more whey, and cut into inch cubes, salting them until they seem a trifle too salty. It can be pressed in a round or square hoop, as you may choose, though there will be less waste in the round. Such a hoop may be a round tin can or one of wood with a follower in both ends of the hoop to hold the curd, and must stand on a draining board. It must be lined with a cloth long enough to twist over the top of the curd before the top follower is put in.

The curd is pressed slightly for eight to twelve hours by placing a weight on the top follower. During pressing, remove the cheese occasionally to trim it and shape it and get its final cap on. Stand the pressed, green cheese in a cool place until dry and then dip it quickly into hot beef tallow or paraffin. Keep it in a cool, dry room not warmer than 60 degrees.

Easily Prepared Dishes

Dock Greens

The curly leaf dock which grows wild in undisturbed fence corners is even more wholesome and palatable than spinach.

Chicken Pilau

The rooster is the foundation for the famous dish of chicken and rice which he French call "pilau." Make it by oiling the chicken until the meat comes off the bones. Season well, and add enough rice to thicken the stew. Cook until the rice is soft. The stew should be thick enough to eat with a fork. Served with a green salad or stewed fruit, this makes a complete meal.

Sausage and Hominy

When you want a hearty dish for supper, spread cold boiled hominy over the bottom of a baking dish, moisten it with milk, season with a little salt and pepper and slice over it some sausage. Bake in a rather hot oven for twenty minutes, or until the sausage is well cooked, and serve hot.

Household Recipes   Easily Prepared Dishes

Potato Pancakes

One cupful of riced potatoes, half a teaspoonful of salt, one egg beaten, one tablespoonful of flour, quarter of a cupful of milk. Mix the above ingredients in the order given, beat thoroughly and bake on an aluminum griddle to save fat.

Pennsylvania Pepper-Hash

Pepper-hash is a good old Pennsylvania relish which is delicious served with oysters, codfish balls, cold meats, etc. It is easily made and keeps indefinitely. The proportions vary, but one red and one green sweet pepper to about a pound of cabbage is a good rule, more cabbage may be added if desired. Also one teaspoonful each of celery and mustard seed, two teaspoonfuls of sugar, a tiny piece of horseradish and enough vinegar to cover the mixture. Remove the seeds from the peppers, cut them and the cabbage into shreds, then chop fine. Sprinkle with a tablespoonful of salt, stand aside for an hour, then drain all water away. Add other ingredients, mix, then pack down hard in a crock or glass jar, pour on the vinegar, diluted if necessary, and cover, but do not seal. In summertime, a few nasturtium leaves may take the place of the horseradish, which will prevent fermentation. During the winter, if sweet peppers are not available, one or two cucumber pickles and a red pepper-pod (containing the seeds) will make a relish which tastes as good as it looks.

Saving Cooked Cereals

No cooked cereal needs to be wasted. Even a teaspoonful can be added to the batter for hot cakes or muffins. A small quantity of oatmeal may be used to thicken a soup.

Grape Foam

Grape foam, which consists simply of the white of one egg beaten stiff and added to two tablespoonfuls of grape juice, is a delightful drink for an invalid. It will quench the thirst of fever and prove nutritious as well. Orange albumen is prepared in the same way, using orange juice instead of grape juice.

Italian Polenta

Polenta, an Italian way of serving corn-meal, makes almost a meal in itself. Put slices of cold mush in a baking dish, cover with a cupful of sliced onions that have been fried in ham or bacon fat; over these pour two cupfuls of canned tomatoes and cover all with a cupful of grated cheese; bake until the cheese is melted and slightly browned.

Buckwheat Spice Cakes

Buckwheat spice cakes offer an excellent way in which to use up your surplus buckwheat flour. Use three-fourths of a cupful of sugar, three tablespoonfuls of fat, one egg, one cupful of sweet milk, one cupful of buckwheat flour, two teaspoonfuls of baking-powder, one-half teaspoonful of salt, three teaspoonfuls of cinnamon, one-quarter of a teaspoonful of cloves, three teaspoonfuls of vanilla. Mix like an ordinary cake. Bake in muffin tins in a moderate oven.

Goldenrod Eggs

Goldenrod eggs make a wholesome and attractive dish. Boil until hard as many eggs as are needed. Cut the whites in narrow strips and place on slices of toast. Pour over this a white sauce, and sprinkle the egg yolks, pressed through a potato ricer, over the top. White sauce is made by rubbing to a paste two tablespoonfuls each of flour and butter, then pour on gradually (stirring all the while) a cupful of hot milk. Bring to boiling point and season with salt and white or cayenne pepper.

Spinach Properly Cooked

Spinach, "the broom of the stomach," ought to be on the spring dinner-table several times a week, and asparagus the other days. Wash well some young spinach leaves, shake off the water, put in a hot stew-pan. Shake and stir until the juice of the leaves comes out and they seem to be melted, then cover and cook slowly for twenty minutes. Cooked thus all of the valuable vegetable salts are retained. It may be eaten with butter or vinegar, as preferred. Slices of hard boiled egg add to its appearance and flavor.

Roasting in Paper

The French have a way of making a year-old fowl as tender as one that is half the age, by wrapping it in brown paper before it is put into the oven, and allowing it to cook in this envelope until it is nearly done. The paper retains the juices, allows the fowl to cook slowly and evenly and grow tender before the outside is browned. At the last the paper is removed long enough to bring the surface of the fowl to the desired color. Young mutton can be brought to lamb-like tenderness in the same way, and roast veal may be cooked thoroughly without the hard outer crust which sometimes spoils this meat when roasted.

Barley Cereal

Barley is one of the grains recommended as an occasional substitute for wheat, and though it has hitherto been principally used for soups and gruel for babies and invalids, it is nourishing and palatable; it builds muscle and makes bone. To use as a cereal, wash four tablespoonfuls of the barley and put it into a double boiler with a quart of cold water and a teaspoonful of salt. Do not stir but let it cook for two hours. It should swell to four times its original size, and be like jelly. Raisins can be cooked with it, if desired. Serve with cream.

Skim-Milk Bread

Skimmed milk used in bread in place of water adds as much protein to a pound loaf of bread as there is in one egg. It gives a softness of texture to bread that adds particularly to the palatability of graham or bran bread.

Potato Muffins

Potato and corn-meal muffins are delicious. To make them use two tablespoonfuls of fat, one tablespoonful of sugar, one egg well beaten, one cupful of milk, one cupful of mashed potatoes, one cupful of corn-meal, four teaspoonfuls of baking-powder, one teaspoonful of salt. Mix in the order given, put in muffin pans and bake forty minutes in a hot oven.

Mixed Breakfast Food

A delicious and nutritious breakfast food can be easily made by mixing together equal parts of bran, oatmeal and corn-meal. A few raisins chopped finely and mixed thoroughly into the porridge just before removing from the fire will add greatly to the flavor and no sugar will be needed. Stir the cereal gradually into boiling water and allow it to cook slowly for about fifteen or twenty minutes. If there is any left over, it can be fried like cold mush.

Boiled Rice

Boiled rice should be prepared thus: Wash a cupful of rice in several waters. Then drop it slowly into a quart of boiling water, salt to taste,

Household Recipes · Easily Prepared Dishes

boil for fifteen minutes, cover and place on the back of the stove where it will finish swelling without burning. Do not stir. Arrange the rice in a ring on a hot dish and place in the center any meat hash, stew, creamed fish or chicken. Nothing more is needed for dinner but a dessert of stewed fruit or a green salad.

Mock Cherry Pie

A good substitute for the cherry pie that you will want to serve on Washington's birthday is made by cooking together one cupful of cranberries, cut in half, a half cupful of seeded raisins, a half cupful of water and a dust of flour. When the cranberries are done, add one teaspoonful of vanilla and sugar to taste. Line a pie plate with good pastry, put in the filling, add the top crust and bake.

Boiled Soy-Beans

Boiled soy-beans form the basis of many nourishing dishes. As thorough cooking is essential, follow these directions: Soak two cupfuls of dried beans for twenty-four hours. Drain, add boiling water, one tablespoonful of salt and a pinch of baking-soda and simmer four hours. To cook the beans more economically, boil for one-half hour and finish cooking in the fireless cooker. The beans may also be cooked with one-half pound of fat salt pork, cut in cubes.

Pop-Corn Cereal

Allow the children to make their own crisp breakfast food. Pop-corn, fresh, warm and crisp, smothered in whole milk and dusted with sugar, is delicious. The children will enjoy this food because they can prepare it and eat it perfectly fresh. Pop-corn pudding will also please the children.

Run two cupfuls of pop-corn through the food grinder, add two cupfuls of milk, one tablespoonful of butter, a half cupful of sugar, an egg beaten light, and a sprinkle of salt. Bake half an hour and serve warm.

Brown Sugar Pickles

If the sugar shortage prevented you from making your favorite kind of sweet pickles last autumn, try these: Cut one dozen plain sour pickles in slices one-fourth of an inch thick. Place in layers in a jar, sprinkling each layer with brown sugar, a few cloves and a little stick cinnamon. Stir or shake every day for one week, when they will be ready to use. One pound of brown sugar is sufficient for a dozen large pickles, and the syrup formed on the first jar may be used to start a second jar of pickles.

Apples and Dates

Apples and dates make a good combination requiring no sugar. To prepare them, steam until tender in a covered pan, one and one-half quarts of sliced apples and the grated peel of one lemon with one-half cupful of water. Add one-half cupful of chopped dates, simmer the fruits together for six minutes and serve cold.

Corn Fritters

Corn fritters are good for a change. To make them, run through the meat-chopper a can of the corn you put up last summer. Add two eggs, well beaten, just enough flour to hold the mixture together, and salt to taste. Beat thoroughly, then drop by spoonfuls into a pan in which there is hot fat to keep them from sticking. Fry a nice brown on each side and serve hot.

Re-Worked Butter

To make re-worked butter, use one pound of butter, one pint of rich milk, one tablespoonful of gelatine and salt to taste. Cream the butter, as for cake, then dissolve the gelatine in a little of the milk, heat the rest of it, pour it over the gelatine. When this becomes lukewarm, pour it slowly over the butter and work in a small churn or beat with an egg beater until well mixed, smooth and thick. Drop by the spoonful in a shallow dish and keep in a cool place.

Boston Brown Bread

Boston brown bread is a good standby. To make it, use one cupful of white flour or bread-crumbs, one cupful of corn-meal, one cupful of Graham flour, one teaspoonful of salt, three-quarters of a cupful of molasses, one and three-quarter cupfuls of sweet milk, three-quarters of a teaspoonful of soda, one teaspoonful of baking-powder. Sift the dry ingredients together then stir them into the liquids. Fill well-greased cans two-thirds full, cover tightly and steam for four hours. Good either hot or cold.

Orange Marmalade

If you have reason to think that your jelly supply will not last until spring, try this recipe for orange marmalade: Use three oranges and two lemons, or for the bitter flavor, one large grapefruit, one orange and one lemon. Squeeze out all the juice and put all the skins through a meat-chopper. Put both together and add three times as much water as the entire quantity of juice and skins. Let stand over night; next day boil for ten minutes, then measure and add an equal amount of granulated sugar. Boil from one and one-half to two hours and put up as any other marmalade. This quantity makes about twelve glasses. If you can get oranges for a fair price, the marmalade will be very inexpensive, and will be a wholesome and delicious sweet for your table.

Measuring Hard Butter

To measure butter without softening it: If half a cupful is needed, fill a cup half full of water, then add pieces of butter until the cup is full. If a cupful is wanted, repeat the process.

Corn Bread With Rice

Corn bread with rice is an economical dish, requiring two cupfuls of sour milk, one teaspoonful of soda (scant), two cupfuls of boiled rice, one cupful of corn-meal, one tablespoonful of shortening. Combine the ingredients in the order named and bake in a greased pan until firm. Serve from the pan with a spoon.

Scotch Oat Crackers

Scotch oat crackers are crisp and good as well as cheap. To make, take two cupfuls of rolled oats, a quarter cupful of milk, a quarter cupful of molasses, one and a half tablespoonfuls of fat, a quarter teaspoonful of soda, a teaspoonful of salt. Crush or grind the oats in the food-chopper and mix with the other ingredients. Roll out in a thin sheet and cut into squares. Bake for twenty minutes in a moderate oven.

Household Recipes   **Easily Prepared Dishes**

Soy Beans with Tomato Sauce

Soy-beans with tomato sauce are good. Prepare the sauce by cooking one cupful of stewed and drained tomatoes with a slice of onion, three cloves and a bit of bay leaf, and strain. Melt two tablespoonfuls of butter, add two tablespoonfuls of flour and when browned add the tomatoes gradually with one-half teaspoonful of salt and a little pepper. Cook thoroughly. To one cupful of tomato sauce add two cupfuls of boiled beans, and reheat.

Delmonico Potatoes

Delmonico potatoes make a good substitute for meat. Cut cooked potatoes into dice, place them in layers in a buttered baking dish, alternating with layers of thin, white sauce and shaved cheese. Allow about one-half as much white sauce and one-fourth as much cheese as potatoes. Sprinkle the top with buttered crumbs and bake in a moderate oven until the top is well browned. Cheese increases the food value of the potatoes, but the latter should be cooked before being combined with the cheese.

Louisiana Corn Bread

Louisiana rice corn bread: Rice improves either plain corn bread or corn bread made with eggs and milk. Use it in both. Corn bread made with eggs and milk is rich in protein and makes a good meal served with a little gravy. This recipe is not extravagant for the housekeeper who has an abundance of eggs and milk. Use three eggs, a pint of milk, one and a half cupfuls of cold boiled rice, and one and a half cupfuls of corn-meal, two tablespoonfuls of melted fat, a tablespoonful of salt, and a teaspoonful of baking powder. Beat the eggs very light; add the other ingredients in the order named; beat hard and bake in a shallow greased pan in a hot oven.

Chocolate Cornstarch

Chocolate cornstarch pudding properly made, is almost as good as ice cream. To make it, put into a double boiler one cupful of milk, one cupful of water, one-half cupful of sugar and one ounce of unsweetened chocolate. When boiling hot add four level tablespoonfuls of cornstarch which has been wet in cold water. Cook until it thickens and the raw taste of cornstarch has disappeared, then add one teaspoonful of vanilla and pour slowly into this the stiffly beaten whites of three eggs. Mix thoroughly, then pour into a mold to harden. Serve with a custard sauce made of three egg yolks beaten with three tablespoonfuls of sugar and added to one and a half cupfuls of hot milk. Cook carefully until it thickens; flavor with one teaspoonful of vanilla and cool. Chocolate should be used largely, as it is both a food and a dessert; it is rich in fat and starches, and a pudding or other dessert in which this product has a part will be welcomed on days when a meat substitute is served as the main dish for dinner.

A Toothsome Rice Dish

is prepared thus: Place hot boiled rice on a platter, cover with white sauce and garnish with sliced hard-boiled eggs and a little finely-chopped ham, if you have it.

Dried Vegetables

can be stored easily, shipped easily, and do not freeze in winter. Small quantities, too trifling for canning, can be saved by drying, and will be found palatable and often better than the canned product. In this season of plenty the wise provider will store large quantities of dried vegetables for winter stews and soups. If she should be pressed for time, and sugar happens to be scarce, she may put up ample supplies of dried fruit for preserves later, or to be used, after soaking over night, like fresh fruit for sauces and desserts. For full directions for constructing different types of dryers and drying fruits, write to the Food Administration, Washington, D. C.

Peach Souffle

is a sugar-saving dessert that is sure to please. To one quart of peaches, canned or fresh, use one-half cupful of honey or syrup and three eggs. Drain and mash peaches through a colander, add the honey or syrup and egg yolks well beaten. Mix thoroughly, then beat the whites of eggs until stiff and fold carefully into the peach mixture. Turn the whole into a greased baking dish and bake in a quick oven for six minutes.

Cottage-Cheese Pie

uses up surplus milk. It requires one cupful of cottage-cheese, one-half cupful of maple syrup, two-thirds cupful of milk, beaten yolks of two eggs, two tablespoonfuls of melted butter, one-half teaspoonful of vanilla, salt. Mix the ingredients in the order given, and pour into a pie plate lined with crust. When baked cool it slightly, cover with meringue, and brown in a slow oven. For the meringue use whites of two eggs, one-fourth teaspoonful of vanilla, four tablespoonfuls of maple syrup. Beat the egg whites until they are stiff, add the syrup gradually, and then the vanilla.

Plain Corn Bread

inexpensive, but good, is made by sifting together one cupful of yellow cornmeal, one cupful of wheat flour, two heaping teaspoonfuls of baking powder and one level teaspoonful of salt. Stir in enough sweet milk to make a rather stiff batter and beat it well. Pour the mixture into a well-greased pan and bake in a moderate oven for about a half hour. The mixture should be no thicker than pancake batter, and if properly made will make a moist, delicious bread.

New Recipes for After the War

Corned Beef Hash

is an appetizing dish, easily prepared. To make it, chop slightly—not too fine—a half can or six ounces of corned beef. Measure this and add a trifle more than the same amount of chopped cold boiled potatoes. Flavor to taste with onion, pepper and salt, and mix all together. Melt the fat in an aluminum omelet pan, put in the mixture, cook until the surface is browned, then turn and brown on the other side; or spread the mixture in a frying pan and, when the under side is browned,

Household Recipes   **After-the-War Dishes**

fold like an omelet. Turn out on a platter, garnish with watercress and serve hot.

Baked Beans With Oil

are found by many to be more digestible than those cooked with pork. To prepare soak one quart of white beans over night. In the morning pour this water off, add more water and parboil them until the skins crack, changing the water twice. Put into the bean pot one small onion, cut in slices. Fill the pot with beans, adding one-fourth cupful of molasses, one-third cupful of vegetable oil, one tablespoonful of mustard, two even teaspoonfuls of salt, one-half teaspoonful of soda, and one-fourth teaspoonful of black pepper. Cover the beans with water, and bake from eight to ten hours, adding water as needed.

Shrimps

are higher in food value than all other shell-fish, and they make a pleasing addition to the bill of fare. Canned shrimps are delicate in flavor, but should be thoroughly rinsed in cold water before using. Placed upon lettuce leaves, and served with mayonnaise or French dressing, they make a delicious salad.

Cabbage and Shrimp Salad

could be served when lettuce is unobtainable. Shred or chop finely enough cabbage to make about two and one-half cupfuls. Rinse one small can of shrimps in cold water, flake with a fork and add to the cabbage. Mix with the following dressing: One heaping tablespoonful of sugar, one tablespoonful of flour, one teaspoonful of salt, one-half teaspoonful of dry mustard. Add one egg, beat all together and stir in gradually one cupful of vinegar and water, about half and half. Cook in a double boiler until it thickens.

Shrimp Sandwiches

are new and delicious. With one can of shrimps use one green pepper, one tablespoonful of lemon juice, three olives, one cupful of chopped celery and one-half cupful of mayonnaise dressing and watercress. Soak the shrimps in ice water for fifteen minutes, drain, dry on towel and break in pieces. Squeeze the lemon juice over the shrimps, add the pepper (cut in thin strips) and the olives, chopped. Then add the seasoning and toss about lightly to mix well. Mix with the mayonnaise dressing, to which is added an onion, if desired. Spread between slices of sandwich bread. A sprig of watercress in each sandwich adds to the flavor and appearance.

Peanut Butter

is a delicacy which we cannot well get along without, as it takes the place of butter, apple butter, jams and jellies, which are scarce many times in the year. The use of it need not be limited to a spread for bread and crackers, for it combines well with a variety of foods, and lends itself to cooking with excellent results.

Peanut Butter Cookies

(sugarless), require six tablespoonfuls of fat, one-fourth cupful of butter, one cupful of molasses, one-fourth cupful of sour milk, one-half teaspoonful of soda, one teaspoonful of baking powder, three and one-half cupfuls of rye or whole wheat flour, or enough to make a stiff dough. Mix the ingredients in the order in which they are given. Roll out the dough, cut the cookies with a small cutter and bake them in a moderate oven.

Peanut Fondu

combines rice and peanut butter in an appetizing way. Use one-half cupful of peanut butter, one cupful of cooked rice, one and one-half cupfuls of milk, one egg, one-half teaspoonful of pepper and one teaspoonful of salt. Beat egg white until stiff. Mix all the other ingredients together and fold in the beaten white. Turn into a buttered baking dish and bake in a moderate oven thirty or thirty-five minutes.

Peanut Bunny

is a nice supper dish. It requires one-half cupful of peanut butter, one cupful of milk, one tablespoonful of butter or butter substitute, one tablespoonful of flour, one-half teaspoonful salt, one-half teaspoonful of mustard, and a dash of cayenne pepper. Melt the butter, stir in the flour, add milk gradually. When smooth, add the peanut butter and cook until it is melted. Add seasonings and serve on crackers or toast.

Banana and Peanut Butter Salad

is made quickly and sure to please. Cut bananas in half crosswise, then lengthwise. Spread the pieces with peanut butter, press them together, place on lettuce leaves, serve with French dressing.

Molded Vegetable Salad

makes a nice company dish. To make it dissolve one tablespoonful of gelatine in one-fourth cupful of cold water. Then add one-half cupful of boiling water, one tablespoonful of sugar, two tablespoonfuls of vinegar, two tablespoonfuls of lemon juice and one teaspoonful of salt. Strain and cool and when it begins to thicken, add one cupful of celery cut in small pieces, one cupful of shredded cabbage and one-half cupful of green sweet pepper, cut in fine strips. Pour into individual molds, set on ice, and when hard turn out on lettuce leaves. Serve with any preferred salad dressing.

50-50 Oatmeal Muffins

are quickly made and quickly eaten. To make them use three-fourths cupful of cooked oatmeal or rolled oats, three-fourths cupful of sifted war flour, one cupful of salt, one teaspoonful of baking-powder, one tablespoonful of sugar, one egg, one tablespoonful of shortening and liquid to make a rather stiff batter—about one-fourth cupful. Sift together, twice, the flour, salt, sugar and baking-powder. Beat the egg until light and add the melted shortening, then the flour mixture, alternating with portions of the liquid until a batter is formed somewhat stiffer than for ordinary flour muffins. Drop by spoonfuls into greased muffin pans until half filled and bake twenty to twenty-five minutes in a fairly hot oven.

Chocolate Bread Pudding

should be served on a meatless day. Use two cupfuls of stale bread-crumbs, two cupfuls of milk, two cupfuls of water, two squares of unsweetened chocolate, two-thirds cupful of brown sugar, one large or two small eggs, one-fourth teaspoonful of salt and one teaspoonful of vanilla. Scald the milk and water together, pour it over the bread-crumbs and allow them to soak for thirty minutes. Melt the chocolate in a saucepan placed over hot water, add one-half cupful of sugar and enough milk (taken from crumbs and milk) to make it of a consistency to pour. Add this to the soaked crumbs with the remaining sugar, salt, vanilla and eggs, slightly beaten; turn into a greased pudding dish and bake one hour.

50-50 Cocoa Cake

is a favorite in The Farm Journal lunch-room. To make it, use one-half cupful of flour, one-half cupful of cocoa, one-fourth teaspoonful of cinnamon, one cupful of sugar, one teaspoonful of baking-powder, one-half cupful of cold water, three eggs and a pinch of salt. Mix the cocoa and sugar together, add the cinnamon and water. Mix until the sugar is dissolved; separate the eggs, add the yolks to this and beat until light. Sift the flour, baking-powder and salt into this mixture; mix lightly together. Beat the whites of eggs until dry and fold in lightly. Grease a Turk's head cake pan, pour the mixture into it and bake in a moderate oven for forty minutes; or bake in greased muffin tins for twenty minutes.

Cocoanut Drop Cakes

are highly patriotic, as they require neither wheat nor butter. The recipe calls for three cupfuls of corn flakes, two eggs, one-half cupful of sugar, one package or cupful of shredded cocoanut. Separate the eggs, beat the yolks slightly, add sugar, cocoanut and corn flakes and mix well. Beat the egg whites very stiff, fold in lightly, then drop by spoonfuls on greased pans and bake twenty minutes in a hot oven.

Recipes for Busy Summer Days

Steamed Brown Bread

is always good, but especially so in hot weather, as it can be cooked on top of the stove with less heat than would be required if it were baked in the oven. It requires one pint of corn-meal, one pint of graham flour, one pint of buttermilk, one cupful of dark molasses, two teaspoonfuls of baking-powder, one teaspoonful of soda, one-half cupful of raisins, one-half teaspoonful of salt. Mix in order given and fill cans or brown bread pots half full and steam for three hours. If the cans are placed in a roaster with other dishes, an entire meal can be cooked at one time over one burner, thus conserving fuel.

Quick Tomato Soup

requires one pint of canned tomatoes, one quart of water, four tablespoonfuls of butter or butter substitute, four tablespoonfuls of barley or potato flour, one teaspoonful of salt, one tablespoonful of chopped onion, and bay leaf or parsley. Mix the water, tomato and seasoning, heat to boiling point, then add butter and flour rubbed to a paste. Boil for a half hour; add two teaspoonfuls of beef extract, then strain and serve.

Cream of Asparagus Soup

Take one pint of the hard, rejected portions of asparagus, cover them with a pint of water or stock, cook slowly for one-half hour; press through a colander, add a pint of cooked rolled oats, a pint of rich milk, a level teaspoonful of salt, a saltspoonful of pepper. Stir until the mixture reaches the boiling point. Add a tablespoonful of vegetable oil, strain through a sieve and serve hot.

Cream of Cucumber Soup

is delicious to serve at the beginning of a fish dinner. Pare two good-sized cucumbers, cut them in slices, cover them with a pint of cold water and

For Busy Summer Days | **Household Recipes**

add one grated or chopped onion. Cook slowly for thirty minutes. Add one pint of cooked rolled oats and stir until boiling hot. Add a pint of rich milk, one bay leaf, a level teaspoonful of salt, a grating of nutmeg and a saltspoonful of pepper. Stir constantly until it reaches the boiling point, strain through a fine sieve and serve at once.

Delicious Corn Chowder

One can of corn, one-half pound of salt pork, two onions, three potatoes, one quart of milk, six pilot crackers, one teaspoonful of beef extract, salt and pepper to taste. Cut the pork in small cubes and fry until light brown. Parboil the potatoes, cut in cubes, return to the kettle and cover them with boiling water. Add the corn, cook until the potatoes are tender. Add the crackers which have been soaked in milk, and lastly add the beef extract. Serve very hot.

Fish Flake Souffle

is an appetizing dish, quickly prepared. It requires one pint of heated milk, four tablespoonfuls of vegetable fat, three tablespoonfuls of flour, one teaspoonful of salt, one-eighth teaspoonful of pepper, four eggs, two cans (or one pint) of fish flakes, one teaspoonful of finely chopped parsley. Make a white sauce by mixing the fat and flour together, then pour on the hot milk, stirring over the fire until it thickens. Beat the egg yolks, add fish flakes and seasoning to the eggs and pour the sauce over them. When cooled off, fold in the egg whites, beaten stiff, and pour into a well-greased glass baking dish. Bake in a moderate oven for thirty minutes and serve at once.

Creamed Fish Cakes

Rub to a paste two tablespoonfuls of vegetable oil and two tablespoonfuls of flour. Add a cupful of milk, put into a saucepan over the fire until boiling. Season with salt and pepper, stir in a half pint or one can of fish flakes. Cover and allow it to stand until hot, then pour over slices of toast and serve with baked potatoes.

Red Bunny

is an attractive cheese dish, quickly prepared. Melt a tablespoonful of butter or butter substitute in a saucepan, add a half pound of American cheese cut in thin slices; melt slowly, then add a can of tomato soup, stirring until thoroughly blended and heated. Arrange squares of toasted bread on a platter, pour the mixture over them and serve hot.

Canned Salmon

has possibilities not always realized by those accustomed to eating it just as it comes from the tin. It is especially nice combined with rice. To prepare it, boil a cupful of rice; when done stir into it a small can of salmon and serve hot. Accompany this dish with a green salad, corn bread, stewed fruit.

Salmon Salad

Remove all bones and skin from a can of salmon and flake it with a fork. Put through a meat chopper one head of celery and a small bottle of olives stuffed with red peppers. Mix all together with mayonnaise or any preferred salad dressing, and serve on lettuce leaves.

Salmon Box

is made thus: Line a well-greased fireproof glass dish or a bread pan with warm steamed rice. Fill the center with cold salmon, flaked and

Household Recipes

For Busy Summer Days

seasoned with salt and pepper. Cover with rice and steam for one hour. Turn on a hot platter and serve with cream or tomato sauce thickened with cornstarch, barley or potato flour.

Tomato Jelly Salad

is especially welcome while waiting for the fresh tomatoes to ripen. To make it, add an equal amount of hot water to a can of condensed tomato soup. Soften one-half package or two tablespoonfuls of gelatine in a half-cupful of cold water. Bring the soup to boiling point, season with salt, pepper and sugar; remove from the fire, add the softened gelatine and stir until this is dissolved. Pour into individual molds, which have been moistened with cold water. When the jelly is cold remove from the molds and serve on lettuce leaves with mayonnaise dressing.

Quick Mayonnaise Dressing

Be sure to have a fresh egg, one with a stiff, firm yolk. Drop the yolk into a deep bowl, add four tablespoonfuls of salad oil, one of cider vinegar, half a teaspoonful of salt and a dust of cayenne pepper. Whip with an egg-beater until it is thick and heavy. Cover and stand away in a cold place.

Kidney Beans and Sausages

make an agreeable combination. The canned beans need only reheating, with added seasoning, if desired. Use half-smoked sausages and put into a saucepan, cover with cold water and bring to the boiling point, when they will be ready to serve. Pile the beans in a shallow dish, arrange the sausages on top and garnish with water-cress.

Grape Jelly

is a refreshing hot weather dessert easily made by combining one of the quick gelatines with grape-juice. Follow the given rules for making the jelly, using heated grape-juice instead of hot water. Whipped cream is a pleasing addition.

Jambolayo

is a delicious dessert, easily made. Fill a glass dish, or individual glasses, a little more than half full of sliced fruits, bananas, dates, canned peaches or cherries; add a few chopped nuts, if desired. Make a lemon jelly, using one of the quick gelatines. When cool, pour over the fruit and stand in a cool place to harden.

Strawberry Delight

To make this, cook a pint of berries until soft, strain and add to the juice enough hot water to make a pint. Pour this over a package of quick gelatine, lemon flavor, and, when cool, pour into a ring mold to harden. When set, turn out on a flat dish, fill the center with fresh uncooked berries, and serve with sugar and cream. Raspberries and blackberries are equally good served in this way.

Canned Pineapple and Dried Apricots

make an excellent marmalade which can be used to eke out last year's supply of fruit, or in sections where fresh fruit is unobtainable. Cook one pound of dried apricots in water enough to cover them, until soft. Then add a large can (size two and a half) of pineapple which has been cut in small pieces. Measure the fruit and add three-quarters as much sugar. Return to the fire and cook until it thickens. This will keep indefinitely.

Use More Fish

The foreign-born residents of the United States are the real fish consumers of the nation, and even then the amount consumed by each person in a year is less than in any other country on the globe. If we are to feed a hungry world, the use of fish will have to become more general, and while the fresh article is not always available, many appetizing dishes can be prepared with salt or canned fish.

A can of salmon or tuna, or a package of codfish, should be found in every pantry, not only for emergencies, but because it saves meat and money, and provides the change in diet which is so necessary, particularly in the late winter.

Creamed Codfish

Creamed codfish with potatoes is a good and inexpensive dish. To prepare it, soak one pound of salt codfish over night, drain and set aside. With a fork, break up enough cold boiled potatoes to equal one-third the amount of fish, and mix together. Heat one and a half cupfuls of milk, and rub together a heaping tablespoonful of flour and a scant tablespoonful of butter. When the milk reaches the boiling point, pour it over the flour and butter and stir well, then let it cook slowly until thickened. When thick, smooth and creamy, set the pan containing it in a dish of hot water. Fifteen minutes before serving time, turn the fish and potatoes into the cream and cook briskly for ten minutes. Break one or two eggs into the creamed mass and stir rapidly until eggs are cooked. Serve with hot toast.

Fish Balls

Fish balls are prepared thus: One cupful of prepared codfish, one pint of potatoes, one teaspoonful of drippings, one egg, well beaten, a dash of cayenne pepper, salt if needed. Wash the fish, pick apart and free from bones. Pare the potatoes and cut into quarters. Put the potatoes and fish into a saucepan, cover with boiling water and boil until the potatoes are tender, but not soggy. Drain off all the water, mash, and beat the fish and potatoes while hot, until very light. Add the drippings and pepper, and when slightly cooled add the egg and salt, if needed, and beat again. Shape in a tablespoon without smoothing very much, slip them off into a pan and fry in smoking hot lard or drippings.

Fish in Potato Cases

Fish in potato cases can be prepared with fresh flounder or with tuna fish. Peel large potatoes and cut into halves. Bake, then scoop out as for stuffed potatoes. Fill the shells with small pieces of fish prepared in white sauce. Put buttered crumbs over the top and return to the oven to brown. Having peeled the potato, the crisp shell is an edible bit.

Salmon Loaf

Salmon loaf: Half a cupful of salmon, fresh or canned; half a cupful of stale bread crumbs, one beaten egg, half a cupful of lemon juice, half a teaspoonful of onion juice, salt and pepper to taste. Mix all ingredients together; put into a greased baking-dish and bake in a moderate oven for about twenty minutes.

Shrimp Salad

Shrimp salad is greatly enjoyed by those who like sea food. It makes an especially nice dish for company, being unusual yet inexpensive. Open a can of shrimps, pour off the liquid and rinse them in cold water. Arrange on lettuce leaves in a salad bowl, or upon individual plates, and serve with any nice salad dressing, or, if preferred, salt, pepper and vinegar.

Christmas Sweetmeats

There are all sorts of candy substitutes, such as stuffed dates, candied ginger, fruit pastes and salted nuts. Not only stuffed dates, but stuffed prunes are delicious. Wash them thoroughly, take out the seeds and slip into each one an almond or a peanut and see how eagerly the children will eat them. Dried fruits, such as dates, figs, prunes and raisins not only have sugar, but are also highly nourishing. Raisins and nuts, if given with moderation, will not prove indigestible.

A half pound each of dates and nuts run through a grinder, softened with lemon juice and cut into squares like caramels make a wholesome substitute for candy.

Use more home salted nuts this Christmas than in previous years. Peanuts, pecans or almonds, if prepared in olive oil or butter, will not go begging.

To candy orange or grape fruit peel requires the use of some sugar, but less than for its equivalent in candy, and you are using up what would otherwise be thrown away. The following recipes require very little sugar:

Peanut bars No. 1: One cupful of granulated sugar, half a cupful of broken peanuts; put the sugar in an iron skillet, stir constantly until it melts to a golden brown. Stir in the nuts and pour at once into a buttered pan. Stir constantly while the sugar is melting, as it burns easily.

Peanut bars No. 2: Shell and remove the skins from one quart of roasted peanuts and chop fine. Beat the white of one egg until stiff, but not dry, and add gradually one cupful of brown sugar, one-fourth teaspoonful of salt and one-half teaspoonful of vanilla. Fold the peanuts into the mixture and spread evenly in a buttered shallow pan. Bake in a quick oven until well puffed and browned. As soon as taken from oven cut in bars, using a sharp knife.

Chocolate caramels: One pint of sugar, one pint of extracted honey (or sorghum), one-quarter pound grated chocolate, one-half cupful sweet cream, one tablespoonful of vanilla extract. Try this often while boiling by dropping a small portion in cold water. When it will form a soft ball, pour about one-quarter inch thick on greased tins. Mark in squares just before it hardens.

Walnut creams: Boil to the hard snap stage one cupful of grated chocolate, one cupful of brown sugar, one cupful of extracted honey (or sorghum), one-half cupful of sweet cream. When it hardens on being dropped into water, stir in a piece of butter the size of an egg. Just before removing from fire add two cupfuls of finely chopped nuts, stir thoroughly and pour on buttered plates to cool, then cut it into squares.

Cracker Jack: One cupful of brown sugar, one cupful extracted honey (or sorghum). Boil until it hardens when dropped into cold water. Remove from the fire and stir in one-half teaspoonful of soda, and when this dissolves, stir in all the popcorn it will take. Spread on greased tins and mark in squares.

Salable Things from Waste Apples

In these days when the world faces an increasingly serious food shortage it is unwise to overlook any resources that will add good nourishing food to the depleted supply. In many States the percentage of cull or cider apples runs fully one-third of the total and it is frequently estimated that thousands of tons of such apples are wasted each year.

While a portion of the larger culls may be evaporated to excellent advantage, the most practical way of diverting this enormous waste into good food is by pressing. Practically all the valuable and nutritive elements of fruits are contained in the juice. The other part consists largely of cellular tissue and is of little value except to retain the juice, which in ripe apples runs as high as ninety per cent.

Products of Apple Juice

A modern hydraulic cider-press will extract an average of a little more than four gallons of cider from each bushel of ordinary culls. This juice is readily converted into a variety of food products that are not only appetizing and nourishing, but most of them are in concentrated form convenient to market and easy to preserve. Cider vinegar, boiled cider, apple syrup, apple jelly, apple butter and pasteurized cider are all in active demand and can be sold at a better net profit than is usually obtained from the apples in a fresh condition.

Even the pomace need not be wasted. It is being used extensively as feed for dairy and beef cattle, and for hogs and sheep. Many pronounce it equal to ordinary corn silage. Pomace also has a distinct value as jelly stock because of its pectin content, which is not impaired by drying. Frequently the pomace is pressed a second time, the resulting juice being used for making vinegar or jelly.

Household Recipes   **Apple Products**

A Temperance Health Drink

Pasteurized cider is highly recommended as a temperance drink by eminent physicians and scientists. It is a tonic as well as a nutrient, containing natural salts and acids of special value in the correcting of stomach complaints and liver and kidney trouble, and can readily be made available as a delightful home beverage the year around. Chemical preservatives should be avoided, but pasteurizing to 160 degrees for two hours, and sealing tight, is effective.

Cider Vinegar

One of the staple food products from waste apples that is in universal demand is cider vinegar. Pure cider vinegar commands a premium on the market.

In the process of transforming cider into vinegar, two distinct fermentations take place. First is the vinous or alcoholic fermentation, which is the changing of the sugar of the cider into alcohol, caused by the action of certain natural yeast bacteria. Second is the acetic fermentation by which the alcohol thus formed is changed

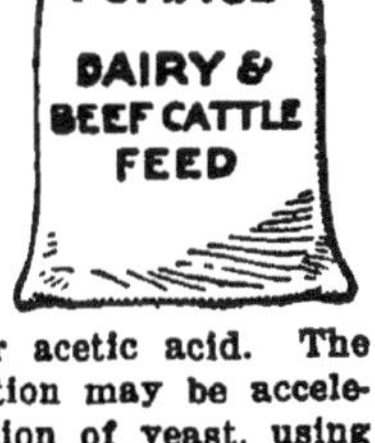

to vinegar acid or acetic acid. The alcoholic fermentation may be accelerated by the addition of yeast, using a cake to each five gallons, dissolved in warm water before adding. The acetic fermentation is also aided by the addition of good vinegar containing some mother of vinegar.

It is important to allow plenty of room for air in the barrel during all stages of fermentation and also to maintain the temperature between 60 degrees and 80 degrees. Care should be taken not to start the second fermentation until all the sugar in the cider is changed into alcohol, otherwise the change to vinegar will be retarded.

Boiled Cider

There exists in this country a potential market for boiled cider that would consume a hundred times the amount now produced if the product could only be obtained. Boiled cider is the fresh juice concentrated by evaporation in the ratio of five gallons reduced to one. In this form it will remain in a perfect state of preservation for years. It is dark brown in color and of a syrupy consistency. It has an extensive use both commercially and in the kitchen, being especially desirable for making mince-meat and apple butter, as well as having a multitude of other culinary uses.

By continuing the evaporating process until the cider is reduced to the ratio of seven to one, the product becomes jelly.

A Home-Made Sugar Substitute

When sugar and sugar products are scarce and high, a practical use of the generous sugar content of apples is especially acceptable. An extensive series of experiments by the Department of Agriculture resulted in the development of a method of making apple table syrup which produces an attractive article of fine flavor.

The process is as follows: Stir into seven gallons of sweet cider five ounces of powdered calcium carbonate—a harmless, low-priced chemical—and boil in a large kettle five minutes. If a large vessel is not available the cider may be boiled in batches. After boiling, pour the cider into glass jars, and allow it to settle until perfectly clear, which requires about seven hours. Return the clear liquid to the preserving kettle, being careful not to pour off any of the sediment. Fill the vessel only about half full, as it foams up when boiling. Add a level teaspoonful of the calcium carbonate for the seven gallons of liquid and boil rapidly until a temperature of 220 degrees is reached, or until it is about one-seventh of the original volume and the consistency of maple syrup when cooled rapidly and poured upon a spoon.

To insure clear syrup the cooling must be done slowly. A good way is to set the jars of syrup in a wash-boiler of hot water and allow the whole to cool. Use this syrup like any other table syrup, and as a flavoring adjunct. Also as sauce for puddings and for making brown bread, fruitcake, candy, etc.

The Baby and His Belongings

Certain things are necessary for the comfort and development of a baby, among which are quiet and clean surroundings fresh air, proper food and clothing. Other necessities are a separate bed, a small bathtub and a basket for baby's toilet accessories. The baby should be trained to sleep by himself from the hour of his birth. Not only will he rest better, but so will the mother, and the chance to form bad habits is eliminated.

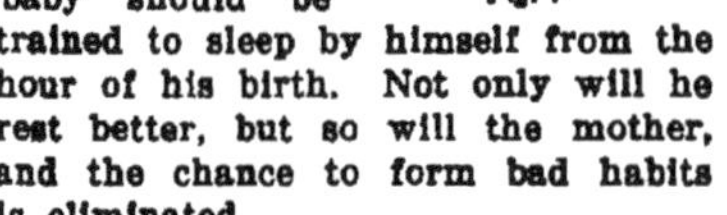

Fig. 1

chairs or upon a low stand, and will accommodate the baby for several months. The bed in the clothes-basket has a bag filled with finely shredded corn husks in the bottom. Over this is placed a folded quilt, and over this may be laid a thick pad of cotton or several layers of silence cloth. This is covered with rubber sheeting, then the small sheets, soft blankets and the white spread, which is made of a yard of crinkled cotton crêpe.

Substitute Beds

A very cozy nest can be made in a clothes-basket or, in a pinch, a bureau-drawer could be used. The basket bed has many advantages. It is easily moved about, may be placed upon two

Selecting a Crib

In choosing a crib, select one having the bars close together, which is absolutely plain (no fancy parts to come loose), and finished, save for the brass knobs, in white enamel. Brass rods across the top are objectionable, for teething babies invariably cool their aching gums upon the cross-bars of their cribs and an enameled rod is preferable for this purpose.

Babies and Children   The Baby's Belongings

A little contrivance for keeping an older child covered is shown in Fig. 1. It consists of two large safety-pins, two brass rings, one yard of wide elastic and as much tape as may be needed to tie around the leg of a bed or crib.

The elastic is cut in half and an end sewed fast to each safety-pin. The other end is fastened to the brass ring and the tape is slipped through, as shown in the circle. The safety-pin is fastened through the bedclothes and holds them securely in place without restraining the child's movements.

Sleeping Bag

For a winter baby, the sleeping bag of outing flannel will keep the little hands and feet warm. The bag is simply made of a piece of flannel, twenty-two inches wide and two yards long. Fold the flannel crosswise, cut a shallow piece out for the neck, cut down the front and face it. Bind the neck, put in a draw string and add buttons and buttonholes down the front. The flannel is then doubled, the edges seamed together, and the seams cat-stitched. The bottom edge is hemmed for a casing and finished with tape and draw strings. Such a bag allows freedom of movement without exposure.

The Bathtub

Quite as important as the baby's bed is his bathtub. For the first few days baby may be sponged upon the nurse's lap, but after that he should be put in the tub for his daily bath, that he may become accustomed to it before he is old enough to fear it. For

Fig. 2

emergency use, a foot-tub will answer and when that is outgrown a small wash-tub of galvanized iron or papier mâché might be used.

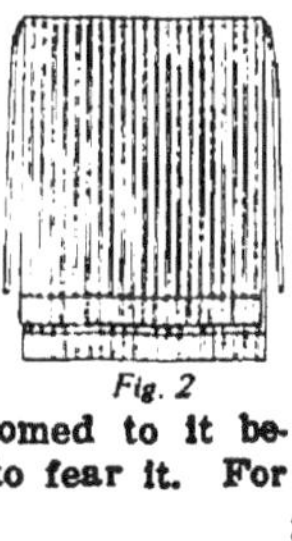

The baby's mother will like to wear the apron shown in Fig. 2, requiring two and one-eighth yards of outing flannel. The ends are hemmed and brier-stitched with white mercerized thread. The upper part is slightly shorter than the under, and a casing is stitched across the top. Cotton tape an inch wide is run through this for strings, and fastened with a few stitches right at the center to prevent its slipping out. In lifting the baby out of his bath, he is placed between two parts of the apron and kept there while he is being dried.

Basket for Baby's Things

A plain basket large enough to hold the things necessary for baby's toilet should be provided. Lacking this, a strong oblong pasteboard box could be neatly covered with pink or blue cambric and, if you choose, white swiss or lawn. The basket should contain talcum powder, a tube of white vaseline, a roll of absorbent cotton in a tin box, wooden toothpicks in a small bottle (for swabs), several sizes of safety-pins, soft towels and washcloths. Soap should be kept in a celluloid soap box.

Early Training

Along with the tub bath and separate bed, baby should be accustomed to using a chair. As soon as he is old enough to sit up, he should be regularly placed upon the "baby's throne" and every effort made to train him to regular habits. A good makeshift can be fashioned out of a box by cutting a hole in it for the seat and building up a back and sides with smoothly planed boards. It may be padded with an old comforter, folded to fit, and a cushion covered with white oilcloth for a seat. A hole through which you can slip a strap or a strong piece of webbing should be bored in each side of the chair, so that the baby can be securely held there. By placing this chair over a vessel it serves its purpose admirably, and its use will save the mother much labor.

Directions for the preparation of baby's food usually call for a double boiler, which is not always available. To improvise one, place a wire dish-cloth in a dish-pan or other large pan. Upon this stand the vessel containing the food that is to be cooked, and surround it with plenty of water. The wire cloth between the pans keeps them from sticking together and the contents of the inner vessel are never scorched.

Sonny's Bath

"Come in!" cheerfully called out the young neighbor, in answer to the old-fashioned mother's knock. "You are just in time to see Sonny have his bath."

"Perhaps I better not," the caller answered, at the same time closing the door behind her; "won't he make an awful fuss?"

"Not sonny," the little mother replied. "He just loves his bath. Why, it's our frolic-time. Eh, little man?"

In answer the baby waved his chubby arms, kicked, smiled, and emitted a series of sweet, cooing sounds.

The visitor was astounded.

"He'll cry before you are through with him, I bet. You're the first mother I ever heard of who spoke of a baby's bath as a frolic-time! My babies always screamed from the moment I took them up to bathe them until I had finished. It was my day's hardest task, and I was always thankful when it was over."

"I don't think he'll cry," was the mother's only answer. "See how good he is while I wash his eyes, nose and mouth."

The older woman watched in amazement. While they were talking the young mother had put a teaspoonful of boric acid into a cupful of warm water. Now she pulled tiny bits from a roll of absorbent cotton. One of these she dipped in the water and carefully squeezed a single drop from it into each eye, quickly wiping the eye with a dry bit of the cotton. The baby gurgled and laughed. Keeping the baby's attention all the time, with deft fingers she squeezed a bit of white vaseline on two more swabs of cotton, twisted them firmly, then carefully cleansed each nostril, using a separate "twist" for each. Again he laughed.

It took but a moment to wash the rosebud mouth. Baby's mother wound a piece of the cotton around the end of her little finger, dipped it in the boric acid solution, and while baby bit at her soft finger, washed tongue, gums and lining of the mouth.

"Well, I never!" the caller said. "I never went through all that for my babies. It's lots of work, isn't it?"

"Yes, it does take extra time, but it's worth it. Baby has never had sore eyes or mouth, and his little nose is so clear he can always breathe through it."

"I wish I'd known that when I had babies to take care of. They always had sore mouths, and sometimes red, inflamed eyes. But we thought that was as common with the babies as cutting teeth. As for the nose, when I saw it was dirty, I cleaned it with a small hairpin. The youngsters always fought against it. I suppose it did hurt." The young mother shuddered at the very thought. "Ah, now he'll cry!" the caller exclaimed, "when he gets the soap in his eyes!" But no soap was used on his face. It was carefully washed with clear water and patted dry.

Until then the baby had been fully dressed. Now his mother removed his clothes—kimono, flannel petticoat, shirt, binder and diapers. "I always take off his nighty, which is apt to be damp, the first thing in the morning, and put on a warm flannelette kimono. He is never fully dressed until after his bath—always at half-past nine."

The old-fashioned mother thought of her babies, who had lain and fussed in their nightclothes until she was ready to bathe them. Perhaps, she pondered, that may have been one reason why they were so cross during the bath.

Sopping a wet cloth with castile soap, she washed first the back and then the front of the baby, and while the caller stared with wide-open eyes, lifted him gently into a tub of water. With the fingers of her left hand spread to support the tiny head and shoulders, she rapidly rinsed off all the soap with a wet sponge, and in the twinkling of an eye had the baby again in her lap, face downward in the large soft towel she had pinned to her left side, and almost enveloped

by the free end of the towel which the mother had thrown over his body.

The visitor gasped. It had all been done so quickly, yet so thoroughly, without a murmur of dissent. Instead, peeping out turtle-fashion from the towel were two bright eyes, gazing at the visitor's red shawl, while their owner contentedly sucked a moist pink arm. A gentle patting with the bath towel, a careful drying of all the creases, a brisk rubbing of the scalp, and then a slight dusting of powder in chafeable spots—and Sonny was ready to be dressed.

Once more the older woman exclaimed, "Here's where he'll cry!" But again she was wrong. There seemed to be no bungling, hard-to-put on clothes. Instead of the tight belly-band which she had always dreaded to sew on, this mother slipped over the youngster's feet a knit band with shoulder straps. The shirt was doubled-breasted and fastened with one small safety-pin. The petticoats were slipped into the simple little dress, and as one garment were drawn over the feet. Baby was turned face downward, and the three garments were buttoned without further disturbing the wearer. He actually enjoyed it.

When at length the little mother brushed back his silky down of hair, and, after wiping her nipple with a piece of cotton saturated with the boric acid solution, placed him at her breast, she turned to the visitor with a happy smile. "Do you wonder I enjoy this hour?" she asked. "Sonny is always like this at bath-time. He is never tired or hungry at half-past nine; I have everything ready so I don't have to make him wait, half-dressed, while I find some necessary thing; the water is always the same temperature—98 degrees—so he receives no shock when I place him in the tub; and most of all, he feels how much I enjoy it, and so has confidence in me. Now he'll nurse and go to sleep."

"It's well-nigh wonderful," the old-fashioned mother replied. "I'd never have believed it could be done if I hadn't seen you do it. Bathed a baby —put it in a tub of water, even—and it laughed and cooed and kicked its legs and waved its arms in glee all the time!"

Indoor Games

Clothes-Pin Game

In this game the only materials necessary are two dozen clothes-pins and two small boxes. Two teams are chosen, each with the same number of players. The players sit on the floor in two straight, parallel lines facing each other. Each line represents a team. About four feet should be allowed between the two lines, and the players should sit close together.

A captain is chosen for each team and takes a position at the end of his line. Each player with his right hand takes hold of the left wrist of the player to his right. This hold must continue throughout the entire game. A box containing twelve clothes-pins is placed beside each of the captains, who, at a given signal, picks out one clothes-pin and passes it to the player at his side. The pin is passed down the entire line until it reaches the player at the other end who places it on the floor by his side. Then another pin is started. Only *one* pin must be in action at a time.

When the entire twelve pins are at the other end of the line start them back in the same manner, and the team getting its dozen pins back in the box first wins the game. If any player releases his hold, it is a foul and the pin must be started again.

The Standing Broad Grin

Line up the contestants in a row, all standing and facing the audience. At the word "go" each must smile and hold the smile. The one who smiles the longest without moving his face wins.

The Continuous Glum

Line up contestants as in the above game. At the word "go" all look glum. The audience tries to make them smile by talking or making faces. They must not touch the contestants. The one who remains glum the longest is the winner.

Fifty Yard Slash

For each heat have four strips of narrow paper, one inch or less in width, and at least twenty feet long. Fasten one end of each strip securely; then the four contestants, each holding the free end of one of the strips, cut with a pair of scissors down the center without running off at the side. The one who reaches the fastened end first, wins the heat. Any one running off, loses.

Chinese Table

The players sit in a circle and each one takes the name of an article used at the tea-table, such as tea, sugar, cream, cake, bread, etc. The one called "tea" begins. He rises, turns around and around in his place, say-

Games and Entertaining   **Indoor Games**

ing: "I turn tea; who turns sugar?"
Sugar turns, saying: "I turn sugar;
who turns milk?" And so on, till
every one in the circle is turning.
They must continue turning till the
leader claps his hands and calls out:
"Clear the table," when all sit down
in their chairs again.

Game of Artists and Critics

A good old game is "Artists and
Critics." Furnish each player with
a slip of paper and a pencil, and direct him to draw a picture of any
sort he pleases at the head of the paper, then write its title at the bottom of the sheet. Usually the less it
looks like what he calls it, the more
fun. He must fold the paper up over
the title so that no one can see it,
then pass it to his neighbor, who
writes what he thinks it is intended
to represent, and folds his title under
and passes it on around the room for
each to add his criticism. When all
of the slips are thus completed, some
one collects them all and, first showing the sketch at the head, reads the
various titles, ending with the artist's
own.

Game of Telegrams

Choose ten letters at random, put
them up where all can see. Provide
each guest with paper and pencil and
have each one write a telegram, keeping the letters in the order given,
and letting each letter be the first letter in a word. Suppose the letters
were A L P O C R G D E H. The result might be, "All lazy preachers
often come running gracefully down
even hills;" or, "All ladies passing
our car receive good dinners eaten
hot." Nonsense, of course, but laugh-
ter provoking.

Heavyweight Throw

Get a half-bushel basket and a
cheap ball. A basket or tennis-ball is
best. Choose sides. The contestants
stand twelve feet from the basket.
Each player endeavors to throw the
ball into the basket. The side wins
that succeeds in getting the ball to
stay in the basket the most times in
a given number of throws.

Game of Biography

The game of biography is also good.
Pass paper and pencils to all of the
guests. Ask them to write their
names at the top, and turn the paper
down across the top so that the
name does not show. Pass the papers
around so that no one will know
whose paper he has. Then ask the
following questions, letting each member of the company write one answer
on each paper by passing the sheet to
his right after he has written and
folded his answer out of sight: Where
was the hero or heroine born—year
and month? What were the child's
first words. How was its youth
spent? What was its first joy?
First sorrow? Present occupation?
Favorite fad, favorite food, favorite
author, favorite book, favorite song,
favorite statesman; politics? Where
will he or she be ten years hence?
Read the biographies aloud before
the company.

Indoor Sports

The hurdle race is for both boys and
girls. Each person is given six
needles and a spool of thread, and the
one who threads them all wins the
contest.

Next comes the standing high jump.
Hang three doughnuts in a doorway,

Indoor Games   **Games and Entertaining**

about four inches higher than the mouths of the contestants. Tie their hands behind them and see who first bites a doughnut.

For a drinking race each player is given half a glassful of water and a spoon. The water must be consumed a spoonful at a time, and the one who finishes first is the winner. If any is spilled that contestant is barred out.

A bun race is great fun. A clothesline is stretched across the room, and from it are hung sugar buns at a height just reaching each player's mouth. The players stand in line with hands behind them, and at a given signal begin to eat the buns. The bobbing of the line makes this very difficult.

Last comes the Rainy Day Race. Each contestant is given a shoe-box containing a pair of over-shoes, and tied with string. A closed umbrella is also handed to each. When the starter counts three the boxes must be untied, the over-shoes put on, and the umbrellas opened. The contestants then walk across the room as rapidly as possible to a set line, remove the over-shoes, replace them in the boxes, tie the boxes, and close the umbrellas before they walk to their starting place. The one who arrives there first wins.

Potato Race

Arrange five chairs at each end of a long room, facing each other. At one end place six large potatoes on each chair. Five persons play at a time. The object is to carry each potato on a teaspoon from the chair at one end of the room to the chair at the other end, without dropping it. The potato must not be touched except by the spoon. If it is dropped, it must be picked up by the spoon, car-

ried back to the chair and a new start made. The one who gets all his potatoes over first, wins.

Nut Game

"Nuts" can be played with small slips of paper and pencil, or the answers given verbally, a nut being handed to the first one who gives the correct answer. Here are the questions and answers:

1. What nut grows nearest the sea? (Beechnut.)
2. What nut is the lowest? (Groundnut.)
3. What nut is the color of a pretty girl's eyes? (Hazelnut.)
4. What nut is good for naughty boys? (Hickory.)
5. What nut is like an oft-told tale? (Chestnut.)
6. What nut is like a naughty boy when sister has a beau? (Pecan.)
7. What nut grows on the Amazon? Brazilnut.)
8. What nut is like a Chinaman's eyes? (Almond.)
9. What nut lives in a pen? (Pignut.)
10. What nut is like a goat? (Butternut.)
11. What nut might be made of stone? (Walnut.)
12. What nut is like a dog's tail? (Wagon nut.)

Potato Peeling Contest

A potato peeling contest is sure to be jolly. Provide a sufficient number of medium-sized potatoes and paring knives and all begin at a given signal. For a second contest allow the guests to carve a face or figure from the peeled potato.

The Blarney Stone

The blarney stone adds one more novelty and is considerable fun. A stone should be treated to a generous

Games and Entertaining   **Indoor Games**

bath of whitewash and be placed in the center of a large table. A round one is best. Tell how the fairies have placed a spell upon it and great good fortune will attend any one who succeeds in kissing it, after having been blindfolded and turned around three times.

A "Thrilling" Pastime

is to seat all guests upon the floor around a muslin sheet, in an absolutely dark room. While some one tells a blood-curdling story, illustrative objects are passed around underneath the sheet from hand to hand. A kid glove, stuffed with bran and soaked in ice water for an hour is as clammy a hand as one could wish. Peéled white grapes, icy cold, make "loose eyes;" cold boiled macaroni and spaghetti will represent muscles; a toy mouse and spider, a hot baked potato and a prickly burr, add interest; while a plaster of Paris skull could be added for the climax.

The Little Dutch Band

A game in which all join is called The Little Dutch Band. The players sit or stand around the room in a circle. The leader assigns to each some imaginary musical instrument—horn, fife, drum, trombone, violin, harp, flute, banjo, etc. Some well-known but lively air is given out, and the band begins to play, each player imitating as nearly as possible the instrument he has been assigned. All continues well until the leader suddenly drops his instrument and begins playing on that of another member of the band. At this the player whose instrument has been borrowed must change his attitude to imitate the instrument the leader dropped.

This continues, the leader taking up the imaginary instruments of the various players, and they at the same time adopting the leader's instrument, the one he started with, not the one he has just dropped.

Indoor Shot Put

Have five pint fruit-jars and fifty beans. Five persons play at a time. A jar is placed in front of each contestant, who receives ten beans. At a given word each contestant drops his beans, one at a time, into the jar from a height level with his chin. The one having the most beans in a jar wins.

Pins and Potato Race

Arrange three large potatoes on a table at one end of the room, and request each member of the party to carry them to the opposite side of the apartment, using an ordinary pin to lift them with. Set a time limit for this feat in every case, and let all who accomplish it in the number of minutes allotted draw for the prize.

Gratitude Game

Each person is given a paper and pencil and told to arrange a list of numbers down the side of the paper, numbering one to fifteen. The players are then to make out a list of things they are thankful for. Only humorous answers will be considered; all others will be ruled out later when the lists are read. The papers are then given in without names attached, and judges are appointed to go over the lists and decide the winning person. The papers are returned, each person taking one at random, so as to relieve him of the embarrassment of reading his own. After the fun is over a small prize may be awarded for the best list.

Going to Jerusalem

The old games never lose in popularity, so you will find "Going to Jerusalem" very good for use for the party you wish to give your class. Arrange chairs in a long row down the middle of the room, placing them so that one faces one way, the next the other, and so on down the line. There should be one chair less than the number of players. Form in line, start the music (a bright march on the talking machine is just the thing), and when all are marching merrily around, stop the music. All scramble for seats, and the one who is left over stands aside, out of the game. Another chair is removed, the music starts up again and then stops suddenly. Again a player is left out, until it gets down to two players and one chair, the one who finally gets the chair wins the game.

Literary Salad

A "Literary Salad" is also appropriate. For this prepare salad leaves by folding and twisting pieces of green tissue paper until they look like lettuce leaves; then paste slips of paper containing familiar quotations on these. The participants of this salad are requested to guess the name of the author of their quotation.

Trees of the Woods

A list of words which suggest the names of trees may be prepared as follows: 1., twins; 2, to languish; 3, well-groomed; 4, a tool, etc., etc. The answers are: 1, pear; 2, pine; 3, spruce; 4 plane. These names are to be guessed. If there are musicians present, such old songs as "Woodman,

Spare That Tree," and "A Rare Old Plant is the Ivy Green" may be sung. For prizes, a dainty apron cut in the shape of a maple leaf and edged with lace, would be pretty for the girls, while a shaving-paper case, also leaf-shaped, would do for the boys, and a book on trees for anybody.

Game of Animals

Sides are chosen, whose members stand closely grouped by their respective leaders, X and Y. X calls out an animal whose name begins with A and counts ten. If Y can respond with another before X has finished counting ten, he does so, and begins counting, and X has to name an animal in A. This is repeated until no more names in A are forthcoming, when another letter is taken. If Y can not give a name before the ten counts have expired, X chooses one of Y's followers, and *vice versa.* When one side confesses its inability to name any more animals, its opponents are entitled to choose as many members of that side as they can give new names, beginning with the given letter. The only duty of the other players is to suggest new names for their respective captains. Before beginning this exercise, it would be well to appoint a secretary to make a list of the different names given, arranged alphabetically.

Sum Contest

For this, each person, standing still, holds a slate and pencil. An easy sum is written out on the slate, the other players start at the word "go!" run toward the slates, do the sum and return to their places. The person doing the sum correctly in the shortest time is the winner.

Game of Sculpture

Here is a suggestion for a party small enough so that all can gather around a large table. Seat the guests. Place in front of each a card having on it an animal's name, five toothpicks, and a small lump of putty, from which each one must form the animal whose name is on the card, using the toothpicks for legs. Give them five minutes to do the work.

Tossing the Smile

The main object of the game is to keep your face straight during the times when you are not "it."

Ask all players to form a circle, either seated or standing. One person is chosen to stand in the center. Suddenly he smiles a broad smile at some one person in the circle, who smiles back, and the two exchange places. None of the other persons in the circle must allow their facial expressions to slip a mite, or a penalty will be exacted. A rapid exchange from the outside circle to the center necessitates alertness and interest on the part of the players who never know when they will be called upon to stand in the center of the circle and "toss the smile."

The following penalties for smiling out of turn may be used: Make the person go around and smile at every one present, individually; have the person smile three times at himself for thirty seconds without stopping. Other penalties on this order may be used.

Game of Wireless Telephone

On rainy days, or at other times when playing outdoors is impossible, the time can be made to pass pleasantly by playing "wireless telephone."

The players are seated in a circle, preferably on the floor, and each one is supplied with a "wireless" outfit, which is simply a newspaper rolled to form a hollow tube about eighteen inches long.

Everything being ready, the starter begins the game by whispering a message through his "wireless" to the neighbor on the left. The message must not be repeated if the neighbor fails to understand it clearly; but it is to be passed on to the next player just as it was received. When the last player has received the message he tells out loud what he was told, and then the player to his right also tells aloud the message as he heard it, and so on back to the starter. This is done to find out why the message did not reach its destination in its original form; and many ridiculous changes will be brought to light. The leader may say, "It is raining hard today;'' but by the time the last player is reached it may be changed to, "Dick is straining lard and hay," with many other funny changes along the line.

"Poor Pussy" and "Rooster"

The one who is "it" gets down on the floor in front of one of the players and meows, spits and cries like a cat. The person for whom this performance goes on must say, "Poor pussy, poor pussy," without smiling. Three "meows" are allowed, and if the person has not smiled, "poor pussy'" goes on to the next one. Whoever smiles is "it." "Rooster" is played in much the same way, except that the players crow, flap wings, scratch the ground, etc. There is no response to the "rooster," the player must only keep from smiling.

Yankee Doodle Kitchen

Divide the company; the girls will then take part in a "Yankee Doodle Kitchen," going through in pantomime the motions of washing clothes, scrubbing floors, churning, washing dishes, etc. The piano or orchestra plays "Yankee Doodle," at first very slowly, gradually increasing in speed, and the workers increase their movements accordingly until they are going at breakneck speed. This, while it sounds very simple, is lots of fun and must be seen to be appreciated. The men act as audience and guess the occupations. They then change places and the men show in pantomime the occupations of the early settlers—felling trees, picking up stones, plowing, sowing, harvesting, building, etc.

Game of Missing Adjectives

Some one writes up a story beforehand; anything will do, but the account of some former gathering, a picnic, drive or walk affords the most pleasure. All the adjectives are left out. When the game is played, some one goes around the room, getting an adjective from each person, in turn, putting them into the blank spaces left in the story. When all the spaces are filled, the story is read and it always proves a success.

A Ring Game

All join hands and form a ring, with the person who is "it" outside. They then move quickly around singing a tune as they do so. The person who is "it" taps some one on the shoulder, when the ring at once stands still; the person who was tapped runs one way, the person who was "it" runs the other; when they meet, they bow to each other three times, then both try to reach the vacant place in the ring. The one who fails to bow three times, or fails to reach the vacant space in time, is "it."

Cotton Contest

Put dabs of raw cotton on the floor and let the players take them up on a teaspoon. Another is to blindfold the players, throw beans or corn on the floor and let them gather them up in teaspoons. The one who gathers the most gets a small prize.

Fish Talk

Each player must talk like a fish. Of course we all know that the finny tribe has no speech, but the person who is "it" calls on each player in turn to make a noise like some certain fish, naming them in turn. The responses are ludicrous, but any one who fails to respond must be "it."

New Form of Stage-Coach

Give to each person the name of a city, town, state or country, then let the one who is "it" call out quickly two of the names, such as "Boston" and "Galveston;" the person so named must quickly change places, while the person who is "it" must try to slip into a chair while the other two are changing. Whoever is left is "it." When the command comes, "All change cars," there is a grand scramble; all change places and the one who is left is "it."

Move the Penny

This is an indoor game that will always find great favor with a company of young people. The whole amusement is afforded by two balls about the size of billiard balls, and a penny. It is necessary to mark out on the tablecloth, with chalk or pencil, a circle about three inches in diameter, and a straight line about two feet from the circle.

Put one ball in the center of the circle and on its top balance a penny. The trick is to bowl from the line with the remaining ball and try to knock the penny out of the ring. Simple as it may seem, it takes a great deal of practice, for nine times out of ten the penny will drop within the circle. The best way to accomplish this is to bowl very slowly, and by knocking the ball very lightly the penny will roll out on the top of the other ball.

Romance

Provide each guest with a pencil and pad. The leader reads the topics aloud, and after each topic the paper is folded over so that no one sees what her neighbor has written; it is then passed to the right, so that each person writes in every booklet. The topics are as follows:

1. Name of the book. 2. Heroine's name. 3. Describe her fully. 4. Hero's name. 5. Describe him fully. 6. Where did they meet? Describe the place. 7. The heroine's impression of the hero. 8. What did the hero think of her? 9. The villain's name, describe him fully. 10. What terrible thing did he do? 11. What effect did his villainy have upon the heroine? 12. What did the hero do about it? 13. What did the heroine do next? 14. How were their difficulties solved at last? 15. What became of the hero? 16. What became of the heroine? 17. What became of the villain? 18. Moral. At the end, each guest reads aloud the book which she has in her hand.

Game of Hidden Cities

Scarcity.	What city has few people?
Duplicity.	What city is full of hypocrites?
Velocity.	What city has many chauffeurs?
Voracity.	What city has greedy people?
Audacity.	What city is for reporters?
Publicity.	What city is for authors?
Sagacity.	What city is for wise people?
Multiplicity.	What city has great crowds?
Eccentricity.	What city has odd people?
Infelicity.	What city has unhappy people?
Pertinacity.	What city is full of office-seekers?
Electricity.	What city is for telegraph operators?
Reciprocity.	What city is for the nations?
Veracity.	What city is full of truthful people?
Loquacity.	What city is for talkative people?

Outdoor Games

Outdoor Fan Tag

Some member of the company is made "it" and receives a palm-leaf fan. He begins operations at once by trying to tag the others, which is done by touching them lightly with the fan. Any one who is tagged instead of becoming "it," as in ordinary tag, takes the hand of the person who tagged him or her and together they try to tag the others. The second person also receives a fan, which he passes to any one tagged by him and constantly playing at his left hand. This goes on as long as any one remains free. The two persons at respective ends of the line carry palm-leaf fans and either end can tag. The fun of the long unwieldy line moving awkwardly around, endeavoring to tag a lively player who is determined not to be caught without a chase, is easily imagined. Any person caught takes the hand of the person tagging him. There is no prize to this game.

Day and Night

The players are divided into two sides, standing about six feet apart. One side is called "Day;" the other side "Night." A boy in the center between the two sides tosses a card—black on one side, white on the other. If the black side comes up, Nights run and Days chase them. Days tag as many Nights as possible. Beyond a line sixty feet away, the Nights are safe. If the white side of the card comes up, the Days run. The number of boys tagged by the Days makes the Days' score; and the same rule applies to the Nights. Repeat this as often as desired, and add the scores on each side; the side having the highest total wins the game. The boy in the center is score-keeper.

"Are You There?"

"Are you there?" is popular on shipboard, but can be played on land as well. Prepare two clubs of straw, winding them round and round with cord so that they will be firm. Blindfold two young men, put a straw club in each one's right hand, then let them lie flat on the ground, clasping each other's left hands. One cries, "Are you there?" the answer comes "Yes!" but the one who answers immediately dodges the blow which follows, for the questioner tries to strike his contestant on the head with the straw club, and tries to locate his head by his voice. Each has a turn, and the one whose head is struck loses.

Follow the Leader

In a company of bright boys and girls the old-fashioned frolic known as "follow the leader" is most enjoyable. One person is chosen as leader. Whatever he or she does is imitated by the rest of the players in a body. The leader may resort to tongue twisters in which he has already acquired some skill, or he may elect to lead the crowd a merry dance, bounding over small obstacles, running around chairs, crawling under bushes and any other mirthful stunt suggested by the occasion.

Church and School Entertainments

Shadow Pantomimes

Shadow pantomimes are very funny and easy to manage. The first thing to do is to select a poem suitable to illustrate. "Lord Ullin's Daughter" is a good one, and there are some excellent things in the "Bab Ballads," by Gilbert. A large curtain of white cotton cloth is stretched across the stage, with a light so arranged that every one who passes between it and the curtain throws a shadow upon the latter. Some one who reads well is selected to read the poem slowly, and during the reading it is acted in pantomime.

An Illustrated Magazine

A large frame of wood, made to represent a picture-frame and covered with gilt paper, is erected on the stage. There is a curtain which can be parted in the middle hung on the inner side, and a screen covered with dark cloth, just back of it, for the background. The magazine contains a poem, a story and the advertisements. A good reader announces the name and the date of the magazine and the title of the poem; he then begins to read the poem, and this is illustrated with tableaux or pantomime, inside the picture frame. The story is read and illustrated in the same way; then follow the advertisements, which need no announcement. Soap, hair-tonic, baby foods, chocolate, etc., anything that can be made into a "picture," can be advertised.

Spelling Game

This spelling game is entertaining: Form the company in a part circle; let some one begin with a letter, an "I," for instance, having in his mind "Independent." The next person must add a letter; but if he inadvertently adds a "t," or even an "n," he is out of the game, as either one of them makes a word. With greater precaution the next one probably adds an "m," with "Immediate" in his mind; but his next neighbor, thinking of "Improbable," quickly adds a "p" and is out; and so on.

A Mother Goose Party

This is pleasing for a home gathering, or for a church or school party. Let each child come dressed as a Mother Goose character. Use cheesecloth for costumes. It is cheap and comes in colors. Even crepe paper may be utilized by clever fingers. There are so many possible characters that, if the party is to have a guessing characters' contest with it, there is no need of duplicating if the children get together and decide on which character each will represent. Boy Blue needs only a blue suit and a horn; Simple Simon, any kind of a suit and a fishing pole. Let the guests guess who the children represent, and when the announcement is made have the Queen of Hearts present the one naming most of the characters correctly with a tart, given on a pretty tray, as a reward of merit.

Church and School | Games and Entertaining

Progressive Initials

A number of tables to suit the number of guests are prepared, and are labeled in order, Fruits, Flowers, Noted Men in American History, Cities of America. In the center of each table place twenty-four assorted letters, face down. These may be pasted on small squares of cardboard, if desired. After guests are seated at each table, give to each a second card on which are written the subjects of the different tables. When all are ready, the bell at the first table rings and the game proceeds. One person turns a letter. If the first table is for fruit, and the letter turned is A, the person who turns it says "Apple," and keeps the letter. Then in quick succession a letter is turned by each person in rotation, until all the letters are exhausted.

The object of the game is to be the first to think of a fruit, flower, noted man or city. When the letters are all exhausted at the first table, the bell rings and the game stops. The two persons who have gained the most letters during the game, progress to the second table, and those at the next table who have the least take their places.

Each person keeps a record on his card of the letters he gets, and at the end of the evening a prize is given to the one who has held the most letters.

Mind-Reading

The so-called "mind-reading" experiments always promote fun. With the aid of an accomplice, one can mystify a roomful of people by offering to guess any number from one to ten which they may choose, while you are out of the room. When you re-enter the room, you go to one after another of the players, placing the hands on each side of his or her head, pretending to get the number in that way. When you reach your accomplice, he compresses his jaw the number of times decided upon by the company. For instance, should they decide upon the number six, your accomplice compresses his jaw six times so you feel it, and so "guess" the number.

Another mind-reading feat is to tell the players that you will guess any musical instrument they may pretend to play. Your accomplice brings the players in one by one, and they go through the motions of playing some instrument, which you will guess, although the player stands behind you and you are blindfolded. You will ask the player to perform a little faster, and when he is going as fast as he can, you can say, "It's coming, a little faster, please. There, I have it. You are playing the fool!" This is all good-natured fun and always raises a laugh.

Game of Advertisements

Pin around the room on the wall, well-known advertisements, as Gold Dust Twins, Ostermoor Mattress, etc. Cut from them all printing and number each one. Let each one write down the numbers on a card, and opposite the number his guess. Whenever possible, pin up several advertisements of the same kind, numbered alike. This enables everyone to see the advertisement without crowding. To the one guessing the largest number give a small prize.

Charades

Charades are always fun, though familiar to almost everyone. Choose sides. Let each side select a word that can be acted out. If the other side guesses the word when it is acted,

Games and Entertaining   Church and School

it becomes their turn. Here are some simple characters:

Mistake: A "miss" steps forward and "takes" something handed to her.

Pekin China: "Peek" "in" a piece of "china."

Bandage: Let all come in pretending to play an instrument "band," and go out doubled over to indicate "age."

Mansion: Let a "man" stand at one side while the others in passing turn aside and "shun" him.

Furlong: All examine first a "fur" and then a "long" string.

To end the evening pleasantly, bring in a board on which is pasted a large rabbit without ears. Blindfold some of the persons present, and in turn hand each one an ear to pin on; turn the person about and give him the other ear. No two ears match, and the results are very laughable.

For refreshments serve coffee and cake, and perhaps fruit. Charge a small sum.

Progressive Conversations

Cards should be prepared with a list of subjects, such as "Should the weather be tabooed as a topic of conversation?" "Do you prefer the mountains or the seashore, and why?" "Which is the greater educator, reading or travel?" "Is music without words capable of expressing emotions?" The gentlemen's cards should be larger than those of the ladies, and each pair, large and small, should bear the same number. They should be distributed haphazard, and the lady and gentleman having the same number are matched for the first topic of conversation. For a fixed time, say five minutes, the first topic is to be discussed. Then at a signal the gentlemen move on one place, leaving the ladies seated as they were. Thus each topic is discussed until the list is

completed. At the close the gentlemen decide by vote who was the best talker among the ladies, and the ladies do the same for the gentlemen. If there is some simple prize, like a small dish of bonbons, it adds to the fun to have the victorious pair share it together.

The Popular Magazine

The contents of a magazine are represented by tableaux. The best arrangement is to have first the frontispiece, some pretty tableau or single figure; then a poem, with human illustrations—the text of course to be read by some good reader; then a story, followed by an illustrated article; and finally the advertisements, which are the most effective of all. The name of the advertised article is to be guessed from the tableau. This entertainment gives unlimited opportunity for ingenuity, skill in making up, etc.

Game of Travel

To play it one of the number must be chosen for the traveler and must leave the room. During his absence the rest attire themselves to represent the natives of foreign lands,—a German takes a book, a Laplander wears a fur, a Chinaman takes two sticks for chop sticks, a Japanese lady has long pins in her hair and carries a fan, a Turk wears a turban, and a native of Holland wears wooden shoes. As the traveler enters, he must pass from foreigner to foreigner, pausing at each to guess of what nationality he is.

Left-Handed Contest

Have the right hand of every guest bandaged before entering the room, and throughout the evening insist on everything being done with the left hand. Another amusing game is to demand something in the way of entertainment from every one present. A good way to manage this is to make a list of old-fashioned or well-known songs. Have four copies of each song ready, write the names of the songs on slips of paper, and let each guest draw one on entering the room. When ready, each group of four must sing the song which they drew. As the groups are sure to be mixed, the result is very entertaining.

Home Parties

Inexpensive Entertaining

There was a time, and it was a good time, when one was satisfied to serve lemonade and cake, perhaps ice-cream, and to be very fine, coffee, nuts and raisins. We enjoyed it because it was good, and the desire-for-something-different microbe had not entered our souls. But now that it has entered, we seek and demand variety, and not to attain it is to confess oneself behind the times.

Put Your Wits to Work

So, when you have a party, you must set your wits to work; and, while your refreshments and decorations may be very inexpensive, you must contrive, in some way, to have the latter suggest the objects of the entertainment or the time of year. If you prefer not to use linen, suitably decorated table-cloths and napkins made of crepe paper, may be had for a small sum. You can also get, for a few pennies, paper cases to hold candies, nuts, etc., and favors suitable for any holiday and almost any purpose. Pretty china is not at all expensive nowadays, and all the glass necessary for the most elaborate entertaining may be had in the pressed ware, a pretty pattern known as "Colonial" being very inexpensive.

A pretty plant, a few ferns, or some wild flowers will do for the bit of green which is now thought essential; while ferns laid flat upon the table are prettier than any embroidered centerpiece ever used.

What to Serve

As for the food, if something substantial is desired, try broiled or creamed oysters, creamed lobster or scalloped crab-meat, if you can get sea-food, creamed chicken, if you live inland. With any of these serve thin slices of brown bread and butter, olives and sweet gherkins. Or you can serve stuffed eggs with mayonnaise dressing and cold sliced chicken or turkey; or a vegetable salad with thin slices of cold ham or tongue. You could also serve sandwiches with various fillings, such as creamed cheese mixed with chopped nuts; sweet peppers, chopped fine, mixed with butter, lettuce and mayonnaise; thin slices of cucumber dipped in French dressing; minced ham, tongue, chicken or hard boiled eggs; preserved ginger cut in very thin slices; chopped dates and nuts; chopped peanuts and butter—and so on indefinitely. The bread for sandwiches must be cut thin and is best when a day old. They must be neatly made to be appetizing.

With the sandwiches may be served a salad. Chicken, lobster, shrimp or crab salad may be used, or a simpler one made of celery and nuts, or celery and chopped apples, or a fruit salad. Olives and salted nuts may be served with any of the things mentioned. Omitting the sandwiches, you may serve crackers and cheese with the salad.

Sweets and Desserts

A sweet dish of some kind usually winds up the menu. This may be that old stand-by, ice cream, or orange jelly with whipped cream; or little individual chocolate puddings with hot chocolate sauce. Still another delicious sweet is made by adding marshmallows, cut in quarters, and halved white grapes, to sweetened whipped cream, served very cold. Small cakes are served with these sweet dishes. Rich cookies, cut in small shapes, or a good layer-cake baked in small tins and iced, will do.

Home-made candy is always enjoyed, and may take the place of the sweet dish, if desired. An old and tried recipe for fudge is as follows: Three squares of grated chocolate, three cupfuls of sugar, one cupful of milk, butter size of a walnut. Boil all together until a drop of it placed in cold water will form a ball between the fingers, then add one teaspoonful of vanilla extract. Beat all until creamy, turn out quickly into a buttered pan and cut into squares before it becomes quite hard. Another good recipe is for Mexican Panocha: Boil together one tablespoonful of butter, four cupfuls of brown sugar, one teaspoonful of salt and one cupful of milk. Cook this until it drops hard in cold water. Then pour in two tablespoonfuls of vanilla and two cupfuls of chopped nuts, either pecans or walnuts, and stir constantly until well mixed. Pour on a buttered plate and cut into squares.

If the weather is cold wind up with coffee or cocoa; in warm weather serve lemonade or grape-juice.

Guessing by Feel

Place on a table about twenty or twenty-five bags, each numbered and containing some well-known article, as, comb, bar of soap, apple, potato, spoon. Each person may handle each bag once, and must handle them all in a given time. Write down each one's name and after it the numbers on the bags guessed correctly. The one guessing most may receive an inexpensive prize.

Smelling Contest

Into each of a dozen or more colored bottles is poured a small amount of some liquid that can be detected by the odor—turpentine, vinegar, witch-hazel, camphor, peppermint, coffee, cologne, bay rum, wintergreen, etc. The bottles are then ranged in order on a table, and each member of the company provided with pencil and paper, passes in turn by the bottles, giving a sniff to each, and then from memory makes a list of the contents of the bottles. For the most perfect list a prize is awarded. This game abounds in fun. Sniffs must be brief; and as perfect solemnity must be preserved for accurate results, the spectacle of the sniffing procession is deliciously absurd.

Home Parties   **Games and Entertaining**

Thimble Club Entertaining

It will be very pleasant to start your club with a social afternoon, and you could play the following games either indoors or on the porch: For a Button Contest, arrange and number your tables. On the one marked No. 1 there are, for each person, fifteen large buttons with thread and needles. On the other tables there are bowls filled with buttons. The person at the head table who sews on her fifteen buttons first, rings a bell and progresses with the one who has sewed on the next highest number, first pulling off the buttons so as to be ready for the newcomers. Make a knot in the thread, sew once into each hole, then fasten enough to hold the button on. Break the thread each time. Every person reaching the head table sews on the fifteen buttons as the first did, the remaining persons beginning over again, and keeping the score. Those at the other tables sew on just as many buttons as possible, while the ones at the head table are doing the requisite number. After fifteen progressions, the cards are collected, the score counted and prizes are awarded the most successful contestants. After this, distribute cards with the following misspelled words written upon them: 1, Dhetar (thread); 2, Yreemgba (emery bag); 3, stbutno (buttons); 4, Nisp (pins); 5, Eednsle (needles); 6, Eatpniel (tape-line); 7, Koosh (hooks); 8, Seey (eyes); 9, Xaebesw (beeswax); 10, Sfeipnatsy (safety-pins); 11, Tgrisn (string); 12, Halck (chalk); 13, Hrteadiskl (silk thread); 14, Ardib (braid); 15, Ssssiorc (scissors); 16, Elbmith (thimble); 17, Debas (beads;) 18, Nragdintoocnt, (darning-cotton); 19, Knodbi (bodkin); 20, Ranettp (pattern); 21, Dugro (gourd); 22, Inkdleeettngin (knitting-needle); 23, Blueck (buckle); 24,

Grsa (rags). Tie a pencil to each card and allow the guests to write the correct word opposite each misspelled one. Simple prizes may consist of articles needed in the sewing-room.

Thimble Parties

Thimble parties are popular this year, when groups of women come together to sew for the needy. Invite your friends to luncheon, which may be a simple affair. Serve clam broth, creamed chicken, rolls, sweet pickles, salted peanuts, chocolate corn-starch with custard sauce, little sponge cakes and coffee. In the center of the table have a work-basket filled with ferns, and, with the dessert, bring on a pretty glass dish of strawberries (emery bags, of course) to be given as favors. The following contest may be arranged on cards, and guessed during the luncheon:

1. What does the farmer do to his sheep—shears.
2. To pick one's way—thread.
3. What is thrown away—waist.
4. A sign of servitude—yoke.
5. A berry—thimble.
6. A blow—cuff.
7. A company of musicians—band.
8. Deep-sea animal and part of his body—whalebone.
9. An exclamation—a-hem!
10. A kind of music—piping.
11. Necessary to hang a picture, and part of the human body—hook and eye.
12. A piece of furniture and a weight—cotton.
13. Money and a derogatory adjective—cashmere.
14. A grassy yard—lawn.
15. Preposition and a fisherman's term—overcast.

Games and Entertaining   **Home Parties**

16. What the cook does to the turkey—baste.

17. A part of an eatable animal—mutton leg.

18. Part of a door—panels.

19. A negative—knot.

20. A prejudice—bias.

Tie a small pencil to each card, using narrow white tape for the purpose, and present the work-basket centerpiece to the person making the largest number of correct guesses.

A Married Folks Party

In a gathering of married people, each man might write a description of his wife's wedding dress, while the wives might describe any difficult experiences they had met with, such as ironing "John's" shirts, making "pies like mother used to make," preparing for unexpected company when the larder was empty, etc. A pretty centerpiece is in the nature of a Jack Horner pie, but may be called "Cupid's Pound Cake" instead. Suitable favors are wrapped in tissue paper and tied with red ribbon or string. These strings are pulled through a hole in the center of the (paper) crust, and one is carried to each place. The sides of the "cake" are decorated with paper hearts or cupids, and at a given signal each guest pulls a string and finds a favor at the end of it.

To match partners for the affair, have a number of pretty cards in a fancy basket; write the name of a lady on each. The men each take out a card and must take for a partner the lady whose name appears thereon.

Hat Making

Provide each lady with a paper napkin and ten pins, with which to make a hat in five minutes. Let the ladies put the hats on and march around, while a jury decides upon the prettiest hat. The one making it receives a prize—a paper of pins, or an equally insignificant article.

A Cat Party

A Cat Party will be funny and novel. Cut out of magazines, advertisements, etc., all the pictures of cats, large or small, which you can find, and draw, paint or trace a cat on your invitations. Cut pieces of cardboard about 6 x 14 inches in size; at the top of each one paste a cat picture and mark it in large numbers. Underneath in large letters, write or print a sentence representing a word which has for its first syllable "cat."

1. A midnight cry.
2. A list of books.
3. An ancient engine.
4. A raft with a sail.
5. A waterfall.
6. A mountain range.
7. One who furnishes food.
8. An accident.
9. One of the finny tribe.
10. One of the furry tribe.
11. To ask questions.
12. A sauce.
13. A common herb.
14. An instrument of punishment.
15. A dupe.
16. An insect.
17. Part of an instrument.
18. A cold.
19. A church.
20. A jewel.
21. One of the feathered tribe.
22. A religion.
23. A common weed.
24. A burial place.
25. Grazing animals.

This list can be added to indefinitely, the dictionary furnishing several pages of words beginning with "cat." Pin or stand the cards in conspicuous places about the room, and furnish each guest with a pad and pencil. The answers are numbered and the person guessing the largest number receives a small prize.

It will be quite easy to obtain suitable prizes for a small sum; if not, a penwiper or needlebook made of black cloth, cut the shape of a cat, will do.

The game of Stuffed Cat is next in order. Two leaders are chosen, they form the company into two lines of equal numbers. One line is called "face," the other "back." One of the leaders throws a stuffed (calico) cat into the air. If it falls on its back the "face" line laughs loudly, the other line keeps still; if it falls on its face, the other line laughs, etc. If any one laughs on the wrong side, he must forfeit a player to the other side, and the game goes on until all the players are on one side.

Pieces of black tissue paper, about 5 x 7 inches, are then passed around, and each player is expected to make a cat out of his piece. He must tear it with the fingers, as no scissors are allowed. When done, they are mounted on sheets of white paper and the guests vote for the best and the poorest. If there is room, the old-fashioned game of Pussy Wants a Corner can then be played, and for those who would rather keep quiet; provide string for playing Cat's Cradle. For refreshments, serve cookies cut with a cat-shaped cutter and Catawba grape-juice.

Progressive Novelty Party

A progressive novelty party is easily arranged and enjoyed by young and old. Arrange as many tables as will be required, seating four persons at each table. At table No. 1 have some sort of a card game, such as Flinch, Old Maid, Snap, etc. For the other tables you can have peanut stabbing, using hatpins to remove the peanuts from a bowl; needle threading; the new jackstraws, using magnets and steel nails, the latter of various sizes, and the game played just as the original jackstraws was played. A miniature game of ring toss can be prepared for another table, using a spool and skewer (the latter driven into a smooth pine board) for the stake, and the rubber rings which come for fruit jars for the rings. Table croquet can also be managed with a home-made outfit,— a smooth pine board with the spool and skewer stakes, crossed wire nails for hoops, mallets made of corks and skewers, and moth balls will be sufficient for a good game. Number your tables, allow five minute intervals of play, and when the bell rings those who have won move up to the next table; and so on indefinitely until the end. Simple prizes may be given to the winners.

After this let all join in the game of Flying Angel, seating all the players save the leader in a semi-circle. Give the leader a white handkerchief and ask him to pass it to the player at one end of the circle. The game is to throw this handkerchief from one player to another without it being caught by the leader, who may stand wherever he chooses or run about from player to player. If the leader touches any player while the handkerchief is in his possession, that player must take his place as leader.

Special Occasions

February Entertaining

The birthdays of Washington and Lincoln, together with St. Valentine's day, afford many suggestions for the hostess. For Lincoln's birthday, the guests might come dressed as members of the G. A. R. and their families. War songs and plantation melodies could be sung, and some one could read the Gettysburg address aloud. Decorate the table with a cheap plaster bust of Lincoln, with streamers of red, white and blue crepe paper, extending to the corners. Serve army beans (baked), salt horse (corned beef), pickles, hard tack (biscuits), cake and coffee.

St. Valentine Entertaining

Decide on your date and send out invitations from ten days to two weeks in advance. If the fourteenth is the chosen day, a roll of red wall-paper, cut into hearts of various sizes, will help along in your decorations, and an occasional arrow of gilt paper may be thrust through the hearts. There are other suitable decorations— favors, napkins, etc., to be had from paper manufacturers for a small sum, and you can use the heart device through the entire evening. Start with a heart hunt, using paper hearts, candy hearts, etc., with a prize for the one finding the largest number, and a mitten for the booby prize. The larger hearts can be cut in two pieces, and these are matched to form partners. Pass tablets and pen- cils, and ask each man to write a proposal of marriage to his ideal, while each girl writes an acceptance to her ideal. These should be well mixed up and then drawn from a hat, and a proposal and acceptance read together. For refreshments serve creamed chicken in heart-shaped paper cases, or heart-shaped sand- wiches, little cakes cut heart-shaped, or a large layer cake with heart- shaped candies to decorate the top, and a ring, a thimble and a piece of money hidden in the cake. The one who gets the ring will be married first, the thimble goes to the one who will remain single, while the money brings wealth.

A Valentine Engagement Luncheon

Last year a girl who was about to announce her engagement, did so on St. Valentine's day. She entertained her girl friends at luncheon, the table being suitably decorated with a cen- terpiece that told the story. Ferns, asparagus vine and red tulle were prettily massed around a black velvet cat emerging from a bag of red silk. Around its neck was a red ribbon, and hanging from this was a white card containing the names of the newly betrothed pair. "The cat was out of the bag" at last. Heart-shaped place cards, strings of red paper hearts draped around the sides of the table and from the chandelier to the corners of the table, and sandwiches, cakes and candy in heart shapes, all carried out the idea.

St. Valentine Games

Follow the heart idea in your games. Old Maids may be played with a set of cards each of which is pasted on a red cardboard heart. Play "Hearts Up" with a tiny heart, just as you play "Up Jenkins." Have the men guests each write a description of the ideal lady of his heart, and the girls in turn write descriptions of their ideal mates.

A Valentine Soap-Bubble Game

Suspend a large sheet across one corner of the room and on it paste three large red paper hearts, numbering them one, two, three. Above each one write a small verse. The first:

Blow your bubble right on here,
And you'll be married before another
    year.

Above the second write:

   To be engaged this very week
   Number two is the one to take.

Above the third write:

A sad, an awful fate awaits the one
   who seeks me,
For he or she will ever a spinster or
   a bachelor be.

On a small table nearby have a large bowl filled with soap-suds and also clay pipes decorated with hearts. Small paper fans should be given to each player, who first blows the bubbles off his pipe, then tries to fan them on the heart where he wishes them to go. Most will try to avoid heart number three.

Your Heart on Your Sleeve

Wearing your heart on your sleeve: One person out of the assembled company retires from the room. Those remaining behind choose a state of mind, such as "Joy." The person outside is called back. When he counts 1, 2, 3, those taking part in the game strike an attitude representing "Joy." The person called in then tries to guess what they are representing. The first person who laughs while the attitude is being assumed is sent out after the player guesses the word—to be "it" next time. Each guesser has three chances; if all three guesses are wrong he goes out again.

Some suggestions for words that can be acted out are: Anger, indifference, jealousy, pity, curiosity, stupidity, pride, expectancy, disgust, fear, self-consciousness, dignity.

To match partners, provide a basket of cardboard hearts and, on arrival, require each boy to punch one of them with a key in his possession. Distribute the punched hearts among the girls. Find partners by matching keys and the keyholes.

An Evening with Shakespeare

Should you be entertaining on St. Valentine's day, heart-shaped programmes with pencil attached should be provided for each guest. Upon the programmes are written the following questions, the answers to which are all names of Shakespeare's plays:

1. Who were the hero and heroine?
2. What mythological characters did they resemble?
3. What did their courtship resemble?
4. Of whom did he buy the ring?
5. What did he write her?
6. When were they married?
7. Who acted as best man and maid of honor?
8. Who were the ushers?
9. What black man tended the door at the wedding?

Games and Entertaining | **Washington's Birthday**

10. What ladies gave a recitation?

11. What three kings (relatives) attended?

12. Where did they make their home?

13. What kingly thing did he do that caused their first quarrel?

14. What did he later say about it?

15. What did her temper resemble?

16. What did he consider his duty after marriage?

17. What did he tell his servant to do?

18. What did she give him?

19. What did their marriage prove to be?

20. What was their daily life like?

21. What man with a Roman nose caused them to forget their troubles?

22. What would you say of their marriage in the end?

Answers:

1. Romeo and Juliet.
2. Venus and Adonis.
3. A Midsummer Night's Dream.
4. The Merchant of Venice.
5. Sonnets.
6. Twelfth Night.
7. Antony and Cleopatra.
8. Two Gentlemen of Verona.
9. Othello.
10. The Merry Wives of Windsor.
11. King Henry IV, Henry V and Henry VIII.
12. Hamlet.
13. King Lear.
14. Much Ado About Nothing.
15. The Tempest.
16. The Taming of the Shrew.
17. Julius! Seize her! (Julius Caesar).
18. Measure for Measure.
19. The Comedy of Errors.
20. Love's Labour Lost..
21. Titus Adronicus.
22. All's Well That Ends Well.

To match partners, each young man is given a heart-shaped card upon which is written the name of some lover famous in history or fiction; the girls are given similar cards, with the names of the ladies to whom these heroes were devoted. Of course each lover seeks his lass and thus becomes her partner. The following are offered as suggestions: John Smith and Pocahontas, Paul and Virginia, Romeo and Juliet, Hiawatha and Minnehaha, John Alden and Priscilla,- Orpheus and Eurydice, Dante and Beatrice, Isaac and Rebecca, Petruchio and Katharine, Gabriel and Evangeline, Pygmalion and Galatea, The Judge and Maud Muller, Touchstone and Audrey.

Valentine Proposals

As the guests assemble, give each gentleman a slip of paper bearing the name of a woman, and the ladies the name of some man noted in fiction as a lover. Thus the one who has Romeo hunts for the lady who has Juliet on her paper. When all know who their partners are, the ladies must evade every attempt on the part of the gentlemen to propose to them during the evening.

A prize is given to the gentleman who has succeeded in proposing, and to the girl who has eluded all efforts of her partner by her wit and sagacity.

For Washington's Birthday

Have a colonial supper for the twenty-second, the guests to be dressed in old-time costumes, the rooms lighted with candles and sperm-oil lamps. A collection of relics may be arranged for, old songs and games indulged in, and a real colonial supper served. Serve tea and old-time cookies, and if it is a money-making affair, sell tea and Japanese china on commission.

For St. Patrick's Day

Cut from green paper a number of pieces approximately representing the map of Ireland. There are as many of these as guests, and to each a little pencil is attached with ribbon. Each player is given one, which he or she is called upon to fill out with the names and positions of the various large cities, rivers, mountains, etc. A book bound in green makes a suitable prize.

Chalking the Pig's Eye

If possible, draw a pig on the floor; if the floor is not suitable, draw it upon a blackboard, or, using charcoal, upon a sheet. Then blindfold each player, turn him around three times and tell him to mark the pig's eye with a cross. A variation on this is to have the players draw the tail, which should be omitted in the original drawing. Still another fun-making scheme is to pass sheets of paper and have each person, while blindfolded, or with closed eyes, draw a pig. Suitable prizes for such contests are the little brown earthenware pig money banks, to be had for a few pennies each.

April First Entertaining

Some clever games for entertaining on All Fool's day may include the "Foolish Walk," for which pile sofa-pillows, books, plants and anything in the way of obstruction on the floor; then tell a certain person to mark each article carefully in mind, blindfold him and tell him to walk across the room. In the meantime, after the victim is blindfolded, the objects have been noiselessly removed, leaving the floor clear. It is amusing in the extreme to see the blind one making his way, and when the bandage is removed the astonishment is great.

This may be followed by a guessing contest. Provide cards and pencils for each guest, with numbers for each course of a dinner menu. Have your courses prepared beforehand, bring each one in separately, and after two minutes remove it. The contestants write down the name of each course as they guess it, and a prize is given to the one making the largest number of correct guesses. The following "dishes" are suggested: Oysters, short-pointed ends of blue crayons (blue-points); soup, small brown cardboard turtles, in a soup-plate of water; relishes, toy or paper red dishes (radishes); crackers, tiny fire-crackers; meat; a toy lamb in a small pan; poultry, a map of Turkey with the name erased; dessert, a curl of hair (lady-lock) or a strawberry emery in a dish of ice (frozen strawberry); cake, the ends of sulphur matches (devil's food); nuts, the iron nuts used in bolts and machinery. Decorate with vegetables instead of flowers, and amongst the refreshments have a dish of chocolates, which are nothing but cotton-batting, dipped in melted chocolate.

Hallowe'en

Old games, old customs, old tricks and charms are appropriate for Hallowe'en entertaining, and the gathering can take the character of a Hard Times Social, the guests wearing their oldest clothes (a prize may be given for the very oldest), and if the company can be accommodated in

Games and Entertaining   Hallowe'en

a large kitchen, so much the better.

Use wrapping paper and the cheapest envelopes procurable for the invitations, and arrange the table with white oilcloth or a colored cotton cover. For a centerpiece fill a toy wooden washtub with rich red apples. Around this arrange candles stuck in potatoes or carrots for holders. Tin pie plates, or the wooden picnic kind, may be used, with paper napkins and tin cups. Refreshments may be simple—sandwiches, salad, salted or plain peanuts, gingerbread, doughnuts, cookies, molasses candy and coffee; or, if a hot supper is desired, you can choose between baked beans or scalloped oysters, rolls, pickles, individual pumpkin pies, coffee and nuts.

Corn husks will hold salad. The nuts should be brought to the table in a great wooden bowl. This should be placed in the center of the table and the guests will be asked to help themselves. However, some will be found tacked to the bottom of the bowl, two of the guests will find their nuts fastened together by means of a tiny wire or thread, and all kinds of confusion will result. When the nuts are opened they will be found but empty shells, the kernels having been removed to make way for small bits of paper on which is printed the fortune of the finder.

Hallowe'en Fortunes

Here are some Hallowe'en fortunes, short and optimistic, that may be used:

For you will come bright, happy days.

You will never marry unless you are suited.

Profit will attend your ventures.

Your companion in life will be ever true.

You have genius, but must develop it.

You will not become wealthy, but you will never want.

Early in life you will know honors.

You will wed the one you love.

Continue unafraid of work—it is not afraid of you.

Of course there are sorrows in your life, but they are balanced by joys.

Never spend money foolishly—you can not earn it foolishly.

You will travel extensively.

Your wealth will come from the earth.

A companion worthy of you will enter your life.

Hallowe'en Suggestions

At one successful Hallowe'en affair the guests entered the house through a cellar door. The cellar was lighted by means of pumpkin lanterns, and a ghost met them and silently motioned them toward the cellar stairs. Instead of ducking for apples, which wets the hair, have two pieces of stick, sharpened to a point at each end, and these nailed together to form an X. On the four points are stuck, respectively, an apple, a potato, a piece of soap and a piece of candle. The X has a piece of string caught into the nail in the middle, and is suspended from the chandelier and set spinning. Then you stand around and try to bite at the apple as the cross spins. If you bite the apple it is a sign of a rich and early marriage; if the potato you marry a farmer; the piece of soap means you marry a poor man, and the candle is to light you sitting up waiting for your husband to come home.

The old tricks never lose interest, but for the sake of novelty may be changed. A piece of candle in the tub of water may take the place of the apples for which young people enjoy "bobbing." Provide a package of the little paste alphabets used in soup. Place these in a bowl, allow each guest to draw a handful and scatter them in a tub of water; the combinations they will form, suggest the name of the wife or husband to be.

For hallowe'en burn all the letters of the alphabet on a big pumpkin, with a hot poker. Then hang it in the doorway, twirl rapidly and have each guest try to stab a letter with a hatpin. The letter hit is supposed to be the initial letter of one's future mate. If none is hit, celibacy is the fate in store.

Costumes for Hallowe'en

The list of costumes is endless. Sheets and pillow-cases, with a white

Witch

or death's-head mask, are easily arranged for a ghost costume for either boys or girls. A witch requires a dark

woollen skirt, a black cape and a wig of coarse hair hanging in strings from beneath a black pointed hat. She carries a broom, of course, and a black cat made of paper may be perched on her shoulder.

Topsy has her face blackened, wears a wig of black hair done in little plaits all over her head, a short-waisted and short-skirted dress of gay cotton, striped stockings and old shoes.

A gipsy girl wears a red petticoat, a black velvet bodice with a silk scarf

Gipsy

around her waist, a gaily colored handkerchief around the neck and a broad banded bracelet on the arm. She carries a tambourine.

With a checked gingham dress, a huge apron and a bandana handkerchief over the head, any one will pass for a colored "mammy."

For a rag-doll costume, take two pieces of muslin each about fourteen inches long and eleven wide, and round the corners. Sew up on three sides; paint nose, eyebrows and mouth on it and cut out places for the eyes; slip this over the head. Wear white cotton gloves, and wear stockings over your shoes and a cotton dress made

Games and Entertaining   Christmas Holidays

with a long skirt. Practice walking in a loose-jointed, floppy way, to carry out the illusion.

Rag Doll

A baby costume is easily fashioned by wearing a rather full nightdress over long white petticoats. A mask representing a baby face, a bib, a white cap and rattle complete the costume.

As for the boys, a slender lad dressed as a girl is always a success. Uncle Sam, Indians, cowboys and pirates are always popular and are easily copied from pictures.

Christmas Holiday Entertaining

Delightful entertainments may be given during the holiday season, when the young folks are home from school or college. A gathering of friends and neighbors, old and young, may be entertaining and instructive and yet inexpensive. The decorations consist of the Christmas greens, bells, etc., and the tree may be the center of attraction. There are many pretty cards at this season which may be used for invitations and place cards, the shops are full of toys and novelties which may be used as favors and prizes, and the refreshments may be very simple,—home-made candies and Christmas cakes playing an important part in your menu.

A Twelve-Month Social

All young people, and some not so young, love to "dress up," and an interesting affair can be made out of a Twelve-month Social. Ask your friends to come dressed or wearing some device to represent the months of the year, and offer prizes for the best ideas. January may come as Father Time; February offers a wide choice with its famous birthdays and the feast of St. Valentine; March offers the hare and "Paddy;" April brings the Easter bride; May is the blossom month and is also sacred to our dead heroes; June brings roses; July is, of course, patriotic; August offers the summer girl; with September comes Labor Day and the "whining schoolboy, with his satchel, and shining morning face." We now remember Columbus in October because of the anniversary of his discovery of America; while November brings us Thanksgiving day, instituted by the Pilgrims; and the year winds up very properly with Christmas and Santa Claus.

Charades

A Christmas or New Year's dinner could be enacted in charade form, the audience to guess the viands as acted. Give each person a copy of the menu, with only the courses written thereon. These are filled out, as guessed, with a prize for the best. The menu is as follows:

Soup, noodle (new-dull); roast, turkey (Turk-key); gravy, giblet (jiblet); vegetables, potato (pot-eight-o); cauliflower, (call-I-flower); succotash (suck-at-ash); jelly, current (currant); dessert, plum pudding (plumb); beverage, coffee (coughfee). The old game of "Consequences" may be varied for the occasion, called "Resolutions," and played accordingly. If the party is held on New Year's eve it may wind up with the birth of the New Year, finding all standing in a circle with joined hands. As the clock strikes twelve, the company sings, "Should Auld Acquaintance be Forgot," and then with a handshake and a greeting for every one, the party breaks up.

Game of Resolutions
For New Year's

Provide guests with papers and pencils. Begin by having ten letters of the alphabet read to the company. These are to be copied down and the guests must choose a new year's resolution of ten words, each beginning with one of the letters used in order in which they have been given out. These impromptu resolutions when read, will cause much amusement.

Things For Boys To Do

Making a Megaphone

Tom and Bill are setting a gate at the far side of the twenty-acre pasture. You want Bill to bring in the monkey wrench, so you swing your hat and bawl at him for fully five minutes. Finally they both drop their tools and come in empty-handed.

It wasn't done that way at the Fort Myer training camp. Our company, split up into platoons, might be drilling all over the huge parade ground; but the West Point instructor, with his megaphone, could control the most distant unit as easily as the nearest one. After I came home —discharged for a small physical defect—I made several useful megaphones.

The lid of a large pasteboard box

furnished the material. I worked out the pattern, Fig. 1. In one corner of the box lid I drove a pin with a string tied to it; by looping the other end of the string around a pencil I drew the two curves at the distance I have figered.

Cut out the piece that is shaded and roll it up; lap the edges about half an inch and glue them so. Then hold it as in Fig. 2, with your thumb and first finger crooked around the small end, which is pressed against your lips to form a sort of mouth piece. A sentence spoken in a firm, distinct tone will carry a wonderful distance

A couple of thick coats of shellac will waterproof the megaphone. A larger megaphone, naturally, will carry farther. Still,

the one shown will answer most folks' needs.

What Boys Can Do   **Things to Make**

Making a Barometer

This will show the air pressure, and thus tell the amount of moisture the air contains. Fill an ordinary quart fruit jar half full of water. Put a small quantity of water into a long-necked bottle and invert it quickly into the jar, as shown. By raising the bottle carefully just enough of the water may be let out of the bottle so that it will stand at B.

When a storm is approaching the air in the bottle will push the water in the neck down to R. When it is to be fair the water will be forced up to D. If the water is down, say to the line A in the evening and still lower in the morning, look for storm. A slight lowering is natural every evening and does not mean anything.

With the aid of this device and a little keen observation as to the direction of the wind, a boy can predict weather for many hours in advance.

Making a Clothes-Tree or Costumer

Wide-awakes can easily make this useful piece of furniture. Any kind of wood will do, but use oak if possible. It requires one upright five feet six inches long and two inches square; one base (Fig. 2) twenty inches long, two and a half inches wide and two and a half inches high; two bases (Fig. 1) each eight and three-quarter inches long, two and a half inches wide and two and a half inches high; and four triangle braces (Fig. 3) four inches across base and eight inches high.

On the twenty-inch base, directly in the center, cut a four inch square, mortise one inch deep, into which fit the upright. Nail it securely. To each side nail or glue one of the side bases. Then nail one of the four triangle braces to each base and against the upright. Screw on four clotheshooks at the top. The tree is then ready for a coat of varnish or paint.

Note: By using small-headed finishing nails, driven below the surface with a nailset, and then puttying the holes, the wood will not split nor will the nail holes show.

Making a Box Kite

Use four light-weight wooden sticks five-sixteenths of an inch square, thirty inches long; four sticks five-sixteenths of an inch square, nineteen and seven-eighths inches long; four strips three-sixteenths of an inch thick, one inch wide, ten inches long; and hemmed chintz or muslin seven inches wide and 124 inches long.

To make the framework, nail the ten-inch strips to the four thirty-inch sticks, two and a half inches from each end. On the ten-inch strips, one inch from each end, bore a small hole. On each end of the

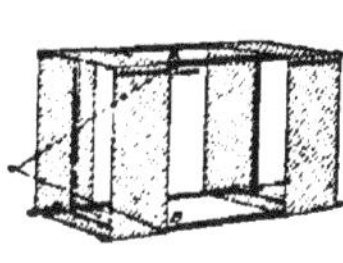

Fig. 1

nineteen and seven-eighths-inch sticks make a round pin (Fig. 2) to go three-sixteenths of an inch deep in the hole on ten-inch strip. Fit these pins in holes and the framework is complete. A larger kite can be made by increasing all dimensions in proportion.

Run chintz or muslin tightly around each end, as shown in Fig. 1.

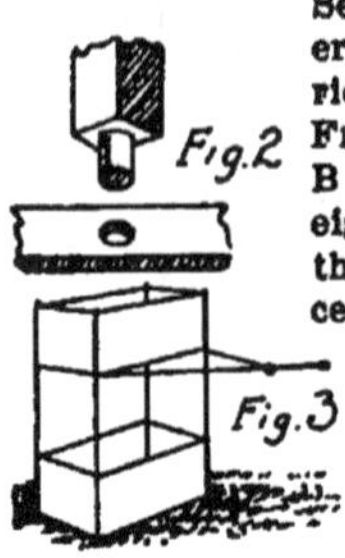

Sew the ends together and tack the fabric to framework. From sticks at A and B tie ends of an eighty-inch piece of thin twine. In the center of this twine fasten a small ring. To this ring attach the kite-flying string.

To fly the kite, stand it on the ground as in Fig. 3. Walk away about fifty feet and gently pull on the string. The kite will rise at once. When it begins to pull or dive, let out more string quickly and give it a number of jerks. This jerking causes the kite to rise. Use strong string. No running is necessary to make it fly.

Making and Using Skis

To make skis, use two pieces of white ash, free from knots and with a straight grain, six and one-half feet long, four inches wide and half an inch thick. Bring one end to a blunt point, starting nine inches from the end (Fig. 1) and curve by steaming

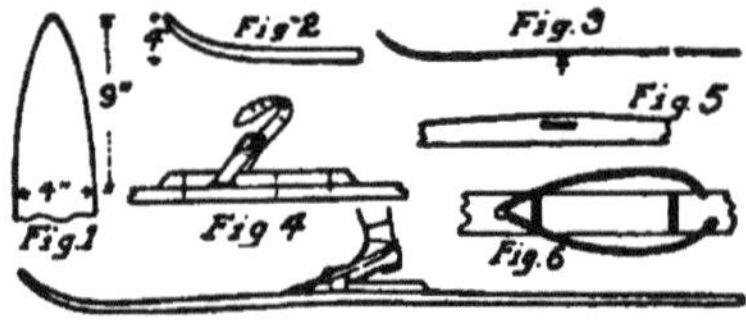

the wood and bending it into the shape shown in Fig. 2. Shave the curved end thinner than the rest of the ski. Make a shallow grove half an inch wide in the center of the bottom, running the length of the ski from where the curve begins. Make

each ski with a "spring," so that when resting on a flat surface the center will be one inch from the ground (Fig. 3). Sandpaper the bottom surface until it is very smooth.

The toe straps to hold the foot should be sixteen inches long and one and a quarter inches wide, with a buckle on one end and holes punched in the other. To attach the strap to the ski, take a board three-eighths of an inch thick, four inches wide and fourteen inches long,

and four inches from one end make a slot on the flat surface large enough for the strap to pass through. Fasten this board to the ski with screws (Fig. 4), the end with the slot toward the front and so placed that it holds the heel-strap in place. Where the foot rests, fasten a piece of rubber matting to prevent the foot slipping.

For skiing, bundle yourself up and wear heavy shoes. Choose a day when the thermometer is not above 20 degrees, for skis will not slip on sticky or "heavy" snow. The snow should be soft; a crusty surface is not good. Select a hill that is not very steep and practice coasting down it. Keep the feet as closely together as possible (skis parallel), the body upright, and knees and hips straight. Lean slightly forward. When you can keep your balance, try steeper hills.

Climbing a hill is known as "herring-boning." Edge your skis and

What Boys Can Do

Things to Make

straddle your points apart until they nearly form a right angle (Fig. 7). The tails are lifted over each other at each step. On the level use a slipping, sliding motion. Jumping should not be tried until you have thoroughly mastered the skis, then try only short ones. A fairly safe jump can be made on a hill as per Figs. 8 and 9. Make an embankment about one foot high to start with, and after practising jumping from that, keep raising it until you reach a height of four feet.

Making a Snow House

We are sorry that all of our boys haven't the opportunity to build a snow house. It's lots of fun, and if the house is properly made it will last for a long time. First of all shovel some snow in a big pile, stamping it down until it is compact and hard. Make the pile at least eight feet high. Then at one side start to hollow it out. The entrance can be small, so that one has to creep in; or it can be made large. If the snow has been carefully packed the roof will not cave in. The snow brought out can be piled at the sides of the entrance as a windbreak. If the nights are very cold, a little water can be sprinkled over the outside of the house. This will freeze and the house will last longer. We'd like to see a good photograph of your snow house and have a full description of it.

Making a Cardboard Lantern

Secure some good tough cardboard. Make base (Fig. 1) twelve inches square. Score it an inch all around, cut at each corner and turn up edges. Make and fasten securely in center a tin holder for the candle. If you can't make this, an ordinary tin candlestick can be used. The top (Fig. 2) is made the same as bottom, except that in the center a hole five inches in diameter is cut. The sides (Fig. 3) are twelve inches square; score and turn in one inch as per sketch; cut out center as shown. Cover openings on all four sides with yellow or red tissue paper. Any design desired can be painted or pasted on the tissue paper. Paint outer side of cardboard black. Glue lantern together or fasten with wire. Attach a wire to each of the four upper corners, join and hang lantern from a pole. See that candle is held securely, and watch out that top or sides do not catch fire.

Making Money—A Shampoo Board

A useful article that every girl or woman in your district will want to buy the minute she sees it, is a shampoo board, made from a board 8 x 10 inches. Stand the board on end and, about an inch and a half from each upper corner, begin to cut down a curve which must be shaped to fit the neck, as shown in the illustration below. Cover the edges of the curve with rubber from an old inner tube.

A brace to make the board stand up may be sawed from a block one and a half inches thick and six inches square, making the cut diagonally across it, as shown in illustrations.

The use of this board when washing long hair avoids the tangling sure to follow when the hair is thrown forward over the head in the usual way. Place the board in position in the sink or a box of proper height. The user leans back, places her neck in the padded curve, letting her hair fall down in its natural position in the sink. The shampoo is quickly given by another person without tangling a single hair—a point not to be overlooked.

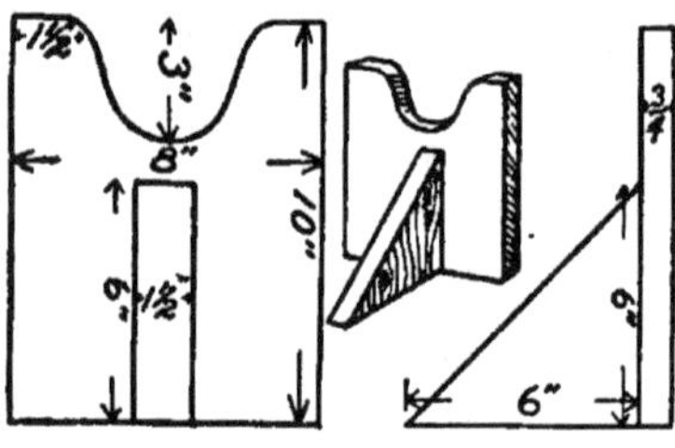

Sailing on Skates

Skating is the king of winter sports and, if a sail is used, the fun is increased fifty per cent. A race with sails will make your blood tingle and give you all a great chance to show skill in tacking.

This sail is very easy to make. Use

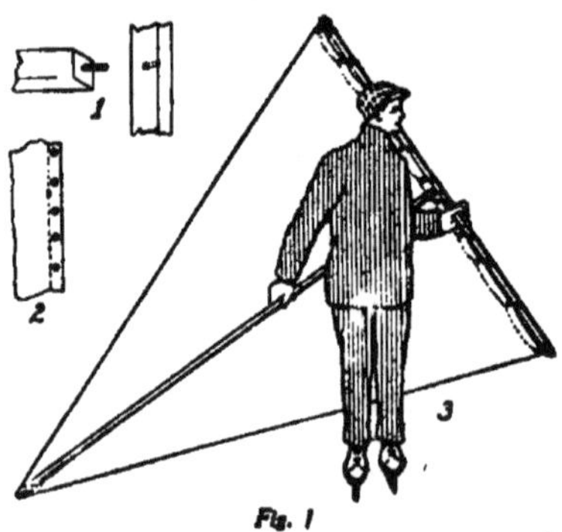

two strips of ash or white pine, 1¼ x 1¼ inches, one about nine feet long, the other six feet. Fasten long strip into short strip by pin, made by driving a heavy wire nail into the end of long strip; file off head of nail. If the wood shows signs of splitting, wrap it with fine wire. Bore a small hole into middle of short strip. (See detail drawing 1, Fig. I.)

To make the sail: Take a piece of unbleached muslin 3⅛ x 9½ feet ("A," Fig. II). Cut it in half (as at "B," Fig. II). Measure nine feet diagonally from each point and cut each piece across (as at "C," Fig. II). Place the two parts together (as at "D," Fig. II), and sew, making the sail about 6 x 9 feet. Hem all edges, which will make it slightly smaller.

Make a two-inch hem and fasten grommets every nine inches on the side of the sail that is to be attached to the short strip. (See detail 2, Fig. I.) Lace with heavy cord (3, Fig. I), and tie sail securely at far point of long strip. When sail is not in use, untie, lay the strips side by side and roll them up inside of the sail.

Study Fig. I closely and note the correct way to hold the sail, remembering that the *left* arm is *around* the long strip.

Tacking: With a little practice you can readily learn to "run before the wind," and to "quarter" and "tack," all of which are nautical terms, and

What Boys Can Do   Tree House

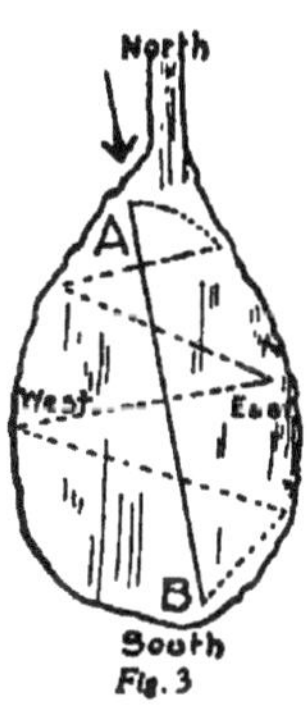

are used because the skate sail is no more nor less than a human sailboat. For instance, if the pond, lake or river upon which you skate is something like Fig. III, and the wind is blowing from the north, as shown by the arrow, you could start at "A," Fig. III, place the sail behind you (holding it as shown in Fig. 1), and speed to "B" at the south end of the pond. This is called "running before the wind." Then, starting out from "B" at the south end of the pond, you will find that by going at an angle with the wind, either right or left, you can make considerable headway. You are now "tacking," and by tacking first to the right and then to the left you will reach the head of the lake or pond once more.

If the wind is blowing from the east, by setting the sail at the proper angle you'll find that you can go up and down this same pond without tacking. This is by far the most fun, and is called "quartering."

A Tree House

There probably is at least one tree on every farm that can, be used as a base for a tree house. A good strong tree, with branches spreading in such

a way that boards can be easily laid from branch to branch, should be used. If properly built there is no danger, and much enjoyment can thus be had.

Build it strong, take no chances. Remember, safety first. Here is a house built by boys out West.

A Handy Twine Holder

Do you know where I can find a piece of string? How often is this question asked? Well, here we have an idea which when properly worked out will give us an ornament, good luck omen, and a good twine holder.

First design the front and back parts of a cat, as in Fig. 1. Trace these on a piece of three-eighths-inch poplar wood and cut along the outline with a scroll saw. Sandpaper both pieces until they are smooth on both sides and on the edges.

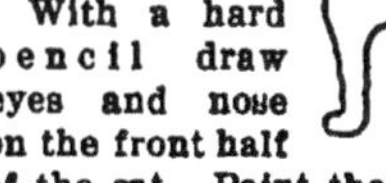
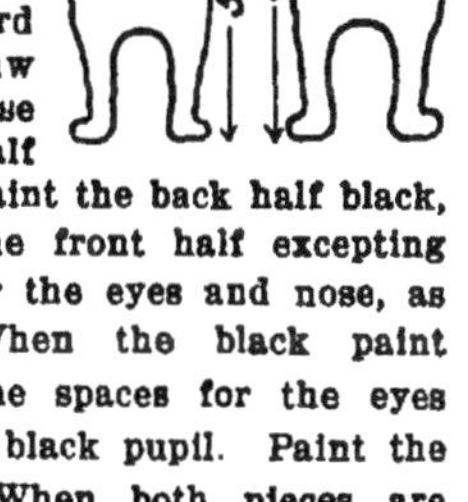

With a hard pencil draw eyes and nose on the front half of the cat. Paint the back half black, and all of the front half excepting the spaces for the eyes and nose, as indicated. When the black paint dries color the spaces for the eyes green with a black pupil. Paint the nose pink. When both pieces are thoroughly dry varnish them with white shellac.

After the front and back halves of the cat are finished, measure a piece of wire one-eighth of an inch in diameter and file or grind both ends to points. This acts as a support for the spool which revolves on it, Fig. 2. Force these pointed ends into the wood by carefully hammering the wood on to the points. This should

be done before the cat is fastened on to the base.

Put the spool in place and wrap on the twine (any color). Make a wooden base for the completed twine holder and fasten by nails or screws which pass up through the feet.

Varnish the entire wooden object and your article is complete. If the body is short the effect will be better than if too long. Paint white whiskers on each side of the nose.

A Tree for Climbing

Climbing is a natural instinct, and a tree for climbing should be in every playground. A straight tree trunk about thirty feet high, with the bark removed and made smooth but not necessarily even, should be used. Plant it *securely* in the ground and protect the top by a platform sufficiently wide not to allow its edge to be grasped by the climber, who must not get on the platform. Only one boy to climb at a time. Take turns and play fair.

The Balancing Tree

This can be easily made for your community playground. Cut down a large and perfectly straight tree, free it of the bark and limbs and round it off; it should be fifty or more feet long. At the thicker end the tree may be two or more feet in diameter. It tapers to an end of four to six inches in diameter, which is to be free to sway.

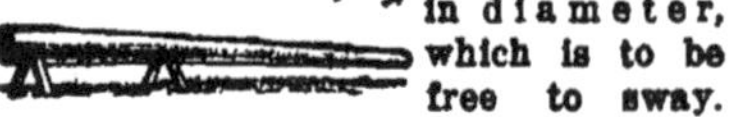

It is supported by two securely nailed wooden feet (which are anchored in the ground and placed as shown), one at the extreme thick end, the other one sufficiently far from the thinner end to allow the thin end free play to bend and sway. The tree is so supported that at its thicker end its upper edge would be three to three and one-half feet from the ground. This tree, as its name implies, gives a chance for balancing exercises on a broad and steady and also on a more narrow surface, which sways and bends. It is not so dangerous as tightrope walking and just as much fun. Always see that the ground on both sides of the tree is free from stones, and never push anyone off. Take your turn and play fair.

Horizontal Bar

Here is a horizontal bar suggested by the Playground Association, and it can be easily made. Two 2 x 6 inch planks seven feet long and a 1¼ inch x 6 foot steel pipe are required. Sink the planks securely in the ground. Bore holes in them to admit the ends of the pipe; the pipe to be left free to be raised or lowered to a higher or lower level by moving it to higher or lower holes. The pipe should be kept from turning by putting a bolt or pin through both plank and pipe at right angles to the other holes in the plank. Your blacksmith will help you out. Instead of pipe, the crossbar can be made out of fine second-growth hickory, thoroughly seasoned and nicely finished. Some prefer steel bars, others wood. In either case use good material to prevent accidents. Under the bar have plenty of soft dirt, free from stones.

 |

Pole-Vaulting

Not very long ago a certain boy learned to pole-vault in the back yard, using a beanpole and a clothesline. Before he was 20 he was asked to go abroad with the team representing the United States at the Olympic games at Stockholm. Not many boys might have such a chance, but here is a way to have some healthful outdoor sport; any boy can make the pole and standards:

The Pole: It should be round and smooth, made of a good straight piece 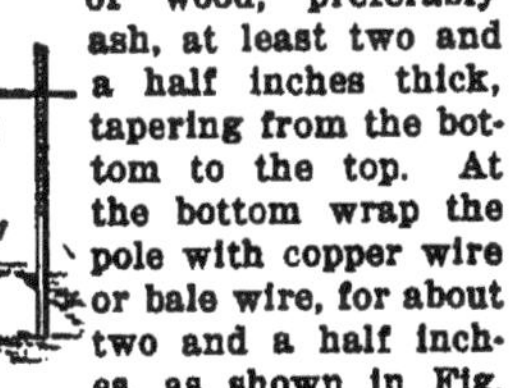 of wood, preferably ash, at least two and a half inches thick, tapering from the bottom to the top. At the bottom wrap the pole with copper wire or bale wire, for about two and a half inches, as shown in Fig. 1. Then drill a five-sixteenth-inch hole at the bottom of the pole, and into this screw a three-eighth-inch bolt about four inches long. Then file off the head of the bolt.

The Standard: For these use any 2 x 2 inch material about ten feet long, with holes bored every inch, beginning about three feet above the ground (Fig. 2). The standards may either be set in the ground or fastened to a base; they should be placed about eight feet apart and a peg provided for each. In this way you can put the crossbar at any height desired. For the crossbar (Fig. 3), any light, straight stick will do.

Take-off and Pit: The take-off (Fig. 4) should be a board two or three inches in thickness, two and a half feet long and four to six inches wide, and should be placed on edge with the top even with the ground and held in place with four pegs. For the pegs, point four pieces of wood and drive them in the ground. Then dig a small pocket in front in which to place the pole.

A pit is needed behind the standards, to land in. Simply dig the ground up and break the clods of earth. Be sure to keep the pit well raked and free from stones.

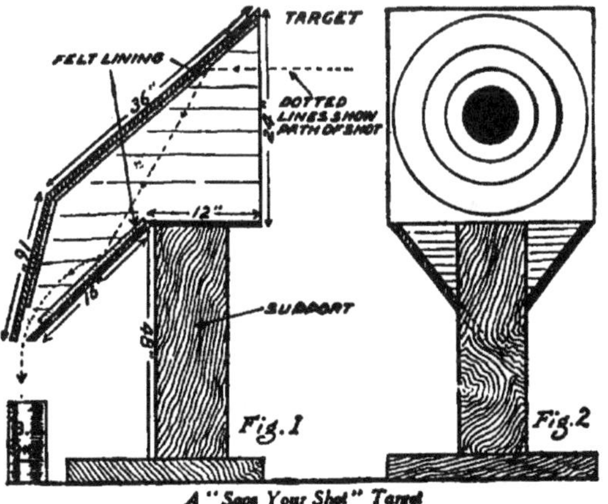

A "Save Your Shot" Target

A "Save Your Shot" Target

This target allows shot to be used over and over, dozens of times. With it the air rifle can be used in the house without any danger of injuring the walls, furniture, etc. Upon a wooden support forty-eight inches high place a box built in the shape shown in Fig. 1. This may be made any size desired. I built this one with the front, or target, twenty-four inches square. The slanting back is thirty-six inches, base twelve inches, and the tapered conductor, which runs the shot into the can has its sides sixteen inches long.

The back and sides of the conductor are lined with felt. The target is a bull's-eye drawn on a cardboard twenty-four inches square, and the cardboard is tacked upon the front of the target box. The dotted lines show how a B. B. shot passes through the cardboard, hits the slanting back, bounds against the conductor side and drops into the can behind the target. The slanting back, with its felt lining, kills the force of the shot without battering it, and so the same shot can well be used dozens of times.

Tub Tilting

Tub tilting on land is equally as exciting and requires just as much skill as tilting on the water.

Secure two barrels, about flour-barrel size, and two poles. Each pole should be from eight to ten feet long, of the lightest possible wood, with a big soft pad on the end. These are spears for attacking. The barrels are set level, exactly at poles' length apart, center to center. Each contestant takes his place on a barrel, and he must try to put the other fellow off. The umpire stands alongside, near the middle. For safety's sake it is a good idea to have some one stand behind each player to act as a catcher in case of accident.

It is counted a foul to push the other player below the knees, to use

the spear as a club, to push the barrel, or to take hold of your opponent's spear with your hand. A foul gives the round to the other boy. A round is up when one boy goes off his barrel. If one drops his spear and can recover it without getting or falling off, it is all right.

A battle usually lasts for about seven or eight rounds. The best players gain their points by wriggling their bodies and keeping in continual motion. There is a lot of fun and excitement in keeping your balance.

Game of Smugglers

The Clan must be divided as equally as possible and a captain chosen for each side. One side takes the part of the "smugglers," the other side the part of the "laws." The game can last as long as desired. One day and one night is a good length, but it can be made longer or shorter. Decide on the time before the game begins. A line must be fixed and two marks made on it about one-quarter of a mile apart. The smugglers camp on one side of the line and the law on the other. Camps should be hidden as much as possible. The smugglers must land their goods across the line between the marks made. The goods of the smugglers can be so many sacks (as decided upon) filled with sand or other weighty material. The players should watch each other's camp and report to their respective officers. The side of the law must (unknown to the smugglers) watch the line between the camps all the time.

The players can report in person, by a companion, or by lights; never by calling, as that would give the position away. A very good line over which to land the goods is a small creek. One-quarter mile in length is about all that can be watched successfully. At night sentries should be changed every two hours, oftener if possible. Two must watch the line at one time. If the smugglers land their goods undiscovered, they win. It is a good game and you must use your wits.

Foot in the Ring

Here is a good game for cold weather; any number can play, but it is best to have squads of eight. For each squad draw on the ground a circle about two feet in diameter. Boy No. 1 comes forward, places one foot in the ring, bending the knee and having the weight of the body over this foot. He then folds his arms and waits the attack of boy No. 2, who, also having his arms folded, hops forward. No. 2 hops around No. 1, who keeps changing his front to where No. 2 is, until he finds a chance to attack No. 1, and while hopping, push him out of the circle. If he succeeds, he wins, and takes the circle, No. 3 coming forward to attack him, and so on. If, however, during the contest No. 2 gets both feet on the floor, he loses, and No. 3 then comes forward to attack No. 1. The player in the ring, so long as his foot is in the circle, may cause the attacker to fall by evading or dodging him. The arms must always remain folded and the pushing must be done with the shoulders, and never with the raised arms. An exciting contest is had by putting two attackers against the one in the ring.

The Snake Chase

It is said that St. Patrick drove the snakes out of Ireland, and here is a game that will help celebrate the event. Any number can play. One boy is chosen "keeper of the snake." The rest are "chasers." The snake is made of a log ten inches long and at least six inches in diameter. Fasten three horseshoes securely on each end (see sketch), and around the log drive spikes. Fasten a rope so that the log can be dragged by the keeper. This snake leaves a trail which is easily followed on ground, but which is much harder to follow on grass or through a woods. The keeper of the snake starts out ten minutes ahead of the chasers. The snake must never be lifted from the ground, but the keeper can retrace his steps and do everything possible to confuse the chasers. Three places should be chosen and known to all as the snake's home, and the object is for the keeper to get his snake to one of these homes before he is captured by the chasers, who must follow the trail, and who cannot leave it to capture the snake.

Giant Stride

The apparatus can easily be made and should be in every community playground. It is the next best thing to flying. A strong old wagon-wheel, a pole eighteen feet long and five inches in diameter at the smallest end, and sixty feet of one-inch manilla rope are needed. In almost every district some one will provide the wheel, and the pole can be cut in the woods. If the wheel has a wooden axle hub, remove the axle from the skein, which is the metal sleeve surrounding the axle spindle to protect it from wear. Shape the top of the pole to fit into the axle skein, then fasten the skein securely in place.

If the wheel has a metal axle, get, if possible, a blacksmith to help you. Have the axle cut off about a foot from the hub and sharpen it to a point. Into the middle of the small end of the pole bore a two-inch hole six inches deep and drive the axle into it. Then have an iron collar shrunk around the pole to prevent it from splitting. An all-metal wheel and axle is better than a wooden one.

Cut the manila rope into four lengths of fifteen feet each, and with copper wire or by splicing attach the four ropes to the hub. Knot each rope every two feet from the bottom for a distance of six or eight feet. Set the pole securely four feet in the ground in concrete. Cover the hub with a tin shield to protect it from the weather. Have the ground around base of stride free from stones.

How to use the giant stride: Catch hold of the rope, start to run around the pole, and the momentum will soon take you off your feet.

Pom-Pom-Pullaway

This is an excellent recess game, as it takes just about fifteen minutes to play it:

A boy stands out in the middle of the long school-yard, or some other open place where two lines can be drawn about fifty yards apart. All the other boys range themselves on one of the lines, or against the school-yard fence. Then the boy who is "standing" yells, "Pom-pom-pullaway!" All the boys on the line run for the other line, past the boy in the middle, who tackles any runner that he thinks he can catch and hold. As soon as he has brought one of the boys to a stop, that boy is compelled to join him in "standing," and these two then call the others who have got by to the other line. "Pullaway!"— the rush is resumed, and probably two more boys are caught; and so on until all are caught; and there is always a terrific struggle at the end, because the biggest, swiftest and strongest are hardest to stop; and by that time these difficult ones have the whole school tackling them. Pom-pom-pullaway is good football practice.

Handy Devices

Leading the Bull Safely

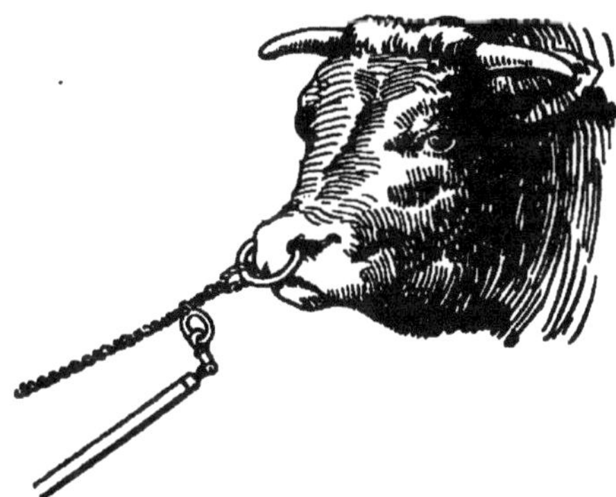

Where a bull is kept on the farm great care must be taken that he has no chance to do any one an injury. No chances should be taken. A rope attached to a ring in the nose serves as an extra hitching arrangement in the stall, but the bull should not be led by this alone. He can charge on the one leading him at will. Put an extra ring in the rope near his nose and have a stick with a snap in the end, and then the bull can be led anywhere in safety, the rope and the stick being taken together in the hand.

Calf-Weaner

Take an inch board and cut it the shape of the head of the calf or steer that sucks the cows; then take a piece one and a half by one and a half inches thick and eighteen to twenty inches in length and put two rows of sharp nails in it; then nail this piece to the narrow end of the board which comes over the calf's nose. Make a halter that fits the calf and nail it to the board, or if it has horns already take a lead-strap and nail it to the end of the board which will come around the calf's nose. Then bore two holes in the upper end of the board and put a rope in it and fasten it around the calf's horns.

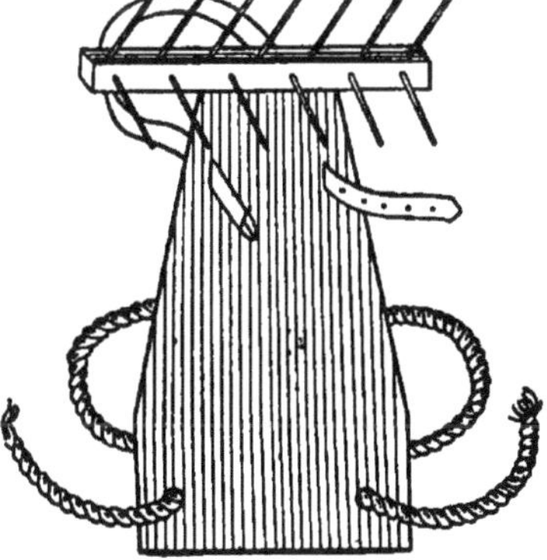

Shelter for Young Stock

Thousands of calves and lambs are turned out to pasture in the spring, and are not housed again until fall, no protection being afforded against cold rain storms and violent showers. This results in great suffering on the part of this young stock and a serious check to their growth oftentimes. Trees afford a slight shelter in case of a passing shower, but for a steady rain they are worse than no shelter. Protection ought to be given, and this can be accomplished

Safety with Bulls

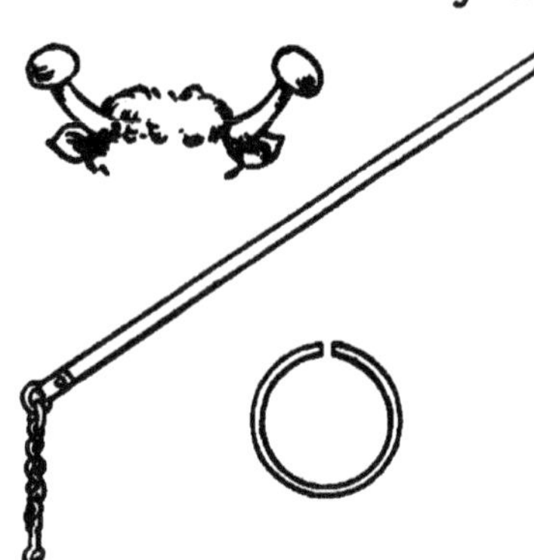

The gentlest bull that ever lived should not be trusted. He may live to an old age, and do no harm; but he cannot be trusted with safety. It's born in him to be inclined toward ugliness and treachery. Put big knobs on his horns—not the ordinary small ones, but such as are two inches in diameter. Put a ring in his nose, and never take him from his stanchion without fastening to this ring such a staff as is shown in the cut. With this he cannot get away nor charge upon his keeper.

Dehorning Chute

A dehorning chute for cattle is easily made as shown. The upper cross-bar is four feet long and of 2 x 4. The levers, clamp-bars and lower cross-bars are of 2 x 4. The two heavy pieces are of 2 x 6. The positions of the clamp-bars and lever when the chute is open are shown by dotted lines. The clamp-bars are four inches apart at the base.

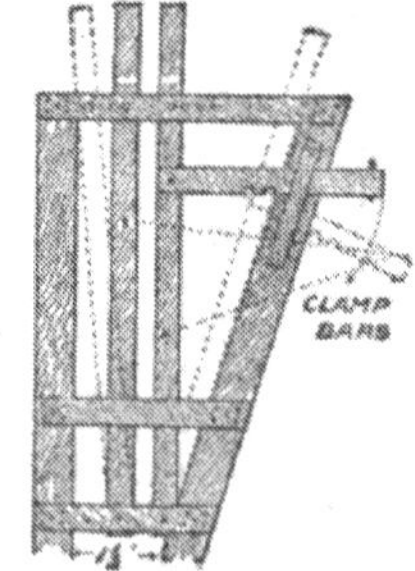

Picketing the Calf

When no pasture can be provided for the calves, they can be picketed so as to have plenty of feed without too cumbersome a rope, by having two picket pins joined by a smooth wire, to which by means of a swivel is attached the calf's rope so that it will slip on the wire. When new grass is necessary move the picket pins alternately. The length of the wire will determine the size of the grazing space.

A Revolving Tether

The calf that is hitched to a bar or stake in the field, is constantly getting "wound up." The cut shows a device —a block of wood and a long sharpened bolt of iron—that will obviate the trouble. The block, with rope attached, turns on the bolt as the calf moves. Other animals can be tethered in the same way, making the bolt longer and stouter as the size of the animal increases.

One More Tether

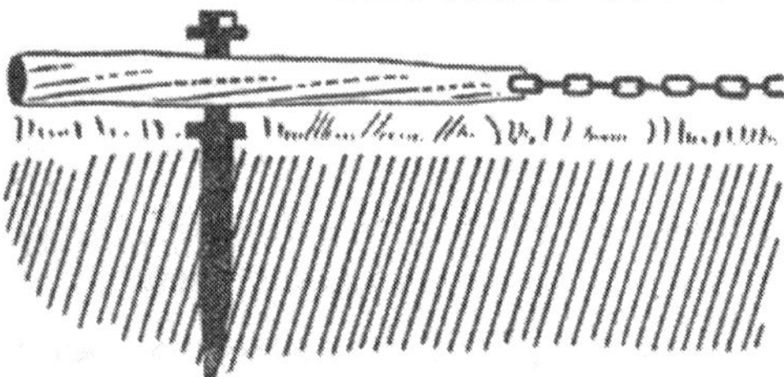

One of the best tether pins can be made of an old axle and the axle box that fits it from an old hub. The axle is cut off eighteen inches from the shoulder and sharpened to go into the ground easily. Then the box is sunk in a stout pole, ten to twenty feet long and a rope or light chain, with swivel, attached to this. An occasional oiling, the same as when the axle is serving on a wagon, will make it all the pleasanter to use. A washer between the nut and axle box may be needed to prevent loss of the nut.

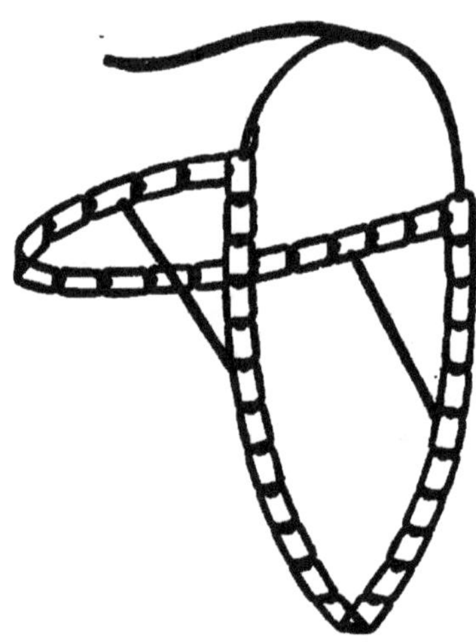

Very Cheap Halter for Cows

Take twenty-two or twenty-three links of an old binder chain; pound them together so they will not come unhooked, to go over the head, and twelve or thirteen around the nose, with the hook sides of the links inside. To the lower end link of one side attach a rope or strap, with a snap in the other end, to snap to the lower link of the other side, to go under the jaw, and to this tie the halter rope or strap. Also brace the nose piece to keep it from sagging too low on the nose.

Drag for Tethering the Cow

The cut shows one of the very best ways to hitch a cow out to graze. Put a head halter upon her and tie the end of the rope to a bran sack in which several smooth, round stones are placed. Put in just enough weight so that the cow can drag it a few inches at a time, or if it is wished to keep the cow in one place put in enough stones to hold her in one place. The beauty of this arrangement is that there is nothing to wind the rope about, while if the cow gets a leg caught, or starts to run, there is no solid anchor to twitch the rope up suddenly and perhaps do injury to the cow.

Handy Devices   **For Stock**

Tether

A very common method of tethering an animal is to set a crow-bar in the ground and tie a rope to it, when the tethered animal generally proceeds to wind the rope about the bar. Add a piece of iron gas-pipe or water-pipe and a bit of hard wood board for it to turn on, and you have a tethering device that cannot be improved. File the lower end of the gas-pipe smooth, so it will turn easily on the smooth board, and an animal cannot wind himself up to the bar.

Squeeze Gate for Treating Sick Stock

These squeeze gates can be put up in the barn, or attached to a fence

In a great many cases, animals suffer more from the excitement and exhaustion of being caught and handled than from any ailment they may have. Young cattle are often chased until they are heated, then roped and thrown. Besides being detrimental to the ailing animals, this often puts them in an awkward position to work upon.

To get around the difficulty, squeeze gates are a great help. They are hinged to the front part of the stanchion, into which the animal is coaxed with feed, and swung around to prevent the animal from jumping sidewise. The gates are especially valuable for milch cows, when treating cases of contagious abortion or infection of the genital passages. Their use insures greater safety for the one who is treating an animal.

Collar for Self-Milker

The drawing shows a simple device to be applied to self-milking cows. Since it is not patented any of our readers are at liberty to use it. It is simply a necklace made from old broom or fork handles strung on a strap and buckled around the neck. It should be fitted to the cow and the sticks made long enough to keep her from putting her head on her side and not long enough to chafe the shoulders or throat when the head is not turned.

A Wagon Crate

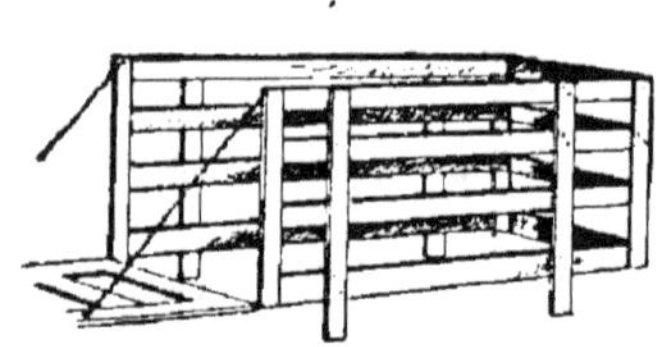

This crate should be about three feet high, and fitted with side pieces extending below it that will just fit into the side irons of the wagon body. It can thus be set upon the wagon bed in an instant, and will be found most useful in moving calves, sheep, pigs, or other stock. It will fit onto a sled in the same way for winter use. It is also convenient when hauling loose material. If this is long, the rear gate can be hinged to open at the side. The slats should be of hard wood, three-quarters of an inch thick.

Breaking the Colt

The cart shown here is a home-made rig useful in breaking a colt to single harness. Its prominent feature is the shafts made of strong hickory poles, extending two or three feet behind the wheels, and braced as shown.

This prevents the colt from rearing and also from backing, for the trainer's seat is so arranged that he can move back and throw his weight on the rear end. A stout strap over the loins from shaft to shaft prevents the animal from kicking. It is important to have the dash board and seat well braced and everything about the cart strong. Combine strength with gentleness and do not break but train the colt.

Rig for Breaking Colts

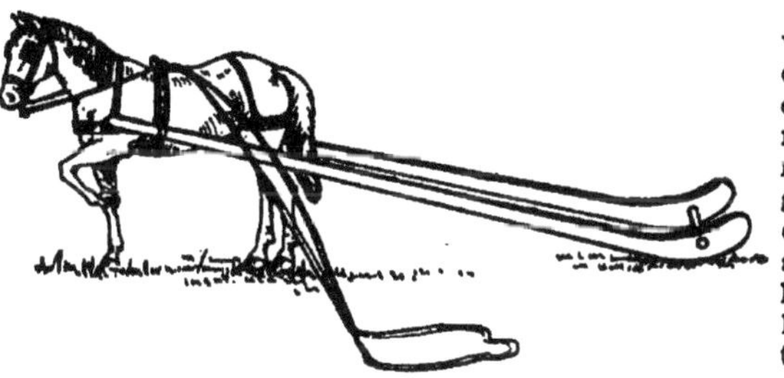

A good rig to use when "breaking" a colt is shown in the cut. It can be made in a few moments from two green saplings. The colt hitched into such shafts cannot hurt anything if he kicks, and can be turned, backed and driven ahead as well as though in a wagon, and that, too, without the risk that is always taken when using wheels at first behind a green colt. When used to the work he can be hitched to a light wheeled vehicle.

Device to Prevent Casting

The intelligent reader will not be slow to catch on to the idea of the illustration herewith. The horse or colt that is accustomed to getting cast in the stall can be prevented from doing so by the use of a strap fastened to a joist overhead so that the animal cannot get its head quite down to the floor. This device is necessary in some cases, and is effective.

Protecting the Horse from Rain

The horse or horses that must be driven in the rain should have such a cover for the neck and mane as is shown in the cut. A horse will soon dry off where the hair is short, but a soaked mane is most uncomfortable and disagreeable, and is likely to subject the horse to a cold. The cover is made of duck, goes under the collar and ties to the top of the bridle.

353

Protecting Dobbin's Head

Here is something to keep old Dobbin's head cool when working in the hot sun. It is simply a piece of pasteboard and a bit of wire, and can be made in two minutes. When bent into position the ends of the wire slip into the long side loops in the bridle. This device allows the air to circulate under the pasteboard, and still keeps off the sun. The straw hats used for horses' heads as a rule lack ventilation.

Another Protector

In the heat of summer the heads of horses should have protection when at work as well as the heads of their drivers.   Get some stout wire and bend it into the shape shown in Fig. 1. Now cover with cotton cloth as in Fig. 2, and the device is ready to slip over the horse's head, letting the lower ends of the wire frame fit into the long leather loops on either side of

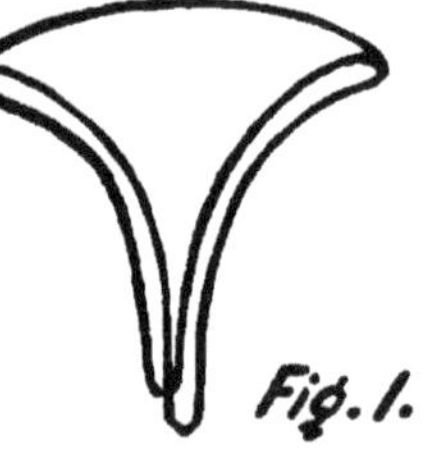

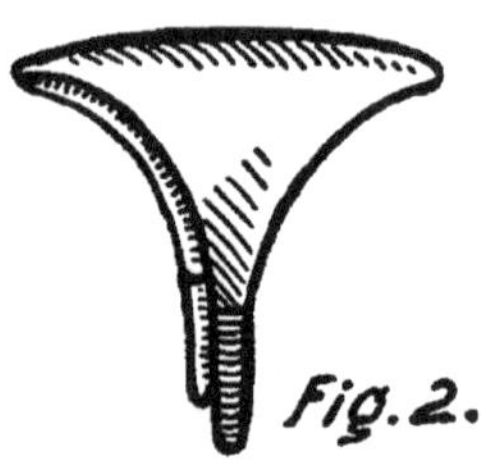

the bridle.   They will set firmly in these, and afford the horse much relief from the sun's heat.

Handy Devices   **For Stock**

Leading Unbroken Horses

The drawing shows how we lead wild and shy bronchos behind a wagon here in Colorado. Take a half-inch rope, lay the two ends together and tie a knot about five feet from the other end where you see A. Now lay it on so that the knot is in the center of the horse's back and his tail over the rope; put the ends through the halter ring at C, one rope being on each side of the neck, and tie the ends together to the wagon. You may be sure that the animal will lead and not pull back the wagon. Horses if tied in this way in narrow stalls can be cured of halter pulling.

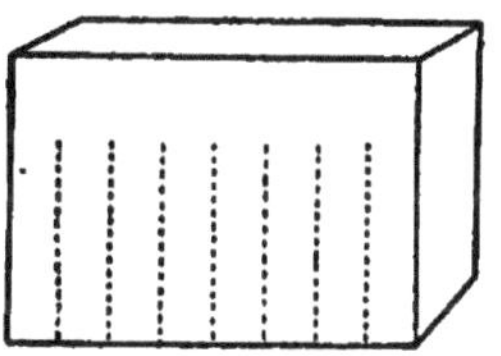

Combs for Manes and Tails

Take a piece of hardwood four by four inches and one inch thick. Saw into it two and a half inches every three-quarters of an inch across, (see dotted lines on first cut) and then whittle each tooth to a point and nail a strap on top and you will have a handy comb like the second cut that will do effective work.

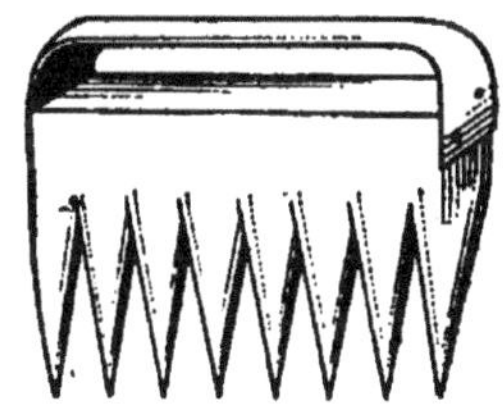

Scraper for Horses' Legs

The drawing shows a device that is better than a scraper whittled out of wood to clean horses' legs with. Have a broom made narrower and shorter than usual, sewing an extra seam in near the end of brush, and trim out a space in center of brush in shape of crescent, two and a half by two inches, like cut. This will do the work thoroughly.

Keeping the Pig's Bed Dry

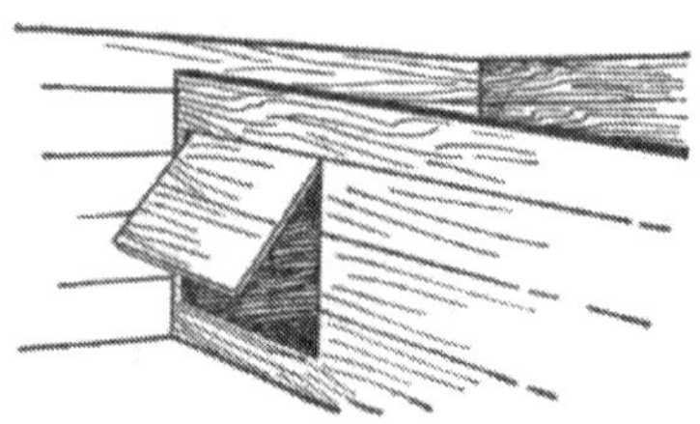

Give the pig half a chance and he will keep clean. In the ordinary pen the bedding soon becomes badly soiled, because it is soon tossed into the wet. Divide the pen and have a swing door, as shown in the cut. Put a thick bed of straw in one end, and the occupant will always have a dry bed. Moreover, in winter, an extra supply of straw can be given, in which the pig can burrow and keep warm. Greater warmth would be secured by covering the top of the straw pen.

Warm Bed in the Hog Pen

The ordinary hog pen is cold in winter. Make in one corner of it such an arrangement as is shown in the cut. The cover is hinged and rests down upon a wide board set on edge. The end has burlap tacked to it. Put straw inside for a bed, and the hog will be very warm all night. The cover can be raised for renewing the bedding or cleaning out the soiled straw. In a small, confined space, an animal's own bodily heat will keep it warm.

Safety Pig Fender

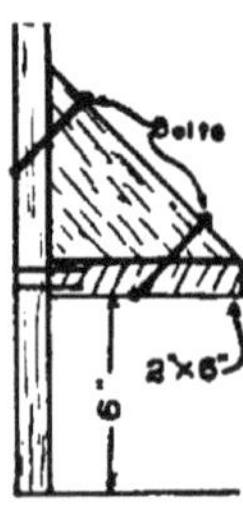

"She is a good brood sow, but she always kills about half the pigs by lying on them." Haven't you often heard that expression from your neighbor? Next time you hear it, tell your neighbor how to prevent such trouble by fixing up a pig fender like the one shown in the illustration. Tack a piece of 2 x 6 on the wall six inches from the floor by means of some four-inch spikes. Then place three-cornered blocks above it every three or four feet, and put quarter-inch bolts through. The pigs will be safe every time.

Farrowing Pen

A farrowing pen should be on every farm where there are swine raised, and here is a design for one. Note the fenders all around to keep the sow from crushing her little ones. The rail is about eight inches from the floor and the same distance from the wall.

Handy Pig Catcher

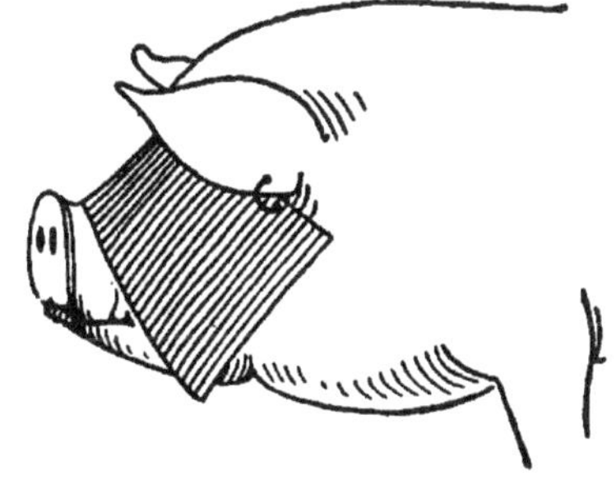

With this very handy pig catcher one can scoop a pig weighing up to thirty pounds. If dropped in front of a running pig he will go into it of his own accord and will not squeal. An old-fashioned hind-leg grab isn't in it with this catcher. To make one, take a piece of three-eighth-inch round iron four feet or a little longer, heat and bend it into a hoop, allowing about four inches of each end to project for insertion into a shovel handle with a heavy ferrule. The net is made of a heavy burlap sack, secured to the round iron frame with heavy twine in such a way that three-quarters of the sack shall hang down and form the net.

Stop Chicken Eating

Here is a cure for that old sow that ate up all those chickens and is now eating old hens. It is a piece of stiff leather wide enough to cover the hog's face within an inch or so of the snout, and secured by a hog ring to the lower edge of the ears. An old boot-leg will do.

Movable Pig Pen

The pieces of wooden rail of which the sides of this movable pig-pen are made extend beyond the boards, and cross each other as shown in the cut. A pin is fitted to slip through each and thus make a firm corner. Such a pig pen can be taken down in a moment, and set up as quickly, and enable one to pasture successive portions of a clover or other patch.

Portable Pig Pen

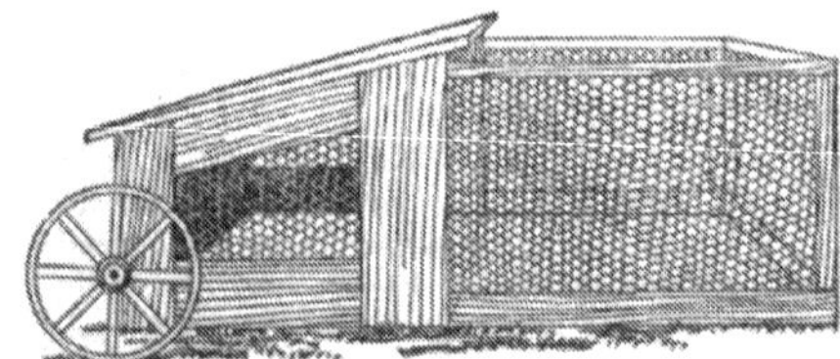

Here is a picture of a movable pig pen, which has a floer and a movable frame on which wire netting may be fastened to make the out-yard. Then the pig can have a nice fresh, clean place every few days. A horse is used to draw it.

Another Portable Pig Pen

Getting the pig out of doors in the summer is good for the pig, and often good for the land on which the pig runs; but the pig's efforts need direction, or damage may be done. A convenient, portable pen is shown herewith. One end protects the occupant from sun and rain.

Two trucks made from a board are screwed to one end, while the handles are placed at the other end. It can thus easily be moved its length every day, and quite a bit of ground dug up and fertilized during the season.

Ewe Pen

Sheep that are about to drop lambs in cold weather, and spring months bring much of such weather, should have a specially warm corner penned off in the sheepfold where they may be placed, both for warmth and to avoid crowding by the rest of the flock. The pen can be just high enough to admit the sheep at the little door, and a grating should extend along one side to give light and ventilation. A couple of such pens will answer for a small flock, and may be placed in some other location than the sheepfold itself, if there is room elsewhere.

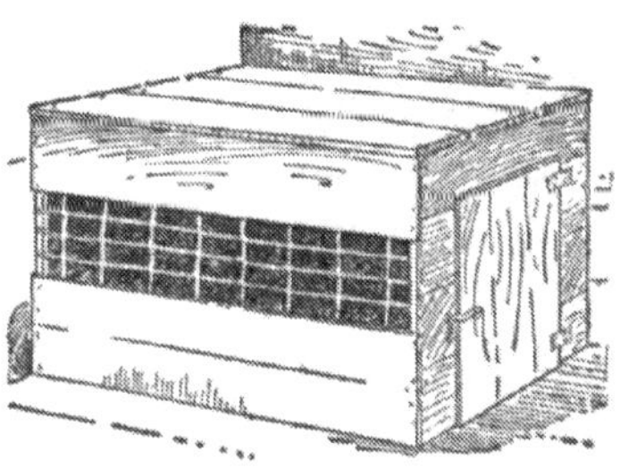

Salt-Ashes Hopper

We have used a box like the sketch for a number of years. It is sixteen inches high, twenty inches long and one foot wide. The mixture that we place in it is equal parts of ashes and salt. It feeds down as it is eaten out from the bottom. Rain sometimes clogs it a little, but that is easily remedied and the contents runs down easily again. A stake in front and one on each side hold the box in place, so that rooting pigs cannot tip it over easily.

Keeping Chickens Out of Barn

When the thermometer is at 100 degrees very few people can keep their temperatures down to normal if they have to chase chickens out of the barn when the doors are open to let in fresh air. By hinging a slat door to the frame as shown, it is easy to keep the chickens from scratching in the stalls and flying into the feed boxes. Screen can be placed on such a door to keep out flies. Hinges can be screwed into the door-posts and removed in the winter, if desired. The outer doors may be kept open.

Another Hen Discourager

Hiram Hogg: "At last my owner has solved the hen problem to my entire satisfaction by hinging the door to my sty so that it will always swing shut. When I leave my house to roam in the alfalfa I push it open with my snout, and need not worry about any fussy old hen and a host of chirping chickens scratching in my nest. Nor will I again waken from my afternoon nap to find that same fussy old hen hovering her brood on my back."

Chute for Holding Hogs

Sometimes it is necessary to hold hogs while ringing or vaccinating. The chute shown here solves the problem so that one man can do the work while the boy drives the hogs. The posts may be 2 x 4 inch stuff. If a small alley is made to lead up to the chute, it will simplify matters. The holes in the upper cross-pieces may be different distances apart, so that different sized hogs can be held. A bolt is slipped through the lever when in place.

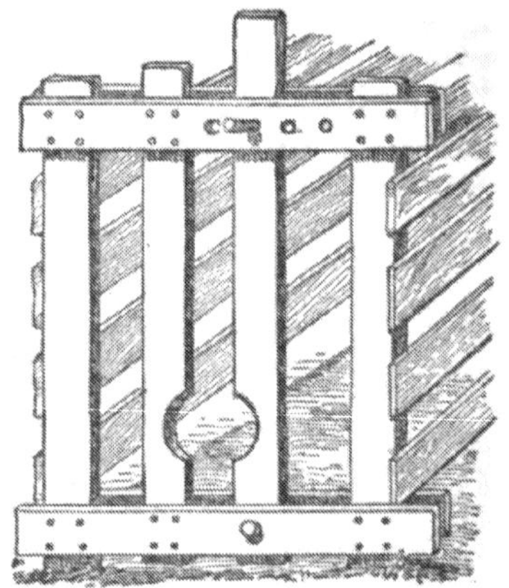

Home-Made Hog Scratcher

Here is a device that will take the lice off the farmer's hogs when he is sound asleep. Drive a stake in the ground, wrap an old rope around the stake and tack with shingle nails. Saturate the rope with equal parts of coal oil and lard once a week, or use one of the commercial coal-tar dips. Drive the stake near the hogs' sleeping quarters. This is so effectual that the hogs will stand in line waiting on their turn.

Space-Saving Grain Bin

A grain bin similar to the cut will prove very satisfactory for its economy of space. Its thickness from the wall out is not more than one foot, yet because it reaches to the floor and can be carried up four feet from the floor, its capacity is large. One-half of the front is hinged so that all the contents can be easily reached, even when the grain is nearly used up. For limited quarters, such a bin will certainly prove very useful.

A Handy Animal Crate

The drawing shows a kind of conveying rack for transporting sheep, calves and poultry to market. It is made of basswood, two feet wide, four feet long and three feet high. Corner posts are two by two; bottom and top frame boards six inches wide and three-quarters of an inch thick. The side slats, the floor, middle deck and roof are made of half inch boards, the slats being two and a half inches wide. A strip one by six is nailed in the center of each floor to strengthen it. The middle floor is, of course, used only in carrying poultry. Both this and the bottom are set in loose so they can be removed when necessary. It will be noticed that the door is in two parts and has two sets of hinges. Two handles should be bolted on each end for convenience in handling.

Water Heater

Have the tinman make a box of galvanized iron, shape of drawing. Have it water-tight (except at top, which is open), high enough to reach above water-line in trough, and large enough so that a small two-burner oil-stove can be lowered into the box and slid into the L part. Fasten box into place in trough with wooden cross-strips. Fill stove night and morning, keep wicks in order and water will be warm at all times. Trough should be in a closed shed.

Water Heater and Food Cooker

Take a barrel, an iron tube about three inches in diameter and perhaps ten feet long, and a wooden plug. Have thread cut in one end of pipe; screw it into right-size hole in barrel, four inches from bottom. Drive plug into outer end of pipe. Put water into barrel. Build fire under pipe (dig trench for fire, if necessary). Can have scalding water in about one-half hour. When cooking feed, it is desirable to have a piece of wire fly-netting fastened over the inside end of the tube.

Another Water Heater

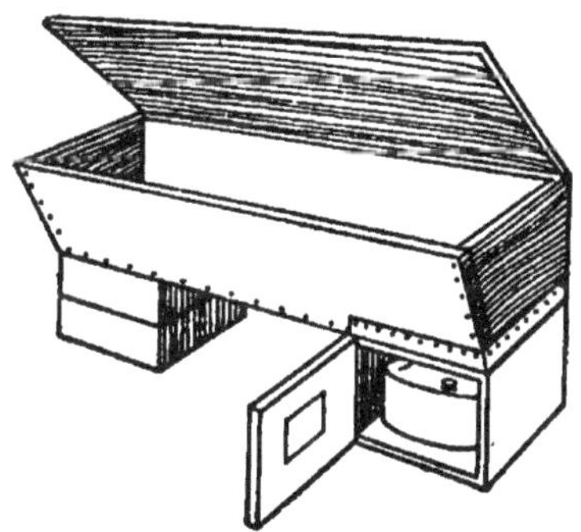

It does not take very much heat to take the chill off the water for the stock, and the chill ought to be taken off. The illustration shows an appliance for using an oil stove for this purpose. The bottom of the trough at one end is made of sheet iron, or better of galvanized iron, plenty of white lead being used to make tight joints. Let the trough be filled with water, the cover shut tight, and the oil stove lighted. If this is done a short time before the cattle are to be watered the chill will be removed to a considerable extent from the water.

A Feeding Chute for the Basement

In feeding a pig or calf in a basement, do not pour the milk down a spout, for the inside of the latter will soon become very filthy and hard to clean. Make a chute like that shown in the cut, and lower a pail down inside. When the pail is in place, raise the front by a cord running up to the first floor. A slide, to be operated by a cord, would answer as well. Without the slide or raised front, the animal would hear the pail descending and would stick its head into the chute.

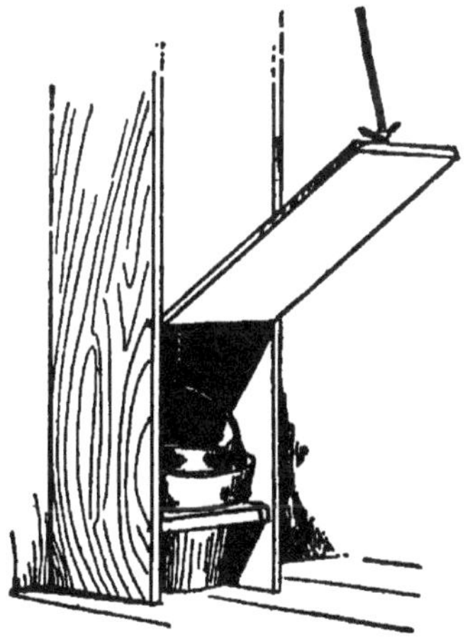

A Pail Holder

Where calves are pastured in orchards, or where cows, horses or calves are hitched out to grass by a rope, the pail of water must be provided. To keep it from being upset and spilled have two small iron rods bent as shown in the cut, sharpened at the other end, and use by pressing them down into the ground by the sides of the pail as shown.

Folding Feed Box

The feed box in the drawing, suitable for hogs (or fowls) is held firmly in place by two iron rods that are bolted to the side of the pen. V-shaped crooks at the ends hold the trough with screws in the sides. When the food has been eaten the trough can be turned up against the wall out of the way and fastened there.

Noon Feed Box

Any one who is obliged to feed his horses during the noon hour at the wagon will find this feed box will take the place of a nose-bag. When the box is hooked over the top edge of the wagon box, the height is just right for the horses. The box prevents waste of grain and provides a better way to feed the horses than in the wagon box.

Hog Feeder

If you use an open trough for feeding slop to hogs, this device will prevent splashing and waste. It is simply a wooden f u n n e l, eighteen inches high, twelve inches square at the top and three inches square at the bottom, and is nailed fast to one end of the trough. Any one can make it who can use a saw and hammer.

Improved Feed Trough

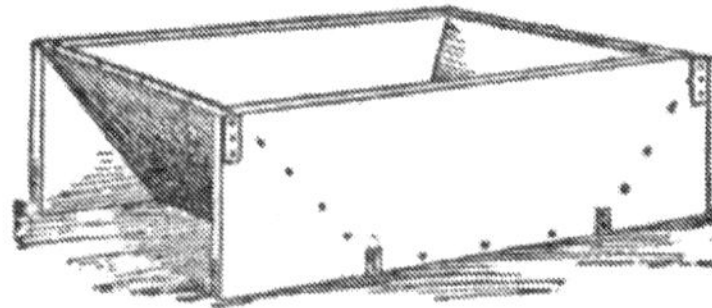

The cut shows a fine plan for making feed troughs that cattle cannot pull apart, and into the corners of which their noses and tongues can easily go. The pieces of strap iron hold the sides firmly in place at both top and bottom. When making a watering trough, let the ends be upright, not sloping, and let the band of iron extend entirely around each end. If an old piece of cart or wagon tire is at hand, have the smith "shrink" it on as he would upon a wagon wheel. This will insure tight ends.

Making a Horse Eat Slowly

Many horses swallow their grain ration without proper mastication. To compel them to eat slowly, try a box arranged like the one shown in the cut. It is long —at least two feet in length and has a sloping bottom with rounded cleats, arranged as shown in the cut. The front of the box is shown cut away, in order to bring the sloping board into view. Scatter the grain ration over the whole of the slope. The horse will eat much of it slowly on its way down the slope, and the small amount that finally reaches the bottom will have to be eaten slowly, since there will be little depth to the grain.

Feeding Trough Protector

If you are having some trouble with your hogs at feeding time, by the stronger ones forcing out the weaker ones, breaking the troughs and spilling the slop upon the ground, then this device will remedy your trouble. The illustration shows the protector over the trough in position after the slop is poured. The device is constructed of any heavy, strong lumber and can be made to turn on pins B, B, which run through the side pieces into the posts. When the device is in position to pour the slop, the end with the heavy slats across it is let down to tne ground and protects the trough from the crowding hogs. After the slop is poured the heavier end is raised and fastened to hook A, as shown.

Feed Box Hopper

Once in a while you get a horse that bolts his feed. Some people try to cure the habit by putting stones in the feed box, but this is a dangerous practice, for the horse is almost sure to bite the stones and injure his teeth. The picture shows just how to make a feed box that compels the horse to eat slowly. The box is built to extend through the partition. The part on the outside is made with a slanting bottom so that the feed will run through an opening in the partition, this opening being just large enough to let the

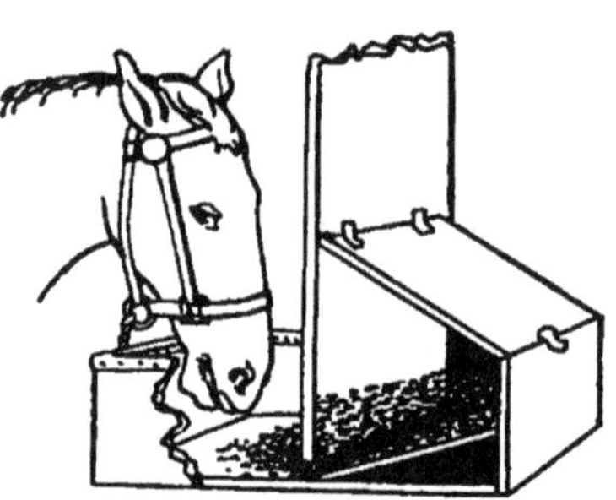

feed through slowly. The horse cannot do anything to make the feed run through this slot faster, so he is forced to eat slowly. A lid is placed on the outside of the box to keep out trash and hens. Often you throw feed into a trough in the dark without knowing that you are mixing it with hen manure.

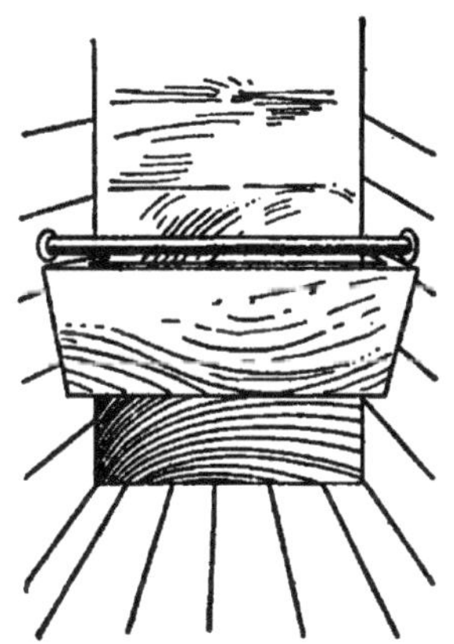

Keeping Horses From Gnawing

There are few horses that will not gnaw the edge of their crib if compelled to stand idle in their stalls for hours at a time. A neat and convenient crib guard is shown in the cut. A piece of two-inch gas, or other iron pipe, is cut at the right length and fitted above the edge of the crib with collars, as shown. This not only keeps the horse from gnawing the edge of the plank but provides also a most convenient place for hitching the horse.

Feeding a Harnessed Horse

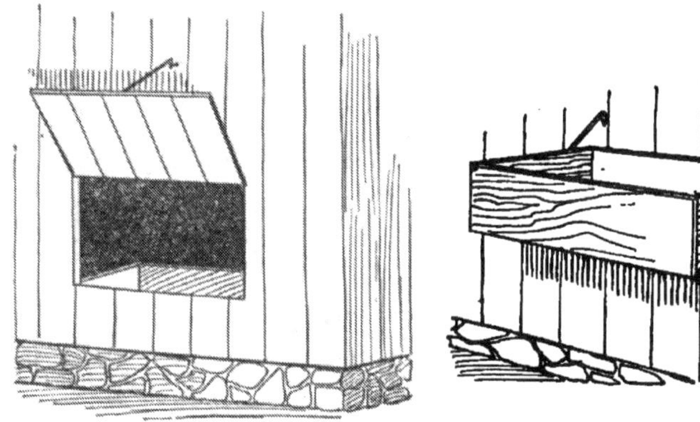

It is often desirable to feed a horse without unharnessing him from the wagon. A handy arrangement for this purpose is shown in cut No. 1. The crib is inside the barn, and is reached by the horse through a raised door, that ordinarily shuts down and gives no sight of its presence. Where this plan is not convenient, have a box that can be attached to the side of the barn, in the manner shown in the second cut, from which the horse can be fed without unharnessing him.

The cut explains itself. The upper part of a section of t h e barnyard o r pasture fence can be thus hinged, provided with chains and let down for the purpose of giving the stock fodder at night in order to supplement failing pastures. A number of light extra stakes are driven down in the line of the fence to hold the fodder in place.

Fodder Rack on Fence

Roadside Watering Place

High watering places by the roadside are becoming more and more frequent every year. They are grand institutions. The cut shows how to make a handsome drinking place. Set a hardwood barrel on end, and convey the water into the bottom. Have an overflow pipe two inches below the top. Put such a framework about the barrel as is shown in the cut—four posts at the corner, and cross pieces. Stretch wire poultry netting around three sides, and plant woodbine about the base. In two years the place will be an ornament to the roadside.

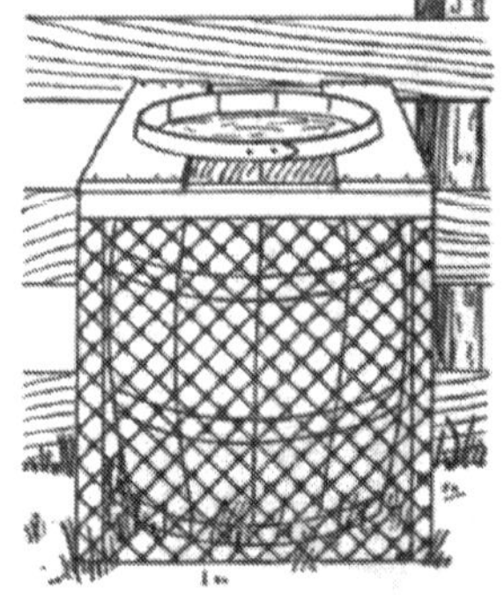

Feeding Trough for Steers

This can be quickly made by any one who can use a saw and hammer. Materials needed are three pieces of two by ten, sixteen feet, for floor of trough; two pieces of two by six, sixteen feet, for sides of trough; seven pieces of two by six, eight feet, for uprights and side braces shown; one piece of two by six, ten feet, for ends of trough and one upright; two pieces two by four, sixteen feet, for side braces; two pieces two by four, twelve feet, for eight cross braces. The illustrations show how to build the trough.

Trough for Water or Feed

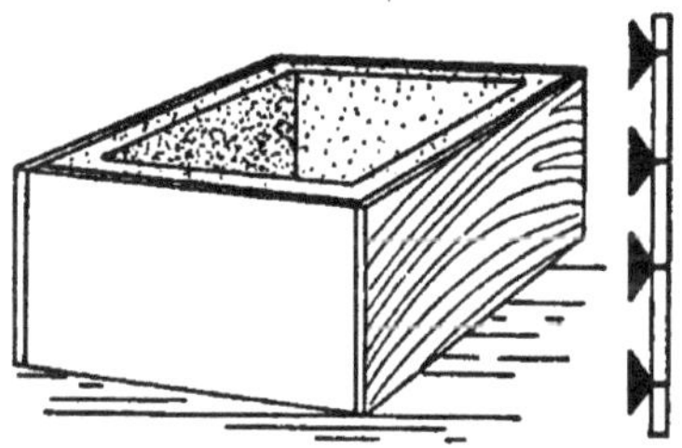

This water-tight trough is much lighter than one of concrete, and no more trouble to build. It may be used for watering stock, or for feeding pigs or calves—the plan of making is the same, though the shape may be changed for different uses. Select a strong grocery box and nail laths over the inside, as suggested at the right of the cut. The laths should be beveled before nailing on. Then plaster the whole inside with cement, an inch thick. The beveled laths will hold the cement firmly in place, and the trough will be water-tight, and easily kept clean.

Swinging Cover for Trough

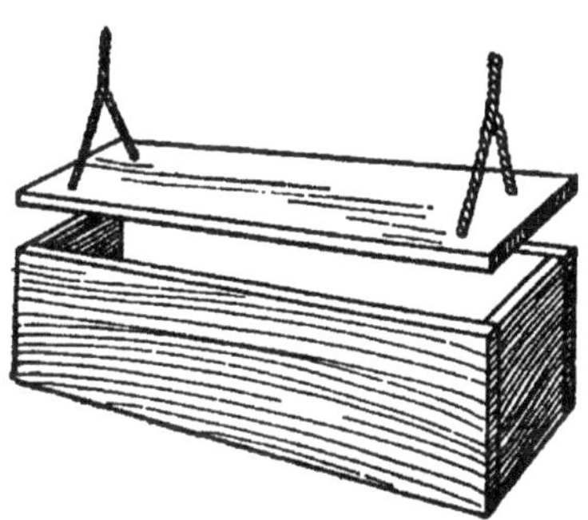

A trough or tub of water standing all day under the rays of the summer sun, becomes very unpalatable for cattle. An automatic arrangement for keeping water cool is shown in the cut. A loose cover is suspended over the trough by ropes. Cattle or horses will quickly learn to press this aside and drink, when the cover will come back over the water again. Make the cover light, and support it evenly as shown.

Cement Lining for Crib

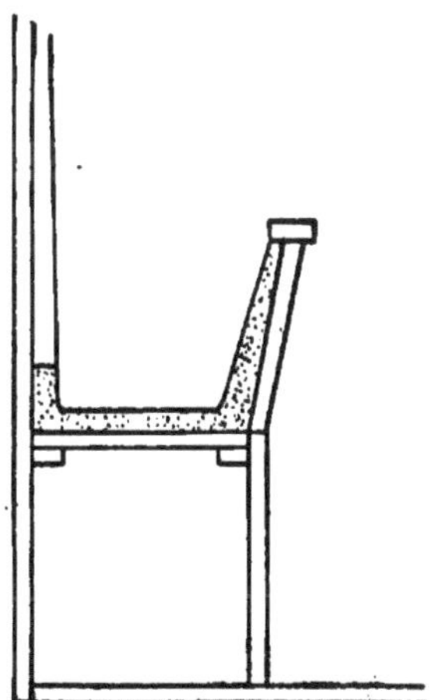

Many horses are continually gnawing their cribs. The rail along the front edge can be bound with sheet iron, but it is difficult to protect the inside of the crib in the same way. Line the inside with Portland cement mixed half and half with sand, in the manner shown in the cut, and there will be no gnawing. The crib will be water-tight, too.

Trough on Pivots

Chopping ice out of a trough, after it has frozen solid to the bottom, was never an agreeable job, even in our boyhood days.   Try a trough like the

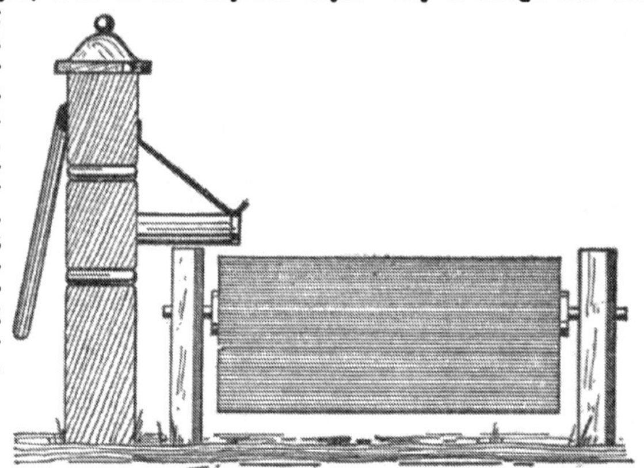

cut, which is hung on pivots placed just above the center, so that when the stock have had what they want, any water that remains can be easily dumped out. By a simple device it might be held upside down, in order to prevent snow and sleet from accumulating in it when not in use.

Creep for Feeding Lambs

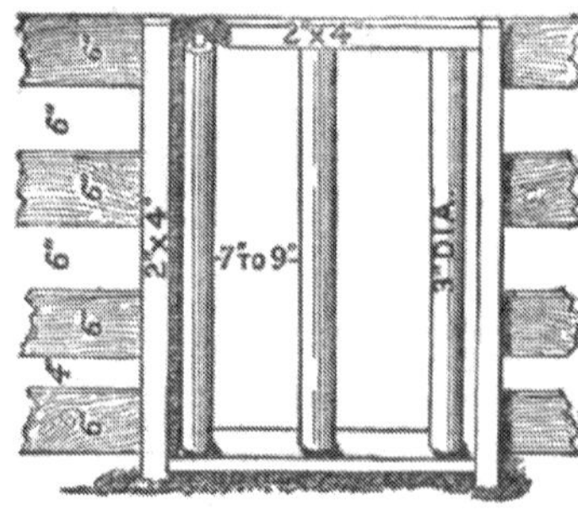

Get some sort of a creep fixed up for the lambs, so they can get some grain. Fence off a little space in one side of the lot, using six-inch boards, placing the lower ones four inches apart and the others six inches, as in illustration. Take three old binder rollers and set them seven to nine inches apart in the opening. If there are no old rollers, take some smooth, straight pieces of cord wood about three inches in diameter and set them in with spikes. This creep lets the lambs through to the grain troughs, but keeps the ewes back.

370

Feed Rack for Sheep

A good feed rack for sheep should save all the fodder and prevent hay-seed and dust from getting into the fleece. To make such a rack place 1x2-inch cleats four inches apart as shown. A twelve-inch board at the top of the rack will prevent hayseed from falling on the sheep's heads. The grain troughs at the bottom are ten inches wide on the inside. A 2x6 beneath the floor of the hay-trough divides the grain trough into two parts. The end of the rack should be closed to keep the sheep out. The lumber needed can be figured from the sketch.

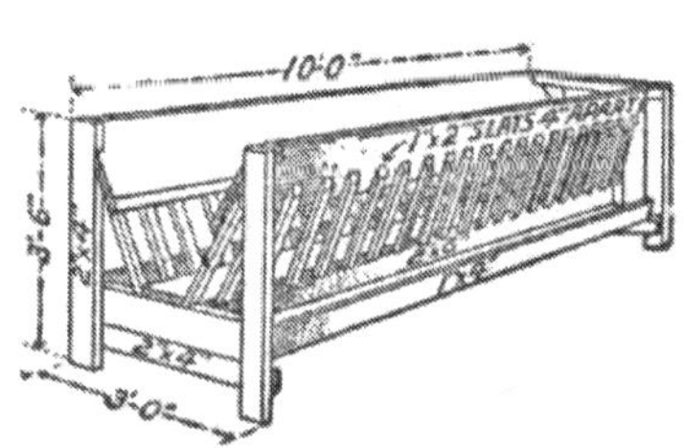

Salt Box for Horses

Salt should always be accessible to the horses in the stall. A place for salt may be made near the feed box, something like the illustration shown. Common salt is preferred unless a very good grade of rock salt is used, for cheap rock salt contains hard substances which will irritate horses' tongues if they lick it a great deal. Draft horses require more salt than those put to less severe work.

Self-Feeder for Hogs

The outer frame can be made of 2 x 6 as shown, and the inner cross-pieces are of 2 x 4. The floor is one or two-inch stuff. A barrel without any bottom is bolted down on top of the 2 x 4s. The diagram shows the feeder as it appears when looking at it directly from above.

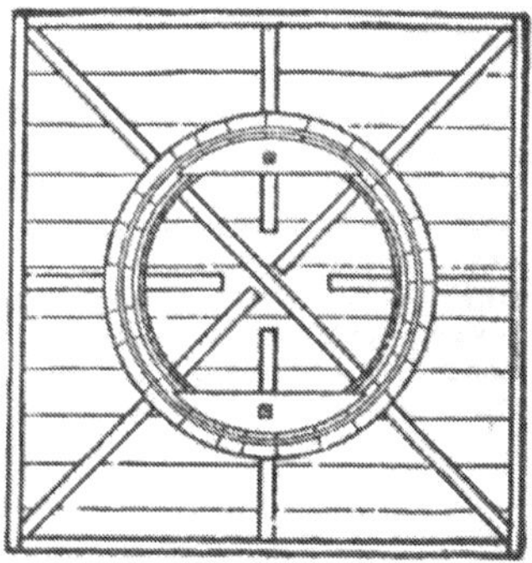

Home-Made Feed Carrier

Where there is a long row of cattle to be fed a grain ration, the device shown in the cut will prove labor saving. Have the feed room at one end of the line of stalls, where the car can be filled. Then with a measuring scoop in one hand the feeder can pass rapidly down the line, pushing the car before him. The whole method of construction is shown in the cut. Any blacksmith can mount the little wheel in a frame with hook attached.

Milk Aerator

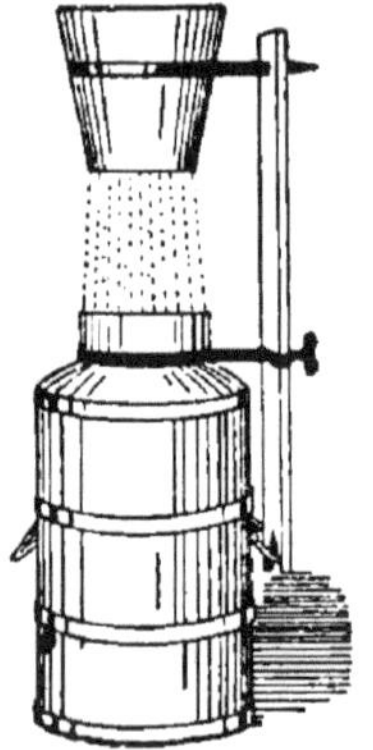

Here is a simple device for aerating milk, upon which there is no patent. Take a pail eight and a half inches in diameter at the bottom and sixteen inches high, drill a row of one-sixteenth inch holes in the bottom near the outside edge, about three-eighths of an inch apart. Suspend this pail over the can with a hoop about one-third way down from the top of the pail, with an arm fastened to it of three-eighths square iron. Make a standard of two by three hardwood, and cut three holes in it to fit the iron so that it can be lowered or elevated as the wind blows. Cut a slot in the bottom to slip on the can handle, and a clamp at the top of the can to hold it from swinging, and you have a complete aerator.

Handy Devices   **Dairy**

Another Aerator, But Different

The aerator, shown in the drawing, is a home-made affair and is not patented. A platform to set the can on is made of two by four scantling, set on edge with a post at one end two or three feet higher than the top of the can. Mortises are made in the post about nine inches apart. Take a piece of iron hoop from an old oil barrel, fasten to a piece of hardwood board to fit in mortises in post; take a ten quart pail, put a row or two of very fine holes around the edge in the bottom, set in the hoop and strain your milk in the pail. Raise the hoop in post according to force of the wind.

Draining the Milk Bottle

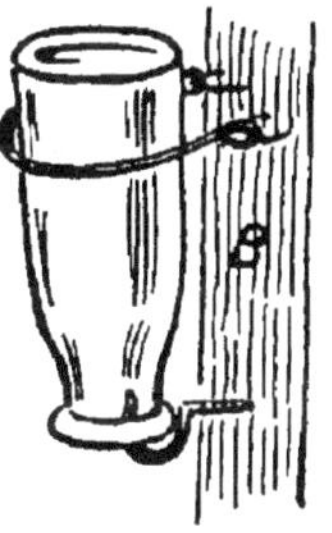

This is the way a milk bottle may be well drained. Bend a bit of wire, horseshoe form, making a small loop at each end (Fig. A). Fasten it to the sink board, using two double-pointed tacks and adjusting a large screw-hook (B) four inches below. When the bottle is rinsed, raise the ring, putting the bottle within it, the neck being held by the hook.

The Strainer Milk Pail

The strainer milk pail figured in the cut is very convenient in form, and those who cannot buy one like it may be enabled to have a similar one made, as any tinsmith should be able to make one from this cut. The flange on the top keeps the milk from overflowing when poured out, yet does not obstruct the top when milking into it, or pouring milk into it, being tilted at such an angle. The hinged cover to the spout keeps all dirt and dust from getting into the spout, and is a great institution.

Milk-Pail Holder for Calves

Bang! went the calf's head into the milk-pail, and the milk splashed in every direction when the pail tumbled over. The man lost his temper and the calf missed a meal. And all the trouble might have been prevented had the man driven four pieces of 2 x 4 into the ground and set the pail inside them when feeding the calf, as shown in illustration. This plan saves milk, time and temper, all of which farmers need to save.

Milk-Can Hoist

Lifting heavy milk cans out of the ice chest at arms' length is heavy work. The cut shows a device for doing the work easily. A carrier pulley is fitted to a rod running over the ice chest. A lever is arranged as shown. With it a can is seized and swung clear of the water, then wheeled along to the table where the milk is drawn off and the cream secured.

Cellar Storage for Milk

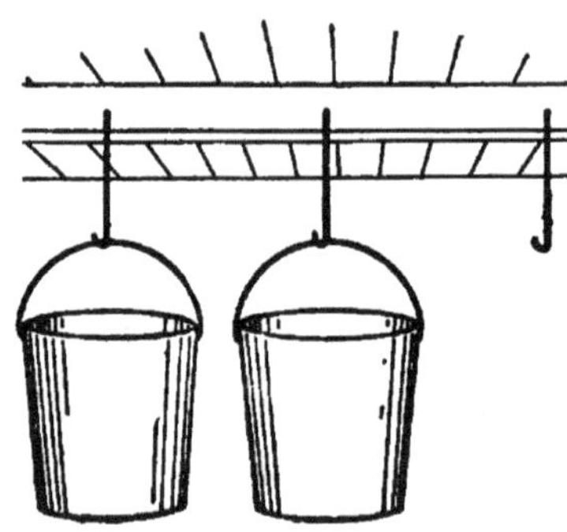

On farms where ice is hard to procure, it is a problem to keep the milk during hot weather. Where one's cellar is cool and sweet (it ought to be sweet at all events) the device shown in the cut will be useful. Cords, or wires, are suspended from a beam and the milk pails suspended from hooks in the lower ends. Rats or mice cannot get to the milk here. Ten quart pails are in many particulars more convenient than pans for holding milk, and cream will rise nicely in them.

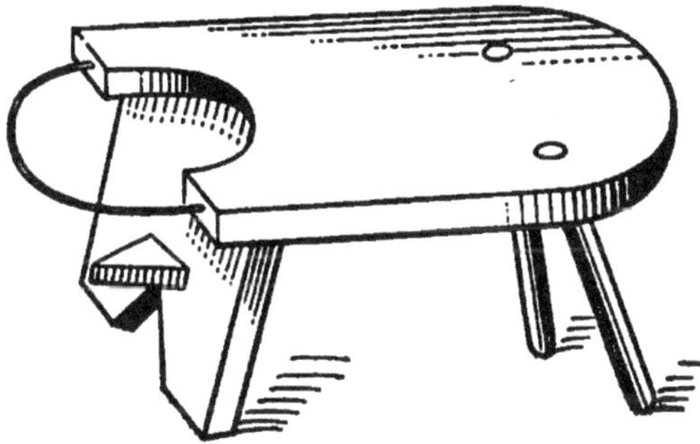

Milking Stool and Pail Holder

A milking stool is here illustrated. It needs no description. It furnishes a seat for the milker and a support for the pail. We commend it to those in need of a milking stool.

Another Milking Stool

If you have not bought your milking machine yet, here is a handy thing to have around the barn morning and night. It is easily made of one-inch boards, put together with screws so as to secure the maximum strength. Then paint it if you want to make a good job, but don't groom the cows with it or you'll wear off the paint.

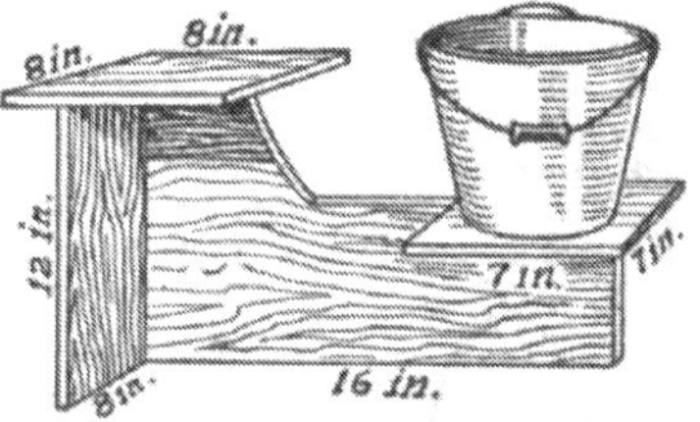

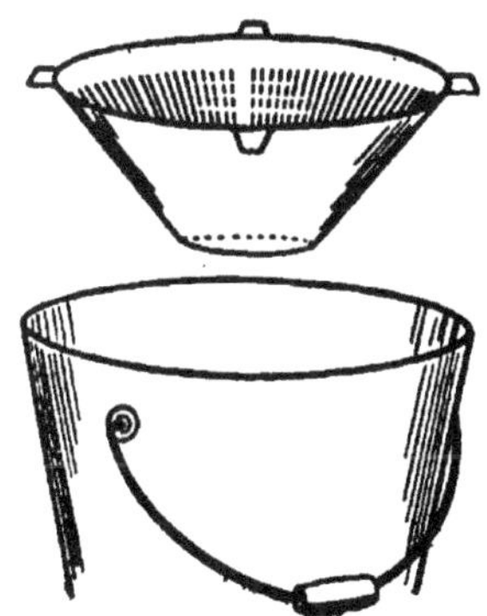

Simplest Kind of Milk Strainer

Good butter making begins as far back as the milking, if not farther. The process of milking must be cleanly if sweet butter is to be made. Fit a cover, with strainer in the bottom, to the milk pail and milk into this. This will keep out much floating dust, and will also assist in keeping the milk closed to odors while it has to remain in the stable. The cut shows plainly how this device is made.

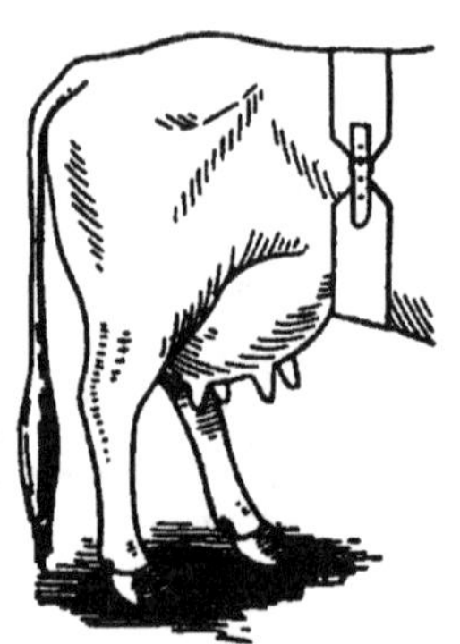

Belt for Kickers

The best plan to discourage a kicking cow is to place a strap as shown in the drawing. Buckle it a little tight, and if she kicks then make it a little tighter and you will find she will not bother you any more.

Fetter for Kicking Cow

To make a fetter for a kicking cow use a piece of 2 x 2 inches about two feet long. In the center drive a headless bolt on which to fasten the two-inch strap, C, which is secured to the board and is long enough to encircle the cow's leg. D is another strap which is joined to C at E. At each end, B, a piece of iron is bent to fit the leg, one above and the other below the knee.

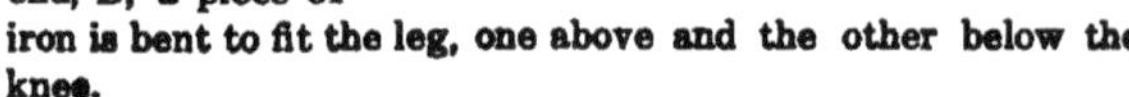

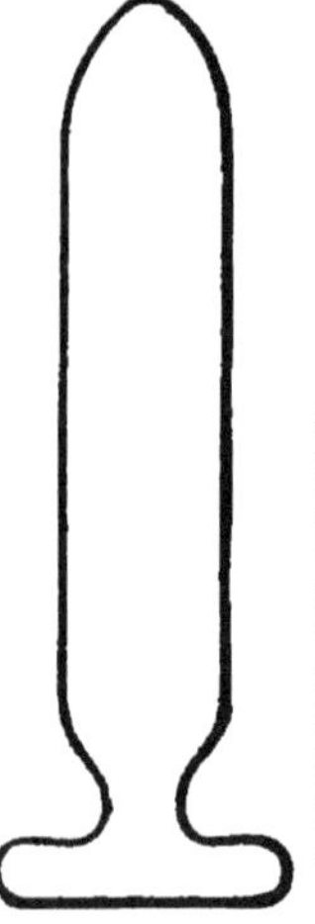

Peg for Hard Milkers

A cow that milks hard is easy to manage. Procure some seasoned oak or hickory and saw it in inch lengths. Split this into pegs and whittle into the shape shown. Remember, the size is one inch; this illustration is enlarged. A broad head must ornament one end. Next it must be a neck not quite so large as the body of the pin. Oil this well and work it gently into the teat that milks hard. The head will prevent it going entirely in and the neck will keep it in place until the next milking. After the cow wears it two weeks, if the milk opening has not grown large enough to permit easy milking, make and use a slightly larger peg.

Handy Devices   **Dairy**

A Simple Tail Tie

Here is a simple way to keep sookey's tail from hitting the hat instead of the fly.   Take a piece of rope about two feet long, make a loop in one end, fasten a small iron ring where the knot is formed; at the other end of the rope fasten a rein snap, put the loop end around the cow's tail, put the snap end through the loop, draw tight; put around her leg and snap into the iron ring.

Another Tail Holder

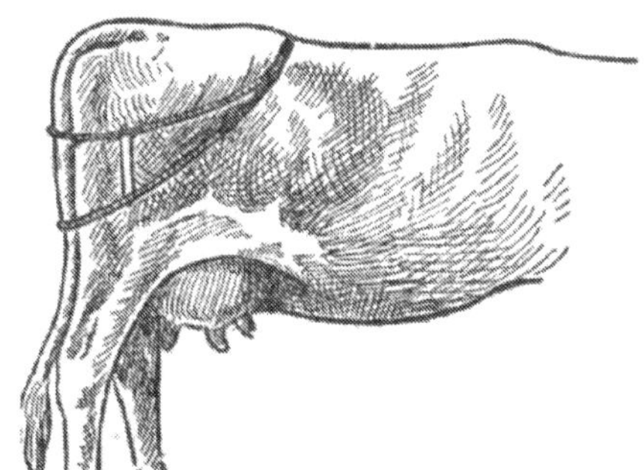

One of the most unpleasant features of milking is the switching of the cow's tail. With many cows this is a trouble, both winter and summer. A single circle of heavy rope laid over the rump helps somewhat; but a persistent cow will soon free her tail. A double rope, kept apart by a bit of a stick on either side, as shown in the cut, will securely hold the tail.   The two ropes should be tied together where they pass over the back.

And Still Another

Here is a device to hold a cow's switching tail: Take a piece of common fence wire and make one end pointed with a file; then make a hook in one end and a loop in the pointed end two inches in diameter, keeping them apart half inch at A. Have fence staples all along the stable in the joist behind each cow; take the wire and run the loop through the hair in the tail and give a couple of twists and hang to the staple.

Single Salt Box for Stock

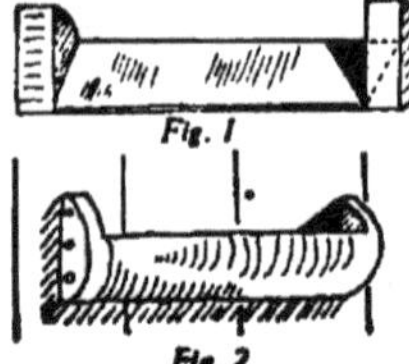

A handy little salt-trough for individual use in the cow barn may be made in the following manner: Take a roller-shaped piece of wood about a foot in length and four inches across. Saw and chisel this down until it looks like Fig. 1. Then toe-nail it on to the wall just above the cow's feed-box, as in Fig. 2. Thus you will have a lasting salt-box. Nothing complicated or expensive about this.

A Gilt-Edge Cream Stirrer

We show a cut of a gilt-edge cream stirrer, or rather the kind of cream stirrer best for putting the cream in the right condition for making gilt-edge butter. The bottom is from six to eight inches across, and the top two to three inches. It may be five to six inches high. A No. 9 wire, galvanized, is used for a handle, with which to push the mixer down and to lift it up in the mess of cream. The effect is to stir the cream and aerate it from the bottom to the top and to mix it most thoroughly. There is no patent that we know of. Everybody who attempts to make butter should have one.

Portable Ice Box

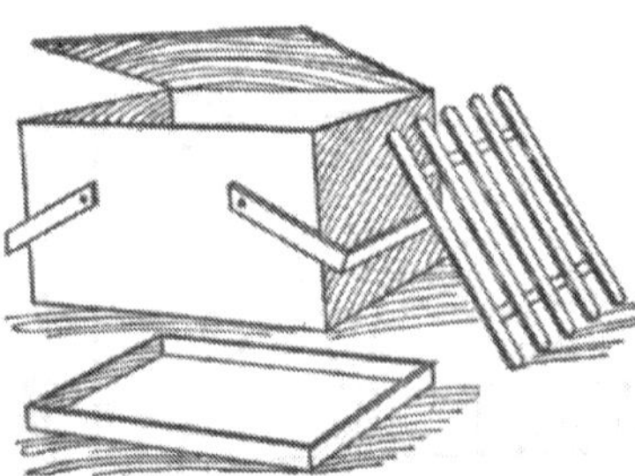

One of the troubles of summer dairying, when butter is supplied to private customers, is keeping the butter nice and firm when going about delivering it. Have a box with a galvanized iron pan in the bottom, with a small tube just reaching through the wooden bottom to let out the melting ice, which is placed, cracked, in the pan. Put in the rack over the pan and place the butter on this. It will come out at each house as firm and sweet as one could wish.

Cow Stall Device

The cut gives a hint that is useful with many cows in keeping them from dropping manure while standing forward in their stalls eating. The horizontal rod just clears the back of the cow as she reaches forward into her crib. To drop manure she must step back so that she can raise her back as cows always do in such a case. Any other arrangement for holding the rod in the right place will answer as well. The same arrangement keeps the cow from wetting her stall except when standing well back.

Cow Stall Design

Here is a sectional view of almost ideal cow stalls, designed to keep the cows perfectly clean and to keep them comfortable. Platform with seven-inch drop; chain to keep cows from backing off platform; swing stanchion; sloping manger, so cows will not strain ahead; partition between cows to keep them from stepping on each other's udders; feed alley in front. Let cows' hind feet rest on rear edge of platform, then not being able or needing to crowd ahead or step off into trench, the platform will always be clean.

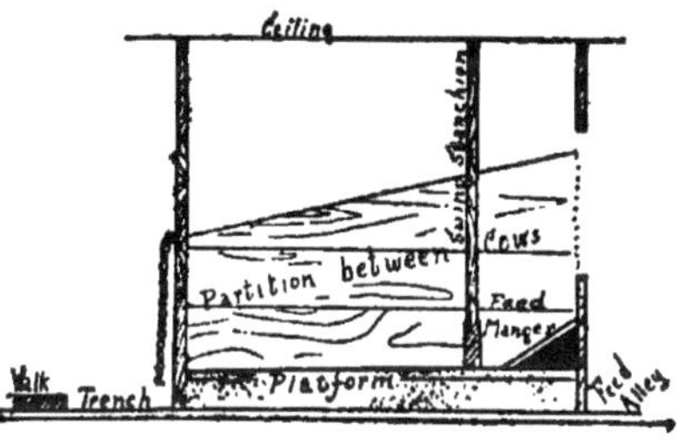

Concrete Lining for Water Tank

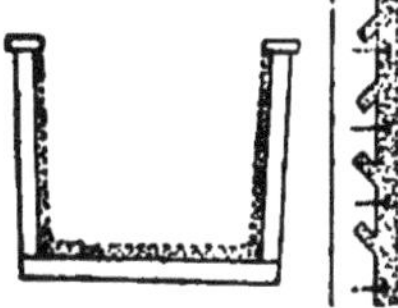

Raising cream in tall cans means ice-water in summer, and a tank to contain the latter. Such tanks are commonly lined with tin, zinc or galvanized iron by a tinman—and at much expense. The cut shows how to line a home-made tank with cement, which can be put on by any one with a trowel. To make the cement hold fast to the sides, bore holes a little way into the inner surface of the sides in different directions, and also drive short nails part way into the surface, as shown in the enlarged diagram of side. This will hold the cement firmly to the sides, and a tight box can be made at small expense.

Feeding Dry Mash in a Box

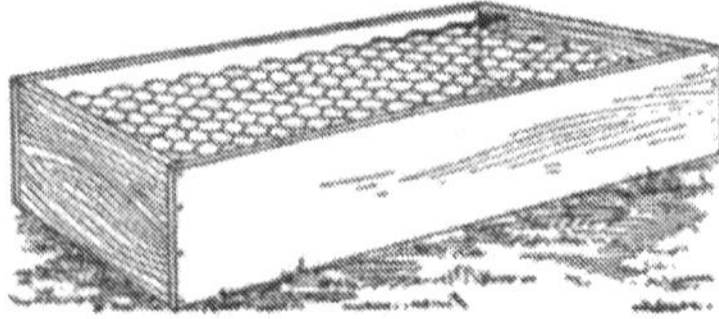

When feeding dry mash, place a piece of chicken wire inside of the box. As the mash is eaten the wire moves downward. This simple device saves the mash from being scratched. The box should be twelve by twenty-four inches, and shallow.

Feed Hopper

While visiting a New Jersey poultry farm, the editor's attention was called to a self-feeding hopper for dry feed. It is easily constructed; the cut makes it very plain. Hoppers of this style are self-feeders, and can be readily kept free from dirt.

Emergency Brooder for Chicks

The cut shows how to keep young motherless chicks warm during cool nights in spring. The roof of the coop is hinged so that it can be raised conveniently. A strong hook is screwed into the roof boards and on the hook a jug of hot water is suspended. This jug should be first wrapped in a piece of old carpet. Thus protected it will retain its heat throughout the night.

Homemade Oats Sprouter

The accompanying illustration will give a general idea of how to make a homemade oats sprouter—one that may be built quickly by any one handy with tools. It is compact, easy to make and portable.

The oats should be soaked for twelve hours in warm water and then spread out in a layer one-half to one and one-half inches deep in a tray, which should have openings or holes in the bottom, or the bottom should be made of quarter-inch mesh wire covered with burlap, so that the water drains freely. The oats should be sprinkled to moisten them, and stirred daily. ·

Chicken Catching Traps

To catch chickens easily, fasten a piece of wire netting 2 x 5 feet, to the side of the fence, by tying at intervals through the vertical center. Fasten the edges to stakes driven about six inches from the fence. Chickens will naturally run along the fence and enter one of the angles between the fence and the piece of wire, where they can be caught readily.

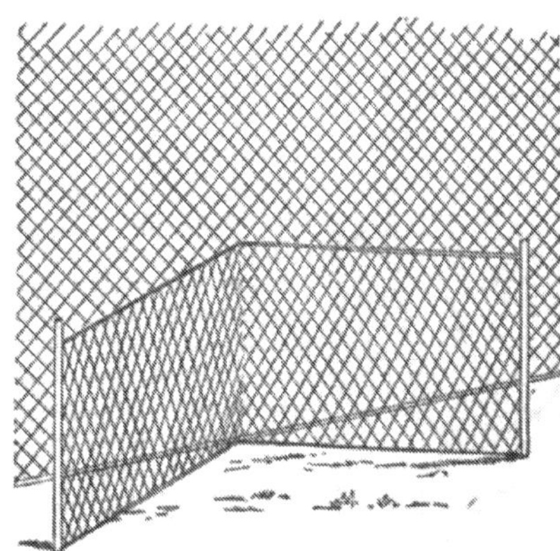

Darkened Nests

This figure shows a style of nests used especially for White Leghorns, as they are rather nervous fowls. The hens enter on the side and go into semi-darkened nests. The eggs are gathered from the front by lifting up the lid.

Air-Tight Foundations

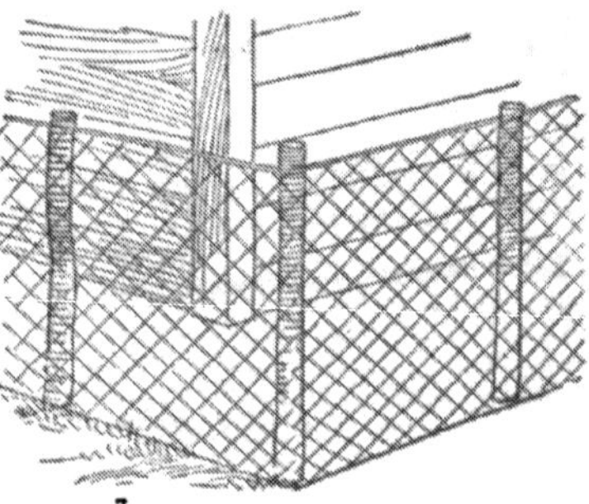

Poultry houses, and many other farm buildings, could be made much warmer in winter by a generous supply of "banking" material. Drive a few stakes about the building and stretch a length of poultry netting over them. Refuse hay, corn butts, etc., can then be pitched in to any height the wire netting may reach, and the building be made secure against hard winds. The netting can be removed in season to be used in the poultry yards when spring opens.

Simple Bottle Fountain

An inexpensive fountain can be made by cutting two pieces of board, one inch thick—one piece nine inches long and the other six inches. Nail these together at right angles, as shown. At the top of the long piece fasten with three staples an open wire ring big enough to fit around a quart bottle, having the opening at the front. Make a T-shaped plug out of quarter-inch wood, the stem about two inches long, and top or cross-piece an inch long, and as wide as will be necessary to keep below the level of an ordinary saucer. Fill the bottle with water or milk, insert the plug; then, with a saucer placed on the plug, invert the entire arrangement; pass the neck of the bottle through the opening in the wire, and place the fountain on the six-inch board or base.

Chaff and Clover Sifter

Where clover hay is at hand a splendid lot of clover heads, dried clover leaves and chaff can be secured to use in the poultry mash. The one shown in the cut was made for just this purpose. An empty box was taken, and a part of one side removed. This can be hinged on again. Tack on a piece of wire netting, half-inch mesh, to form a sieve. Rub the chaff over the top, and the seeds of clover

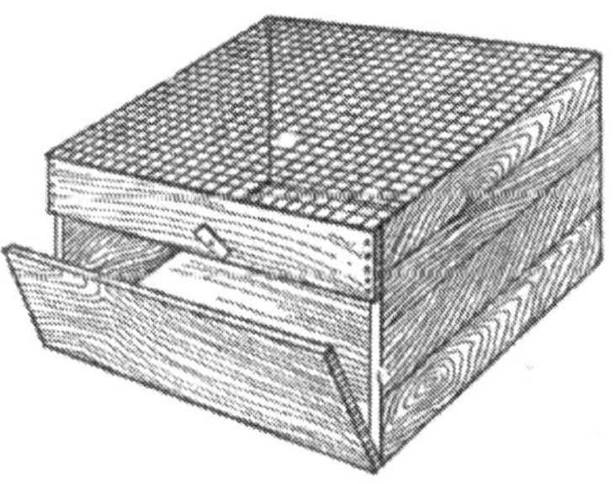

and leaves will fall through, the coarser part being left on top. The sifted portion can be emptied out at the side. This material forms half of the morning meal.

Garden Protector--Try It

To make this clever device take a forked stick like the drawing, and from three to ten inches long, according to breed of fowls kept; attach to the leg just above the knee. When Mrs. Biddy puts her foot forward to scratch, it will act as a back stop and walk her right out of the garden.

Door Catch

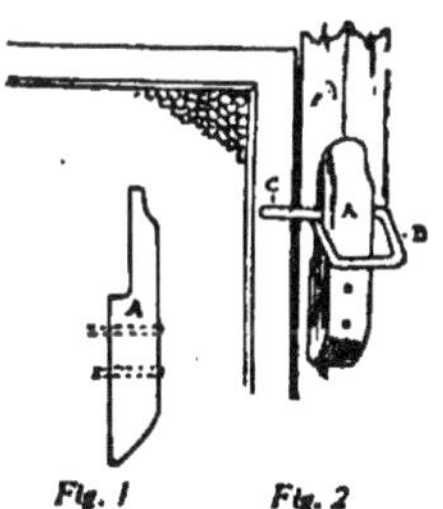

A very handy door catch for a cage or coop can be made out of a small block of wood and a piece of wire. Fig. 1 shows the block as cut out. It can be made three inches in length, about an inch in width, and a half inch in thickness. Shape it out in the manner as shown in illustration. Then take a piece of wire and bend it to make a loop, as shown in Fig. 2, to drop over the wooden catch, and attach to the door of the cage or coop. A represents the wooden catch; B, the wire fastener; C, the door of the cage.

Feeding Coop for Chicks

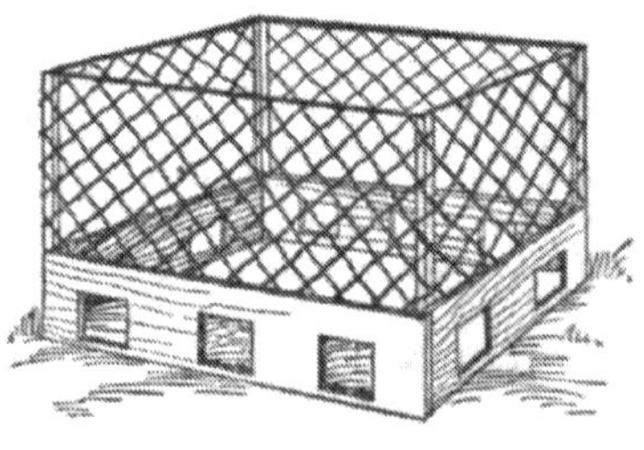

A separate feeding coop for the youngsters is essential to their comfort and thrift. If feed is thrown to a promiscuous flock of all ages, the larger will get the lion's share, besides injuring or, it may be, maiming the weak ones. There are many ways of making such coops, but our illustration shows a good one. Four corner posts, four pieces of foot-wide boards, and a strip of twenty-four inch netting with a few nails and staples are all the materials required to construct it. Make the entrance holes large, and for small chicks tack a piece of thin board over the side of holes, and chip off a little or move the piece as the growth of the chicks require.

The old birds are not likely to fly over the low sides of such a coop, but if they do, it can easily be covered by nailing strips across from post to post and laying boards. For shade and the protection of the feed and the chicks from rain, a covered feeding coop is always to be preferred.

Another Trough with Pivoted Cover

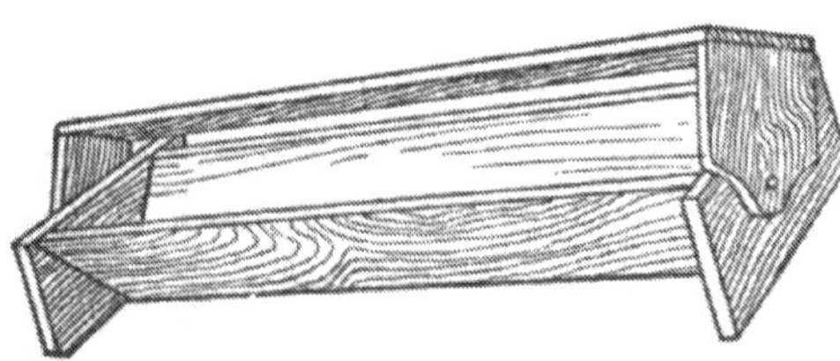

All poultrymen know the disadvantage of having the fowls get into the feeding trough with their feet, taking up room unnecessarily, making the feed dirty and wasting it. The usual rigid cover interferes with throwing the mash in the trough, but the device shown in the cut, herewith, avoids this difficulty. The top board is movable and is thrown to one side when the feed is put in. It is then brought to an upright position and fastened with a peg, allowing the fowls to get their heads only in the trough.

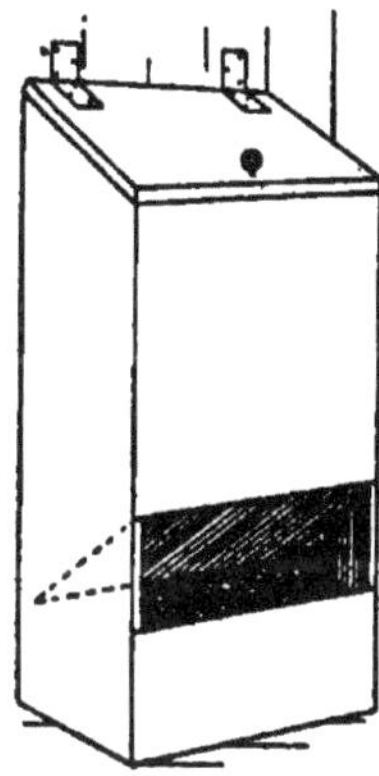

Shell Hopper

This box is for holding a year's supply of shells. It is two feet tall, one foot wide and eight inches deep. The lid is slanting so that it is impossible for the fowls to perch on it. The opening is eight inches from the bottom and is six inches wide. A slanting board extends from the front to within an inch of the back and allows the hopper to keep always full as long as the supply lasts. Fasten it against the wall inside of the poultry house.

Drinking Dish

A convenient drinking dish for poultry is made by taking two old three-quart tin pans with flat rims, as shown in Fig. I, and putting them together, as shown in Fig. II, after four square openings have been cut in one, the edges of which are neatly turned over. At one side the rims are fastened together with a rivet, on which they turn apart to allow filling and cleaning. A wire nail passes through two holes on the opposite side. Neither the fowls nor dirt can get into such a drinking dish. Worn-out cooking dishes can be thus utilized.

Simple Water Protector

The illustration represents an arrangement I use in my poultry yard for furnishing drinking water to the flocks. It consists of an old stave basket with every other stave broken and turned over any suitable vessel for holding water. I put a brick or piece of rock on top to keep the basket in position.

Roost and Dark Nest

This shows the roosts, the dropping board platform and the nests below, the last so constructed that the fowls lay in the dark—a cure for egg eating. The method of construction is too plainly shown to need description. The writer has built three of these and finds them the best combination of roosts, dropping board and nests that he has ever seen.

Protected Roosts

In cold climates, especially when the temperature goes down to and below zero, the fowls must receive extra protection at night while on the roost. The illustration shows how the roosts can be boxed in, and a muslin curtain dropped down at night. This will not only prevent drafts, but the inclosure will conserve the animal heat and there will be no danger of frost-bitten combs and wattles.

Protecting the Chicken Exit

A convenient storm door over the exit hole in a poultry house will keep out driving rain and snow and also afford good protection against drafts. The illustration tells its own story. Such a storm door can be made out of any thin lumber. About an inch in thickness would make it stronger.

Louse-Proof Nests

This entire nest is constructed of tarred paper. A frame is made of shingling lath and the bottom, sides, back and partitions are composed of nothing but tarred paper. This nest is set upon a board platform measuring about 3 x 4 feet. The illustration plainly shows how it is built.

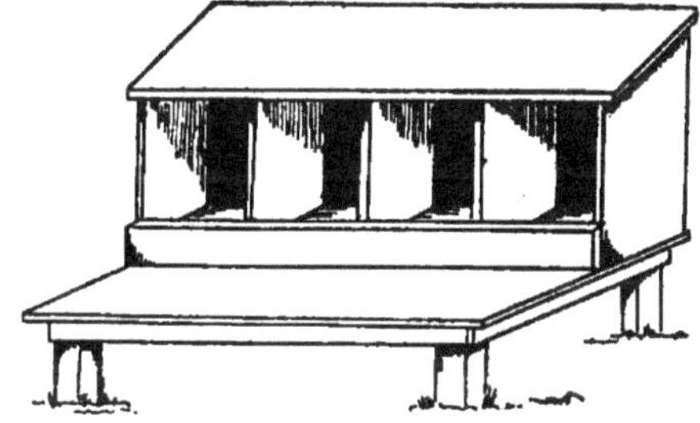

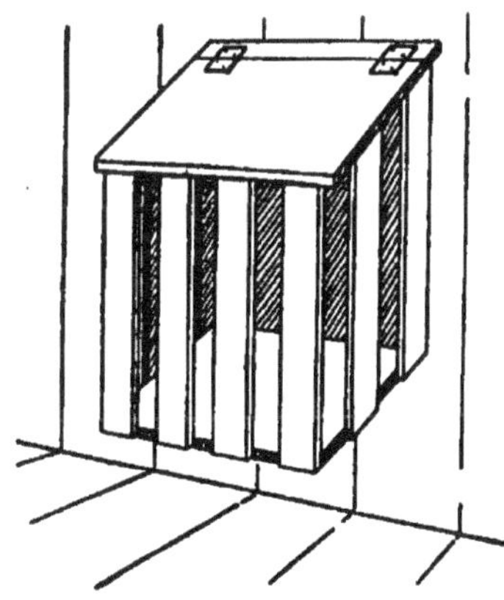

Vegetable Rack for Poultry

We could never make the often advised plan of hanging cabbage or other roots up with a string for the fowls to peck at work in a satisfactory manner. Throwing them on the floor in the litter, or in the dust and filth, is not to be tolerated. A slatted rack like the cut with sloping roof and sloping bottom keeps them clean and holds them just where the birds can make the best use of them.

Poultry Door Construction

The swinging door between two poultry pens should extend to the floor, so a wheel-barrow can be run through; but, if the door goes to the floor, the litter in the pens interferes with its opening and shutting. Let the lower part of the door be made to swing in or out, as shown in the cut, and it will ride up over the litter, but come back into place when the door swings back into position.

Partition in Henhouse

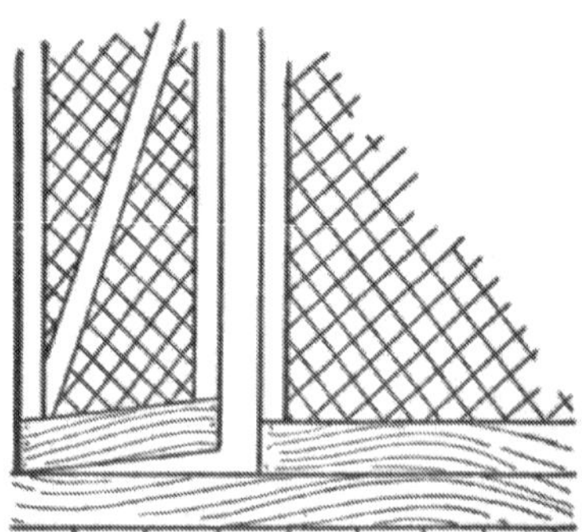

Fowls do better in flocks of twenty-five or so each. If the house is long, better divide with wire partitions, having two boards at the bottom to prevent fighting. The bottom board runs beneath the door, as shown, so that the door can be opened without hitting against the deep litter with which the floor should be covered to induce scratching. Such a door should have spring hinges, then no latch will be needed. A walk in front of the pens takes just so much room from the hens. With doors arranged in this way, it is easy to walk directly through the pens.

Shade for Poultry Yard

Where there is no natural shade in the poultry yard, make a frame and stretch wires across it, as shown in the cut. For a permanent shade set woodbine at the base of the frame; for temporary use plant the seeds of morning glory, gourd or other quick-growing vines.

Feeding Pans

If you do not use hoppers and need some device to hold food, hunt up your tin pans that are past mending, find a board that you can place two or three pans on in a row, then take some little pieces of leather and several small wire nails and tack through the leather and pan, using three nails for each pan, and you will find you have a most satisfactory feeding place for your chickens, either for mashes, scraps or grain, and one which cannot be easily tipped over.

Following Nature in Setting Hens

The illustrations show how the barrel nest and run are constructed. The former consists of a flour barrel laid on its side in an excavation sufficient to bury the sides of it about two or three inches. The dirt secured in digging this ditch is spread on the inside of the barrel, and upon this a nest is hollowed out and filled with tobacco stems.

On the top of the barrel is fastened heavy roofing paper. If this paper is nailed to cleats and fastened by wire it can be readily removed after hatching season and used in this way for years.

At night a wide board is placed in front of the barrel with an opening above sufficient to afford ventilation. The nest being on the ground, the eggs receive plenty of moisture, and a better hatch is assured.

The run is constructed of whole length plastering laths and made in sections, as shown in picture

Keep Roosters from Fighting

Take a stiff cord and make a loop long enough to hang from the bird's neck nearly to the ground. Fasten end of cord around his neck. When he makes a jump for his antagonist his foot catches in the loop and over he goes, and he soon concludes he don't know anything about fighting.

An Airy Chicken Coop

When chicks get to weigh two pounds, or even less, apiece, the small coops in which they were first placed become much too little for them. The cut shows an excellent coop for growing chicks—one in which they can be kept till cold weather in the fall. It has roosts and a slat door front, with burlap curtains to let down in case of storms or cold winds. Here the chicks will be safe from harm, have pure air, and will grow rapidly, not being crowded.

Coop and Run

If you raise chickens in the old-fashioned way you will need a number of small coops with long runs. You can make these yourself and the boys can help you and be happy about it.

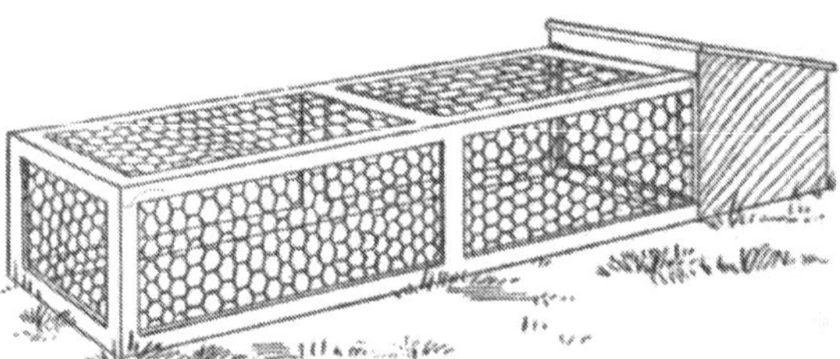

Chicken Coop Double Roof

A double roof over the chicken coop will not only keep the hot sun from distressing the hen and her brood, but it will also prevent rain from driving into their sleeping quarters. Make a framework above the coop, as shown in the cut, and cover with cloth or building paper. The air space between this and the coop will keep the coop very comfortable, while the protection in front will make grateful shade for both hen and chicks. The chicks will not thrive unless they are comfortable.

390

Colony House

One of the most convenient devices ever invented for the poultry farm is what is known as a colony house. It is invaluable in raising young chicks, or keeping growing cockerels, or placing surplus hens. This house measures four feet square, three feet high on the sides, and is raised about a foot from the ground. On many farms there are a dozen or more of these houses. In some cases they are scattered about the orchard, being located 200 feet apart, in which case no yards are required.

Coop and Run in One

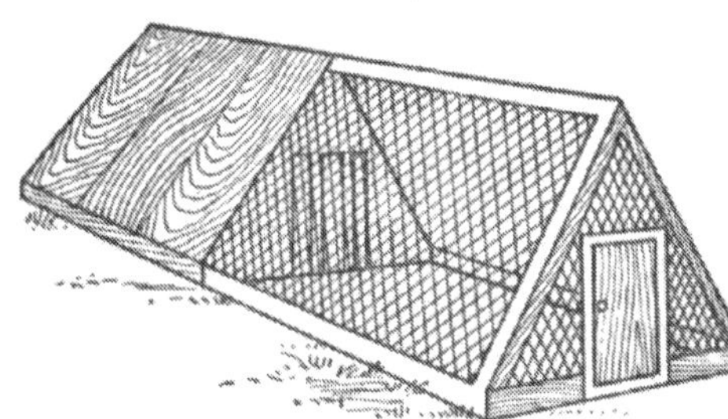

The coop and run here shown is an improvement on an English coop advertised in foreign papers. A door in the rear gives access to the hen. The chicks can come into the yard as soon as it is light. They can be given their liberty by the little door in their run during the day, if desired. Such a run gives security a n d out-door air to the chicks before their owner cares to arise in the morning.

Simple Protection for Coops

Hawks and crows will often s t e a l young chicks early in the morning if they are a l l o w e d free run as soon as awake. The protected chick run here-with illustrated, not o n l y saves space, but it is easy to handle in case of remov-

ing the chicken coop to a new location. The illustration so plainly suggests the construction of these runs that no further explanation is needed.

Hanging Sectional Coop

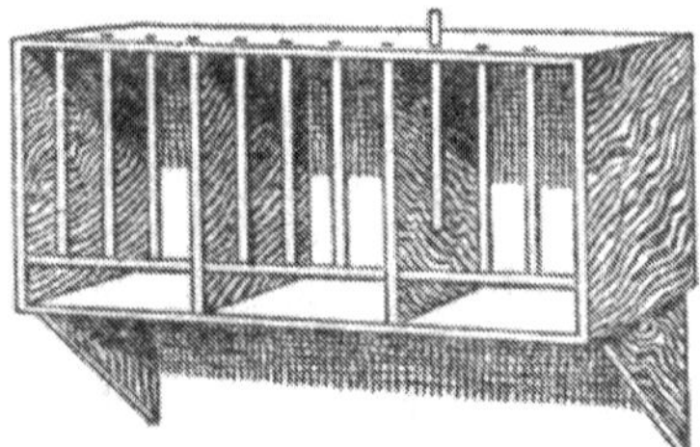

This is one of the most convenient devices about a poultry house. If it is desired to carry over an extra male he can be put into such a coop. If a hen gets broody here is a place to put her until broken up. If a fowl is to be fattened the coop stands ready to receive it. Made as shown it can be cleaned out in ten seconds. Dishes for food and drink can be hung on hooks at the side. Set this sectional coop up against the wall of the poultry house and you will wonder how you ever got along without it.

Fence Construction

If other plans fail to help the hens from flying over the yard fence, try the plan of bending the top of the netting over toward the inside of the yard, as shown in the cut, bending in six or seven inches. In scaling a wire netting fence, fowls are inclined to fly up against the fence and then with feet and wings mount up to the top, where they perch for an instant and then fly off. With this turned-over fence they find their upward way barred.

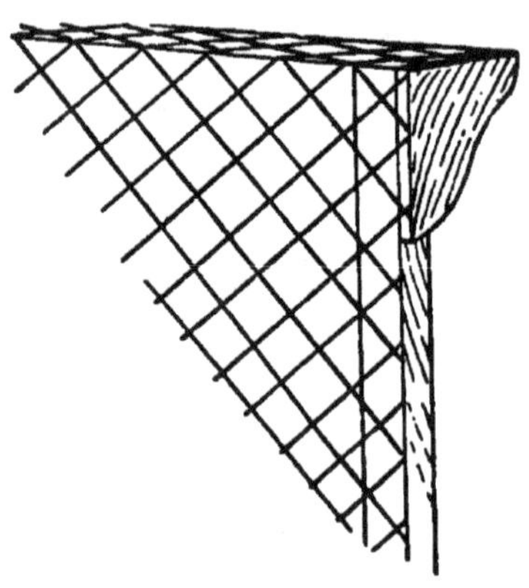

Chicken-Proof Fence

A chicken fence that will turn both grown fowls and small chicks can be made by using a foot-wide strip of inch mesh wire netting at the bottom, and the ordinary two-inch wire above this, as shown in the drawing. A bit of fine wire here and there will hold the two nettings together. The ordinary netting will not hold chicks much before they are five or six weeks old.

Double Fence

This figure shows a very good method of fencing to prevent male birds in opposite yards from fighting through the wire-mesh. A piece of scantling about two and one-half feet long is nailed to each post, and on this scantling is tacked two-foot wire mesh, raised a little from the ground. This makes it impossible for males on either side to reach each other, and they soon give up in disgust.

Saving Posts in the Wire Fence

An economical method of putting up a wire-netting fence is as follows: Ordinary two-inch mesh is used. Posts are placed fourteen to sixteen feet apart, and a heavy wire is tacked to the top of each post. This wire is then drawn down to meet the top of the fence (as shown in the illustration) and fastened with pliable wire. This holds the netting firmly and saves quite a few posts. The usual way is to have a post for every ten feet of fence, but by the use of this top wire, from two to four posts are saved each stretch, which is quite a saving in cost.

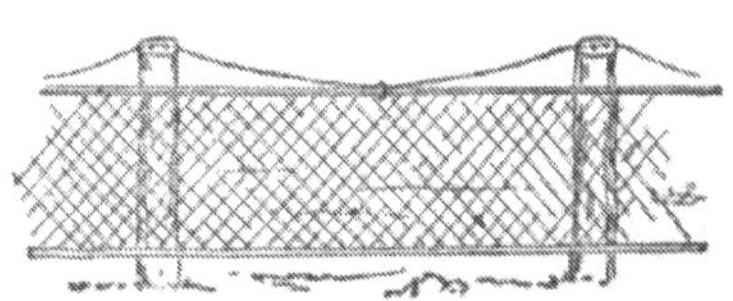

Skeleton Coop

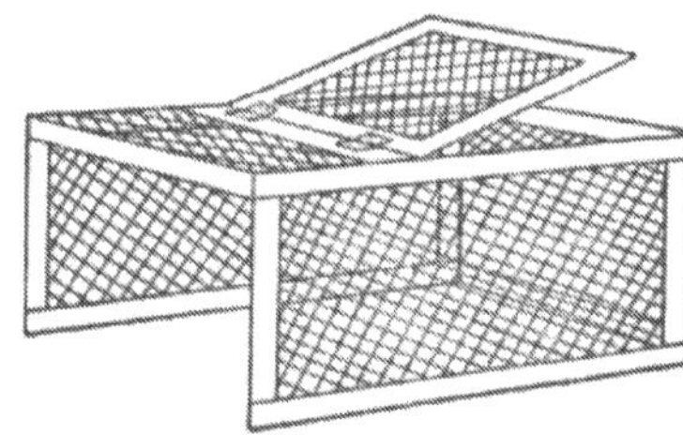

This skeleton coop of lath and wire, when placed in front of the brood coop, makes a nice little run for the chicks while young or in stormy weather. In stormy weather an old piece of oilcloth or other temporary cover may be used for the brooding hens, removing the door of the other coop and placing food and water and a dust box in it for the hens' use.

One-Way Poultry Gate

A neat thing to get fowls to stay in the yard after they come in, and at the same time to permit others to come in, is a little slat door suspended where they come in usually and swinging in only. They soon learn to put their heads between the slats and push to come in, but when they try to go out so, it does not work. A movable pin may permit the door to be set either way. Screw eyes make good hinges.

Raising Early Chicks

Here is a useful home-made attachment to a wood-house I saw the other day. A brood of early chicks is reared in a wood-house where a stove is kept, and on warm, sunny days a little slide door is opened and the brood runs out in the air upon a tight shelf boxed in at sides and top but with net in front. No pesky cat can catch an early broiler, and the chicks seem to thrive and do well.

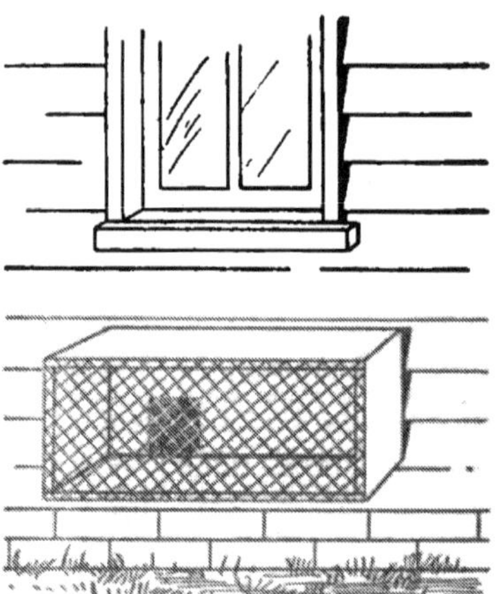

Portable Dust Box

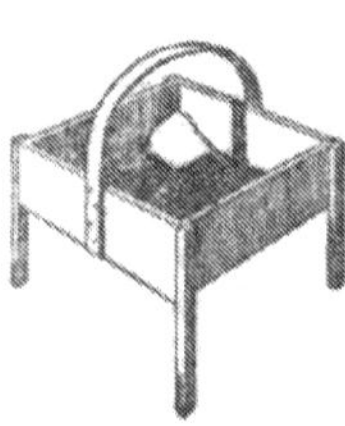

A cheap dust box can be made from a wooden box, a barrel hoop and four pieces of wood for legs, as shown. If this box is filled with road dust or sifted coal ashes, the hens will delight in flying up into the box and taking a good dust bath. This is Biddy's manner of fighting lice, and it is very effective, too. The box can be made any size desired, so that it can be conveniently carried about. Every pen of fowls should have one.

Keeping the Chicks Under the Tree

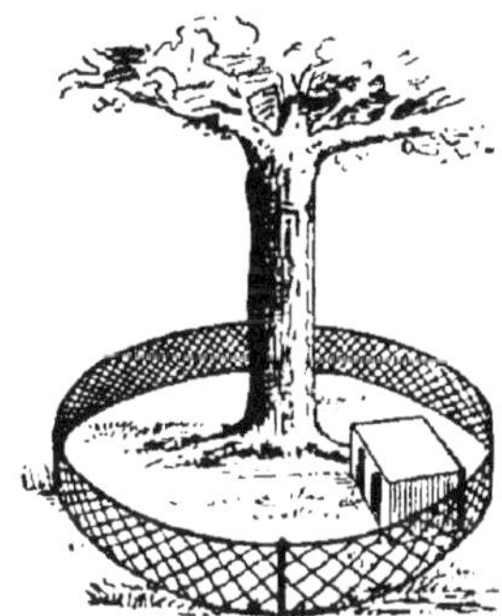

No two things about a farm go together better than fruit trees and chickens. Make them acquainted early. Put the newly-hatched brood in a coop under a tree and surround the whole with a circle of two-foot inch mesh poultry netting (as shown in the cut), it will stand alone in the form of a circle. The shade is good for the baby chicks. The chickens are good for the tree. As they grow, they will scratch the soil and thus cultivate and enrich it, besides destroying many insect enemies.

Summer Poultry Houses

It is often advisable in summer time to establish a colony of fowls on new runs, away from the farm buildings and from the general flock. For this purpose light, portable and cheaply constructed houses are needed.

It is also desirable to get growing broods out of their crowded coops and up off the ground as soon as they are large enough to use perches. The structures described below can be adapted to either purpose.

Number one is a handy and inexpensive "colony" house for hens or growing chicks, and is so light that it can be moved about anywhere. There is no window, but an overlapping portion of the roof can be raised for light and ventilation, the open space being covered by wire netting. Such a hen house needs only a light bit of frame—four pieces for sills; two A pieces for the ends and ridge pole. A panel removed from the door and covered with wire netting would be an improvement.

Number two. Nothing elaborate, is it? And yet it answers every purpose, and not only protects from storm, but from skunks and weasels. It has a tight bottom, and each side of the roof is hinged to the side, so that it may be opened to catch a chick or to clean it out. A piece of coal sieve is the door and ventilator, through which air can pass, but no animal. The handles pass through the house, making low perches for the birdies. When moving it raise one end and drag the handles at the other end. A bit of chain keeps the roof together,

and in very hot weather permits the insertion of a chip to hold it open half an inch or more the length of the ridge to let the heat escape. Every farm ought to have half a dozen.

Number three shows a practical coop for a summer colony of hens that are to be given free range in a pasture. The top is the roosting room, access to which for cleaning can be had by letting down a hinged door in the rear.

The space below can be used during rainstorms, the hinged front protecting this and also the roosting room from rain, while giving free circulation of air. A row of nests is placed in the lower space, access being had by a hinged door at the end. Half a dozen colonies of hens in such coops can be scattered about in rough land, such places being especially enjoyed by poultry.

The floor extends along the rear side under the perches and the hens gain access through the open space. By leaving out the nests and providing cleated boards for stairways to the upper story, this house is equally well adapted to young stock.

Shading the Windows

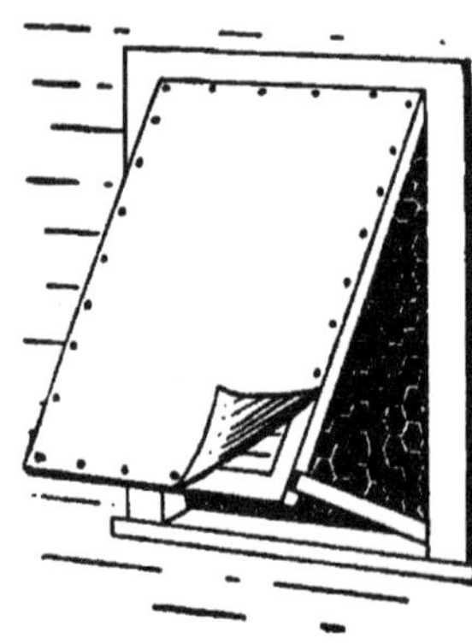

Is the sunshine pouring in your poultry houses through the windows? This is fine in winter, but in midsummer not so fine if the chickens are shut up. The cut shows how to stop it.

Half-Doors for a Duck House

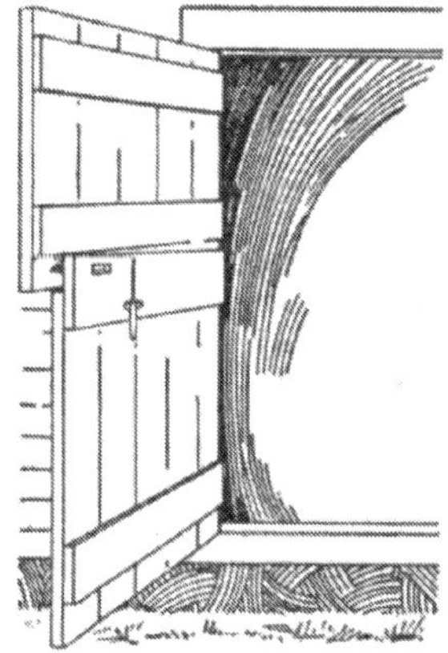

A door divided midway from top to bottom is very useful in a duckling house, when it is desirable to keep the youngsters indoors during inclement weather. It is not best to expose ducklings, while they still have their down, to heavy rain storms.

A double door permits the top to be open for ventilation, without allowing a draft to sweep in across the floor. The cut shows how to arrange such a door so that the top can be stapled to the lower half, and the whole used together as one door when desired.

A cleat covers the midway cut, but is fastened only to the lower half. A staple in the upper part runs through a slot in this cleat, and is secured by a pin that hangs on a string. Bevel the cut between the two parts, to shed water.

Incubator Cellar Roof

An incubator cellar can be greatly improved by having a double roof. The openings on the sides and ends between the roofs allow a circulation of air that will keep down the heat caused by the sun shining directly on the roof. The illustration shows a plan that will keep down the rising temperature in the incubator house when the sun is shining. Late in the spring or summer it is often difficult to keep a uniform heat in the machine, as the house becomes overheated from the effect of the sun's rays. The inner roof can be covered with cheap boards and roofing paper, with lath battens, but the outer roof should be shingled, as a black roof absorbs the heat readily.

Attaching a Scratching Shed

Since poultry need an open scratching shed in winter if they are to do their best in the way of egg production, the illustration shows how a low shed can be attached at slight expense to an existing poultry house. Doors, with some glass or curtains, could be provided for the shed, to be used in stormy weather if desired, but for many sections of the country this simple open shed will be found just the thing for the purpose desired.

Piano-Box Poultry Houses

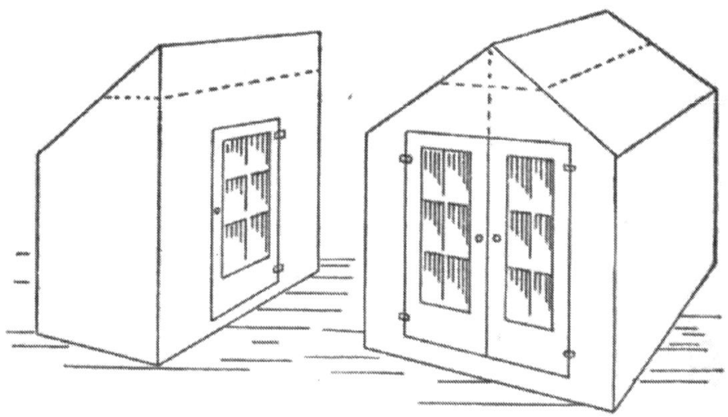

Fig. 1   *Fig. 2*

Piano or organ boxes make good poultry houses, either for half-grown chicks or for a small flock of hens. The cut shows two ways of finishing them. In Fig. 1 the part above the dotted line is added to complete the roof. A combination door and window is also added, and any covering placed upon the outside that may be desired. Fig. 2 shows two piano boxes placed together, and the roof completed. Two combined doors and windows are placed in the end to give sufficient light. The walls and roof may be covered with roofing felt.

Chaff Lining for Poultry House

We show here a cheap way to keep out the cold. Get heavy, red building paper (the heaviest grade) and tack to the studding inside, beginning at the bottom. As each strip of paper is put on, fill chaff loosely in between the studs, as shown. Let the top of each strip of paper lap over the strip next above. When the paper is all in place, nail laths to the studs, as shown. Where hens or small animals will come in contact with the paper, use a strip of wire netting about the base, as illustrated.

Ventilating Window

The question of properly ventilating the incubator room has frequently been asked by Farm Journal readers. Noting the different methods used by poultrymen visited, the editor observed that the simplest and most effective method is to have the windows hinged at the bottom instead of at the top or on the side. The illustration plainly explains itself.

Side-Hill Poultry House

If the land slopes to the east, dig into the bank so as to make a level floor. Dig a trench and fill with loose stones for a foundation. On this build a concrete wall or one of cemented rough stones, as shown. Then cement the floor. This makes a very warm pen.

The cut shows a section of the earth, the straight dotted line indicating the position of the cemented floor, and the other dotted lines the stone foundation and the cemented stone work. The open scratching shed is now the standard among poultrymen. This shed would be improved by lath screen doors to confine the occupants on windy or stormy days.

Poultry House on North Slope

An entirely suitable site for a poultry house is not always available. In this case the poultryman must do the best he can in the situation. The cut furnishes a hint of what may be done on a northerly slope in the way of securing the sunlight, a warm roosting room, and a sunny and comfortable scratching

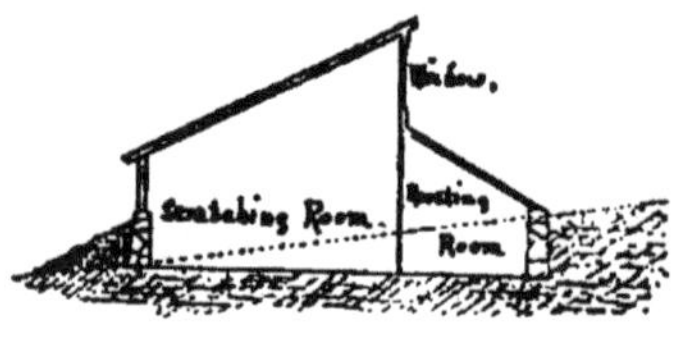

shed. The earth excavated to secure a level floor may be banked up against a northern wall. This may well be made of stone or brick, but if of wood, it should be double boarded, having building paper between to make it wind-proof. Care must be taken to provide drainage for a house so located, or rain and melting snow-water will find their way into it and make no end of trouble.

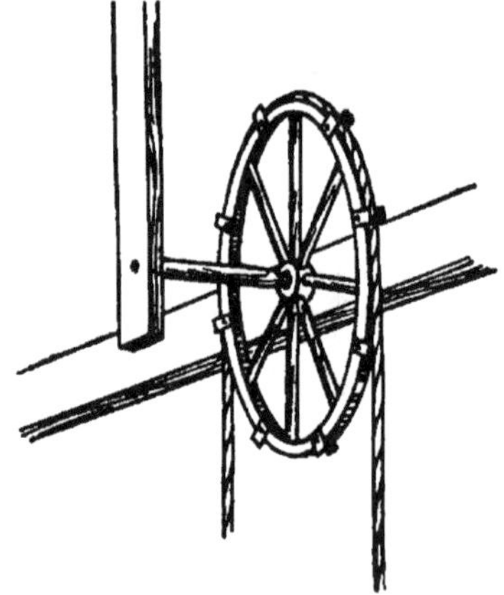

Hoisting Pulley Wheel

An old wagon wheel, with side clips of wood screwed to the rim, and its hub extended to form a shaft, can be hung in the manner shown in the cut, when it will be found exceedingly convenient for hoisting articles, for weighing and other purposes. The shaft should have a stout iron rod through the center and passing through the hub, being firmly fastened so it cannot revolve in the wheel or shaft. Fit one end to revolve in the side of a beam, and nail a piece of joist to a rafter as a support for the other end. The cut shows plainly the details of construction.

Rope Loop for Weighing Hay

It is often desired to weigh out small lots of loose hay, and when this is the case the arrangement of a rope in the form shown herewith is also convenient. Make a loop of a long rope and stretch it out thus doubled on the barn floor. After the hay has been thrown upon it, gather the two ends of the loop and pass one through the other as shown and hook the upper part into the hook of the steelyards.

Hay Distributor

In wide mows or bays when the fork delivers the hay in the center in the usual manner it must be forked laboriously to the sides. To remedy this build a board platform, B, six by nine feet, nailed to a four by six inch piece turned at the ends and pivoted at A, in blocks nailed to the plank, C. A rope is fastened to each end of the platform so that it can be tipped to either side as desired. Planks can be laid across each

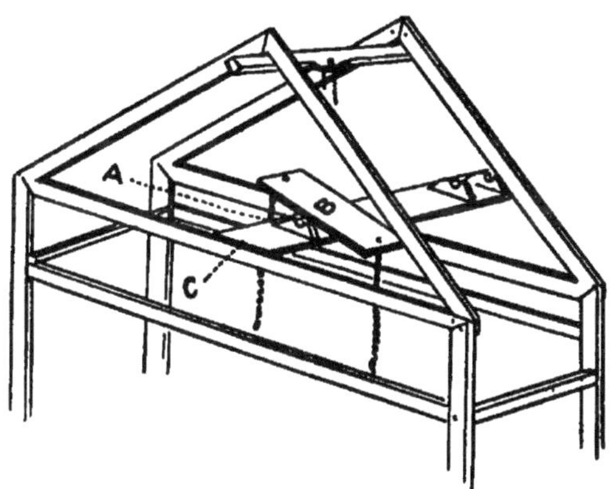

bent and sockets fastened on each one, so that the dumping platform and its attached roller can be moved easily. In order to be effective the hay must drop every time near the center of platform. On a wooden track this can be effected by boring a hole in the track and putting an iron pin through it. On an iron track the same end can be secured by a clamp screwed on.

Safety Device on the Grain Chest

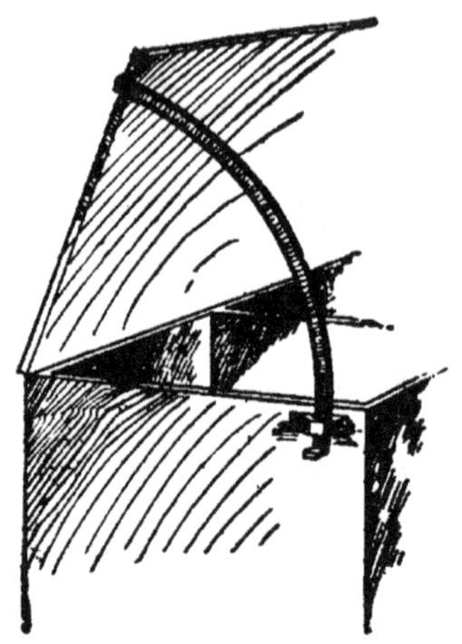

If you find you cannot trust your own memory to be careful, or that of the hired man, put a curved iron strap upon the end of the stable grain chest, as shown in the cut, then the cover will not be left up, and a cow or horse, getting loose, will not lose its life from over-eating grain, as has so often happened. The cut shows the way to attach the iron strap. The cover cannot be raised so as to stand open itself, so must always close after one has taken grain from the chest. A strong strap or chain would do as well.

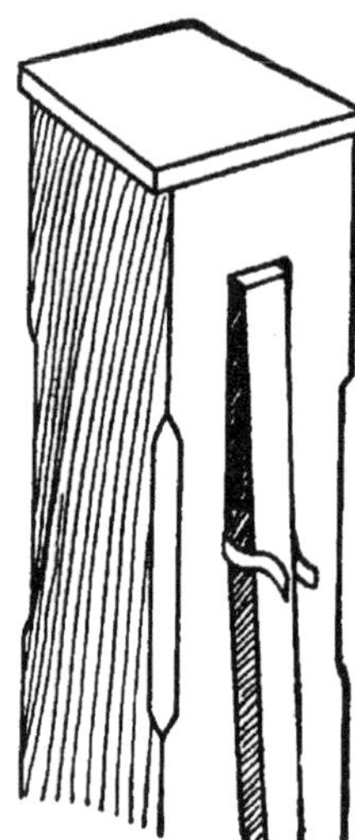

Pump Handle Holder

Unless all pump handles are held in such a position that when you wander out after dark you will not be in danger of dislocating your jaw or cracking your shins, fix them as shown in the drawing. A piece of spring steel bent as shown in the cut and fastened to the pump with two screws will do the business. When done pumping, push the handle down and the clamps will hold it there.

Grain Chest from Piano Box

At almost every village music store there are to be had empty boxes in which organs or upright pianos have been packed. These boxes can be utilized very nicely for grain chests, after the manner shown in the cut. The narrow shelf gives accommodation for grain measures and feed boxes.

Improved Hitching Ring

Rings are frequently inserted in the sides of farm yard buildings so that a team can be hitched there. The horse commonly rubs his headstall against the building to the headstall's great injury. Many horses also use their teeth upon the boarding or shingling. The cut shows a curved iron rod with a snaffle that is hinged to the building's side, which effectually prevents the horse getting his head near the boarding, and keeps him also from twisting about from side to side. A blacksmith can do the work in a few moments.

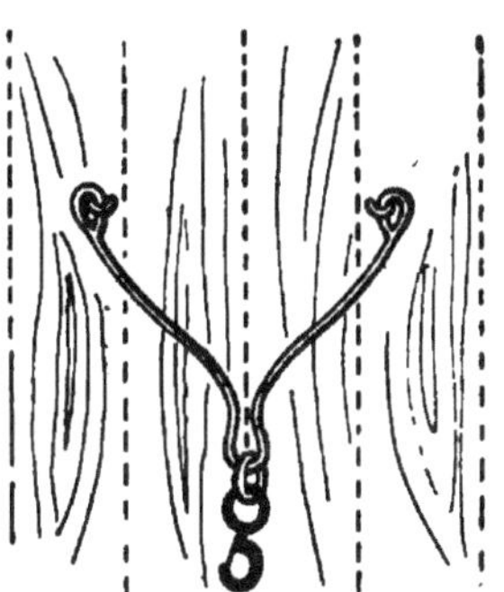

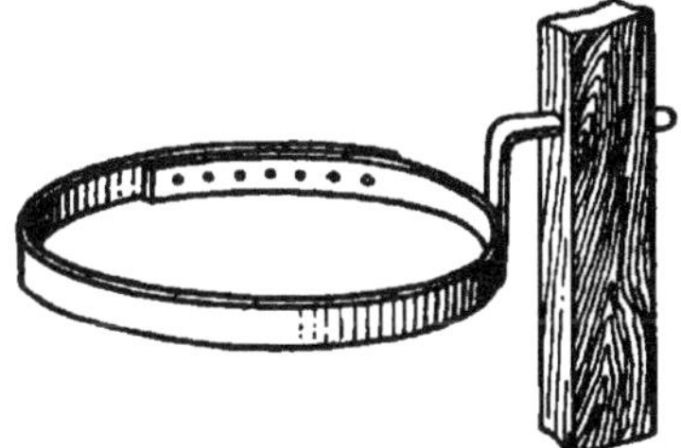

Adjustable Bag Holder

There are bag-holders and bag-holders, but one that is adjustable to all sizes of bags is somewhat of a rarity.

Protector for Hay Chutes

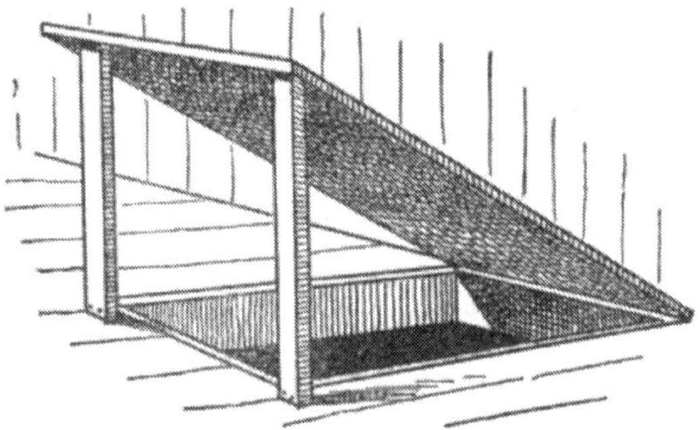

It is a wonder that even more people are not injured by the unprotected feed chutes that extend from many upper floors down into horse and cattle cribs. There is a constant liability of one stepping backward into one of these openings, or of children falling into them when playing in stable chambers. It is a simple matter to protect such openings in the manner shown in the accompanying illustration. Such an arrangement is one of the ounces of prevention that are worth many pounds of cure.

Portable Bag Holder

This handy, portable bag-holder is made with a piece of two-inch plank thirty inches long for a foundation. Holes twenty-four inches apart are bored in this with an inch and a quarter bit, to hold the upright standards. These should be about three feet nine inches long and about as thick as a man's wrist, and must be tough and springy. Shave these sticks to fit into the holes in the plank, and taper the tops for about ten inches so they will fit into a three-quarter inch hole. The two holders, one of which is shown at top of cut, consist of blocks of wood two inches square and four and one-half inches long, a three-quarter inch hole in the end of each and a semi-circular piece nailed on the other end. The points of three small nails in each curved piece and the spring in the standards hold the bag up and wide open.

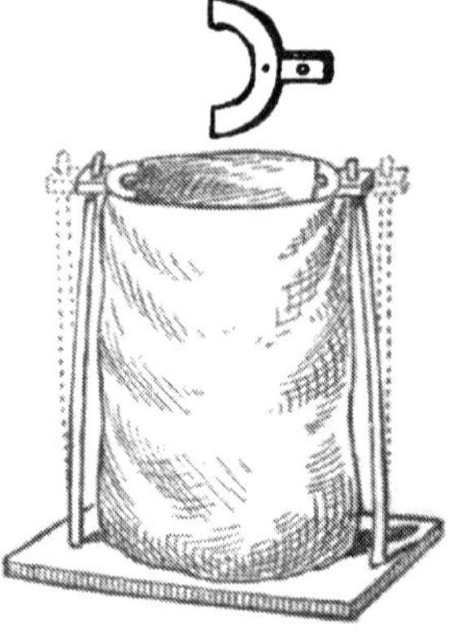

Safe Lantern Bracket

It is being on the safe side when one resolves never to carry a lantern into a barn; a stumble may mean the loss of the barn and its contents. Cut a panel in the big barn door, hinge and chain it as shown in the cut. One can then let down the panel, set the lantern upon the shelf thus formed, and light the interior very plainly. If the cattle quarters are to be visited, have a panel there, also.

Storing the Wagon Box

No system of ropes and pulleys to directly raise a wagon box or rigging is satisfactory unless the ceiling be sufficiently high to permit the suspension out of reach of the heads of men and horses, and usually of top carriages. A better plan requiring less tackling is a pair of heavy brackets on the side of a shed or hay mow with a rope and single pulley in the side wall or barn frame above where the box or rigging will reach when turned on edge upon the bracket. On removing the rigging, etc., drive close beside the brackets and attach the rope to its opposite side. Having caught one edge on the brackets, draw up on the pulley and turn the awkward thing

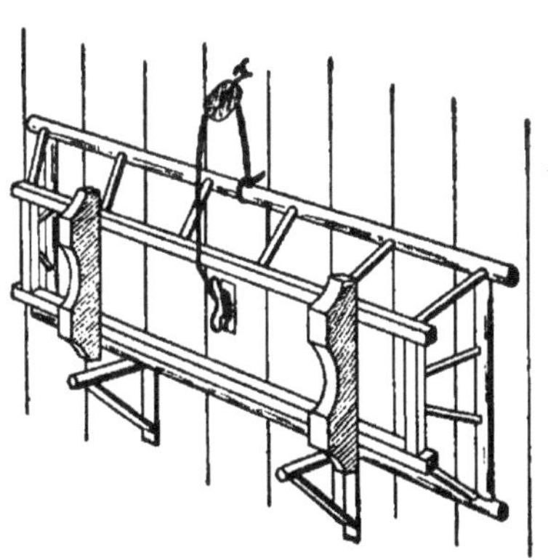

up out of the way. A short rope or chain will hold it where it is put. The brackets can be made of two-inch plank as shown, or natural knees may be cut in the woods and hewed into shape.

Work About the Barn   **Handy Devices**

Seed Corn Protectors

To keep rats and mice away from the seed corn, the bag of chestnuts, etc., when hung in the garret, punch a hole in the bottom of three or four old tomato cans and slip them on the rope on each side, and the thieves will not be able to pass them.

Straw Rick Root Cellar

The cut shows how to make a room under a straw rick, in which to keep turnips and potatoes. It answers as well as a root cellar, as contents are accessible at all times. Cost is very small. Cut t h r e e good-sized posts with forks at one end, length about nine feet, put them in ground deep enough to be firm, then take two heavy poles about sixteen feet long and place them in forks of uprights. Place on each side good-sized poles close together, the ends resting on ground, set a little below surface to keep them from slipping. Against the south "end post" nail an old door frame, with door, so when filled it can be kept under lock and key.

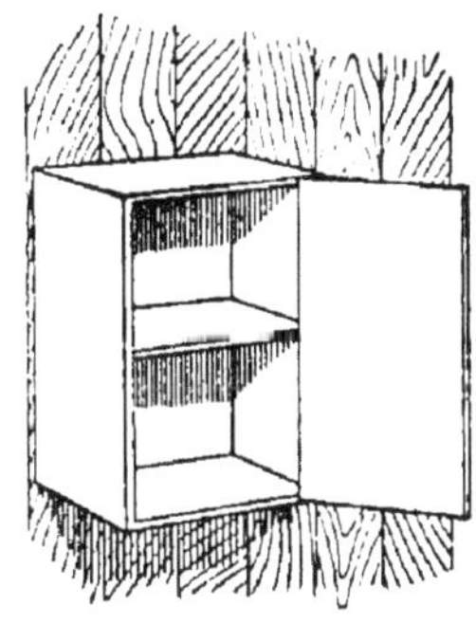

Simple Closet

Small closets are indispensable about the barn, stable and workshop. A ready-made one is shown in the cut. Get a box at the grocer's in which fancy crackers come and screw it against the wall as shown. It has a cover already hinged on. Put in one or more shelves, according to the use that is to be made of the closet.

Lantern Hook

The farmer who knoweth what is good for him does not set the lantern down on the barn floor, where it may be accidentally upset. And here is the proper way to secure both safety and convenience. Back of the stalls, and elsewhere in the barn where needed, he strings an overhead wire at a height within easy reach. On each wire are one or more notched bits of wood. Then all that's necessary is to hook on the lantern and slide it along to just where it is wanted.

Hitching from Rear of Stall

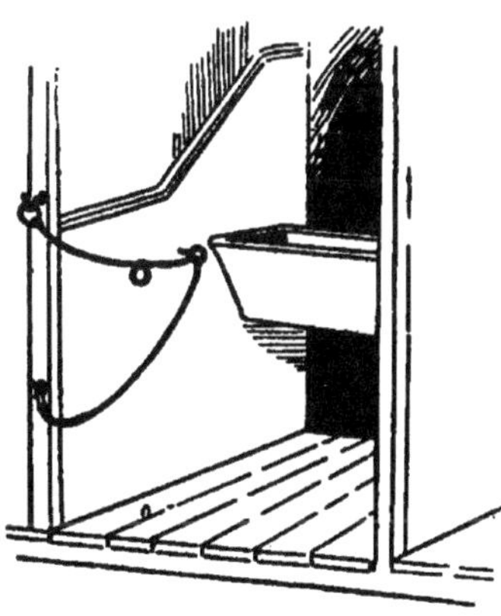

It makes some horses nervous if one enters their stall beside them. Others are so nettlesome that they jump if one's clothes touch them in crowding by. But especially is one liable to get a "horsey" odor on his "best suit" if he goes into the stall with a horse on coming home from church or elsewhere. The cut shows a device for hitching the horse, and unhitching him, without entering the stall. Hook the snaffle on the lower rope into the halter head on the horse. And as the horse enters the stall pull the upper rope till the second ring will slip over the hook. Reverse the process and bid the horse back, when it is desired to take him out.

407

Storing the Ladder

A good place to hang ladders is shown in the cut. When building a long hen house the roof was carried out behind to afford shelter for long ladders that are so hard to accommodate, when not in use, about the farm buildings. Hooks long enough to accommodate several ladders can be placed in the rear wall at distances that will hold up both long and short ones.

Steelyard Rigging

Again and again one sees farmers hanging the steelyards upon a stationary hook, and lifting the article to be weighed up to the hook of the steelyards, be the article light or heavy. Use such a device as is shown in the cut, which explains itself, except the bit of chain hanging on a peg. If the article is low, this chain will lengthen out the steelyard hook, so that no lifting will be required.

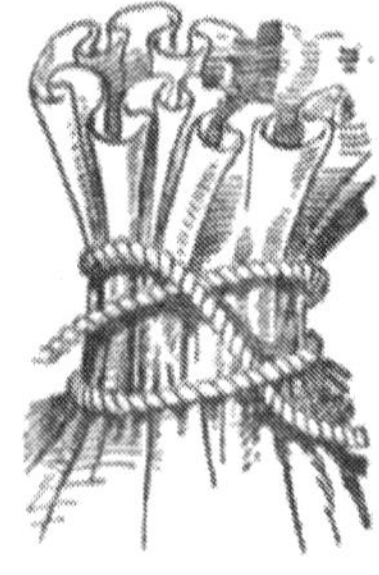

The Miller's Knot for Grain Sacks

The miller's knot for tying grain sacks is one of the most effective. It is the knot used in tying cement sacks, and they seldom come untied. It is almost necessary to cut the string to open a sack, for the knot sticks so tight it is hard to untie.

Straw Bundle Carrier

With this device you can carry enough straw to bed six cows and three horses,—without scattering or losing it even when the wind blows. It beats arms or fork.   Take a piece of sacking eight feet long and three or four feet wide; put a three-inch hem at both ends; slip a two-and-one-half-inch slat in each hem, and tack slats securely in place.   Put a ring  at one end and a cord and stick at the other, as shown. Lay the carrier on the ground; fill with straw; bring the two ends of carrier together, around straw, and button stick in ring; hold by the stick, and swing bundle on to back.

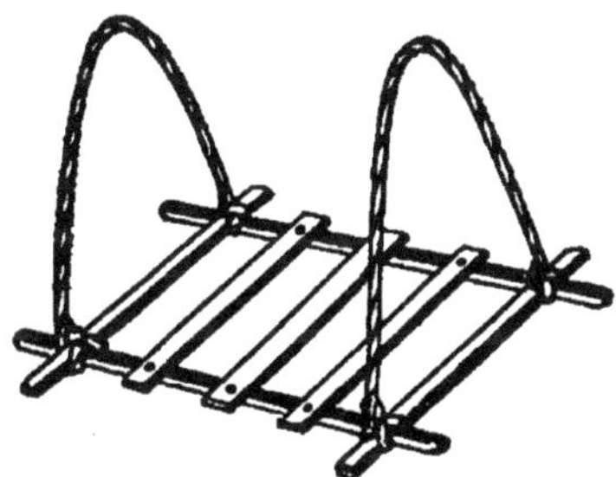

Frame for Weighing Hay

If you will use once a platform like the cut you will never again try to weigh hay with a rope.   Four poles are fastened together at the corners by ropes and light pieces put across as illustrated.

Note—The ropes should be longer in proportion than they are shown by our artist.

Sectional Ventilators for Mows

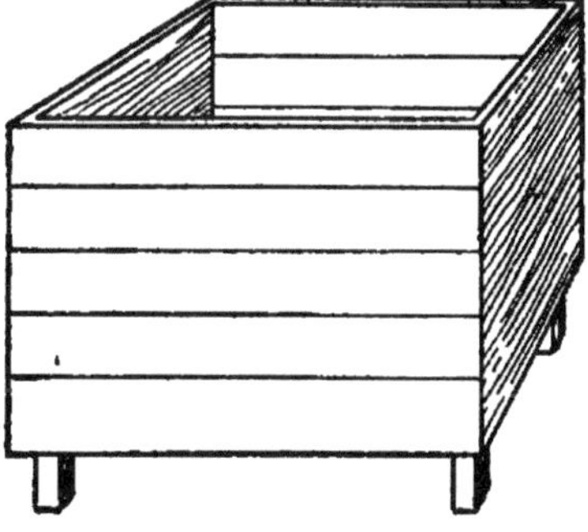

A profitable job for rainy hours is the making of ventilators in sections to use in mows and stacks where corn is put up. These are eighteen inches square, two and a half feet deep and without a top or bottom. The corner cleats inside hang down six inches below the bottom and thus set into and connect one with another as high as wished.   Any rough material, even slats, will do. If there are openings in the sides, so much the better.

Storing and Using Liquid Manure

How the French preserve manure is shown graphically in the sketch. From a cistern where the liquids run, a pump carries to the broad, flat pile that may burn unless it is soaked through it. Almost all of us may build a cistern and save the stable liquids which are the most valuable part of the manure, and though we may not sport a tank on wheels to spread it, we may, like the people of France, whose tillage is noted the world over, pump it frequently upon the drying heap of solids. Economy of manure is what we need to learn.

Filling Grain Sacks

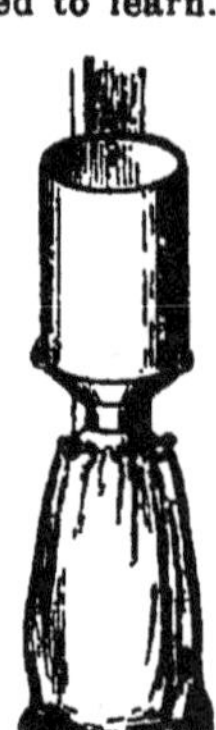

By the use of the following device one man can fill sacks with ease: Take an old five or ten gallon milk can and remove the bottom. Then punch a half-inch hole in the rim near the bottom. Around the neck of the can punch three small holes at equal distances apart, insert a hog ring or belt hook in each hole and clinch. Hang the can on a nail in a post or wall at such a height that a grain sack fastened to the three hooks will just touch the floor. The grain is then shoveled or dipped into the milk can and the sack easily filled.

Handy Rack for Implements

At one side of the shop have a rack for tools and garden implements. The garden fence is no place to hang a hoe when not in use. The sketch given shows how to make a serviceable rack. The two compartments marked I are for small tools, such as hammers, trowels, dibbles, or the like.

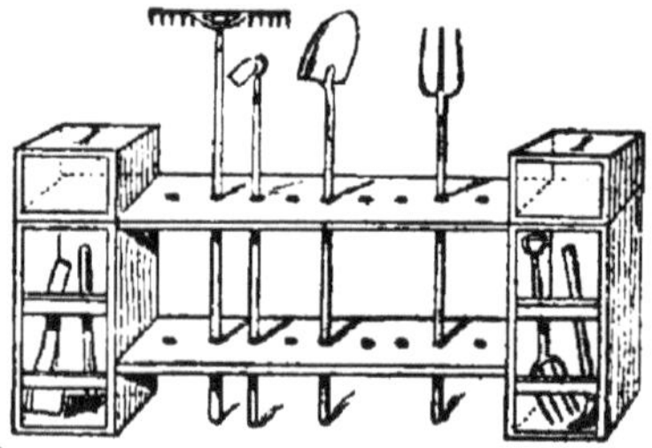

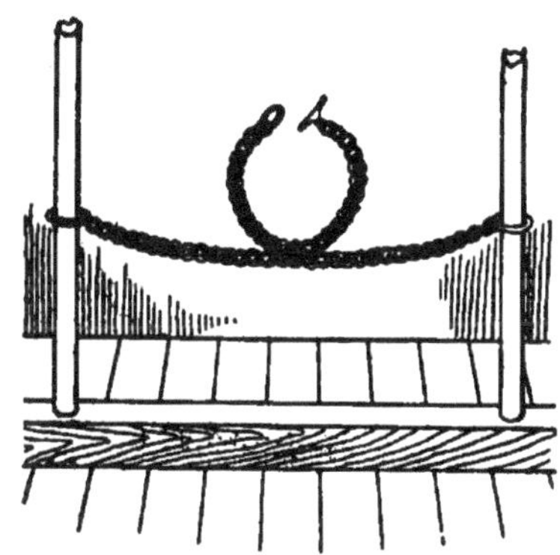

Cattle Tie

If stanchions are not used, one of the best cattle ties is shown herewith. A cross chain slides up and down on two uprights, and has the neck-chain attached at the center. This gives perfect freedom of movement to the animal's head, and still keeps the animal in much the same position always.

Truing the Grindstone

When a grindstone gets out of true, a half-inch rod of soft iron held to the stone like a turning chisel, while the stone is turned, will true it up nicely. If this is not at hand, a spade with the handle against the ground and the edge like a turning tool will soon put the stone in good shape. If the stone becomes glazed over, nitric or sulphuric acid will cut the glazing down to the grit.

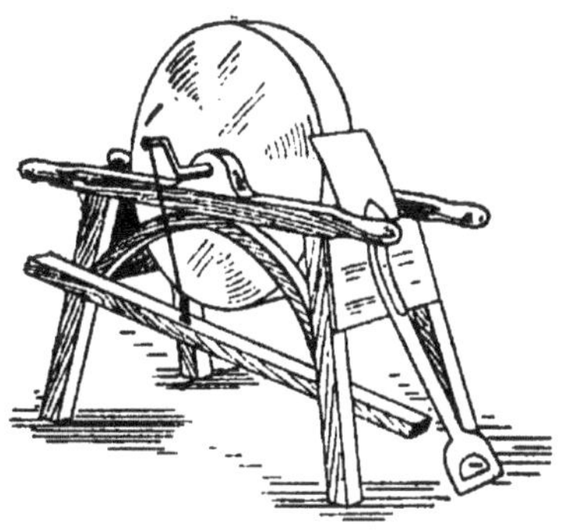

Pump Spout Drain

Why have a muddy puddle about the pump in summer and dangerous ice in winter? Make a trough under the spout, as shown in the cut. It projects beyond the edge of the spout an inch and is about two inches below it. It leads the dribble from the spout, which makes the ice or the puddle, back into the well through a hole cut in the box where its lowest edge joins the box. The trough does not interfere with filling a bucket at the pump.

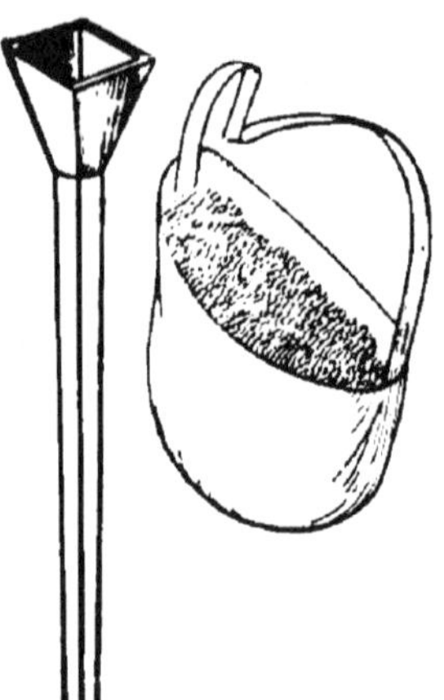

Seed Dropper

Where one has to drop the seed by hand and does not possess a patent seeder, a home-made device will answer about as well. The cut shows a pouch to sling over the shoulder to leave both hands free—one to hold the square (or round) conductor in the right position over the "hill" and the other to drop in the seed at the spreading top. The tinsmith will make such a tube for a trifle, or it can be made at home of laths, with a sheet-iron flaring top.

Husking Seat

A husking seat that avoids backache or stomachache is the device shown in this little sketch. Sit across the end, with bundle of corn across the board at each side of you; the pegs at end prevent bundle slipping away from you.

Side Draft Attachment for Plow

If there is any plowing yet to be done close to fences or trees have your smith put extension rods on the plow and a clevis to hitch the team to, like the cut. If the beam of plow is adjustable to draft, it should, of course, be adjusted before the rods are bolted to the handles.

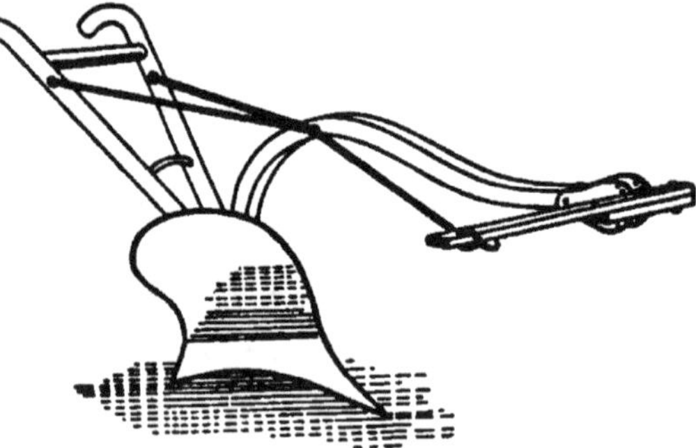

Burning Out Stumps

Dragging out stumps by expensive machinery is slow and costly work, and in many cases may be avoided by drawing away the soil from the stumps so they will dry in the sun and wind, and then firing them in the following manner: With a two-inch auger bore through a large root near the ground on the side toward the prevailing wind; then with a squirt-can moisten the inside of the hole with coal oil and get fire started so the wind will draw through it. In this manner, if a draft can be maintained, the stump may be gradually consumed. Two men can fire and watch to keep the burning going, night and day, and as quickly clear an acre as by the more forceful machinery or dynamite.

An Aid in Tying Corn Shocks

As an aid when tying corn shocks, procure a hook having shank with hole in which to attach rope. A blacksmith can make one. Fasten it to the end of a heavy cord five to six feet long. Take five or six iron rings about an inch in diameter. Slip one on rope to center and tie knot to hold in position. Fasten the others on in a similar manner at intervals of six or eight inches. At the opposite end of the rope tie a two-inch harness ring. To use: Throw hook around shock, pass it through the end rig, draw tight, and hold in position by hooking into one of the central rings while shock is being tied; after which remove.

Binding Corn Shocks Firmly

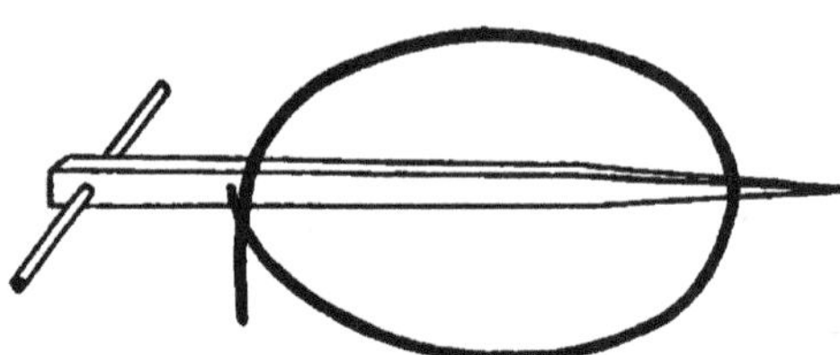

To have corn stand up against the fall winds it is important to bind the shocks firmly near the top. In tall corn, such as many grow, it is difficult to do this without a device like the one here illustrated. It is made of 2 x 2 inch hardwood stick four or five feet long, rounded to within a foot of one end and shaved to a point at the other. At eight inches from the large end a half-inch hole is bored for a rope, and near the same end a three-quarter-inch hole for a pin to go through. A knot fastens one end of the rope in the hole. To use the tightener, thrust it through the shock, adjust the rope as shown in cut and twist until the shock is snugly drawn together. The band of twine or straw can now be put on at leisure and as firmly as necessary, and the implement be withdrawn.

Bench for Corn-Husking

To make a bench for husking shocks of corn, take two cultivator wheels and make an axle about three and a half feet long and fit the wheels on it. Spike two twelve inch pieces to the axle, one near each wheel and to these spike two, two by four pieces fourteen feet long. At the other end nail a cross piece and legs to keep the bench at a proper height. At the axle the uprights are tied to the long pieces by braces. Higher wheels would make the uprights unnecessary. The drawing explains how it is constructed.

Double-Handle Saw

Here is the way I use a cross-cut saw alone: Put one handle down as far as I can, the other in its proper place. Try it. Take one handle in each hand.

Saw Attachment

When one must use a cross-cut saw alone in fitting wood for the stove in winter, such a device as is here shown will save labor. A heavy piece of timber is suspended to swing easily on a hook. Insert one end of the saw at the lower end and begin to saw. The weight of the timber will act like a balance wheel and greatly aid in running the saw. The long hook permits the saw to fall as the log is sawed.

Making a Round Stack

In stacking hay in round stacks I use a center pole which is set in the ground about three feet and allowed to extend a little higher than the stack. Near the base of the pole I nail several short pieces in order to brace it and form an air chamber. Across these are laid other short pieces of boards and the stack bottom is then laid in the usual way. The accompanying illustration shows the stack cut down through the center. A, is the pole, B, B, the braces, C, C, weights hung on D, D, fine wire which is attached to a ring E, of heavier wire large enough to slip easily down the pole. I use five or six weights. In weighing a rick I drive down along the ridge large wooden pins to hold the rings and weights in place.

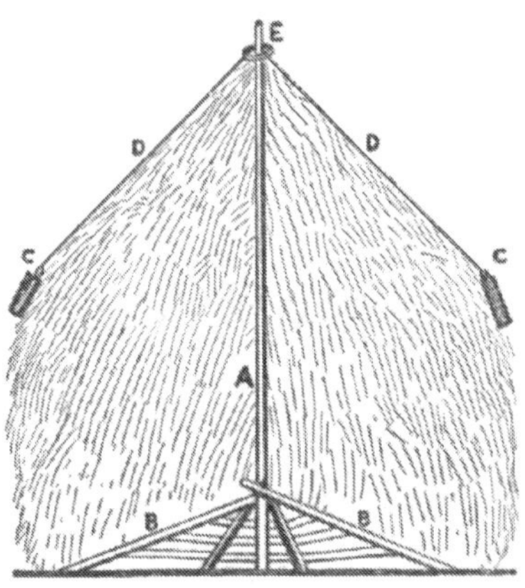

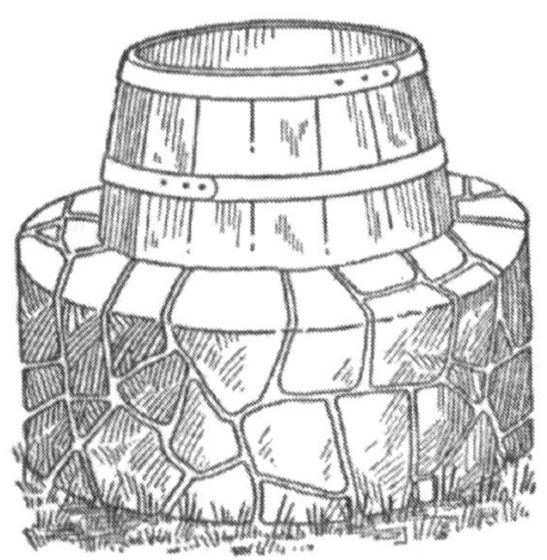

A Barrel Protection for the Spring

The ordinary spring is apt to be little more than a mud hole, where cattle resort to it. A cemented wall can easily be placed about it, bringing the water up much higher and keeping it always clear and sweet. Place a barrel over the spring and build a wall up about it, as shown in the cut. Lift barrel out when work is complete. This will make an attractive wall, and labor and expense will be light.

Another Form of Protection

The ordinary pasture spring is generally trampled and muddy, and often filthy. Make use of the plan shown in the cut. Dig out the spring and sink a tub or barrel in it. Around this drive four stakes and put in crosspieces, as shown. This will not only keep the cattle from trampling in the watering place, but will permit the weaker to drink with the stronger, without any fighting.

Hay-Rack Unloader

To unload the rack drive between the posts at the lower end. The crossbars in the rack will touch the poles, which are about twenty feet long, at A. As the team walks ahead the rack will clear the front bolster at B. At this point the rack will almost balance and the lower ends of the poles can be easily raised so that the rack will clear the wagon. When the wagon is pulled away the lower ends are dropped and the holes fit over the pegs driven into the posts. When reloading, raise the lower end and back the wagon under from the upper end. Posts are far enough apart for a wagon to go between, and the height varies with the height of the wagon wheels.

Two Lumbering Helps

The following devices will save many a backache and the help of at least one man: A famous rig for hauling over rough roads, mud holes or sideling places in the woods, can be made easily and at short notice. Fit a

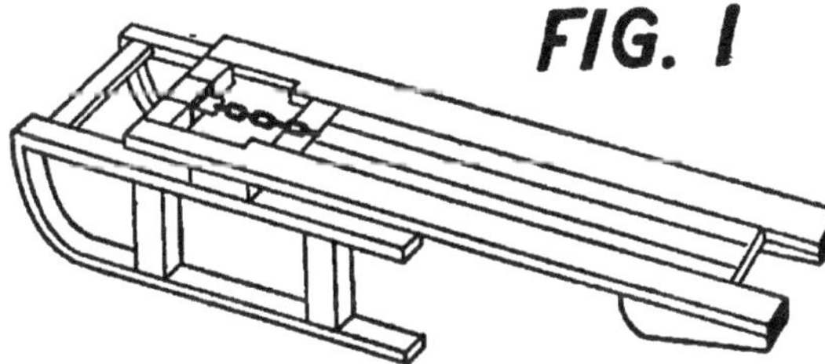

heavy bolster on a bob sled and mortise two blocks, each five inches high, to it in such a way that side pieces to drag will fit on, outside of them. These are fourteen feet long, (Fig. 1) and make the frame of a drag

rack. They are framed together at each end and drawn by a chain connecting the forward cross piece to the sled's king-bolt. Plenty of stake holes should be bored, and low shoes fastened to the hind end of the drag rack as often as they wear out.

No lifting of logs will be necessary because the rack is on the ground. The sled can pitch and dive without exciting the rack, and if the road be sideling the hind end slides around down the hill and braces the load so it never topples over. It glides serenely over mudholes and logs where a hind bob would get stuck forever. It also climbs upon and goes over deep snow better than any other rig, turns easily and goes around any tree past which you can get the nose of the sled. For all these reasons, heavier, larger loads can be averaged, and for sledding out manure in winter it has no equal. When hauling sap or manure put on bottom boards.

When sawing wood from the tree, or desirous of raising a large log or the whole trunk of a tree, no little homemade machine can supply so much power in a small compass as the one shown in the drawing. (Fig. 2). It furnishes a convenient fulcrum for a tremendous leverage and will hold a log waist high for sawing, so the hard work of cross-cutting is obviated. Saw off two pieces of two-inch plank six feet long. It should be ten inches wide. These are bolted together at each end, except that a strip of the same thickness and four inches wide holds them apart. Half-inch or five-eighth inch bolts are none too strong. Next bore two rows of three-quarter inch holes through both planks, and have the rows

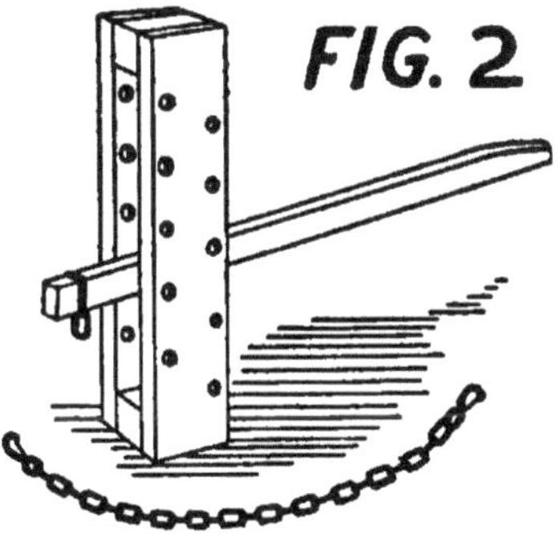

five inches apart, as in the engraving. Put an iron bolt in each row of holes. Let it be as large as will slip in and out easily if the plank gets wet. These two pins alternately act as fulcrums.

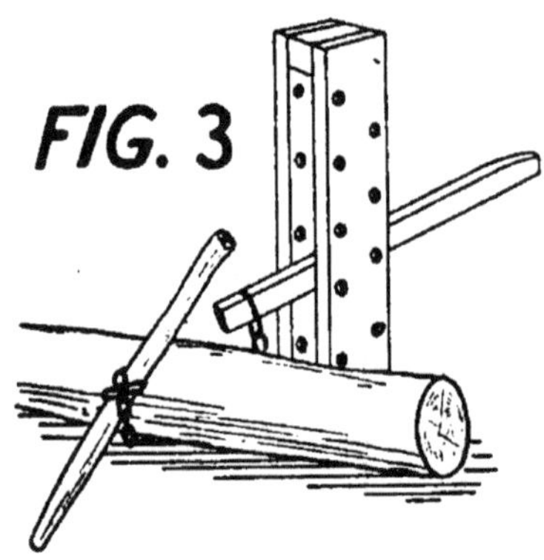

A lever of hickory or other strong wood, eight feet long, must be shaved to go between the planks, and must have a link suspended from it by a band of iron. The link is shown suspended from the end of the lever. This is to make it clear. It hangs from the lever between the planks in reality. Before trying to raise the tree trunk, cut a sapling three to four inches thick and lay it at right angles to the tree, on the side opposite the jack. (Fig. 3). Let its butt rest on the tree. Now pass a chain under the tree and hook it to the sapling. Attach other end of it to the link on the lever. Each time the lever is raised or lowered a pin can be put higher and the motion will make the link rise steadily, weight and all. A boy can work it.

Ridger for Irrigating Alfalfa

Here is an idea which may be acceptable to some of the alfalfa growers. It is a cheaply and easily constructed machine for throwing up levees or ridges for checking irrigation land. It is very effective, easily drawn by two horses, and steered by simply shifting your weight from one side to the other, standing on the side you wish the machine to go. Running a float alongside of levee will give the job a finish hard to beat. It is not patented.

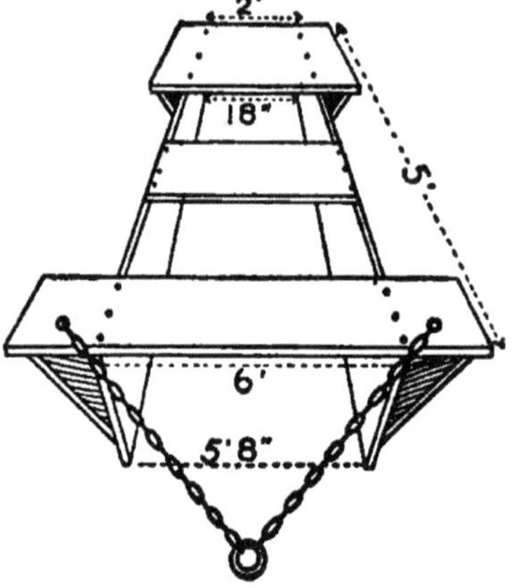

Stone Boat and Lift

The cut shows a contrivance for picking up stones that would otherwise take strenuous lifting. An iron brace is bolted to the rear cross-piece, as shown, to which is hung at the middle point a wooden lever, long enough to give good leverage. The inner end of this lever has a short chain attached with four loose grappling hooks that any blacksmith can make. The stone boat can be hauled alongside the large rocks, the irons put around each one in turn, and each one easily swung upon the drag.

Piling Cord Wood

Thousands of tiers of wood are being piled up every day in winter, and many of them are falling down at each end. A simple plan for holding the ends of tiers straight and firm is shown in cut. The green "withe" firmly holds the end upright when the wood is piled upon it.

A Simple Field Marker

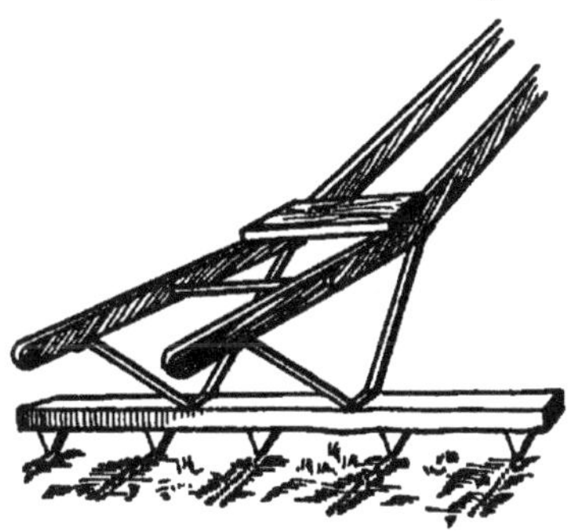

The advantages of a marker in planting corn, potatoes or other hoed crops, is that the rows can be made to run in both directions, which permits the cultivator to do nearly all the work during the subsequent growth of the crop. A one-man farmer can thus raise much larger crops than would otherwise be the case. The marker shown can be made at home and ironed by any blacksmith. The shafts and handles can form one straight piece, if desired, in which case they would have to set a little higher from the marker.

419

Field and Wood-Lot   **Handy Devices**

Another Simple Marker

This cut explains itself. An old pair of buggy shafts are set firmly into a piece of two by three inch stuff, from which stout markers extend below. The handles are from a broken-down plow or cultivator, or may be simply straight, round pieces of wood, inserted in the frame at a convenient angle. Get the first row straight and all the rest will follow suit. The row marker makes it possible to cultivate the rows both lengthwise and crosswise.

Level for Ditch Digging

An implement of great simplicity and usefulness may be used to give an even and

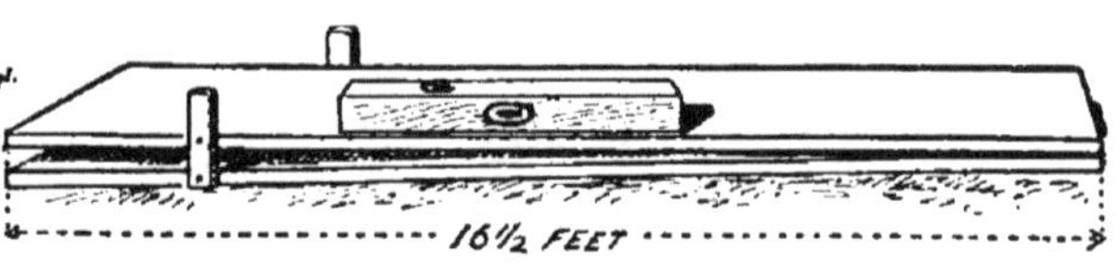

uniform grade to the trench bottom for the tiles: Two pieces of stiff, well-seasoned lumber, each one rod long, are hinged together closely at one end. As illustrated, affix a strong slat to each side of the bottom board one foot from the opposite end. When a grade of one inch per rod is desired, open this end of the contrivance just one inch and drive nails through the slat to hold it firmly in position. Lay this flat in the bottom of the ditch, open end down hill, and on it lay the spirit level. When this indicates a direct, horizontal position for the top plank the bottom must slope just one inch in the rod, and lying flat on the trench bottom indicates that to be the same. Be sure the lumber is not warped nor sprung and that it has a uniform thickness.

Leveler on the Cultivator

After every use of the cultivator the top of the soil is left in ridges and very loose and open. The drawing shows how to use the cultivator and yet leave the top smooth, so as to avoid rapid evaporation of moisture. Simply bind a strip of board to the back teeth of the cultivator as shown, using pliable wire. Make a trial and set the board just as low as needed to accomplish the required result.

Stubble Breaker

The stubble breaker in the drawing is simply a log twelve feet long, dressed to four by six inches and a pole and braces attached as shown in the illustration. The proper time to use it is when the ground is frozen. The drag covers three rows and breaks off the stubble close to the ground. It will be more convenient if it has handles on it so as to steady it and lift it in position when turning.

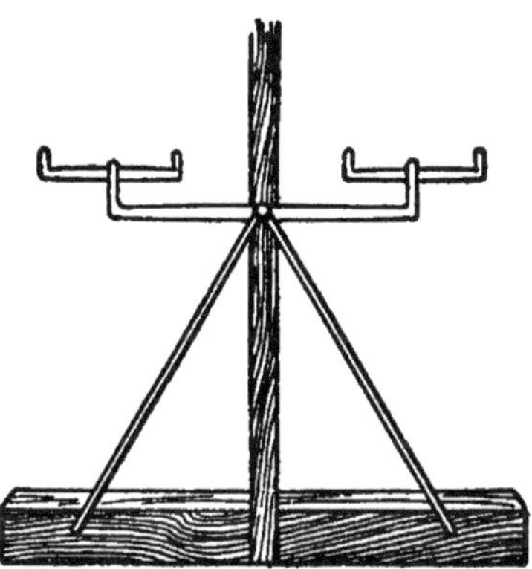

Clod Crusher and Leveller

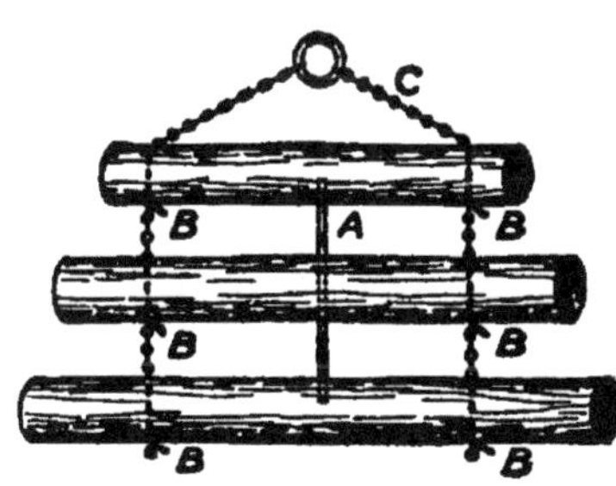

Here is a drawing of a home-made clod crusher and soil leveller which any farmer can easily make. Use three logs, about ten inches thick, and six, seven and eight feet long, respectively. A in the drawing is a four and one-half foot bar which goes through center of middle log and extends three inches into upper and lower log. C is a one and one-half inch chain which extends through holes in the logs and is held in place by plugs B.

Scare-Crow

Make a flaring wooden box, without top or bottom, and to the sides fasten bits of broken looking glass as shown. Suspend this by a cord so it can swing freely. It will constantly keep flashing and dazzling in the sunlight, and will inspire much awe in the minds of crows and other predatory birds. Bright tin cuttings from the tin shop hung up in the field serve much the same purpose, but must be renewed as they rust at the cut edges.

Covering for Tile Joints

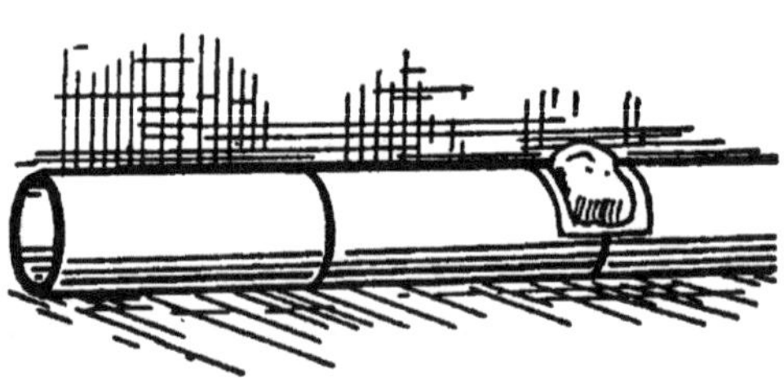

When tile is first laid there is much danger that earth will work down into the joints. Many put paper over the joint, to last until the earth gets firm about the tile. But this soon decays, and water trickling down carries earth in at the top of the open joint. Put on a square of tarred paper, and over this lay a handful of cement, smeared over the joint, as shown in the cut, then you will have a permanent protection when the paper has decayed, and it will always stay in place.

Plow Shoe

A handy contrivance for taking the plow to and from the field is a simple crook cut from a sapling, say two and a half feet long, with a notch cut in the upper side for the plow-point to come up against when in use. Raise the plow upon this runner, bring the point into the notch, hitch the team to the clevis on the plow-beam, and go on. Most farmers know about this, but some don't and this item is for the latter.

Felling a Tree Where You Want It

Cut two poles to help control the tree you are chopping. Set one under a limb, say, twelve feet from the ground, and stand it in a notch in the other, as shown in the drawing. Now raise the lower one and set a short chunk under it. From time to time raise it a bit higher, thus urging the tree to fall in the right direction for safety and easy work.

Long Handled Pruning Saw

For pruning trees it is a help to fasten a handle to a good saw with two-edges of different fineness, s o t h a t branches out of reach otherwise, may be also cut off. This will save climbing through the trees or mounting a ladder. This device may be bought, ready made.

Another Type of Pruning Saw

The cut shows an excellent tool for pruning limbs in the tops of trees without the necessity of climbing into the tree. A pole of any length required has a small, narrow-bladed saw fitted into the end. This is strengthened and supported by an iron rod that is riveted to the end of the saw, curves over its back and is firmly attached to the pole. With this implement a person can stand upon the ground and reach any limbs in the top that need removing.

Barrel Header

A barrel header, such as shown above, is a handy implement to own. But if there are only a few apples to be barreled, it may not pay to buy a press. One can be rigged very quickly by using a plank or scantling with one end under a stud reaching to the shed plate and temporarily nailed in place. The barrel to be headed forms the fulcrum. Be careful not to press the apples too hard.

Another Barrel Header

This barrel-header works to perfection; blacksmiths will make it for seventy-five cents. The parts marked A A A are made of a small wagon tire with hinges at C C. D D are rods of half-inch round iron riveted to the frame three inches above hinges on each side, but left to turn freely as a hinge. R is a piece of two-inch plank nearly the size of barrel head. Place the head on the barrel, then the header in position. Loosen the top hoops, bear down A to press the head in. Drive down the hoops and the head is in.

Mice Protection for Trees

The writer has proved to his own satisfaction the real worth of old newspapers wrapped about the base of young fruit trees as winter protection against mice and rabbits, and sun-scald. An added protection is shown in the cut in the form of an old tin can wrapped about the base pressed down into the earth. A lot of old empty cans can be put into the fire and the solder melted off. The sides can then be pulled apart at the seam. If newspapers or cans are not at hand, use strips of lath placed upright close together and tied into position, or strips of wood veneer, or fine wire metal screen, or odd pieces of tin, sheet iron, etc.

Wire Protection for Trees

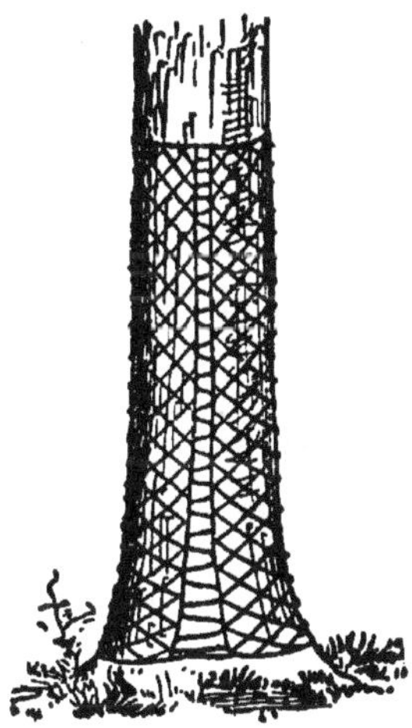

A very neat and serviceable protection for tree trunks against the gnawing of bark by animals is shown in the cut. Wire poultry netting comes in all widths from six inches up. Select a width equal to the circumference of the tree to be protected, and cut off a strip to reach to the required height. Bring the selvage ends together as shown in the cut, and lace them with a piece of fine wire. If the tree grows, loosen the lacing. Many shade trees about one's home are ruined by the hitching of horses to them.

Wooden Protection for Trees

Roadside young trees, and all trees in fields that are to be pastured, should have tree protectors, and the picture shows how they can be made and put up. The side boards are ten inches wide and an inch thick.

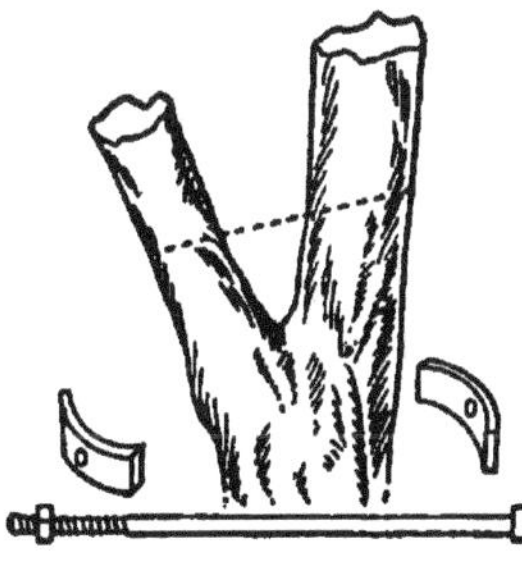

Saving a Tree Fork from Splitting

Many trees are practically ruined by having a big limb split down. Don't wait for it to come, but get an iron rod with curved washer straps and a nut, and put it through the limb and trunk in the position of the dotted line. Many a fine shade or fruit tree can be saved thus at a trifling expense. A long, small-bore auger will make the hole for the rod.

425

Label for Trees

Procure a thin piece of zinc about six inches wide from which cut strips crosswise three-quarters

of an inch wide at one end and tapering to one-eighth inch at the other. Odd pieces of old zinc, stove boards, etc., may thus be utilized. After being cut the pieces should be put in vinegar to allow them to corrode, after which an ordinary lead pencil will complete the business. Either or both sides can be written upon. And the writing will last for years, too, so that "he who runs may read." The diagram explains the idea. Simply wrap the small end of the label loosely around a limb of the tree—never around the trunk. As the limb grows, loosen the label somewhat, or change it to a smaller limb. On the reverse side of the label it might be well to write the name of the agent or nurseryman who furnished the tree; then later, if the variety proves untrue to name, you will know whom to blame.

Moving Large Tree

The arrangement illustrated is very simple, and a big tree may thus be taken up and replanted. First dig around the tree, preferably when ground is frozen; don't injure the roots by going too near; take up a large clump of dirt. Place connecting piece of the standards against the tree, to which fasten with ropes, winding a cloth around to prevent barking the tree. One or two horses hitched to the rope will easily raise the tree with ball of earth, and swing it on to the waiting stone-boat, on which you may haul it to the planting place. There dig a hole sufficiently big, set the tree, and fill and ram the earth securely into place. Then trim the tree branches a little—to make up for any roots lost in the moving operation. If the situation is exposed to severe winds, it may be necessary to anchor the tree for several years. This can best be done with guy ropes, being careful to protect the bark from chafing. Also, it may be necessary to water the tree several times during the first one or two summers.

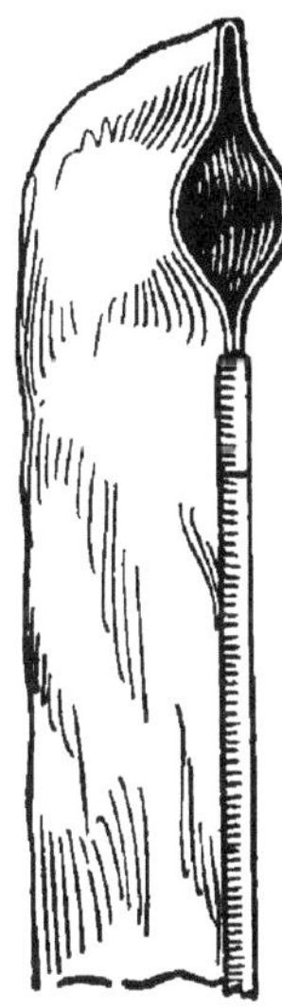

Fruit Picker Which Prevents Bruising

The apple gatherer shown here can be depended upon not to bruise the fruit, for as each apple is picked off, it gently runs down the cloth "spout" that is attached to the pole, extending down to its lower end. The apple falls into the hand of the operator and is placed in the basket, while with the other hand another apple is being hooked off.

Hanger for Fruit Basket on Ladder

A basket hung from the rounds of a ladder is very inconvenient to reach. Fit an iron in the shape shown in the cut, and you can then have your basket at the side of the ladder, in the handiest of all positions. A few wooden pegs up and down the ladder will keep the iron from slipping. A blacksmith can make it in a few minutes.

Another Good Fruit Picker

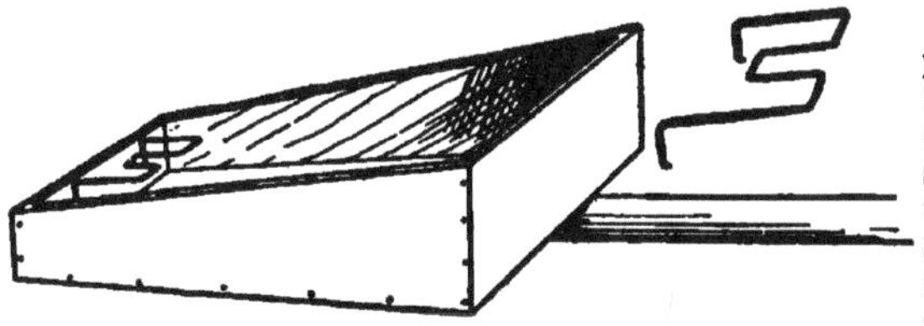

A very good fruit picker can be made thus: On the end of a long pole nail a small, light box with sides cut down, as shown. At the front edge rig a strong bent wire in the manner suggested. Line the box with burlap, and you have a fruit picker that will take off choice fruit without bruising, and bring down a dozen specimens at a load.

Cherry Picking Bag

Here is a contrivance to help in picking cherries. Make a ring of wire that will be about six inches or less in diameter. The basket part is made of cloth; part of a flour sack will do; this should be about four inches deep. Fasten a ring to the basket, as shown. A strap is placed around the last three fingers and through the ring. This contrivance is light, and, being on the left hand, the one usually holding the limb, is always in easy reach of the right hand.

Hook for Fruit Picking

This little hook may be made of a crotched stick or of heavy wire. The important point about it is the slide at the lower end. This is a piece of wood with a pin in the outer end to hook on to a branch. It is made in such a manner that it will move and slide freely up and down the rod. To use the hook, draw the limb containing fruit down to a convenient position and put the slide over a lower limb to hold it. The side strain keeps the slide from slipping on the rod.

Another Hook for Fruit

What is the use of picking the small, knotty, wormy, along with the good fruit? None at all. Let the scrubs stay on the trees or pick themselves. It is much less work to pick them up off the ground for the stock or for cider, later, than to pick them from the tree. Or, if you want to, finish the job as you go along. To do this, have a hook made like that in the cut to hold two baskets, instead of one as is commonly done. Such a hook is bent by a blacksmith from a piece of quarter-inch round iron. Sorting the fruit in this way causes less handling, and therefore, less bruising.

Storage Trays for Apples

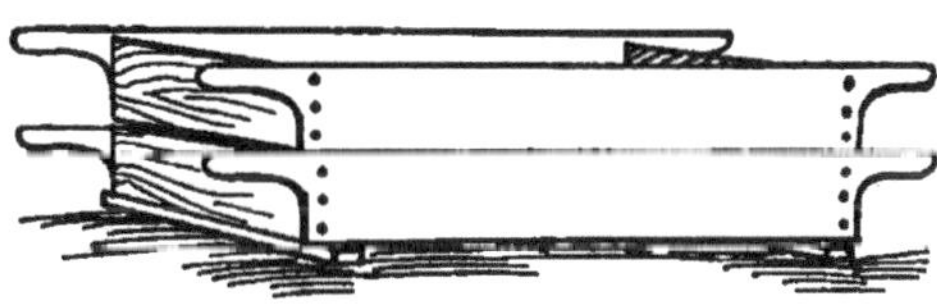

Deep bins for storing ap-
ples have their disadvan-
tages.   So have barrels.
The cut shows one of the
best fruit storing arrangements. Shallow trays with
end handles, and spaces
at the ends for the circulation give ventilation and
ready access to the fruit. They can be piled up several trays deep, and each
tray is instantly accessible and removable.

Apple Barrel Press

We have kept apples through the Winter by burying them,
and have had them come out of the ground with a more
natural flavor than when kept in cold storage. They must
be in good condition, be gathered at the right time, and
carefully handled. Each apple is wrapped in paper, packed
closely in box, with enough newspapers over the top to keep the
fruit dry. We then pile enough earth on top to prevent freezing. Those stored in this way kept well and had no taste of the
soil nor the loss of flavor of cold-storage fruits. A barrel press
is a great help in packing apples.

Gooseberry Picker

When the time comes for gooseberry
picking, this little device is just the
thing for this work: Take a piece
of thick lath about twenty-two inches

in length (or any kind of stick about that size will do), and a stiff wire not
more than half so large as common fence wire and about twelve inches long.
Bend the wire in the center and fasten it on to the stick with small staples.
Have the ends of the wire extend out from the end of the stick three or three
and a half inches (see cut), and just far enough apart to strip off the large
gooseberries on a branch and leave the smallest ones and the leaves. Use
a dish pan, or something that is large around at the top, for catching
the berries.

Pruning Hook from An Old File

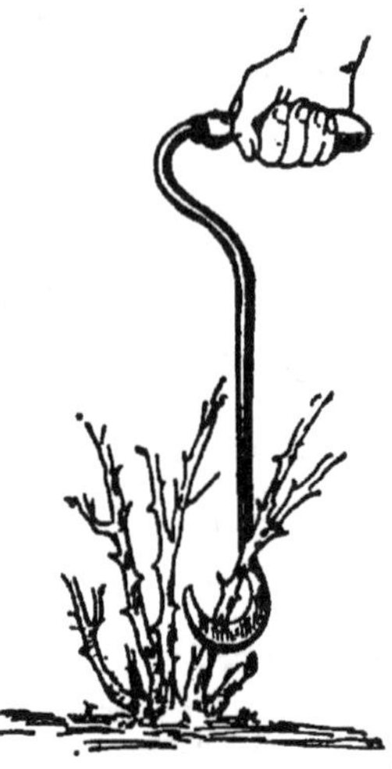

The question often comes up how to cut out and remove the prickly canes of blackberries and raspberries. Try having the blacksmith transform an old file into a hook, weld an iron rod to it, and fix a handle on the other end, making a tool as here shown. Keep the hook very sharp, wear gloves, and you can work with this tool for days very comfortably. Don't wait till spring to take the old canes out, for it ought to be done in late summer.

Cane Knife--"Better Than a Hook"

The drawing shows an implement you will most likely need for thinning out raspberry and blackberry canes. It is simply a chisel having a curved blade and long handle. Instead of pulling the canes against your face (as in using a hook pruner) and breaking off half of those you wish to leave on, you chip them off as slick as a whistle and leave them stationary or push them away. It is well to have two of them with blades one and one-half inches wide and one inch wide. Use the narrow one among currants and gooseberries, the other among blackberries and raspberries.

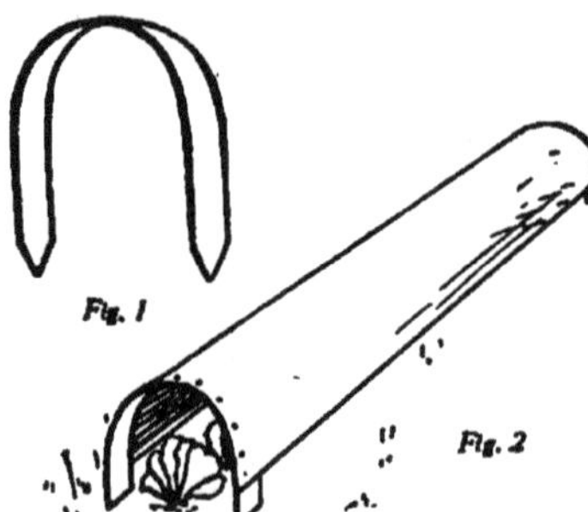

Protection for Young Plants

Instead of the make-shift protection of old newspapers (which are continually blowing off), take long strips of cotton cloth and tack them at each end to such hoops as are shown in Fig. 1. When stretched over newly set cabbage or strawberry plants, the device will show as in Fig. 2. Successive strips of cloth can be sewn together, making the whole as long as desired. If quite long, put extra hoops at intervals between the ends. Such protectors can be used year after year and are always ready.

Window Stand for Seed Boxes

Many people have to start all their early plants in the kitchen windows, and the space is usually rather restricted. The cut shows a stand with a series of boxes, one above another. Each box is pivoted by screws through the side pieces into the middle of the ends of each box. The boxes can thus be tilted toward the window to get the full sunlight. The next day the box can be turned about, and the boxes tilted the other way, as the sun draws the plants to one side and to the other. Small pegs hold the boxes in place when tilted. On cold nights the whole stand can be removed from the window.

Newspaper Shade for Plants

W... en transplanting plants duri... warm weather, shade then with a little tent made instr...ly of a sheet of old newspaper. Fold the paper over the plant, as shown and secure the edges by dirt or stones, or by bits of lath with a stone in the middle. The paper turns the sun's rays, while the open ends permit a free circulation of air.

Reel for Garden Line

Make the garden rows and flower borders straight, by using a long line. And make a reel to wind it quickly upon when not in use (see cut). The free end of the line is attached to a long peg, and the other end to the reel, which serves as another peg—the reel part being firmly hooked when enough line has been paid out.

431

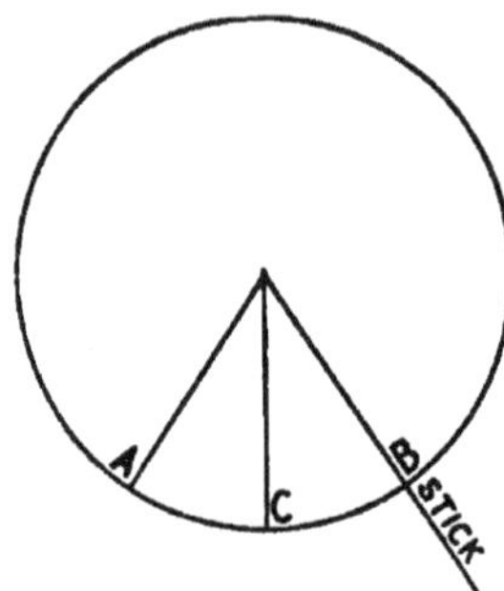

Simple Shade for Young Plants

Here is a shade or cover for newly-set plants which can easily be made of any kind of heavy paper. Cut paper in circle, size wanted, and fold over as marked in drawing—folding A over on to B, with a crease at C; turn the flap down and tack it fast to a stick, which may then be stuck in the ground. Get a circle of paper and a stick, and you'll soon get the idea; although hard to describe in words, it's very simple. By cutting all circles as nearly alike as possible, the covers may be stored away when not in use, taking up very little room, as they fit one into another.

Protecting Young Plants

This device not quite the same as above, is made of stout paper, cut in the form of Fig. 1. It is folded together and tacked on the seam to a light stick, Fig. 2. When tomatoes, cabbages and other plants are first set out the little hood (with its air vent open at the top) is pressed down over the plant, to the earth, as shown in Fig. 4. In a day or two it can be raised, as in Fig. 3, to get the plant accustomed gradually to the light and sun.

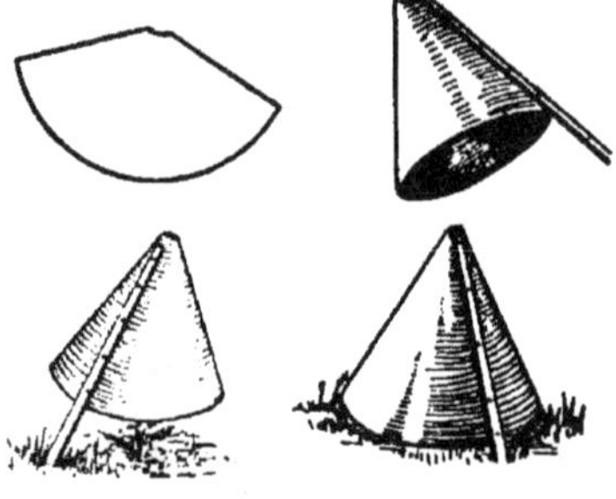

Propagating Case

When a few of the first early plants are needed they may be grown in a portable propagating case like the cut shown herewith. The outer wooden case is three by four feet and two and a half feet deep. In this case is a lamp to supply heat and a galvanized iron water tray; over this a large box or several small boxes of earth and a sash covering the whole. There must be a door in the side of the lamp chamber to admit air and to gain access to the lamp.

Sub-Irrigation System

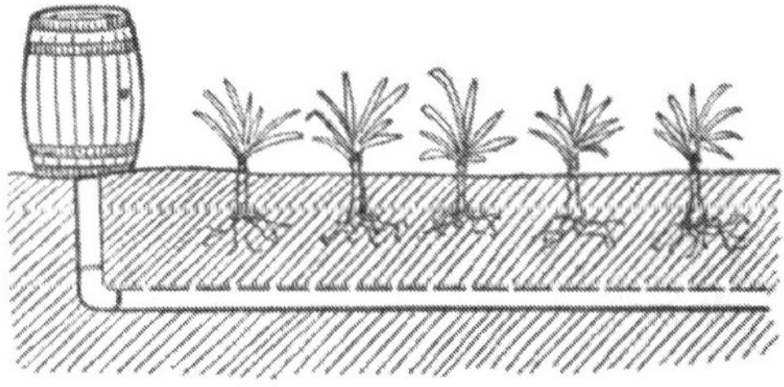

The cut illustrates a method of sub-irrigation which can be practiced in a small way. The outfit consists of a three-fourths inch iron pipe, bent as shown, and a paint keg fastened on the upper end. The lower end is plugged a n d one-eighth inch holes drilled in the pipe eight inches apart to let out the water poured in the keg. The pipe may be fifty feet long and have a fall of two inches in that distance. Put moss or loose earth on top of pipe. A sub-irrigation plant, where tile is used, can be made on the same plan.

Garden Marker

The cut is intended to represent a garden marker. The markers are little wheels of wood beveled on both edges so they will make a smooth V-shaped drill.

Wheelbarrow Garden Marker

To mark the places for setting out plants when transplanting, fasten an old wide leather strap about the tire of the wheel in an ordinary wheelbarrow. To this strip

nail previously little blocks of wood, the proper distance apart. Wheeled along the rows this will not only mark the right spots for the plants but will firm the soil just where the plant is to be set. This is specially desirable if the device is used to mark the spots where beet and other seeds are to be dropped.

Another Garden Marker

Cut a circle from a board for the wheel. Split a bit of ash for the handles, as shown in the cut, and you have the body of a garden marker. A side strip at right angles forms a shaft for the wheel to turn upon and for the marking pegs to be inserted in. A little brace holds this rigid.

Garden Roller

The picture shows a garden roller, made out of a fifty-pound lard can filled with concrete. A three-quarter-inch iron bar forms the axle, the frame is made of 2 x 4 studding and three-quarter-inch washers prevent the roller from grinding against the frame. To give rigidity, the cross-pieces of the frame are let into the side-pieces and held in place with five-inch nails. It weighs just under 180 pounds, and makes a useful light garden roller.

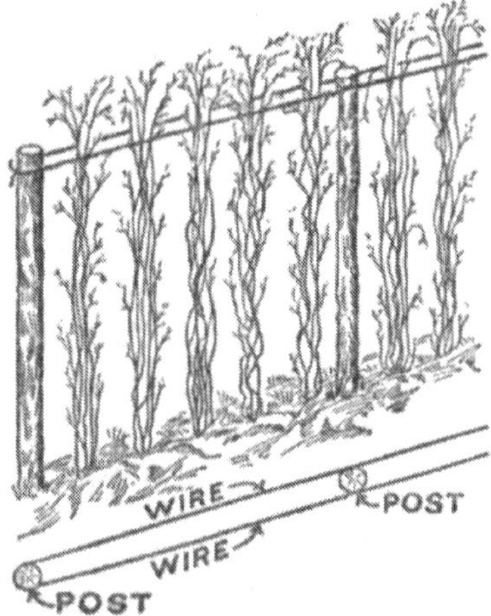

Berry Bush Supports

In many localities raspberry and blackberry bushes must be tied up to prevent heavy snows from stripping off the new and tender shoots that will produce the fruit of the coming season. Fig. 1 shows an excellent plan for supporting the bushes. Fig. 2 plainly shows a sectional view of the same stakes, wires and bushes. This method keeps the bushes erect and gives a clear, open space between rows for the cultivator and for the pickers.

Tomato Vine Support

Tomatoes need a bench-like support, so that the vines can spread out to the sun and air, and yet be held up from the ground. An excellent plan is shown in the cut. A low, wooden support, like that shown, is placed at intervals of eight feet along the row, and across the top is stretched two strips of twelve-inch wire poultry netting, leaving space between for plants to grow up through.

Another Tomato Support

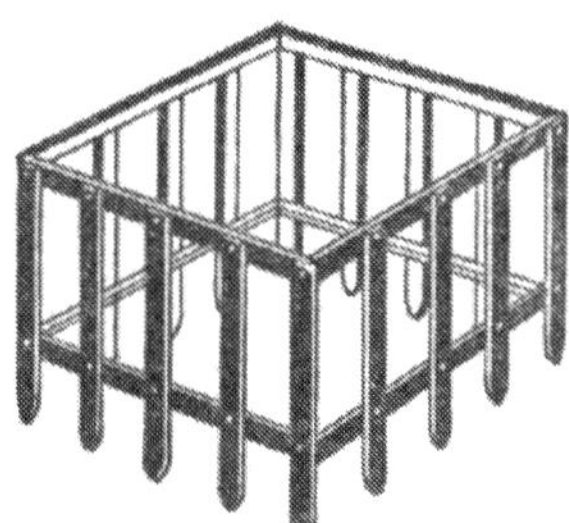

Keep the tomatoes off the ground, on supports of any kind. All that is needed are twenty-four laths two feet long. In making, take four laths, place them as shown and nail two cross pieces. Then fix all the rest in the same way and nail the four sides together so as to make a square. Sharpen the ends so that they will go down into the ground easily, and then set the rack over the tomato vine and pound the ends in the ground about three inches.

Transplanters

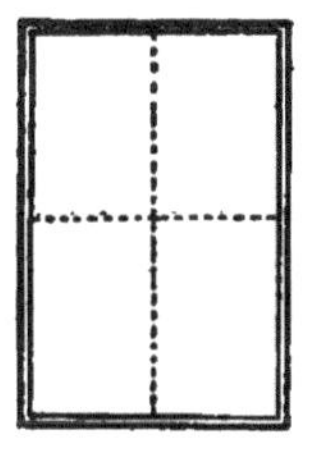 

The drawing shows a handy and successful method of transplanting small fruit and vegetable plants and even shrubs and small trees. For the former tin tubes of different sizes are cut from sheets of tin; for the latter tubes of sheet iron with stout wire around the top. When the pieces are cut out of the sheet the ends are turned so as to lock together, which is better than soldering them. To take up a plant, draw the leaves together, slip the tube over it and press the latter down a few inches. You can lift the plant without disturbing the roots, set it on a tray or wheelbarrow and carry it where you wish. Now set the plant in position, draw the earth around the tube and plant and pour in water. In a few minutes the tube can be slipped out as "sleek as if greased" and there is your plant not knowing it has been moved from its bed.

Hinged Garden Trellis

We expect to grow choice table grapes on our experimental f a r m when we get it. As grapes are somewhat tender it is best to protect them in winter. This is easily done by having trellises like the drawing. Take six eight foot pieces of two by four inch scantling and bind each three pieces together with three boards as illustrated, set posts in ground and hinge the pieces to them with bolts, bevel top ends and lean together tent-fashion and fasten. When winter comes unfasten top and let trellis and vines drop to the ground.

Trellis for Peas

Here is a description of a trellis for sweet (or garden) peas, which is in successful use. The idea consists of using two-inch-mesh wire netting, eighteen inches wide, supported in the form of shelves ten or twelve inches apart. The supporting posts should be driven into the ground about eight feet apart, along the pea row. (The drawing roughly shows the idea, although the posts are pictured too close together and the pea vines are too scanty. This form of trellis has the advantage over the usual up-and-down wire trellis, of giving the vines a chance to grow separately instead of massed together in a narrow bunch, thus allowing each vine more elbow-room, air and sunshine.

Rake for Grape Vines

The drawing shows a simple device for removing the loose vines from a vineyard after pruning has been done. Take an old heavy wagon tire and have a blacksmith weld or bolt on four teeth about twenty-six inches long, shaped like the teeth of hay rake, but of heavy iron like the tire. Put on wooden cultivator handles and a hook to which to attach a horse. No patent applied for.

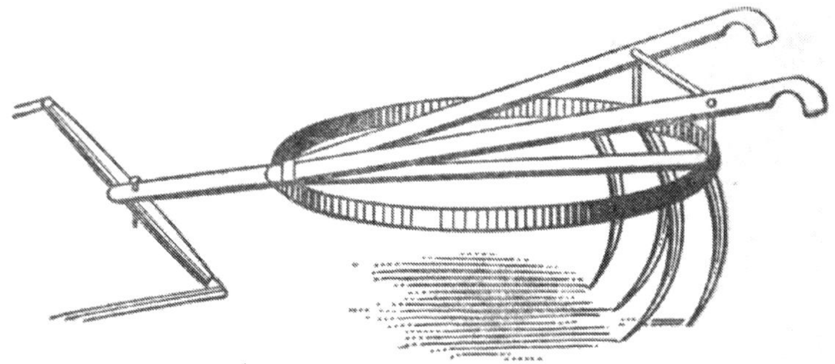

Carrier for Berry Baskets

Basket carriers that will be needed in the berry patch next spring can be made in the workshop during the winter. The drawing illustrates a simple and easy way of making them. Fig. .1 represents an inch board six inches wide. The wood should be of a kind that does not easily split. Saw into pieces following lines of cut. Make the base of blocks wide enough so that they shall be eleven inches wide after cutting off the points at the ends. The sides and bottom may be of very light material (plastering lath will do), and the length sufficient to take in six or eight quart boxes. The top strip for handle should be strong enough to bear necessary weight and firmly nailed to end pieces. A clip of light tin over the joints will be a great advantage.

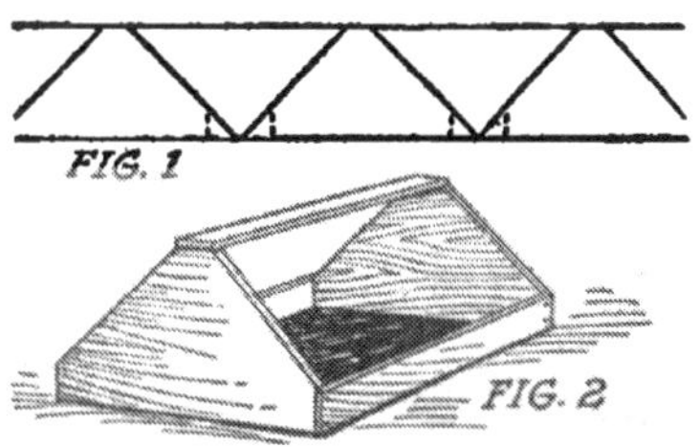

Vineyard Post Repairs

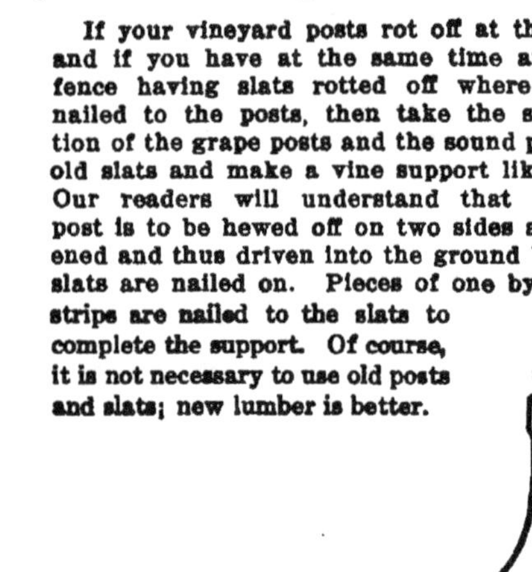

If your vineyard posts rot off at the ground, and if you have at the same time an old slat fence having slats rotted off where they are nailed to the posts, then take the sound portion of the grape posts and the sound part of the old slats and make a vine support like the cut. Our readers will understand that the short post is to be hewed off on two sides and sharpened and thus driven into the ground before the slats are nailed on. Pieces of one by two inch strips are nailed to the slats to complete the support. Of course, it is not necessary to use old posts and slats; new lumber is better.

Home-Made Seed Sower

A very simple contrivance for sowing small, smooth seeds, is shown herewith. It is a bottle with a cork in and a large quill in cork. In order to sow with greater regularity, an ounce of seed may be evenly mixed with a half pint of sand.

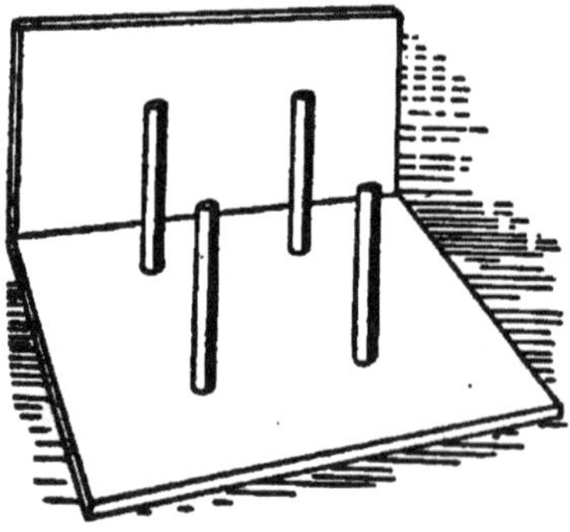

Asparagus Buncher

All seedsmen sell asparagus bunchers. Which of the many patented kinds is the best we cannot say, but here is the old-fashioned home-made buncher that anybody can make for themselves with two pieces of boards and four pegs.

House for Early Vegetables

Where early vegetables can be grown with a profit, a start must be made under glass. A cheap little house is shown in the cut. It is made of matched siding, is not more than six or seven feet wide, and as long as one may wish. The inside arrangement is shown in the cut. Have mats to place over the glass at night and keep the heat up at that time with a small oil stove.

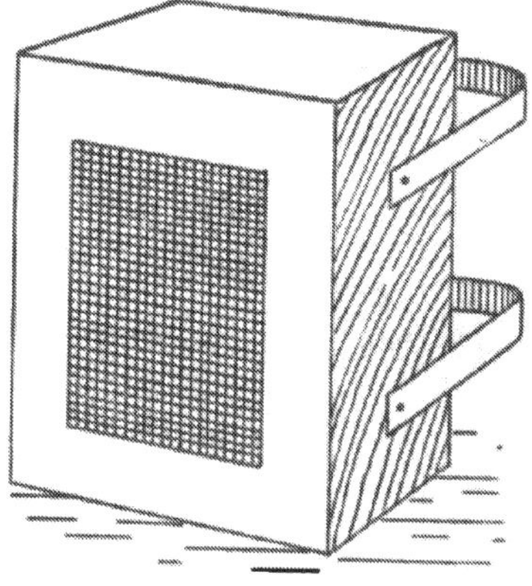

Box for Vegetables

Make an opening in the bottom of a grocery box, and cover it with the stout wire screening used for cellar-windows, having about a half-inch mesh. Nail on two handles made from old hoops, that will swing down out of the way at the ends. Use this box when gathering vegetables from the garden, and wash these by immersing in a tub of water or by pouring water over them.

Simple Plant Support

Here's a simple support for tomato plants, peonies, etc.: Simply get two stakes three feet long and about one inch thick and two inches wide; sharpen them at one end, and get the hoops of a nail-keg for the rings. Place the lower one eighteen inches from the points, on the outside of the stakes, and the upper one at the top, on the inside of the stakes; this gives a flaring top and a support that will hold any vine, and also one that is easy to get around for cultivation, and one that does not take up much room in storing for another season, as one support can be slipped inside the other.

A Ladder Truck

Two people can carry heavy ladders with comparative ease, but the extra helper is not always at hand. With the device shown in the cut (which any one can make), one man can move the heaviest ladders with comfort. The ends of the ladder fit into the little blocks (which slide along the shaft to fit the ladder's width). Then taking the other end of the ladder, one can wheel the whole away to any desired point. The wheel can be cut from a stout piece of plank.

Low Truck

The cut shows an improved form of a device for moving heavy bodies in the house or barn—stoves being handled with special ease by the use of this little platform on very low, broad castors. The rear end is so low as almost to touch the floor. Tilt up the article to be moved and back the platform in under it. It can then be wheeled anywhere.

Another Ladder Wheel

One person can take a heavy ladder anywhere alone, and that, too, with great ease, by taking advantage of the device shown in the cut. Two trucks are sawed from plank and holes bored through their centers, and through the ends of all the heavy ladders on the place. The rod and trucks can be slipped into place in a moment and held by pins, as shown. The other end of the ladder is then seized and the whole wheeled off.

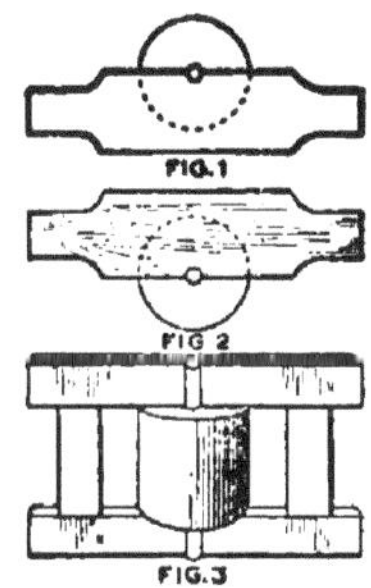

Trucks for Moving Heavy Articles

Make several small trucks to use when moving heavy articles, such as timbers or boxes that are too heavy to lift conveniently. Used in the position shown in Fig. 1, the weight rolls over the roller of the truck. By turning upside down as shown in Fig. 2, the device can be used as a truck on which the weight is moved. By lifting one end of the weight the truck can be removed. Fig. 3 shows how the frame is made of 2 x 6-inch plank; the roller is about six inches in diameter. A piece of heavy iron rod will do for the axle.

Another Simple Truck

To make the truck shown in the drawings, which is especially useful at housecleaning time, obtain a board 18 x 12 inches. Two inches from each end nail two blocks as long as the length of a large spool, as shown at "A." On each side of these blocks nail a strip of inch lumber one foot long and two inches wide, as indicated at "B." For wheels fasten four spools between the cross-pieces "B" by using a large wire spike for each axle. Use wire that will just fit the holes through the spools, and drive it through the cross-pieces about one-half inch from the lower edge. On this home-made truck you can draw a tub or pail of water, and many other heavy articles. To draw it about a rope may be attached, as shown.

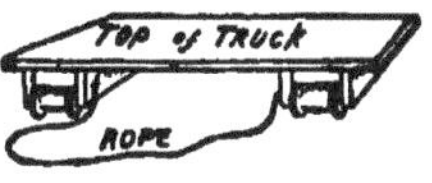

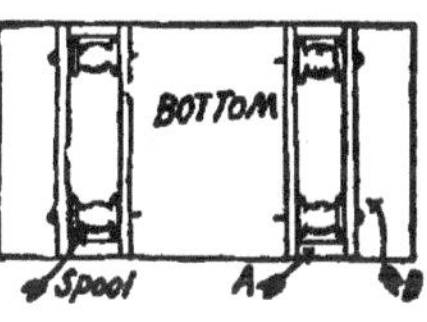

Runners for the Harrow

Some men went past with a harrow dragging on its own runners, and it spoke for itself as a good thing. When they began work

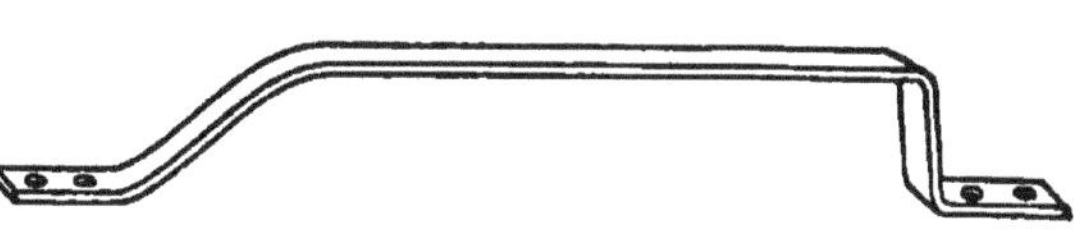

with it I followed to inspect it closely. It was a square harrow built to fold double and the ingenious owner had taken a heavy old cart tire, heated it and bent it into rough runners, which he had bolted on to the top of one section. The other part was folded on to it, and thus it carried itself to the field attached behind a wagon. The sketch shows the shape of one of these runners.

Utilizing the Worn-Out Ax

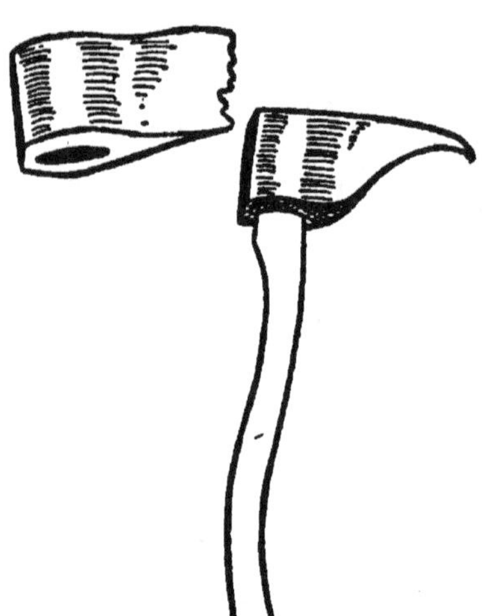

The old ax shown in the cut can be worked into a new form by a blacksmith. Its new form is shown in the second figure. This is in reality a light pick-ax, and will be found very useful when digging in the ground, or when chopping ice from a watering trough in the winter.

Railway Type Truck

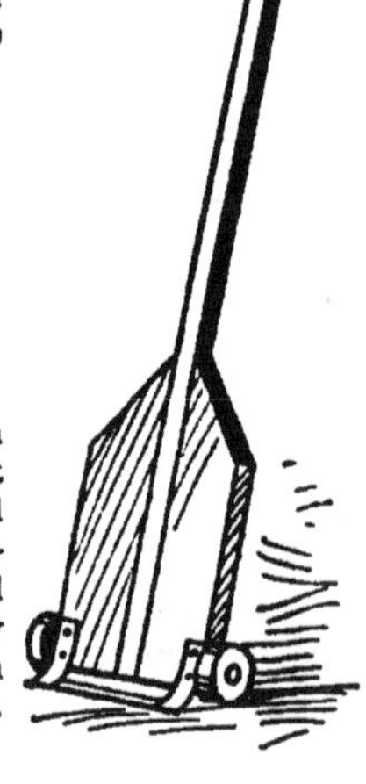

One often sees about railroad freight stations an affair similar to the cut, but not having the convenient shovel handle. The lip of iron at the bottom is placed under the edge of a heavy barrel or box, the whole balanced over the small wheels and the whole easily wheeled away. The shovel handle makes the wheeling away much easier. Such a device will be found very useful on the farm. Make it of hard wood, with wide iron trucks.

Improved Saw-Horse

The advantages of the improved saw-horse shown in the cut are apparent at a glance. A stick of cord-wood length can be laid in the arms, and sticks of any length can be sawed, while the whole lies firmly held by the arms of the "horse." These arms will shut up together like a pair of scissors, effecting a saving of room, while a strap across the bottom will keep the horse from spreading more than may be desired.

A Rake and Stone Box

Ground can be easily cleared of small stones by the simple device shown in the cut. An iron garden rake, and a box with one side removed, gives a "broom and dust pan" arrangement that makes the picking of stones an altogether different affair from the old-fashioned finger and basket method. The box has handles to permit emptying it into the cart which is to haul the stones away.

A Wood and Cement Roller

Cut out two circles from a wide board and run an iron rod through the center, attaching a handle as shown in the cut. Nail very thin strips from the edge of one circle to the edge of the other all the way around and upon these, at regular distances of three or four inches, nail an inch and a half thick strips of wood, with edges beveled, as shown in the cut. Fill in the spaces between with cement, smoothing surface at ends neatly, and you will have a fine garden roller. The cut shows the construction plainly.

Light Garden Roller

Garden soil is generally very light, frequently too light and porous to afford the best conditions for the sprouting of very small seeds. The cut shows a home-made wooden roller that can be used effectively in preparing the seed-beds, firming the soil, fining the lumps, etc., also in going over the rows when the seed has been planted, thus pressing the soil down about the small seeds. Saw off a short piece from a round post, eight or ten inches in diameter, and make a frame and handle, as shown in the cut.

Stone-Weighted Roller

Take a box and saw out a part of one end and a part of the bottom, as shown in the cut. Insert a round stick of wood for the roller and add handles and a bit of board over the roller where the dotted lines show its position. Stones can then be put in the box until the required weight is secured.

Bag Holder

The cut shows a bag holder that will fit any size of bag, easily made and not in the way when not in use. The three-sided frame that with little hooks at the top, holds the bag in place and open, drops down against the side of the grain bin when not in use. It can be hinged to the bin with bits of leather. One should be placed in front of each grain bin to be handy.

444

Snow Plow

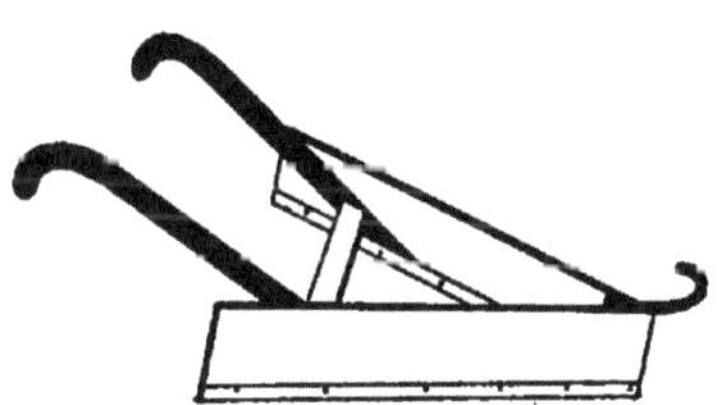

The drawing shows a one-horse snow plow that has been in use for twelve or fifteen years; it has saved many a day of hard labor. It is made of two boards, 1 x 12 inches x 4 feet long, set four feet apart at rear end. The triangular block in the front angle should be two inches thicker at the bottom, so as to give the boards a slant outward at the bottom. The 1 x 12-inch brace near the rear should be sawed to fit. The picture explains other details.

Snow Shovel

A snow shovel is a necessity during the winter months. Here is a good one made in the home work shop. The handle of a broken fork is utilized, as shown, while the blade is made of three-eighths inch stuff strongly nailed or screwed together. Bend a strip of zinc over the front edge, and make the back, in which the handle fits, of hard wood.

Plant Setter

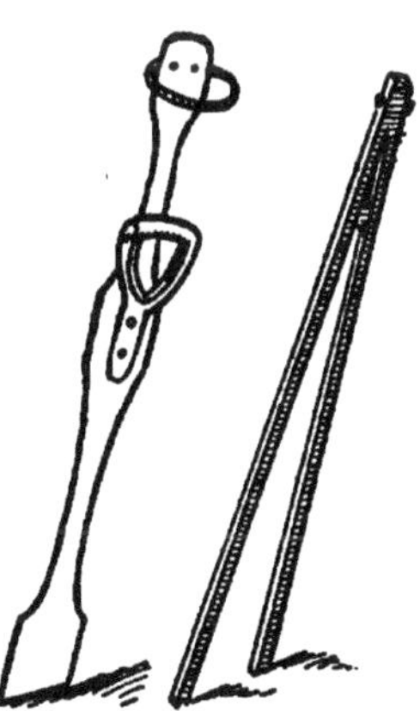

The two implements shown in the attending cut are in common use for setting sweet potato plants. The first is a dibble four feet long, made of a straight slat one by three inches, with a strap of leather at top to fasten around the arm of the user, a fork handle nailed to the side and the lower end beveled on both edges. The second is a pair of wooden tongs made of pine wood slit to near the top, a screw put through to keep it from splitting, and a wedge a little further down to give the proper spring. The tongs are one and a quarter inches wide and about three feet eight inches long. Make hole with dibble, grasp plant by the root with the tongs and thrust into the hole, give a pat with dibble and walk on. A good operator will set 20,000 plants a day.

Scraper Made from Hoe

An old worn-out hoe, riveted to a piece from a broken cross-cut saw blade, makes a very handy scraper for barn or garden use, or for scraping off plat-forms that are covered with snow. Round off the corners of the blade that no one may get hurt upon them.

Utilizing the Broken Rake

When an old garden rake has had so many of its iron teeth broken out as to make it worthless as a rake, fit it to a strip of board as suggested in the cut, fastening it firmly to the board by fence staples as shown. The useless rake then becomes a useful im-plement for leveling off beds for vegetable or flower seeds, and for any kind of grading. In the barn it will be useful for heaping up chaff, or other litter, or for clearing floors of such material.

Staple Puller

The handle is iron, ten inches long, one-half inch thick, seven-eighths-inch wide, welded to a tool steel head three and one-half inches long. This head is round, one-half inch thick; the hook end is flat-tened to one-eighth inch at point.

To use puller, place the hook end back of staple, tap on opposite end with hammer, pull, and staple will come out straight and in good shape to be used again.

Another Staple Puller

Any blacksmith can make this tool from a broken hayrake tooth or any suitable piece of steel. To operate, drive the point behind the staple (by hammering on the bend of the puller), then raise the lower end of the tool—and out comes the staple.

Swamping Hook

A swamping hook (like illustration) for pulling out old, partly decayed stumps and rocks, will come in very handy. When the ground is soft, after a rain, a stout team can be made to pull out quite large rocks with it. Any blacksmith can make one out of some heavy bar of iron. If desired, a five-foot-long wooden handle may be bolted to the upper side, by the use of which it will be easy to hold the hook in proper position while team starts pulling.

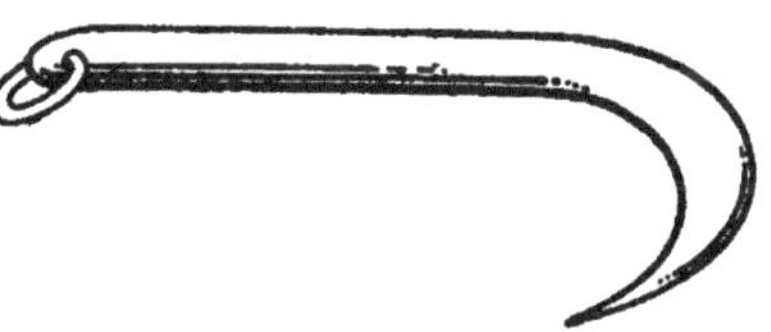

Tire-Saver Jacks for the Automobile

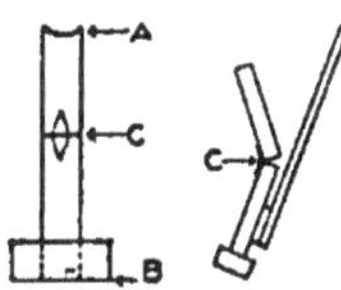

During spare time the handy man can construct a set of four tire jacks which will greatly save the tires of his automobile. During the winter when the car is stored, or during wet times in the spring months when not in use the car's weight should be removed from the tires, and they will thus last much longer.

If soft wood is used the upright piece should be of 2 x 4 inch material, smoothly planed. This should be mortised to the base piece, and fastened with two bolts. Base should be at least twelve inches long. The upper end of upright is hollowed out to fit hub of wheel. Measuring from A to B the upright should be two inches longer than the distance from under side of hub to floor when tires are tightly inflated. Tack a piece of rubber to upper end of upright to prevent slipping.

Now saw upright in two at point C and securely fasten a strong hinge on the back. Get a hardwood lever thirty inches long and bolt it to lower half of upright, first inserting a block one and one-half inches thick to off-set the lever so it will not strike end of hub before upright is in vertical position.

To use, the base of jack is placed on the floor directly beneath the hub, the rubber covered part A is placed under the hub, and the lever is then pushed toward the car.

Handy Wagon-Jack

Every farmer needs a handy wagon-jack. Here is one that can be made in the home workshop. It will fit any axle, high or low, and can be operated instantly. Give it a good-sized base, to secure stability when the weight of the wagon is upon it.

Another Type Wagon-Jack

A long piece of 2 x 3 inch stuff, to which is hinged a short piece of the

same material, makes a handy wagon jack that will lift any wagon, high or low. And with the long lever it is easy to lift even the heaviest of wagon wheels. Slightly sharpen the lower end of the short strip to keep it from slipping. The folding arrangement makes the jack occupy but little space when not in use.

And Still Another Wagon-Jack

A wagon jack is a handy thing to have in the barn. Here's a picture of a home-made one that is easily constructed: Base, about 8 x 12 inches; two inches thick. Two uprights, two inches apart, mortised to base; center block at top and bottom, as shown in cut; uprights 1½ x 3 inches x 3 feet, of hard wood, with holes for bolt through lever. Lever of hard wood. four feet long, notched. Stout wire for holding lever in position. Simple, isn't it?

Home-Made Windlass

The wheel from a worn-out horse power, or hay press, mounted as shown, in the loft of a building, will prove a very great convenience in raising articles from one floor to another. This arrangement is also convenient for weighing any heavy body, the scales being put on the end of the rope about the shaft, when the article to be weighed can be raised and its weight taken.

Hay and Straw Hook

Here is a drawing of a handy hook for pulling hay or straw from the stack.

Hand Cart

The best home-made hand-cart on record can be made without employing a blacksmith, and without bending the axle. Take old wagon clips, fasten two strong handles to the under side of the axle, letting them project six or

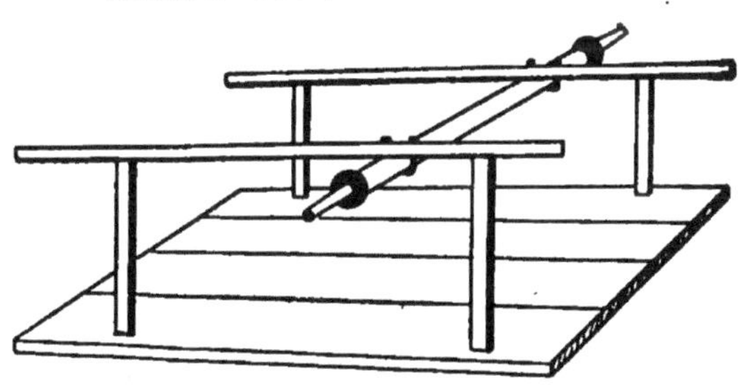

more inches beyond it. Now, run bolts through the projected ends to within eight inches of the ground, and two more of the long bolts of the same kind nearer the person who is to use the cart. These four bolts support a platform suspended upon them. To prevent it from swaying, brace it with sticks to the handles.

Tripod Ladder

You can make this unusual ladder yourself on rainy days, in your shop, which, of course, you will also have. Being a tripod it will stand firmly on uneven ground and will not sink into loose soil. The stiles, one of which is shown in cut, are in truss form, the parts bolted with one-fourth inch carriage bolts. The steps are of one-inch-lumber, one and a half and two inches wide. The foot has a spread of thirty inches, the top fifteen inches. The leg of ladder swings on a five-sixteenths-inch iron rod. It is known in California as the fruit growers' ladder.

Using Ladder on the Roof

A roof ladder may be of the utmost importance if a fire breaks out, while under ordinary circumstances a ladder that can be quickly applied to a roof is often of great convenience. Any ladder, long or short, can be so applied by the addition to it of the simple device shown in the cut. Any blacksmith can make it in a few moments, and it is but the work of a moment to apply it to a ladder. The cut needs no explanation.

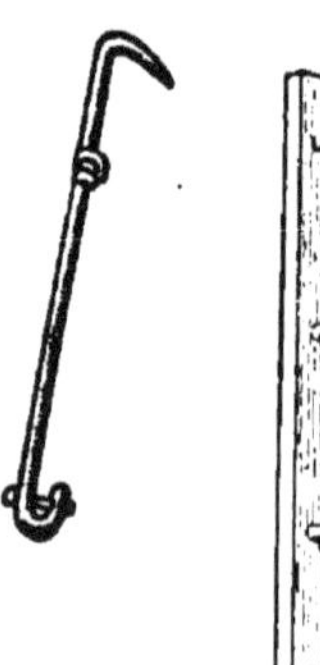

Double Ladder

Here is a ladder for picking fruit from the lower part of the tree. It is employed in picking pears and apples from trees of medium height, and is very satisfactory.

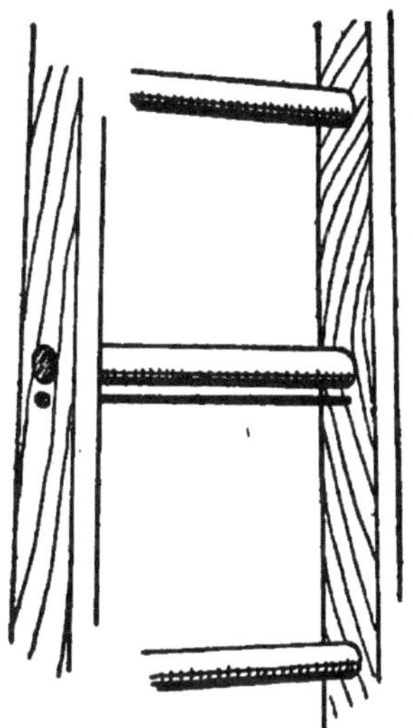

Making the Ladder Light

A ladder may be made at the same time light and strong by having the sides thinner than the usual ladder sides, but wider. Have the rounds stout and of tough wood. Then set three little iron rods, with head at one end and nut at the other, and put them through the ladder as shown in the cut—one at each end and one in the middle. This will keep the ladder from spreading—a source of much danger.

Garden Tool From Broken Fork

About most farm premises can be found broken forks that are no longer useful, and cannot be made useful as forks without paying the village blacksmith for repairs as much as a new one would cost. But the smith can cut the other tines off even with the broken one, sharpen them all and bend the shank, and so turn the worthless thing into a handy garden implement. The drawing shows how.

Useful General Tool

Take a piece of octagon tool steel of the best quality, three-quarters of an inch in diameter and three feet long, and have your smith make you a small crowbar or jimmy, shaped like the accompaning drawing. One end should be flat and cut for nail puller, the other end square pointed and sharp. With this tool you can pull the most stubborn spikes, nails, fence staples, hinges, hasps, and take old buildings apart and thus save much good lumber. Also useful in adjusting heavy machinery. An ordinary nut large enough to slide freely on the iron bar may be put on before it is bent into shape. It increases the leverage of this tool wonderfully and makes it very much more useful. The nut must slide freely and can be pushed out of the way when not needed.

Another Universal Tool

Here is the handiest tool (with few exceptions) that a farmer can have around the house. It is made from a piece of wagon spring about a foot long, with one end turned up about an inch, and a hole in the other end by which the tool can be hung up when not in use. The turned-up end is, of course, well sharpened. With this tool you can open boxes, scrape paint, pull carpet tacks, and do a hundred and one other things with it. Better try the idea.

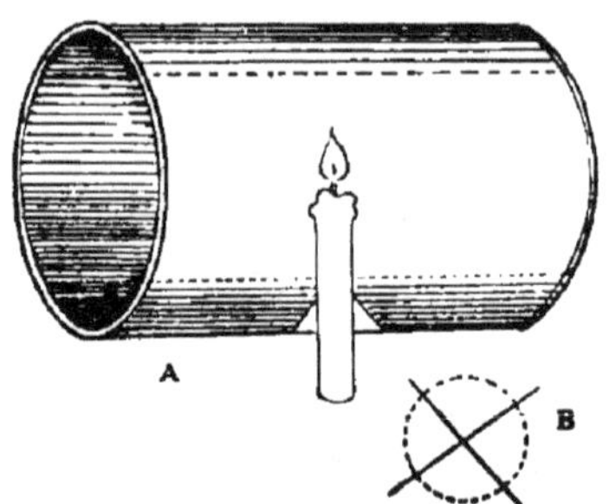

Tin-Can Lantern

Here is a home-made lantern that will serve in emergencies and help to follow safely dark and dangerous ways. It is made out of a cast-off fruit can, and the illuminator is the old-time candle. One end being melted off the tin can, a cross cut is made at A, in the manner indicated at B, and the points are turned inward sufficiently to admit of forcing the candle through the opening, and there is your lantern.

Combined Marker and Dibble

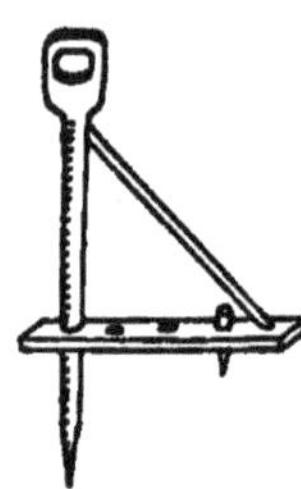

A very good dibble and marker for making holes and marking distances for setting out tomato, cabbage and pepper plants, may be made from a broken spade or shovel D-handle. It is necessary to point the end for a dibble. The marker is made by attaching a light strip of three- or four-inch board at a point about six or eight inches (according to the depth of the hole it is desired to make) from the bottom end of the handle to extend at right angles to the handle, as shown. A brace strip should extend from near the outer end of the marker board to near the top of the handle. Auger-holes may be made in the marker board and fitted with a removable plug, giving a means for the change of distance between plants. The only difficulty is in finding a spade-handle long enough, as the pointed end with which the planting holes are made in the ground should be straight.

Convenient Tool

A convenient tool for many purposes, including the destruction of bushes and the cutting of ice, can be made in a few minutes on the anvil by using the axle of a worn-out carriage. It need not be rounded or changed in any way except at one end where a plate of steel three-sixteenths of an inch in thickness is welded to it in the shape shown in the cut. The edge of this steel plate should be drawn and tempered.

Potato Sifter

The sifter for sorting and sprouting potatoes does good and rapid work. Get a box of suitable size, take out the bottom and in place of the bottom nail on round sticks (old broom handles will do) putting them as far apart as the size of potatoes not wanted. Put handles on the one end and pieces of rope on the other. Fasten the ropes to a support of some kind, have man to shovel in the potatoes and swing back and forth a few times, then dump. By having handles at both ends, and two men to operate it, the potatoes can be dumped directly in a cart or wagon.

Plank Clod Crusher

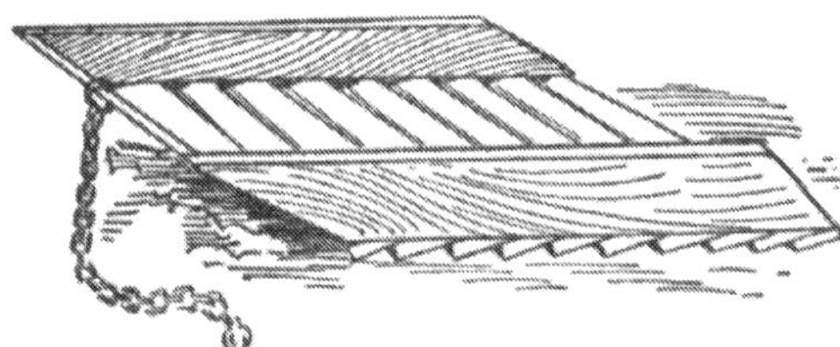

When one has no roller, a substitute can be made of planks fastened to sled runners as shown in the cut. As a clod crusher it does excellent work, even better than a roller in some cases. The planks do not lie flat, but are edged a little so that the corners of the planks scrape the ground.

Double Barn Broom

A single broom does not sweep wide enough a swath in the big barn floor, but two old brooms, put together as shown in the cut, make quick work. Two old brooms with broken handles can thus be utilized, making a handle for the combination out of a broken shovel or fork whose handle is intact.

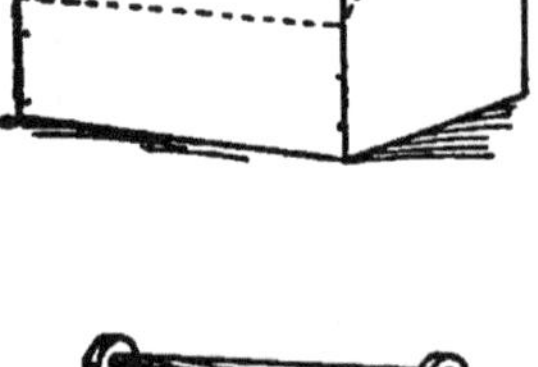

Carrier from Grocery Box

The cut shows how an empty grocery box can easily be converted into a neat carrier with a handle. The sides and ends are cut away as suggested by the dotted line, holes bored in the ends and a round bit of wood inserted. Such boxes are especially handy for use in picking fruit, vegetables, etc.

Fork and Shovel Hanger

All pitchforks, shovels and spades used about the farm should be hung up when not in use. This will prevent stock becoming injured on them, as well as preventing the handles from becoming broken. The illustration shows an easily made hanger. Take a piece of 2 x 8 and bore a two-inch hole in it for each tool. Make the holes a little larger than the size of the handle. Then saw a notch into the hole to pass the handle through. Nail the board on top of a 2 x 4 on the wall of the barn. The tools are hung with the sharp ends up. A broom holder for the housewife may be made in a similar manner.

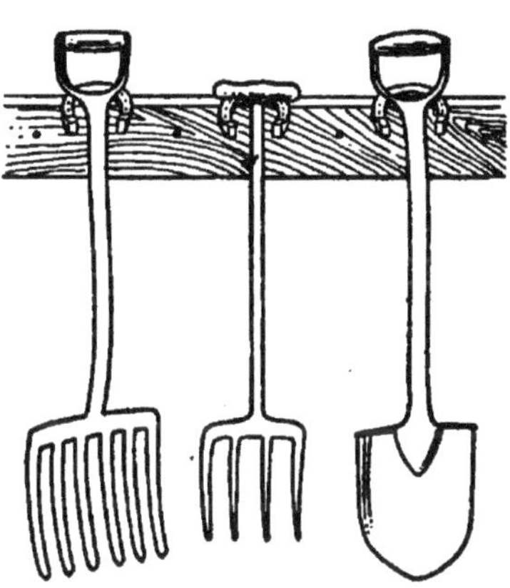

Hanging Tools on Horseshoes

Here is an English idea for utilizing cast-off horseshoes for hanging up farm tools. A strip of wood is fastened to the side wall, and on this the shoes are nailed by two or three wire nails. The main point of the illustration is to have a place for the tools, and to keep them there when not in use.

Rein Holder

This rein holder for the carriage post is home-made. The village blacksmith fashioned it from a piece of steel. It is seven inches long, half inch wide, the lower three inches one-fourth inch thick, the upper part one-eighth inch thick with an out-curve at top. Two screws hold it on while it holds the reins.

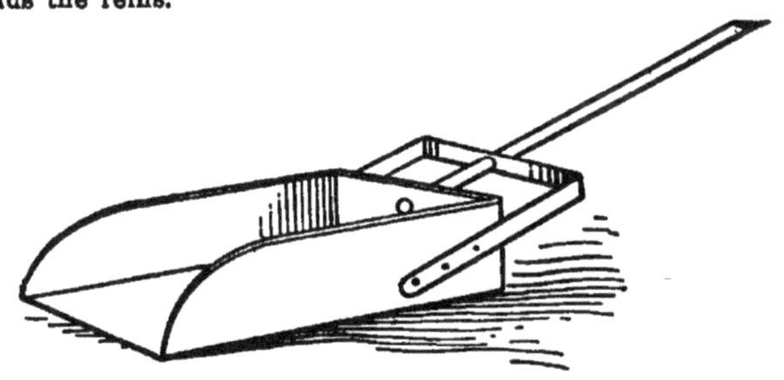

Chaff Shovel

To make a chaff or leaf shovel take an empty box and cut it down to the shape shown in the cut, sharpening the bottom at its front edge. Insert the handle of a broken fork or shovel and get the blacksmith to add an iron brace like the one shown here. A similar but more clumsy brace of wood could be used that would answer the same purpose, by nailing on two side pieces and attaching a cross-piece to these.

Home-Made Root Cutter

This home-made cutter is excellent either for cutting up roots of stock or for trimming the tops from roots when harvesting the crop. For the latter use, the slide-way that is figured will serve admirably. As the roots come down the slide before going into the root cellar, the tops are cut off and fall through the opening into a basket,  while the root without its top is tossed along and goes on down the slide.

Disintegrator

Take a land roller and put in spikes or teeth, six or eight inches long, a foot apart in the row and rows six inches apart. If we wish to mellow a piece of ground we do not take a pick and draw it slowly through the earth. This would be too laborious, and is not the best way to do the work. We do with the pick just what the spike in the roller does, go down a certain depth and put on the leverage, thus lifting and breaking up the soil. See the point?

Land Measure

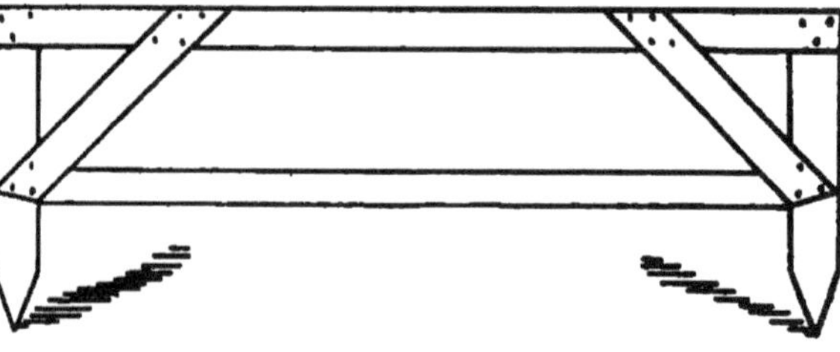

This device is made of three-quarter inch pine or b a s s w o o d strips two inches wide and eight and a half feet long, nailed to uprights two and one-half feet long and three inches wide, braced as shown in cut. The pointed ends of uprights are eight feet three inches apart, or exactly half a rod.

Powder Sifter

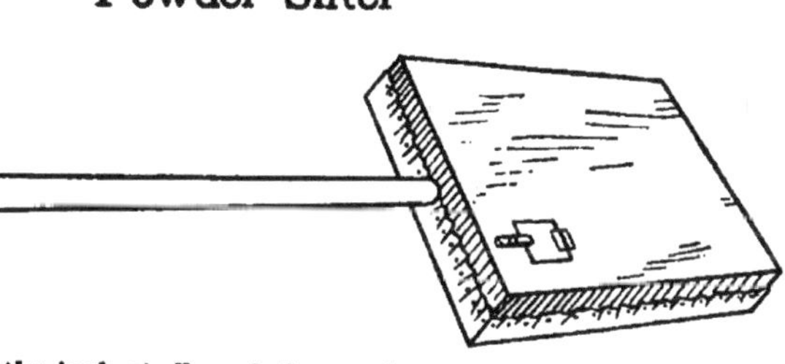

A home-made contrivance for sprinkling powder upon small fruit, plants and bushes is shown in the cut. The box is about nine inches square and an inch and a half deep. The boards are three-eighths inch stuff, and the top has an opening for inserting the powder, which is closed by a stopper. The bottom is covered with a piece of cotton cloth, coarse enough so the powder will readily sift through when the box is shaken. Fasten in the handle firmly.

Flexible Harrow Frame

Where the ground is very rough, the ordinary rigid harrow touches but little of the ground as it passes over it. The cut shows a plan for frame that gives great flexibility, and is specially advantageous on rough and uneven land. The harrow is hinged together with bolts that are linked together with eye heads, as shown in the cut.

Convenient Dump Cart

A convenient dump cart can be made when one has a light pair of wheels after the fashion of that shown in the cut. A blacksmith will bend the axle while the body is easily constructed. The handles are like those of a wheelbarrow, making the cart very convenient to handle. The tail-board is easily removed for dumping the load. Have the body short enough to make dumping feasible.

Reversible Stone Drag

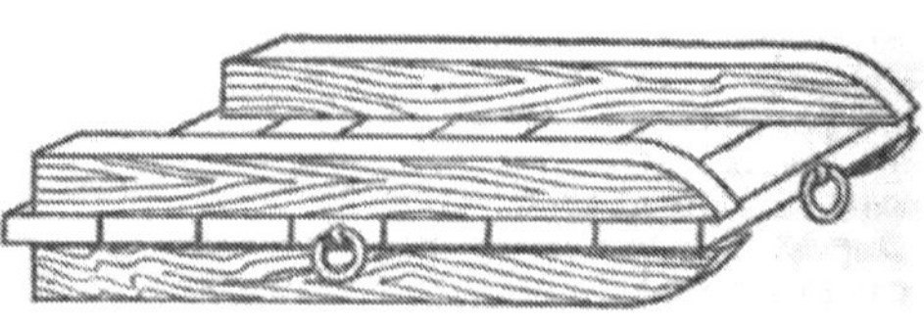

Over a large part of the United States the farms are more or less infested with stones of various sizes. It is doubtful if one of our boys likes to handle a load of stones twice, when by dumping it by the aid of a team once handling will do as well as twice, besides saving time. Here is a drag that is easier to pull than the old stone boat, having runners six inches wide to slide on. But the best thing about it is it has runners on top and bottom, so that when it is turned over in dumping it is still right side up for business. The runners are timbers that are chamfered off to slide over obstructions. They make a strong drag that will stand rough usage. The team is hitched to a ring in the side to invert the load.

Simple Stone Boat

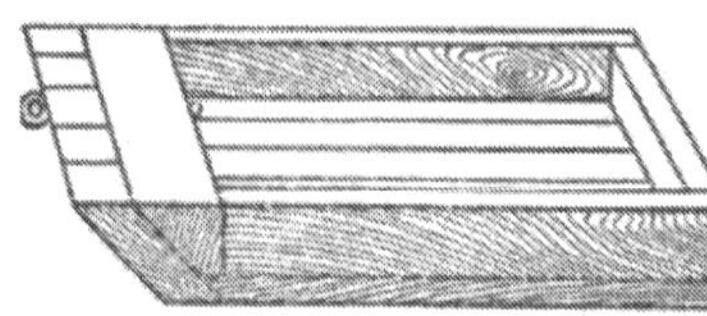

A stone boat or "drag," without planks sawed so that they will turn up at one end, is shown herewith. Specially sawed planks are not always to be had. The forward end can then be made in the fashion shown in the cut. Short pieces of plank, their ends sawed with a bevel, as shown, are spiked on to the short cross-piece in front and to the ends of the bottom planks. The drag will then readily ride up over stones and other obstructions.

Rock Carrier

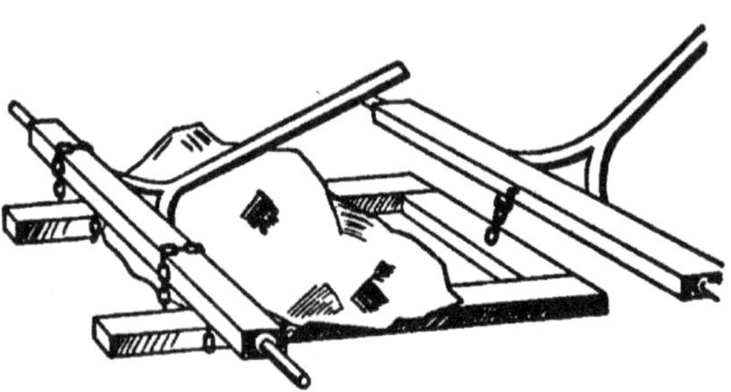

The best gear possible for hauling big rocks is a rough wooden frame slung at the rear of an axle by chains so it will just clear the ground. The forward end rests on the ground when loaded. When the rock is in place the forward wheels are backed up and the pole raised high in the air so as to hook on to the forward end of the frame. When the pole is brought down the end of the frame will, of course, be lifted. The tongue or pole of the rear wheels can then be lashed to the forward axle and the load driven away. The frame can rest wholly on the ground when being loaded by making the same arrangement for raising the rear as for the forward end. The drawing shows the axles without the wheels, for clearness.

Sled for Corn Shocks

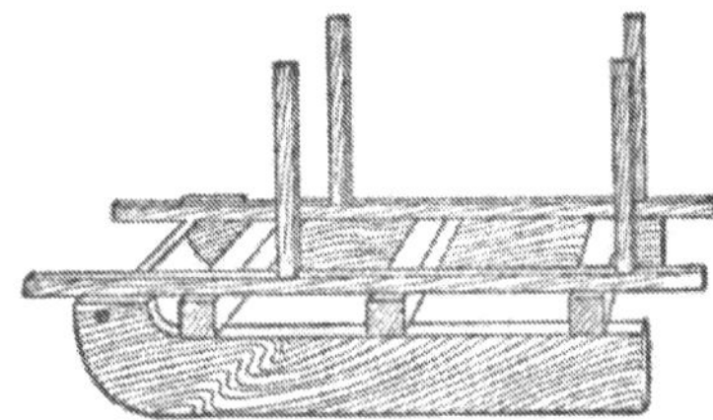

Here is a sled for hauling corn shocks to the barn floor and may be found useful if cold weather catches you with much corn unhusked. On cold, very windy or rainy days, it is not nice work to husk in the field, and such a sled as this will enable you to go on with the work under shelter.

Sled for the Water Barrel

A small sled for hauling a water barrel will often be found useful on any farm. A double headed barrel should be used and it should be laid on its side. A large funnel is useful in filling. A chain is passed around the barrel and under the hind axle of the sled, and a twister put in to hold the barrel on in going over rough places. The runners should be made of two-inch plank and the cross pieces mortised in to give strength.

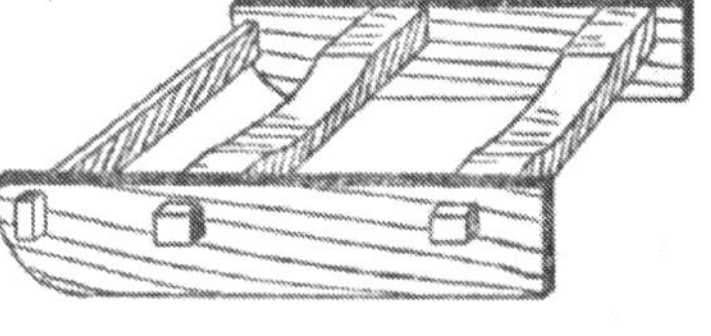

Dump Sled

A dump-cart is often as much needed in winter as in summer for hauling out manure, etc. A sled can be fitted up in the way shown in the cut so that the body can be dumped as easily as when on wheels. The construction is plainly shown in the cut. Braces from side to side can be placed beneath the cross-support to give greater stiffness if desired.

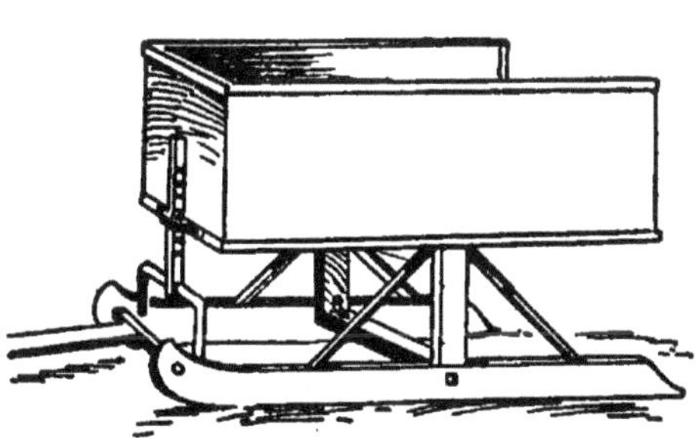

Cheap Sledge for Colts

This cheap sledge has been serviceable in breaking colts this past winter. It is only a pair of hickory poles, trimmed smooth and hewn quite thin, so as to bend in front of the sledge proper, and at the same time answer for shafts. A couple of blocks sawn from the cross-pieces raised them enough for the road. The cross-pieces and blocks were fastened to the pole-runner by four white oak pins. Slats were fastened to a box nailed to them for a seat, and a board bolted on to hold the whiffle tree. An hour's work saves risking your road cart later in the season, for our colts are thoroughly broken to harness, and when time is least valuable.

Balanced Wheelbarrow

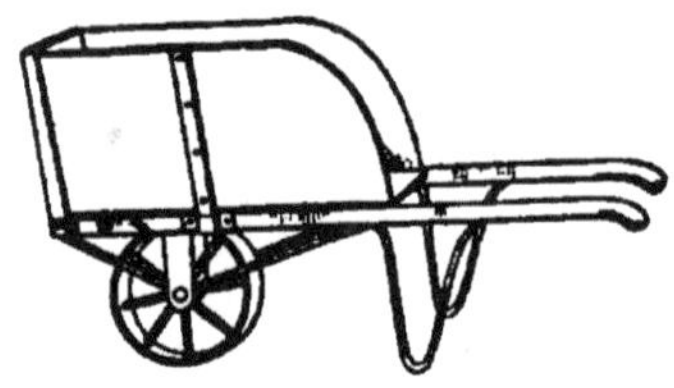

A wheelbarrow has already been designed with two wheels placed under the body of the barrow. A much more convenient form is that shown in the cut, which has one wheel exactly under the center of weight when the handles are raised. This barrow can be tipped over sideways to empty the load, as with a common barrow. This style wheel receives all the weight of the load upon itself; and the barrow takes up much less room in a shed or stable.

Light Wheelbarrow

For light but bulky materials a contrivance like that shown in the cut will prove convenient. The worn out buggy wheel in front takes the place of a second man between the front handles. One man can thus easily handle small crops alone. The device 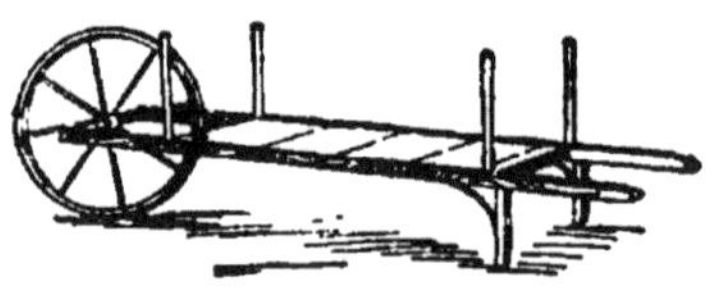 can be made very quickly. Fitted with light sides and end boards a big load of light, but bulky articles, like cabbages, can be wheeled in.

Spring Wheelbarrow

We have springs on wagons, why not on barrows? The springs afford a wonderful relief to the arms in passing over rough places. Get your blacksmith to take an old carriage spring, heat it, bend it and bolt it on as shown in the cut. Delicate fruits can be carried in such a barrow without injury.

Front Extension Wheelbarrow

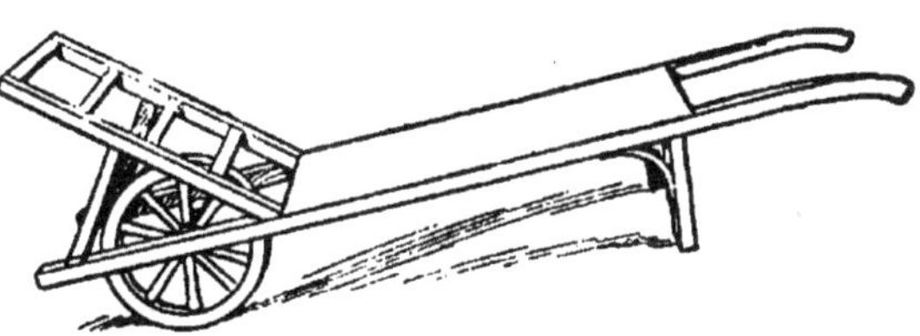

The ordinary wheelbarrow fills up too quickly when corn fodder, grass or bulky trash is to be moved by it. The cut shows how to make such a barrow as will get around this defect. The load, as may be seen, is thrown mainly over the wheel. The slats may be covered and side boards put on if earth or like material is to be moved in it.

Rear Extension for Wheelbarrow

The cut tells its own story as to the manner of lengthening the barrow. The use for such an enlarged implement will be found in the desire to wheel large loads of light but bulky material, such as corn fodder, pea vines, etc. Make the lengtheners of ash or similar tough wood.

Feed Barrow

This is a feeding barrow for the barn floor. The feed, either dry or wet, is placed in this barrow and wheeled along in front of the stalls. With a shovel the proper amount is given to each animal. The shape of the barrow makes it most convenient to shovel up the feed. It can be made at home, the two side wheels

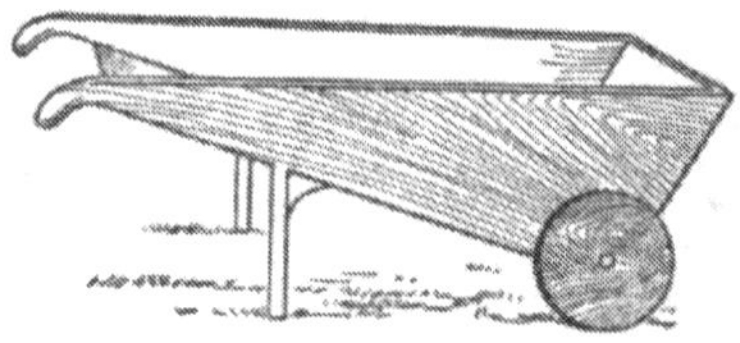

being cut from a hardwood board and then firmly secured to the sides. The extension of the sides forms the handles.

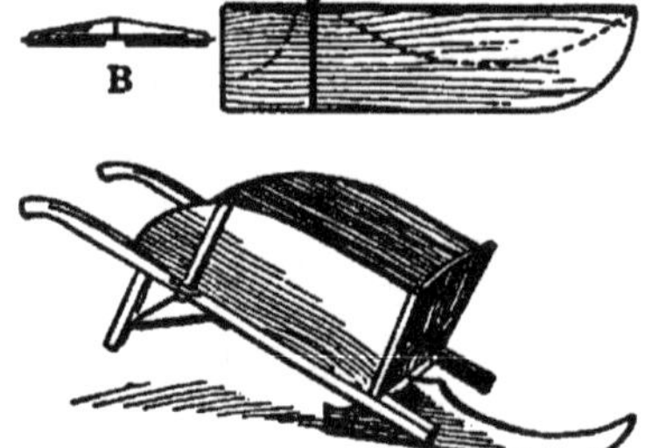

Single Runner Barrow

The runner is one and a half inch plank, twenty-six inches long and twelve inches wide. The shoe is a piece of band iron fastened with four screws. The spindle B, is set back under the body and the runner is fastened to this by a half-inch bolt run through ten inches from the back. The cut shows how the spindle and runner are made, and how they are attached.

Low-Down Truck

Every farm should possess a low truck for drawing fodder or shock corn from the field. This can easily be made from the front part of an ordinary wagon. A strong oak reach about a foot long replaces the longer one. To the rear end of this is bolted an iron clevis that holds a cross-piece, as shown in the cut. For the platform, two

poles fifteen feet long are used. At two feet from the upper end holes are bored and the poles are pinned to the cross-piece mentioned above, the ends resting on the bolster about two inches from the standards. The rear wheels are fifteen inches in diameter, put on an iron axle, the whole taken from old farm machinery. Being so near the ground it is best to board up the lower end of the poles for five or six feet.

Detachable Wide Tire

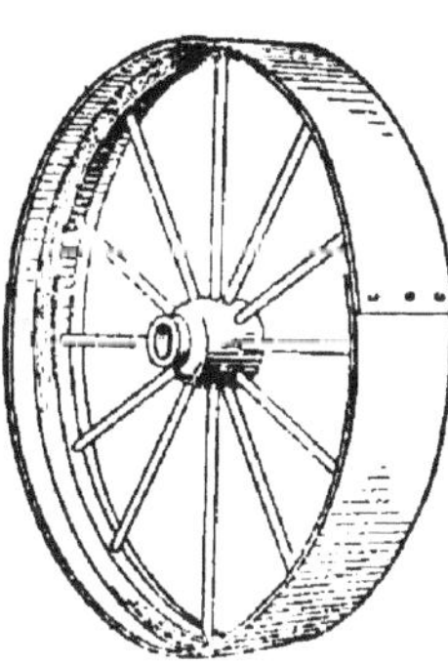

The drawing shows how to put a four-inch tire on a common wagon and take it off at will, by removing the two half-inch bolts shown in the lap; the tire is kept on by means of six small pins riveted into the tire, three on each side. There are times when a narrow tire is needed and times when the broad one is best. In this way we can take our choice.

A "Cut-Under" Hay-Rack

Hay-racks are easy to make if one is handy with tools. The cut shows a diagram of the bottom framing, with the outside sills severed to give the front wheels a chance to "cut under." The two inner sills are a trifle heavier than the outer ones. The cross strips should be of hard wood. The second cut show how the side uprights are put in around the open space. The two uprights have their ends in the inner sill. and their tops in the top rail as the others have.

Improved Tail-Board Fastening

When a cart is filled, with the load pressing hard against the tailboards, it is difficult to take these out if fitted behind an upright at each end. The cut shows an improvement. One end has the upright and the other has the device that is plainly shown in the illustration. A tap with the shovel liberates each board in succession.

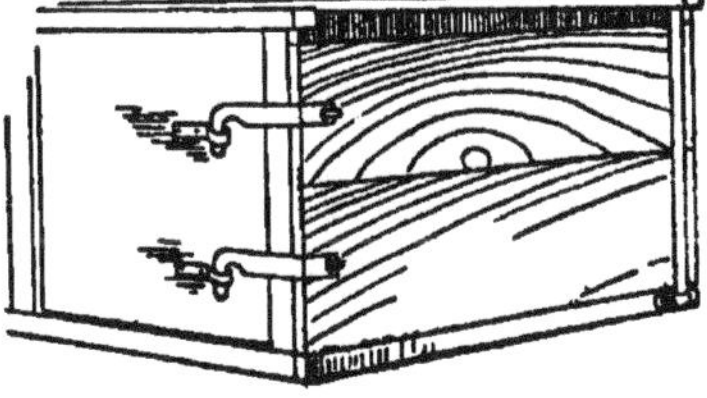

Wagon Tail-Board Platform

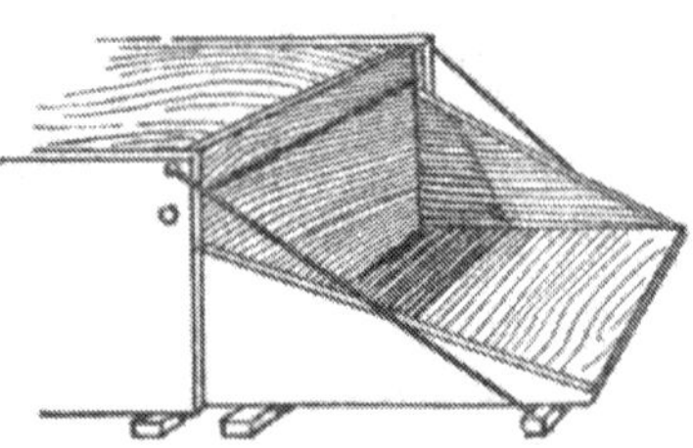

The drawing shows a handy end-board and platform to attach to a wagon when potatoes, corn or other similar products are to be shoveled out. The end-board can be let down, and work begun at once in a most convenient form, without the material falling on to the ground. The cut given herewith explains itself, showing the end-board in two positions, as closed and fastened by an iron rod and when it is let down ready to begin to unload.

Ladder for Wagon Loading

In loading farm wagons, whether from the field or potato or apple bin, there is needless lifting to empty box or basket up over the high side, or end. Have a short step ladder, with hooks at the end, and when loading a wagon hook the step ladder over the end as shown in the cut, and walk comfortably up to the edge and

turn the basket's contents into the wagon. It is straining work to lift a basket full of vegetables or fruit as high as one's shoulders over and over again.

Spring Seat

The seats to farm wagons and wagon boxes on runners are usually springless and very hard. To ride under such conditions is almost as wearying as walking.

Make a seat after the fashion shown in cut. Two boards are separated at the corners by stout furniture springs—the stoutest that are used. This can then be laid into a seat in place of a cushion, or may form the seat by being laid across the top of the wagon box. In this case it should have a cleat on the under side at either end.

Another Spring Seat

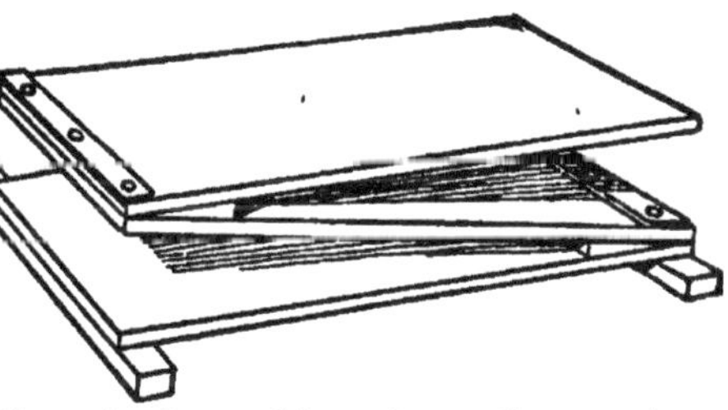

Here is a spring-seat board for a farm wagon that looks as if it might help us over the rough places when we ride. To make it, take three boards of proper length and width. Fasten ends together at a sharp angle, putting blocks in the angles to get the required spring. Fasten the joints together by quarter inch bolts through pieces of hoop iron. Screw nuts up tight. Put cleats on under side to keep it from slipping sidewise.

And Still Another Spring Seat

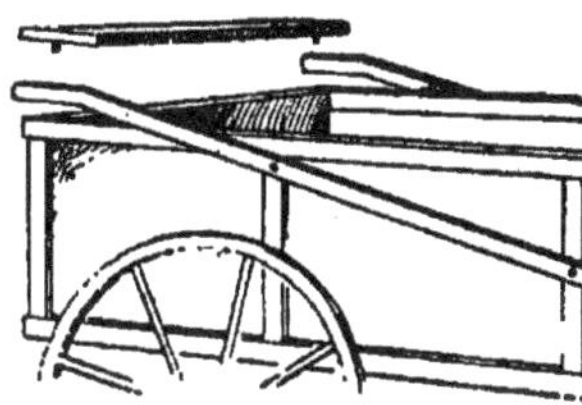

Farm carts and wagons that are without springs give hard seats for the driver. An easy way to make a spring seat is shown in the cut. The cross-board has round pegs beneath that fit into holes in the side spring-pieces. These pieces would best be made of ash.

Movable Wagon Seat

Men usually provide a good seat on their summer carts and gears—usually a cast-off mowing machine seat—but sit in most uncomfortable and cold positions on the winter sled when returning from a logging or wood-hauling trip. The cut shows a way to utilize the summer seat, slipping it out when ready to load and laying it on top of the load. When returning the seat is slipped into place and a pin inserted. A blanket can then be thrown over one, and warmth insured.

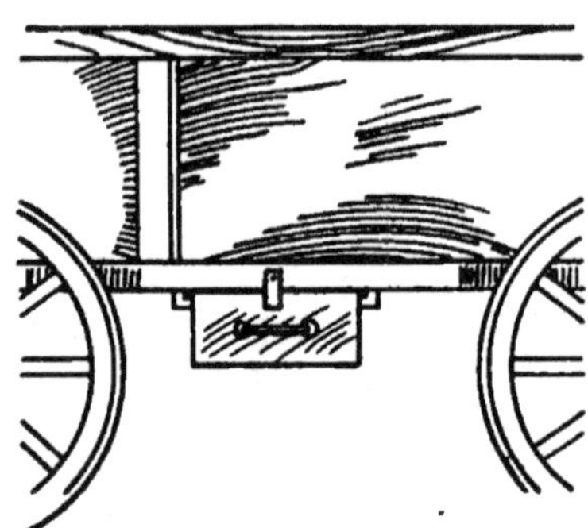

Tool-Box on Wagon

Fit a small drawer to slide beneath the farm wagon, or cart, in the way shown in the cut. In this should be kept extra nuts, bolts, straps, strings, a hammer and other small articles that one needs so often when away from the farm buildings, and a break occurs, either in cart, harness or some farm machine. The presence of such a drawer will often save much time and no little vexation of spirit.

Counter-Weighted Hitching Strap

Some horses, unless hitched with an uncomfortable short rope, will paw the air and get one foot over the rope continually. The cut shows a way to keep the halter rope continually taut. The weight board is just heavy enough to take up the slack of the rope and does not pull upon the halter. The weight can be adjusted till it is right. Have a peg to hold the weight at a certain height, as shown.

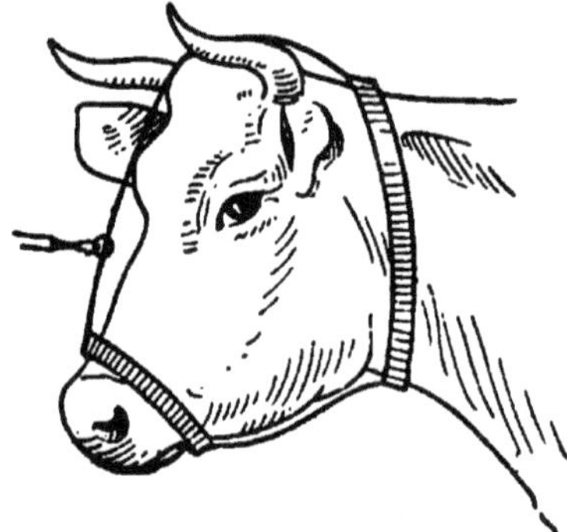

Halter for Cows

A useful halter for cows, and one they seem to like is employed by stock owners. It consists of a heavy strap about the neck and another around the nose connected to it by straps running from one to the other on the face and under the chin. On the face strap a ring is strung so' it plays from nose to horns as the tie strap which is snapped in it is tightened or loosened.

A Rope Halter

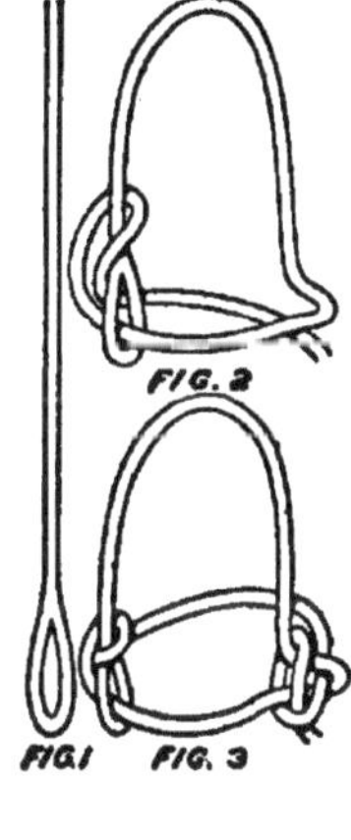

Make a rope halter for horses intended for sale by taking a piece of rope with a loop on one end as shown in Fig. 1. Then fit the rope over the horse's head in the shape shown in Fig. 2. Fig. 3 is the third operation with knots drawn tighter. A small piece of light rope can be attached in front if desired so that the halter will not slip back on the horse's head. These knots untie easily.

Three-Horse Evener

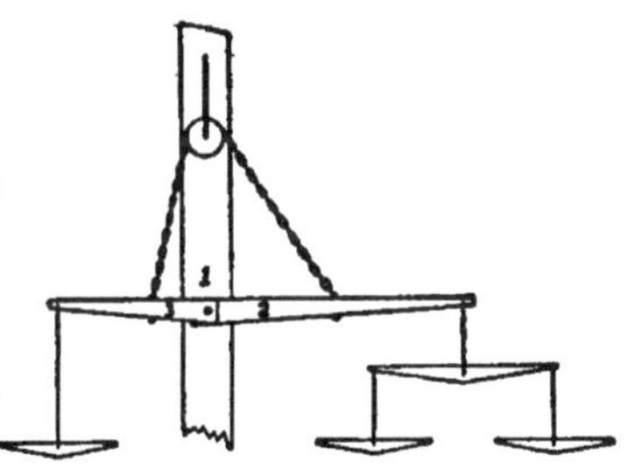

This three-horse evener is made of hickory two inches thick and three inches wide. In the picture, No. 1 is the tongue, and two and three the evener—jointed at a pivot bolt through the tongue as shown. A strong chain with small links connects the two pieces after passing over a pulley farther back on the pole. This chain should be attached to the one-horse side exactly one-quarter of the distance from the pivot to the singletree chain; on the other side just one-half way between. The single and doubletrees are attached equal distances from the chain. It is all hung on the under side of the pole. The rig works evenly and the team is comfortable.

Springs on the Whiffle-Tree

This is nothing more or less than sixteen inches of the end of an elliptic wagon spring, bolted on the back side of the evener, and the clevis made longer. The clevis pin goes through the bolt hole of the spring. The spring is composed of two "lifts," one and three-quarters inches wide. The out curve of the spring is about two inches. A horse must do his best to bring the spring up against the wood so that in ordinary work the spring gives from a half inch to an inch. If every teamster and farmer would use such a starter for the team, there would be far less shoulder-sore and blemished horses than are seen at present.

Hold-Back Whiffletrees

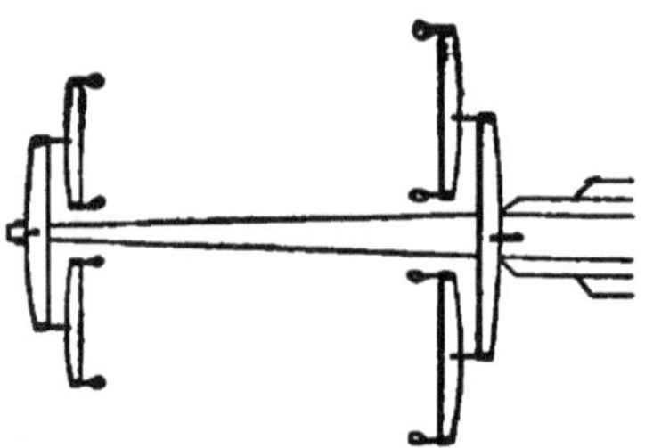

A team is frequently seen with the pole straps fastened direct from the collar to a short neck yoke, or to the end of the pole piece, where a rude substitute for a neck yoke is permanently fixed to the pole. In cities where immense loads are drawn, and the comfort of the team is made a subject of study, it is becoming customary to give each horse freedom in the triple neck yoke. This is a contrivance closely resembling evener and whiffletree. The hold-back straps instead of meeting under the body of the draft horse, extend from the breeching, through the girt on both sides and fasten to the single-tree in front. This permits the horses to walk more steadily without anything to alter their motions, and gives them the opportunity to spread apart, or approach near together when conditions require it. We cannot be too careful of our faithful dumb friends during hot weather, and any farmer's boy of fair ingenuity can rig up this addition to the hay wagon with but very little expense at the blacksmith shop.

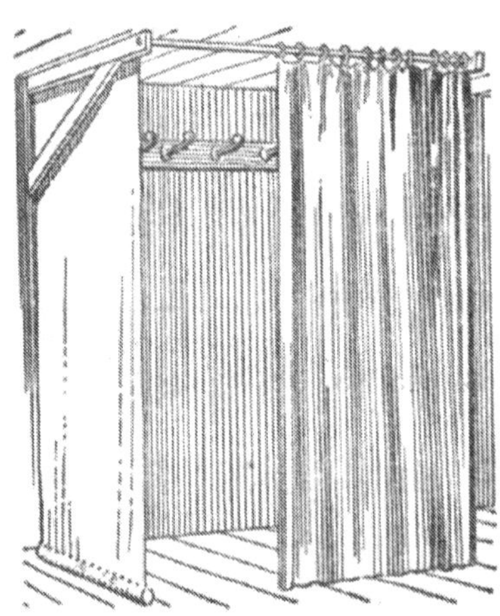

Harness Protection

Harnesses ought to be protected from dust and dampness when not in use—that goes without saying, but a great many people never find the time to build a regular harness closet. The cut shows one that can be built in half an hour's time—a short cut to something that will answer the purpose very well. The sides are burlap, supported as shown. The front is a curtain of burlap, supported by wire rings, running on a wire stretched across the top of the front. It is a simple closet, but it will protect the harnesses.

Wire Fence Crossing Ditch

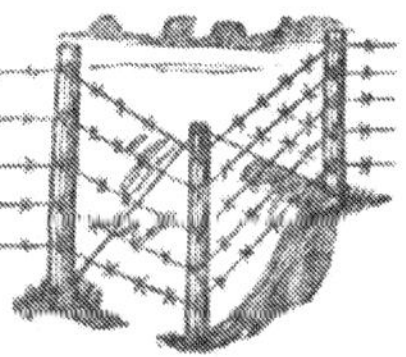

Often, when building a wire fence, it is necessary to cross a small ravine or depression in the field. In such cases it is sometimes a puzzle how to anchor the post in the lowest place, so that it will not "pull out" when the wires are tightened. The accompanying drawing shows a good method of solving the problem. A wire (doubled to secure strength) is stretched from the bases of the two posts near the ravine edge, over the top of the lower post.

Pig Fence

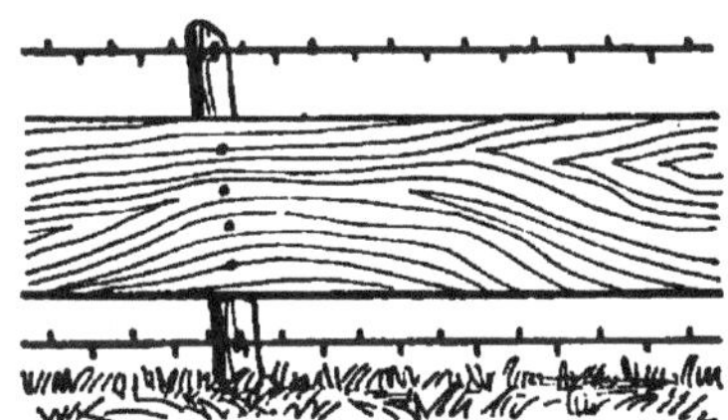

Where pigs are yarded out of doors an economical fence must be had if any extended run is to be given them. The cut shows a good pig fence—barbed wire at top and bottom, and a wide board between that offers ample protection from injury from the wire. The pig cannot root under nor climb upon such a fence.

Movable Sheep Fence

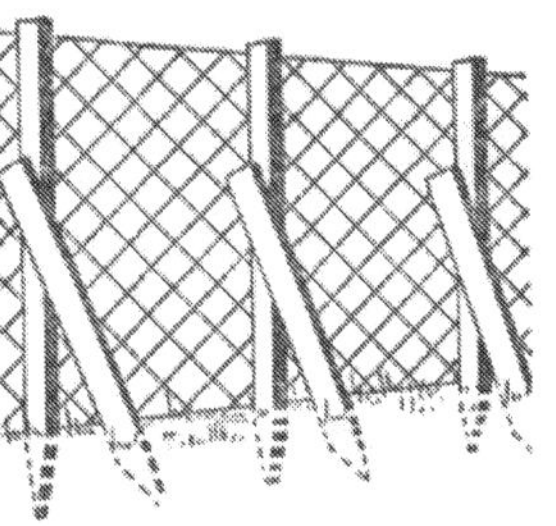

It is often desirable to fold the sheep upon a crop of turnips or other rank growth, where indiscriminate feeding would result in great waste. To keep the sheep confined in a small space that all the crop may be eaten up clean, requires a movable fence, light but strong, inexpensive and easily put up. The drawing shows a panel of such a fence, any number of which can be made as required by the size of the flock. Each panel consists of three light but strong posts, with an anchor post attached to each. A man at each end of such a panel can move it to the desired position, when a dozen blows of a maul will enter the stakes in the ground sufficiently to turn sheep, but not so far but that they can easily be pulled up when it is desired. A rod and a half of poultry netting or the same length of any woven wire fencing may be used for each panel. The anchor posts are, of course, on the side away from the sheep. A very light fence arranged in this way will turn sheep.

469

Sheep-Proof Wall

To make a stone wall sheep-tight use the improved method shown in the cut. Remove a few of the top stones where the stake is to stand and incline it across the open space as shown, having the top of the stake come on the inside of the pasture wall. Replace the stones and nail boards along the upper ends of the stakes as shown. Sheep have no foothold for climbing over such a fence as this, and it is moreover very permanent.

Another Sheep-Proof Wall

"Top-poling" a wall is not a safe fence for sheep. They dislodge the poles, and even go over the poles unless very high ones are used. Wide boards are expensive. Foot-wide wire poultry netting can be bought for quite a low price per running foot, with discount on large lots. Stretch it along the wall as shown

in the cut. It is better than boards, and costs about one-third as much.

Portable Fence

Surely it is unnecessary for one to enumerate the advantages of a few panels of portable fence on the farm. They will come "into play" a dozen times during the season. The cut shows how to make such a fence. The posts are made of narrow strips of board, as shown, the opening being wide enough to admit the ends of two panels. To turn a corner, use a hook on the end of one panel and a staple at the side of the post—or better, two hooks and two staples.

Light Portable Fence

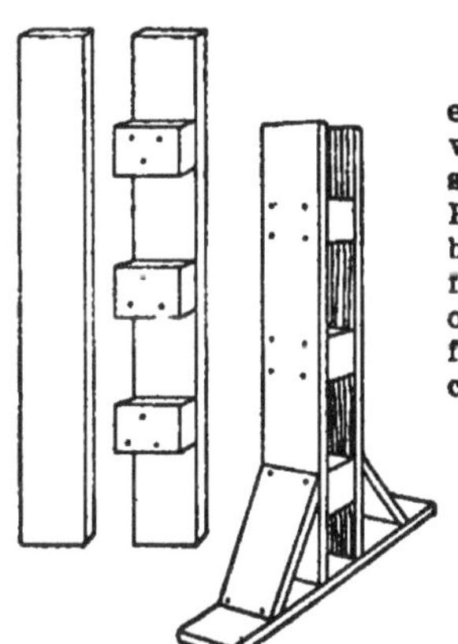

A piece of fence that one can move about to enclose small areas for short periods of time is wonderfully convenient on a farm. The cut shows how to make the posts for such a fence. Put up in a small rectangle, the corners will brace each side of the fence endwise, or, if preferred, the cross piece and braces at the bottom of every other post can run lengthwise of the fence. The whole fence, except the light rails, can be made of four-foot pickets.

Wire Splicer

A is a piece of stout strap iron 1 x 6 inches, with three holes in it as shown. The wire to be twisted is slipped through one of the holes and twisted about the other wire by means of the splicer, making a neat, strong connection. Where the fence wires are some distance apart the end holes may be used, but where the wires are close the middle hole is handier.

One More Portable Fence— This Time Wire

The triangular support on the left takes the place of a post. The cross is made of two two by two pieces four and a half feet long, fastened by a bolt at top and a four-inch board three and a half feet long, nailed at bottom. Two notches are cut four inches wide, one at top and the other at the center of bottom board. In these notches the ends of the panels rest. The panels are made of two by three pine and netting or wire to suit stock to be confined. The two railings are sixteen feet long, the uprights which are placed four inches from end of rails are four and a half feet long and the braces seven feet long, all bolted together with quarter-inch bolts. Make it in winter or other leisure time. Two men can move and set up one hundred panels in half a day.

Strong Wire Splice

Often, on the farm, it is necessary to splice two pieces of wire together. Much depends upon the strength of the splice, and, of course, the strength depends upon whether the job is done in the proper manner. This is how the telephone men do it—and they know their business.

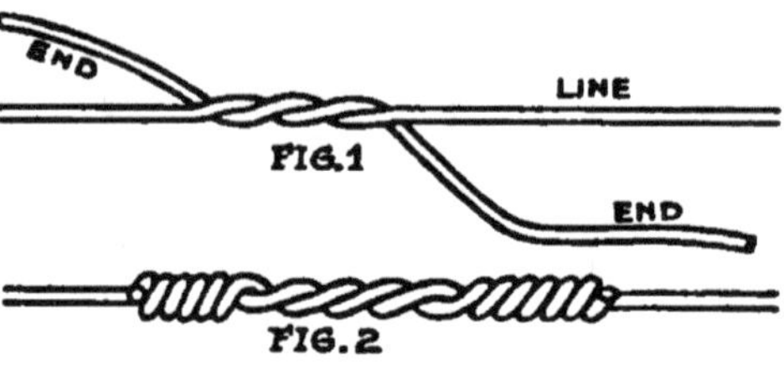

Wire should have at least three inches between the "buttons," this space to be so twisted as to form two and one-half turns, as shown in Figure 1. The buttons are formed by turning the ends about the line wire six times at each end. (See Fig. 2.)

Hurdle Fence

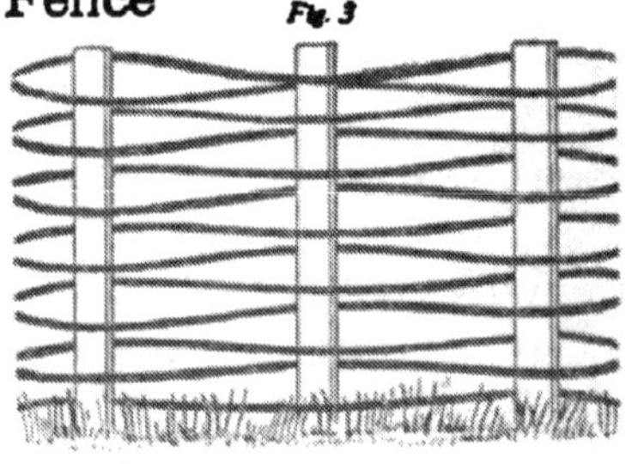

In locating where large timber is scarce there is sometimes a good deal of small sapling stuff, just the thing to make a hurdle or waddle fence, but of little use for anything else. Cut of any good lasting stuff that is three or four inches in diameter, stakes five feet in length. Sharpen these to a point, Fig. 1, and set them in line four feet apart where you want the fence to be. Use a wooden peg, Fig. 2, to make holes for the stakes. The peg is driven twelve or sixteen inches into the ground and withdrawn. The stakes are then driven down with a maul to the proper height. Now cut any small slender saplings that may be about the place, that are from one to an inch and a half in diameter. None should be larger than two inches. Have them as long as you can get them, the longer the better. Weave these saplings in between the stakes like basket work, putting them as close together as they will go, till you get to the top. You will then have a fence looking like Fig. 3, that will turn any animal.

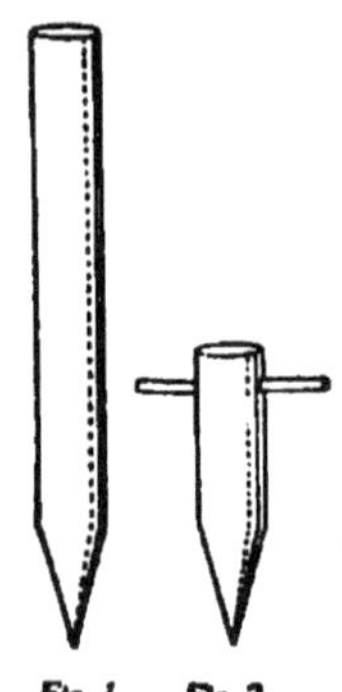

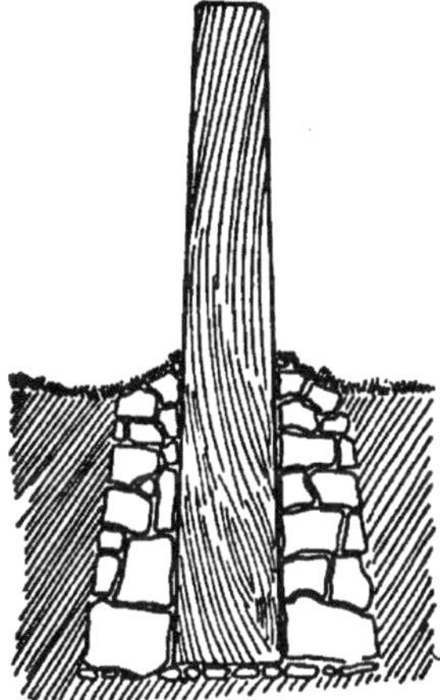

End Post Set in Masonry

The end or corner post of a wire fence has much responsibility placed upon it, hence the need of great firmness. The cut shows a plan that will give satisfaction. Dig a large opening and set the post in the middle of it. Fill in about it with loose stones and pour cement into the spaces between the stones. When the cement has set the post will be in the middle of a massive boulder, and will be exceedingly hard to "budge." Make the sides sloping inward, as shown, that the frost may not lift the whole bodily.

Double-Stake Posts

Where one must use very small stakes for fencing through lack of larger, the fence may be greatly strengthened by putting the stakes in pairs, and binding them firmly together, and to the boards, with the wire that can now be bought very cheaply.

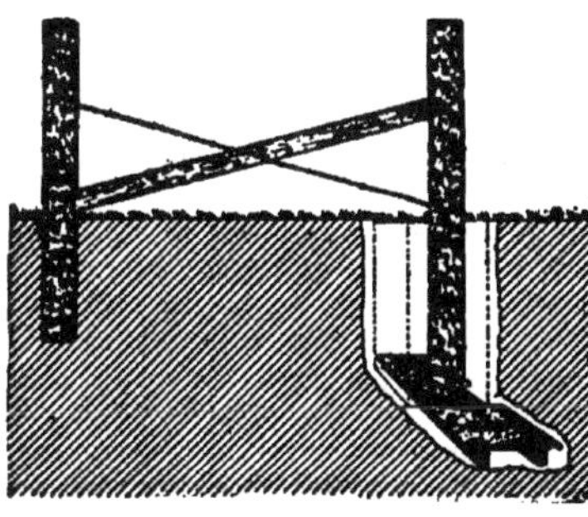

Anchoring the End Post

Everybody knows that a chain is no stronger than its weakest link. And it's the same with a wire fence. Everybody knows, too, that the "weak link" in such fences is usually the end post. Well, here's a method of setting an end post, using wood only. Looks strong, doesn't it? It is strong. Such a post won't pull up nor "give," nor wobble, and it is not in the class of weak links.

473

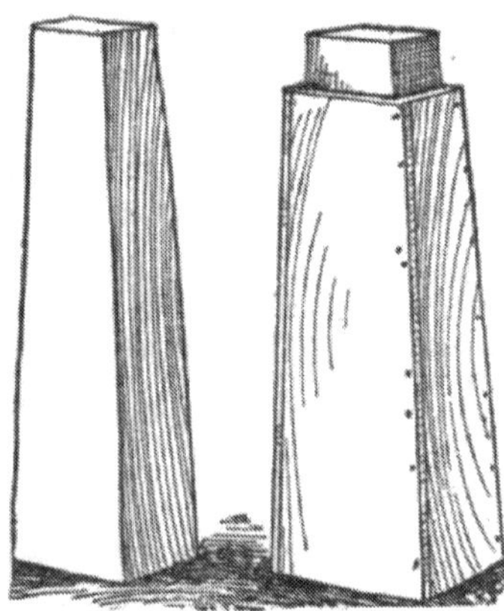

Frost-Proof Foundation Post

A permanent rock foundation for buildings is not always feasible, and stone or concrete posts must be used. Their chief failing is the likelihood that they will be lifted by the frost. Make the post a little smaller at the top, as shown in the first cut, with smooth sides; then box it in as shown in the second cut, allowing the box to slide upward freely. The frost will lift the box, with no chance whatever to lift on the sloping post within.

Utilizing Old Posts

Posts that are too short or which have rotted off at the ground may be lengthened as follows: Take two of them and saw each halfway through twelve inches from the

end. Then split off the twelve-inch piece on each. Put the posts together as shown in the cut, and nail with spikes. You then have one long, serviceable post.

Preserving Fence Posts

The best preservative for this purpose is coal-tar or creosote, boiled into the part of the post to be placed in the ground. The picture shows a convenient way to do this.

Any tank or large can of sheet iron which will allow the liquid to be heated to the boiling point, and which is deep enough to allow the posts to be covered with the liquid to a height of thirty inches will do.

Posts to be treated should be thoroughly seasoned, and treated for about two hours.

If you can do no better you can certainly put on some creoso'e with a brush, letting the liquid soak into the wood as much as possible. Charring fence-posts is not so effective.

Concrete Fence Posts

A stone post will not rot, and this concrete post is the nearest thing to stone. On ground alternately wet and dry, wooden posts do not last long. Make an open box for a mold, of the shape and size desired for the posts. Bore holes in the bottom where holes are desired in the posts. Insert long wooden pins as shown. Make the sides and ends of the mold slightly flaring, that the post may be gently turned out. After the box has been filled and the cement has "set," the pins can be pulled out, leaving the holes in the post, through which fence wire can be passed to bind the boards or stakes, as shown. Dry gradually, out of the sun, to prevent cracking.

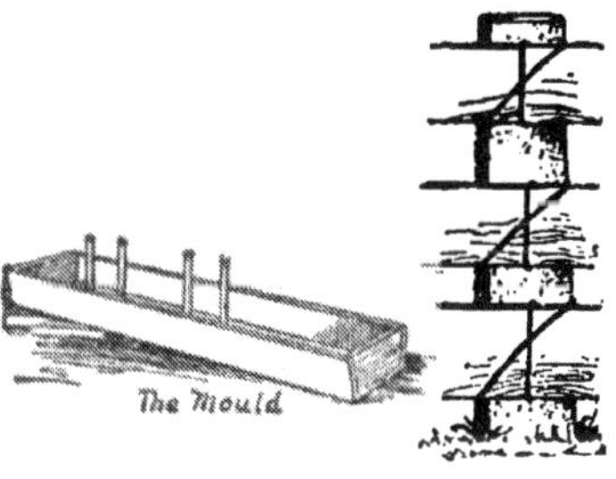

Sharpening Drive Posts

Those who drive their posts can hold them for sharpening by a tripod made like the cut. It consists of three small posts bound together a little above the center, with a chain or rope. To do good work have a large, solid chopping block and sharp, thin-edged broadax.

Pulling a Post

For a handy device to pull old fence posts out of the ground, find a crotched tree limb or fork like the letter Y, about three feet long. Lean the fork against the post; loop a log-chain around the post at the ground, then up the post, through the fork; hitch team to the other end of chain, and you have a lifting power here that will pull any post as fast as you can hitch to it.

Another Post Puller

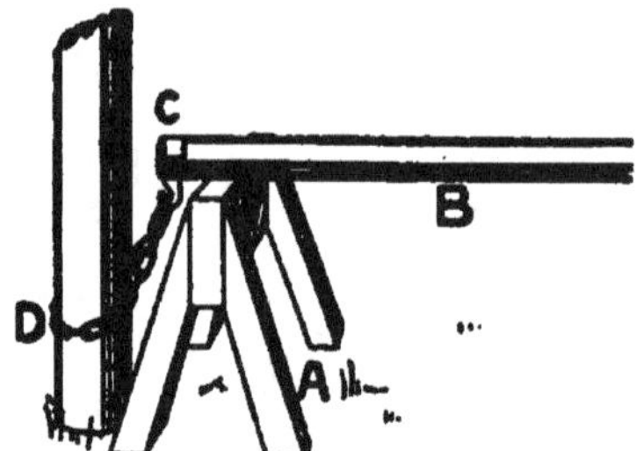

A is a small trestle, two feet high, one foot long; B, a strong lever six feet long; C, a hook; D, a small chain. Simply set trestle A along side post and place lever B on trestle; then take chain D and loop on post and hook one of the links on C, bear down on end of lever, and out comes the post. This device is easily carried from one post to another.

Bracing Tilted Posts

The annual spring fence repairing finds many stake fences tilted to one side or the other. To right such fences and keep them in place a stake is usually driven in as a prop on one, or both, sides and nailed to the upright post. If the ground is at all soft, the pressure of the wind or cattle drives the small prop into the ground and the fence is tilted over again. A better plan for supporting such fences is shown in the illustration, which tells its own story. A support like this at every third or fourth stake will help greatly in keeping the fence upright and firm.

Emergency Wire Stretcher

Take a wheel from a spring wagon, fasten end of wire to rim of wheel, put one side of hub against post, then hold and turn wheel so that the wire will wrap on hub. Had a wire stretcher and broke it, and found this to come in handy.

Fence Post

In many soils a fence post rots out with great rapidity. The cut shows a way to obviate this trouble. A hole is drilled in a broad, flat rock and an iron bolt inserted. A hole is bored in the bottom of the post which is then slipped down over the bolt, or set the bolt in a concrete block. Such a post is practically indestructible.

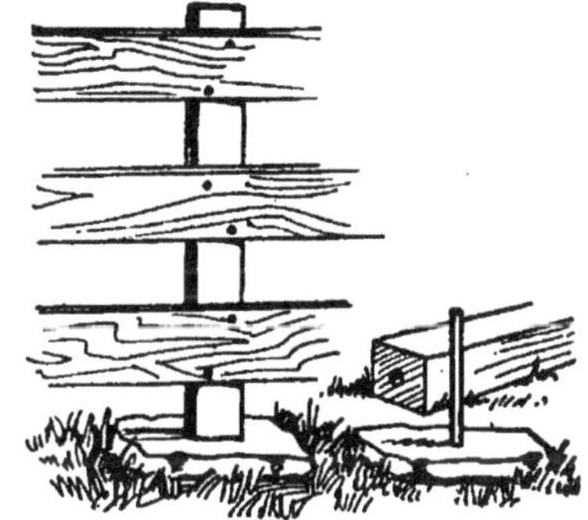

Fooling the Pig

The cut shows how to fool a pig and keep him out of the pasture, at the same time allowing the cattle to pass in and out at will. A wide board is nailed on the posts and another is nailed to short posts set about a foot away. The pig invariably goes in the narrow passage and out the other end and does not attempt to jump over.

A Handy Gate

A handy gate can be made at different parts of a fence around the pasture lot, or any other enclosure, which is much more convenient than to climb over the fence every time one is at a point where there is no regular gate. Saw off two boards and screw over the cuts two strap hinges. Nail on the up-and-down pieces and saw off the other ends. A button completes the gate. (See cut). Where a whole gate is to be made in a fence, it can easily be made in this way after the fence is built. Nail on the up-and-down pieces. Saw off the boards at one side and screw on hinges. Then saw off the other side.

Pathway Through Fence

Where one must frequently pass t h r o u g h a fence that encloses cattle, the plan shown in the cut will be much more convenient than a gate that must always be carefully latched. The center post is held in place by the blocks at the bottom, the openings allowing a person to pass through readily, but not cows or horses. One can never forget and leave this passageway unlatched.

Wire Fence Pathway

Here is a method of opening a passageway in a wire fence, obviating all necessity of braces for the posts and keeping the wires as taut as you want them, and there's no gate to bother with. The accompanying picture explains the idea fully.

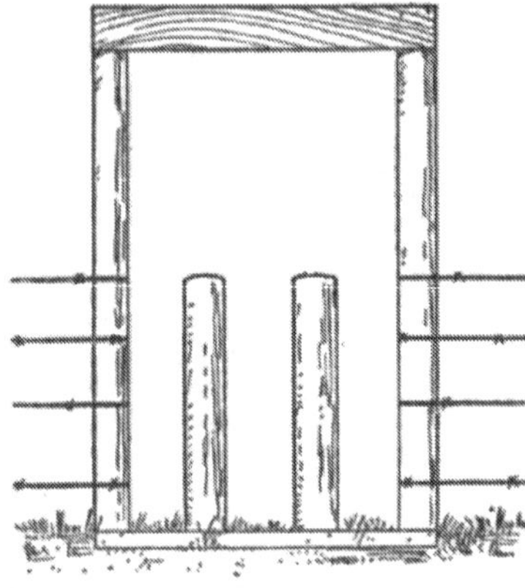

A Sagless Gate

Have uprights (a) the size desired for gate, then three cross pieces (b), letting them into the uprights equi-distant, by mortising. Then have a piece for a brace (c) to reach diagonally across the frame from the right upper corner to the lower left corner; saw a notch in this brace into which fit the middle cross piece. Fit in two smaller braces, as shown in the diagram. Nail on the slats and hang the gate.

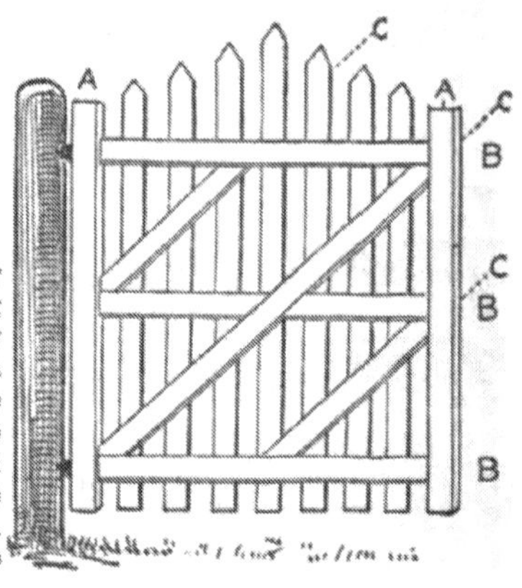

One More Sagless Gate

Farm Journal folks always want things about their homes looking neat and trim, so the farm gate must not sag down. A gate built like the one in the illustration will not do this. This is only a common sense gate hung upon strap hinges, but the bracing effect is the secret of all.

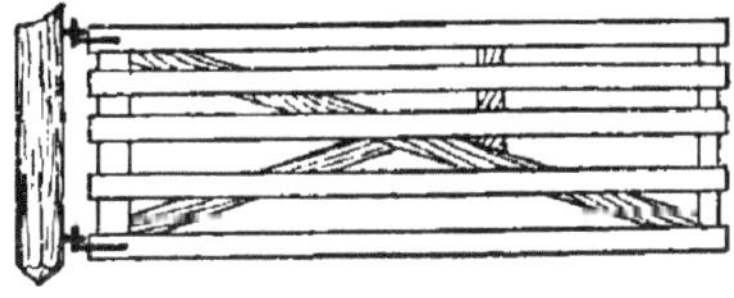

Counter-Weighted Gate

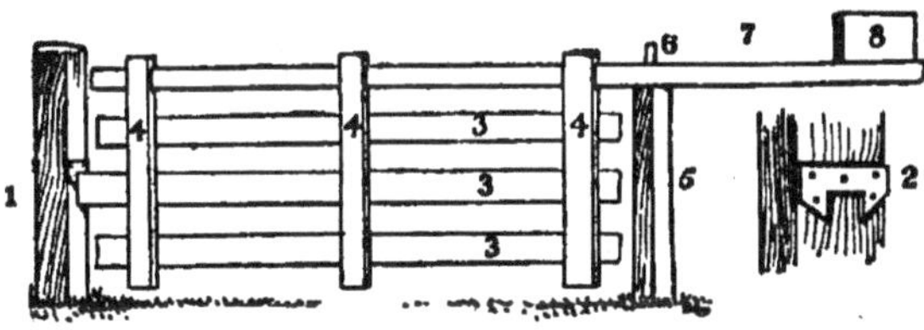

The drawing shows the kind of gate that needs no bracing. The post, 5, is stout and firmly set. In the top of this set a two-inch hardwood pin, 6, on which the gate is hinged. The top piece, 7, is four by four inches and sixteen feet long. The uprights 4, 4, 4, are mortised into the top piece. The slats 3, 3, 3, are five inches wide, the upper and lower are ten feet long, the middle one being a little longer to form a latch to enter the slot on post number 1. The slot is shown enlarged at 2. The gate is balanced by a weight at 8 and will swing both ways.

Wire Fence Gate

It is often desirable to drive a team through a barbed-wire fence, where it may not be worth while to build a regular gate. A passageway can easily be made after the plan shown in the cut, two posts being hitched together by an arrangement of hooks and eyes. To keep the wires taut, though the fence is cut open, run wires from the top of each second post on each side of the bottom of each "gate" post, as shown.

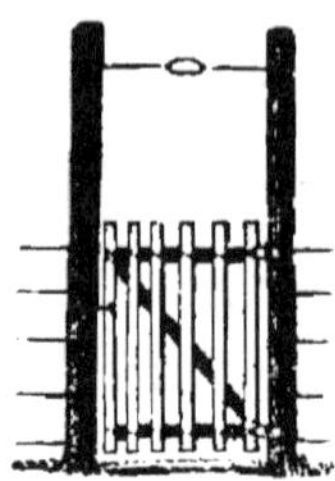

Gate in Wire Fence

Here's a method of building a neat little gate anywhere in a wire fence, and no posts are required. A double wire between the two gate posts, at top, keeps the fence taut. This wire is twisted tight, as shown. Any farmer will know how to twist it, using a stick as a twister. The picture explains itself, and the plan is simple and practical.

An Above-the-Surface Gate

A gate made like the cut cannot sag. It is a Yankee invention, of course, and especially useful where the subsoil is solid rock and a post hole must be excavated by blasting. Both posts can be framed in the same manner, and thus render the whole thing movable at will.

Hingeless Gate

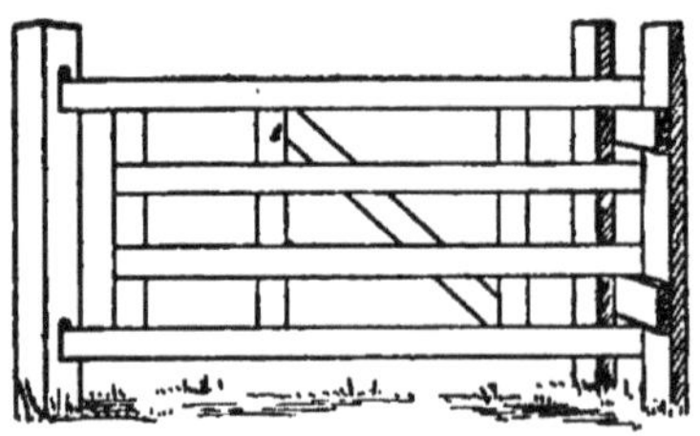

If this gate is made of dry pine, one by five inches it is light, and any boy big enough to drive pigs can handle it. To open it just slide the gate back until the ends of the top and bottom rails come out of the mortises of the left hand post, and then carry it around as far as needed. The two right hand posts should be set far enough apart so that the gate will slide freely between them. As there is no strain on the posts as in ordinary hinged gates, they may be light.

Sliding Gates

The cut shows how this excellent gate is made. Its merit consists in the way it is hung. The hanger is a clevis, holding a roller on which the top bar runs. The clevis is pivoted, and the pivot bolt passes up through a cross piece fastened to two posts, the bolt being held by a nut on top. The roller is three inches in diameter, having a flange half an inch high on each end, and may be  made of wood or iron. The posts where the gate swings must not be set opposite, but one about three inches in advance of the other. To open gate roll it back until nearly balanced, and then swing it around. It will swing in only one direction.

Light Gate for Pasture

Here is a drawing for a light, convenient and cheap farm gate for pasture lot. It's easier to lift off the whole gate than to draw a lot of bars. Gate is made of four one-half inch pipes and two iron uprights one and one-half by one-half. Each gate hangs on two hooks at each end on the fence posts, and generally costs less than $3.00. Second-hand pipe can be used.

Double Bars

Perhaps some of our folks will want to fix a gateway where they do not want to go to the expense of hanging a regular gate. The bars here shown will fill the bill. They are placed loosely in the posts and can be pushed either way and not fall to the ground. The long space is twelve feet, for teams, and the short space is four feet, for stock and people to pass through. 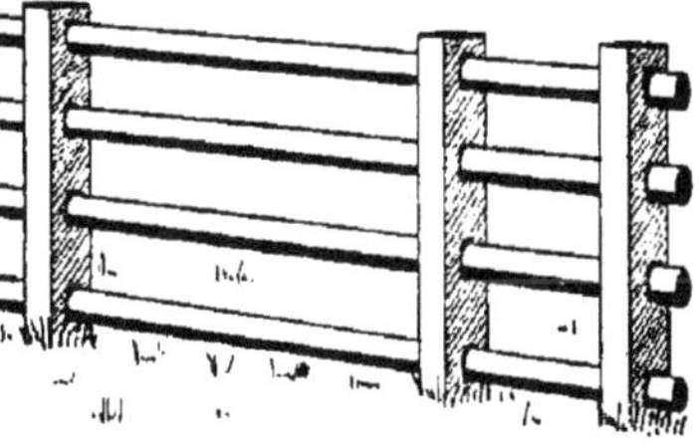

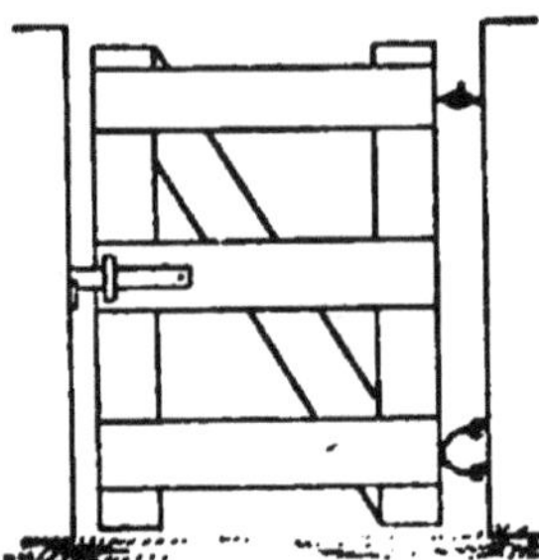

Self-Closing Gate

Here's a gate for the front yard that neither the dog nor the hogs can open, nor the boys leave open after them. The upper hinge is simply an ordinary hook and eye. B is the lower one (enlarged), which does the shutting. It has two curved branches, with a half circle at the end of each, which shut against two square staples, driven into the post a little each side of the middle. The gate opens either way, and when open rests on one or the other of the curved branches, and its corresponding staple; when shut it rests on both. The latch is simply a straight piece of iron, pivoted at the back end, that it may play up and down, and when the gate "slams," slide up over the sloping end of the catch A, and drop into the notch in the middle.

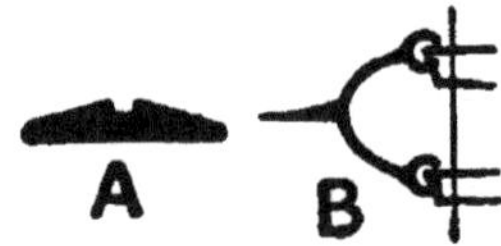

Another Self-Closing Gate

A swinging gate like that shown in the cut is always shut, yet always open. Careless boys cannot leave it so the stock will get through into the cornfield or out into the road. As will be readily seen this gate is for people and not for vehicles to go through. The loose end swings from

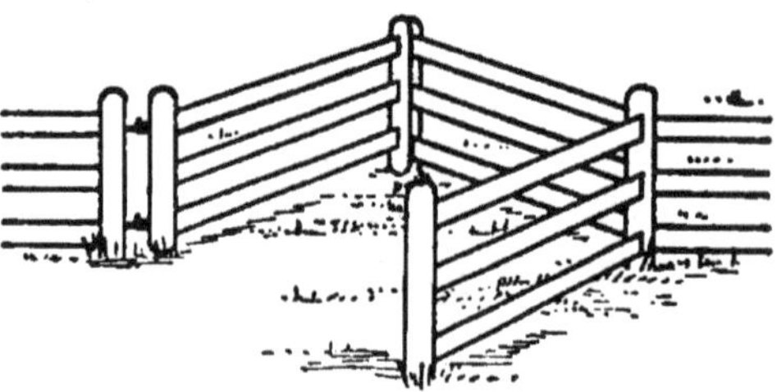

post to post of the fork in the fence. If made of light material and hinged properly it may be lifted off to allow the passage of stock or a barrow, as occasion requires.

Making the Heavy Gate Light

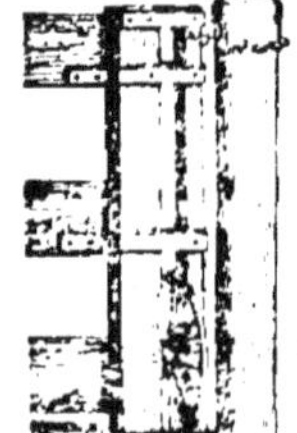

You don't like to carry a heavy gate any better than we do. Fix up the gates with some old cultivator wheels, like this illustration.

Spring-Closing Gate

The necessity of keeping certain gates constantly closed is manifest. The cut shows an excellent plan. A long spiral spring that can be had in different sizes at the hardware stores, is attached to a rearward projection from the top of the gate, as shown. The gate can thus swing in both directions—a great convenience. A latch catches whenever the gate swings to, from whichever side it comes. This is one of the best self-closing devices yet illustrated.

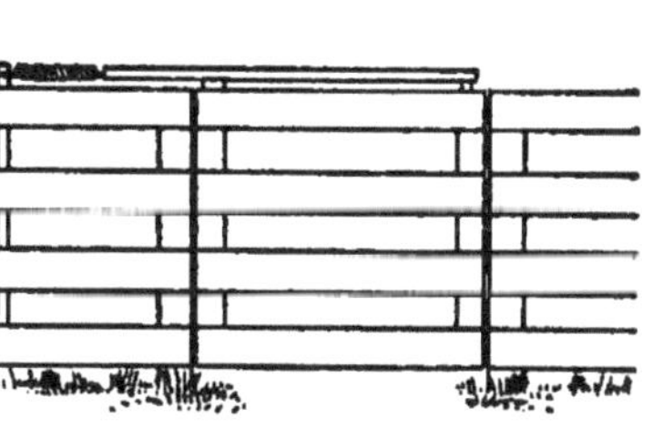

Spring-Locking Gate

To make a self-locking gate take an old hay rake tooth, make holes in the ends for bolts, fasten one end to the gate and the other to the sliding bar as shown. The spring of the tooth holds the bar in place when the gate is closed, and the latch is more likely to fasten than if we depended on gravity only.

Gate Support

The illustration shows a way of keeping a gate in place. Wire netting in time pulls heavily upon gate-posts, often spreading them so much that the gates do not keep shut. After trying all methods of preventing this, the editor hit upon a plan that has ever since proved successful. Besides the braces on each side of the gate-posts, a shingling lath is tacked over the top (the posts, by the way, being a foot higher than the fence). This holds the fence in position. Where these gates are used directly in the chicken yards, heavy wire can be used in place of the shingling lath, which will prevent fowls flying on top.

Gate Latch

There is nothing more important about the poultry yard than a good gate catch. Where ordinary wooden buttons are used, heavy winds are apt to blow them off, as constant use weakens the hold of the screw in the button.

Some years ago the editor had a man working on his farm who was a regular genius. Seeing the trouble we were having with gate fasteners, he bent a piece of wire into the shape shown in the illustrations, and attached it to the gate. It was a success. We are using the plan to the present day. Figs. 1 and 2 show how the catch is made.

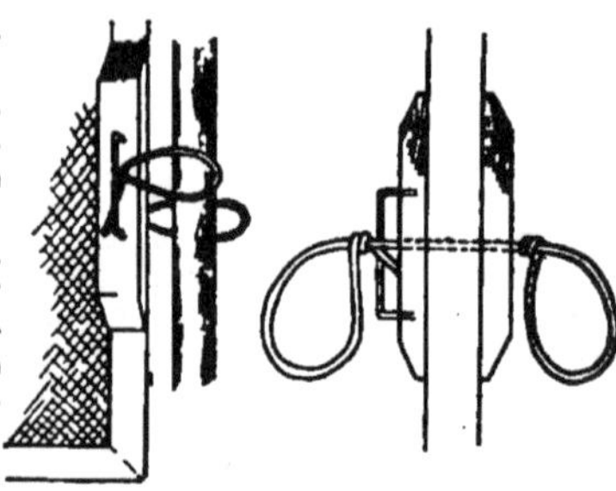

Fig. 1   Fig 2

Simple Gate Catch

A reader sends us a gate fastening that he says is cheap, easily applied, effective, and will "catch" on no matter how much the gate sags. It is simply a piece of iron bent as shown at (1) and stapled fast in a notch of the post. When it falls to a horizontal position it clasps the gate as shown at (2).

Stile for Fences or Wall

Many old fashions are worth renewing. Our ancestors used to get over fences by means of stiles, one of which is figured in the cut. They are quaint in appearance and very convenient where a gate may not be desired. They are also useful in the case of stone walls, where the installing of a gate would entail much labor.

Wire Fence Stile

Every one has experienced the nuisance of trying to cross a barbed wire fence without tearing his clothes or scratching himself. The cut shows a safe way. Let one post be longer than the rest, and nail two cross-pieces to stakes on either side of the fence, as shown. One can now step up one side, over and down the other, with ease and safety. Two bits of board are nailed over the upper wires, as shown to prevent the wires catching into one's clothes. If you must cross such a fence occasionally, try this plan.

Another Stile for Wire Fence

To get over a wire fence easily and safely, make a device like that shown in the drawing. The high posts are to hold on by as you step over. The upper wire has a guard of wood to keep it from catching your clothes. By making such a stile as this you avoid the necessity of crawling between or over wires.

Gate Latch Lock

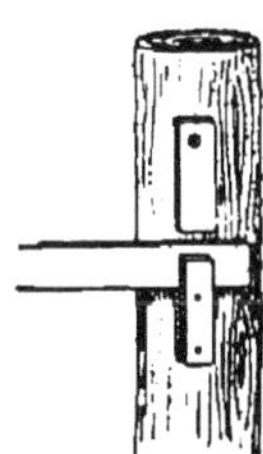

It often happens that the cow or horse will learn to work the common gate latch by lifting it, the cow with her horn, the horse with his nose, and thus go where they should not. A neighbor of mine has thus solved the problem: He takes a block of wood, say an inch wide, three-fourth of an inch thick, three inches long, which he places against the gate-post in an upright position, nearly touching the latch. Then he drives a nail through the upper end of the block (see sketch) and into the post, thus allowing the block to turn on the nail so that the lower end may be moved out of the way of the latch when it is to be lifted.

Plank Culvert

The ordinary stone culvert is apt to be a failure. The sides become pressed together, the top settles, and soon the opening is almost completely choked. An excellent plan for making a culvert is shown in the cut. Two sticks of timber, pine hackmatack or cedar preferably, are placed one above another on each side, and spiked together. A bit of plank at either end embedded under them, with a crosswise planking over the entire top, keep the sides from coming together, even in the slightest degree. The planks above and below are firmly spiked to the timbers. Such a culvert will be more satisfactory in every way than one of rocks could possibly be, and where flat stones are not at hand, will be found much easier to build.

Rot-Proof Sidewalk

Whether it be for the farm or for the roadside, the sidewalk should be so constructed that moisture cannot be retained beneath the planks. The life of the average plank sidewalk is shortened one-half because of this trouble. Where the planks rest upon broad sleepers moisture is long retained between the two surfaces and decay soon begins, especially if the wood be hemlock. Let the sleepers be of the shape shown in the cut and the walk will at once dry out after every rain. Experience has proved the value of this idea.

Foot Path Construction

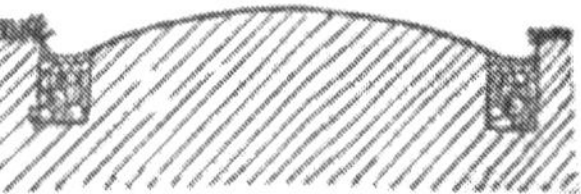

The cut shows an excellent walk for one's grounds or for the path between the house and barn. The rounded surface has a little trench at each side filled with loose stones, with the smallest stones at the top. The water can not wash such a path, for it settles at once to each side of the path, and sinks into the rocks, where it is conducted away without doing any damage. A furrow at each side makes the trench that is to be filled with stones.

Wood-Block Supports for Boardwalk

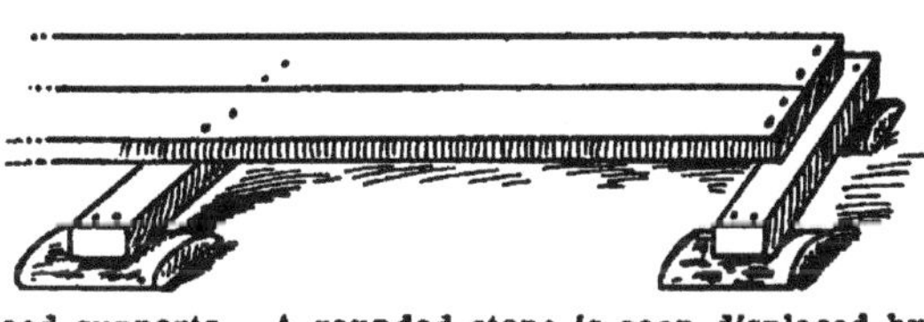

The cut shows one of the best plans for laying a wooden sidewalk, whether for highway or farm use. The cross-pieces are usually supported by stones, and these are rarely so flat and even as to make good supports. A rounded stone is soon displaced by frost and the sidewalk comes to grief. Get cedar or pine log butts at the mill, and split each one in two. Lay them as shown in the cut. It makes the best and firmest support imaginable.

Stiffening the Bridge

There are comparatively few farms in the East, at least, that do not need some bridging of streams. Some are deep, some broad, some shallow and narrow. The cut shows a way to support the middle of a farm bridge where it is not easy to place supports below. The iron rod and the inclined timbers form a truss that is wonderfully strong.

Subway for Stock

A subway for stock will save much bother where a highway divides the pasture. Build the under passage where the highway passes across a dip in the land, making the walls of stone or heavy timber, as shown.

Post-and-Pole Bridge

Four heavy posts eighteen inches in diameter and as high as you want. The approaches to the bridge are set on the bank, back from the current far enough to avoid its rush in freshet and ice. Or if the bridge is to be used especially hard, double these posts. Trees selected at the right points may be used if well rooted. Now notch in deeply on the water side as high up as the water ever rises, and fit a pole (2) of one foot diameter at butt to match the notch. This must

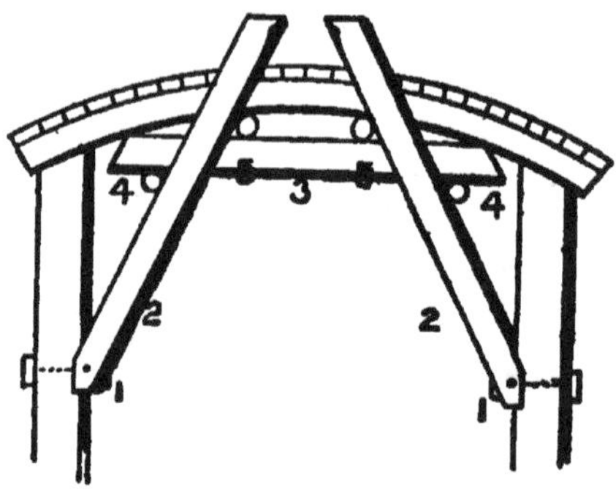

be long enough to reach up past the surface of the bridge. A heavy bolt must join it to the post at 1. Figure 3 is a similar pole laid across upon cross poles, 4, 4, and under other cross poles, all of which are bound together with iron bands or short chains where they touch. All the weight and strain of the span comes upon these poles, which form a truss. The same thing may be put up on a light scale for a footbridge. The poles 2, 2, must not be spiked to the bridge frame, but the bridge may be fastened to 5, 5, and to 3, upon which it lies.

A Simple Stone Abutment Bridge

A wide and shallow brook makes a plank bridge both expensive and hard to build. The ground plan given here shows walls of rough stone laid out from either side of the stream, leaving just enough space for the passage of the water in the center. Within the walled space earth is solidly filled to make a roadway, leaving only a short span over the center to be bridged with wood. The cut shows the idea plainly.

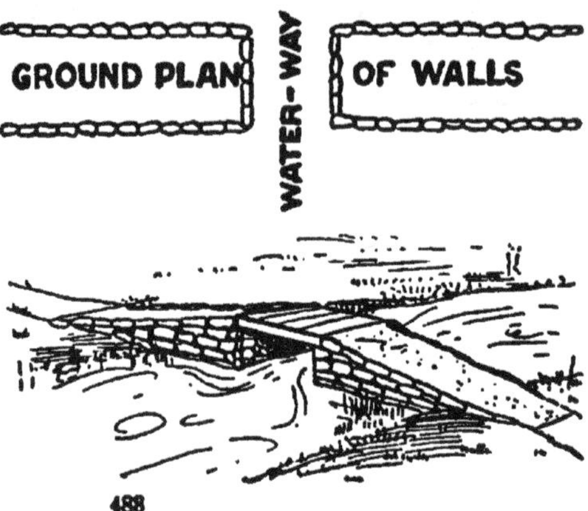

Farm Bridge

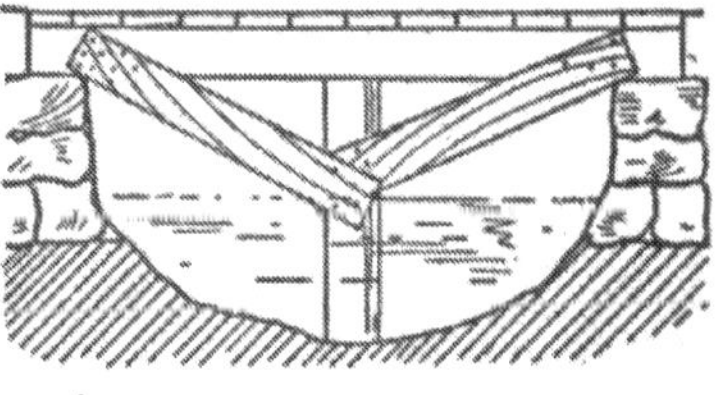

Most farm bridges are across small streams where a single length of stringer will suffice. But the middle of the bridge should be supported to guard against accident with extra heavy loads. Set posts in the middle of the stream beneath the stringers, and brace these with planks spiked on as suggested in sketch. Spike a plank also from post on one side to post on the other, two feet below the stringers. The side braces will virtually form a truss on each side, and will add great support to the center of the bridge.

Joining a New Building to an Old One

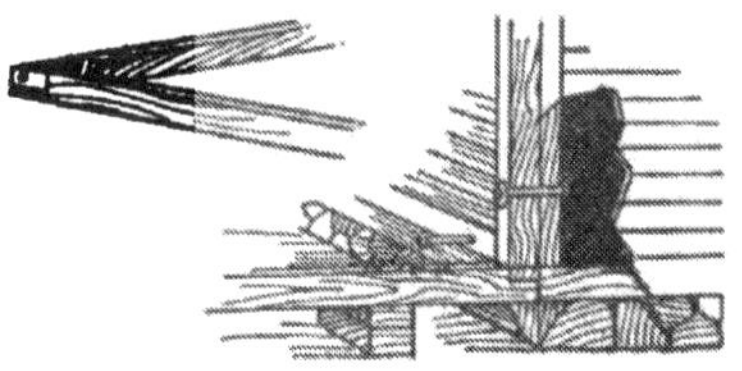

It often happens that one desires to attach a new building to an old one. If not properly done the addition is likely to pull away from the older structure. Take away the old boarding and bolt the new frame to the old, as shown in the cut. The two buildings can then never come apart. It is well also in framing corners to "dog" the two timbers together as suggested in the figure. The frame can then never open at the corners.

Protecting the Barn Door Frame

The hubs of the hay-rack wheels and those of other farm vehicles are constantly in danger of hitting against the side of the doorway when being driven in, unless great watchfulness is exercised. The driver on his big load of grain, hay or other fodder cannot always see the wheels. Have on either side a contrivance like that shown in the cut, then the hubs cannot hit the sides. It may be made of wood with straps of iron across the face, and movable if the doors swing on the outside.

489

Roof for Manure Pile

Many barns have manure heaps under the "tie up" windows, and no protection from the rains, or the eaves overhead. It is very easy to make a covering like the one shown in the cut. A simple framework is spiked to the side of the barn, and covered with battened boards. Many farms in New England have these manure protectors, and their fields show the effect of the manurial elements thus saved.

Improved Smoke-House

There are smoke-houses of several types for curing meats, but none so safe and satisfactory as the one in which no fire is ever put. It has a six-inch tile running from a fire pit in the earth three to eight feet from the house and a trifle lower.

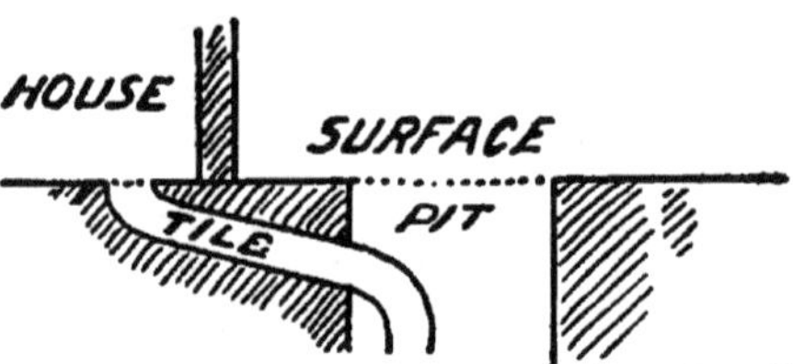

The smoke comes in at or near the bottom of the house and reaches the hams and bacon perfectly cool. Another advantage, the meat may be smoked without unlocking the smoke-house.

Walls of the Ice-House

Too many ice-houses are so roughly thrown together as to make them very unsightly. Besides, rough boarding often lets in much heat. An excellent plan that secures a cool house and an attractive appearance outside is shown in the cut. Inside the frame, plain rough boards are used, while clapboards are laid directly upon the studding on the outside. Building paper can first be put on both sides of the studding before the boards and clapboards are put on, if the tightest wall is desired.

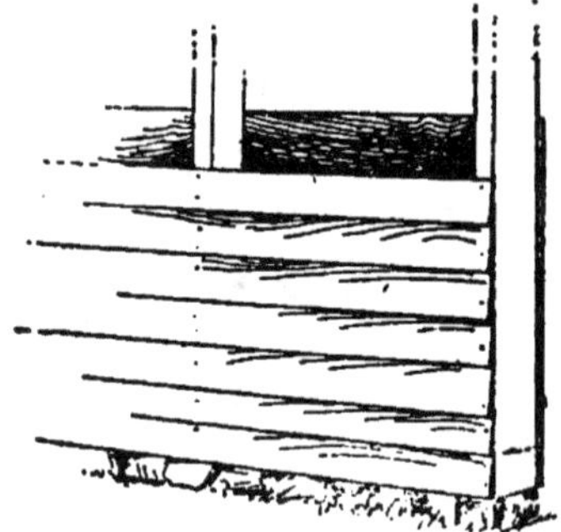

Hog Shed

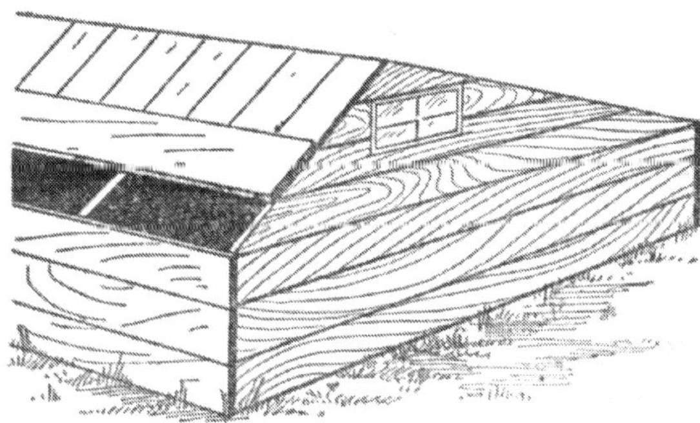

Hogs are not particular about fancy quarters if they be warm in winter and cool and out of the sun in summer. This hog shed is just the thing to build between two long strips of clover that may be fed by hurdling the hogs upon it. It is only three and a half feet high at the rear and four feet in front, and is roofed by tongue and groove boarding, up and down. This shed may be made as long as the number of compartments demand. It is entered by the stock at the rear through swing doors and the feeding is done in front. Here a long board is hinged to let down during very cold or snowy days, and in summer to protect from the sun.

Simple Well House

This is the simplest form of a well house that can be built, and provides for the trough inside the house, with two swinging doors to give access to it. Under such conditions the well will not freeze, while the water will be much more comfortably pumped on a cold, blustering day, than when one is standing on an open platform. The doors also afford protection to the stock from the wind when drinking. They are tightly closed when the stock has been watered.

Pump House

On some farms the pump and well stand out exposed to the heat of summer and to the freezing weather of winter. A well house is a great advantage at all seasons of the year, and can also be made an attractive feature. With it the water will keep cooler in the well in summer, and winter freezing will be prevented. The cut shows a good design with windows open in the summer. Sashes can be inserted in winter. Get a vine growing over such a house and it will look very attractive. Of course such a house is really necessary if you use an engine for pumping.

Glazing Hothouse Sash

Double pointed carpet tacks are better every way than the zinc points formerly used for glazing purposes. For the proper way to use them see the illustration given herewith.

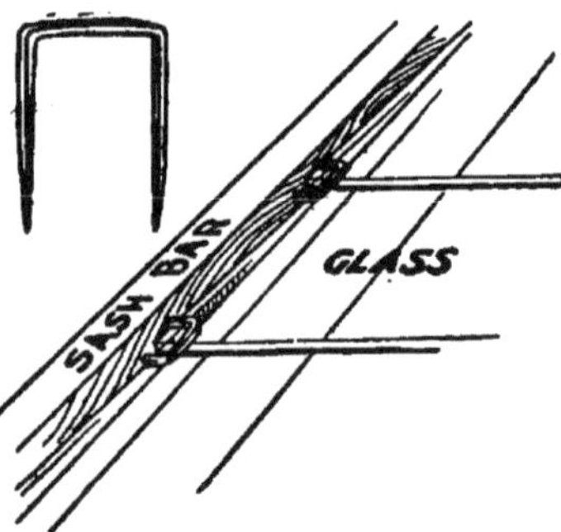

Glazing with Sheet Lead

The drawing shows a method of glazing by means of strips of sheet lead. The sash bar must not be grooved, but have a plain surface. The lead strip is tacked on and bent as seen in the cut on the left. After the glass is laid in position it is bent down as shown in the cut on the right.

Keeping Wind Out of the Foundation

Many farm buildings permit the wind to sweep under them because they have no tight foundation. Such a condition causes much suffering to the animals confined inside. A simple way to bank such a building is to lay down a strip of the stout, red building paper, that is now sold so cheaply, in the manner or put on laths along the upper edge, shown in the cut. Tack one upper edge and lay a narrow strip of board along the edge upon the ground.

Stone and Mortar Foundation

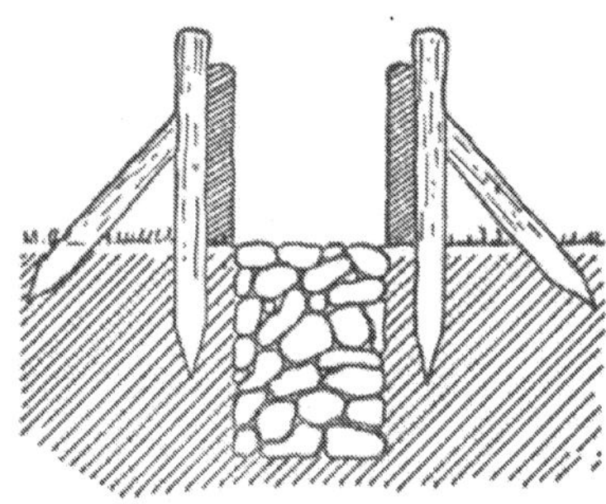

To build foundation walls if concrete is not available, dig a trench to the frost line. Fill with loose stones. Now set up a plank on each side and hold them in place by stakes, as shown in the cut. Fill in now to the top of the planks with loose stones and soft mortar—soft enough to fill all spaces between the stones. Allow planks to remain until the mortar has set, then move along and build another section. When the wall is hard lay a little soft mortar along the top and imbed the sill in it. The wall will then be airtight.

Foundation Lattice

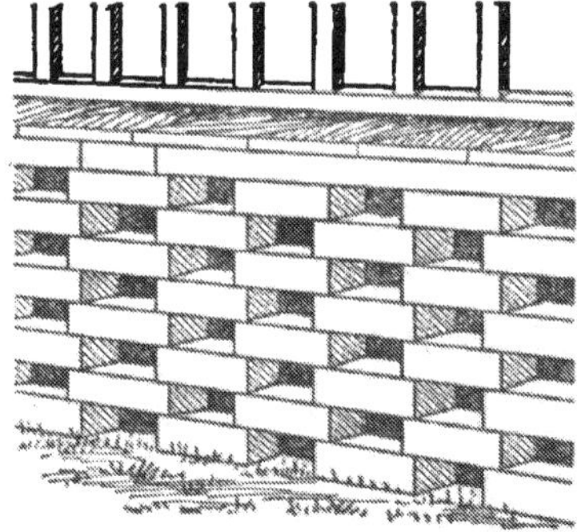

The cut shows a novel and desirable way to build a lattice work beneath piazzas and in the foundations of buildings, where ventilation beneath the buildings may be desired. The ordinary lattice work of crossed laths soon rots out and falls apart, while the brick-work lattice shown in the cut will practically last as long as the building lasts, and will always present a neat and tidy appearance.

Air-Tight Foundation

To keep the cold winds from getting under the sills of the poultry house, a 2 x 4 inch timber may be built solidly into the eight-inch foundation wall and the siding boards nailed directly to it. The cut illustrates the idea so clearly that no further explanation is needed. It may be well to incorporate this suggestion in the new house you are about to build.

Cow Stall Drain Construction

Have the cow platform made of joined planks. Let these project at the lower end over a trench eight inches deep, that is lined, as suggested in the cut, with an inch of cement, made of one part Portland cement and two of sharp sand. The cement should not come up more than two or three inches at the sides of the trench, being capped by an inch board above. It will thus be less likely to break on the sides. With this arrangement all the liquid manure can be saved.

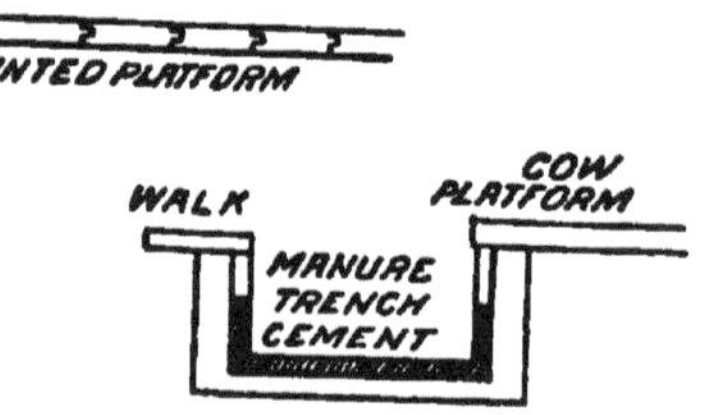

Drain for Horse Stalls

The cut shows a shallow, curved metal drain that can be placed behind a row of horse stalls and the liquid manure led away to a convenient manure heap. This shallow drain can be washed out with a pail of water at any time. A narrow, removable board can be fitted at the rear end of the platform so that no solid matter can fall into the drain. Do not make the openings between the ends of the planks large enough to let any solid matter through.

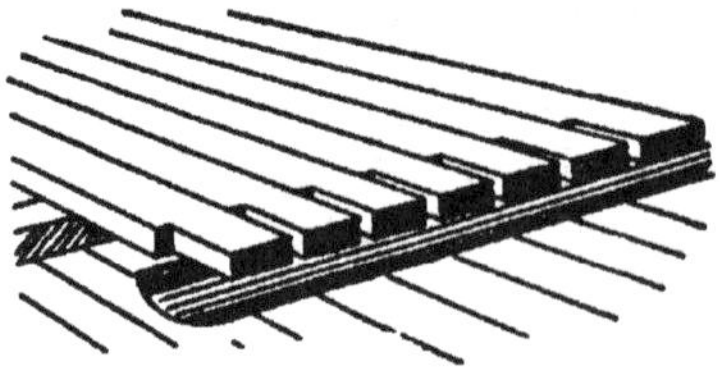

Paper Walls for Warmth

Almost every home has at hand the means of making farm buildings warm. Tack on coat after coat of old newspapers. Then board, shingle, or clap-board directly over them. Air simply cannot pass through successive layers of paper. If desired red building paper can be put over the newspapers in the manner shown in the cut, so if water gets through the boarding, it will be carried down to the ground by the resin sized paper.

Weather Strips for Farm Buildings

Here is a device for keeping a great amount of cold air from blowing in around the doors of farm buildings this winter. The cut is a cross-section of a door and the door posts. When the

door is shut, nail a strip of wood all around inside the edge of the door, close up to the door-frame, nailing the strips to the door. Then nail similar strips to the door-frame. These strips are represented in the cut by solid squares, and the plan will be readily seen. The device simply makes a double door-jamb and will make a building much warmer.

An Outdoor Coal-Bin

An outdoor coal-bin constructed after the fashion shown in the illustration, is a convenient receptacle near the kitchen door. A door made on the bin roof makes it easy to fill the bin. The slant floor throws the coal toward the front of the small chute from which the coal is shoveled. A box four feet square, four feet high in front and five feet in the rear, with the slant floor, will easily hold a ton of coal. The chute in front is one foot wide and four feet long, and about eight inches in front, running up in a slant.

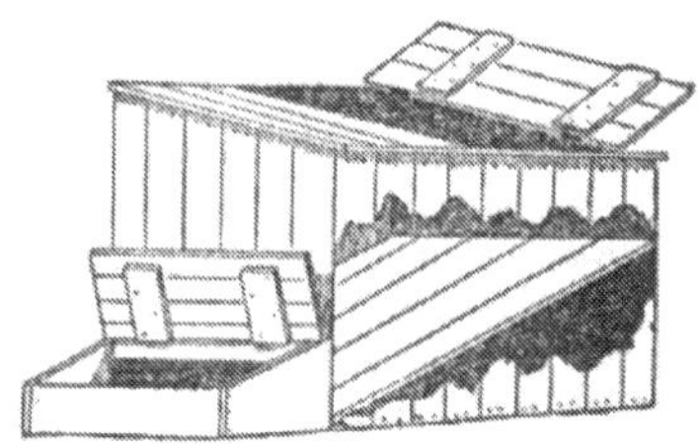

Roof on Hinges

A queer shed, isn't it? But a very useful one nevertheless. It is in daily use in New England. It is so near the big barn doors that they will not open unless its front is dropped down from the position it usually occupies, shown by the dotted lines. It is kept up by two large hooks that act like braces on the underside. During driving storms, also, it is convenient to lower the front to keep out rain and snow.

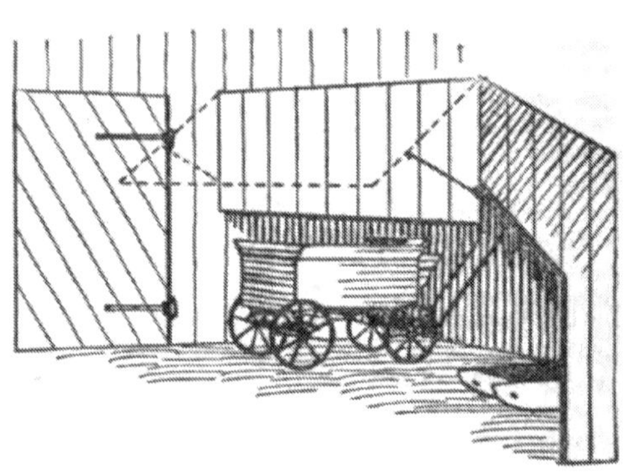

Keeping the Cellar Dry

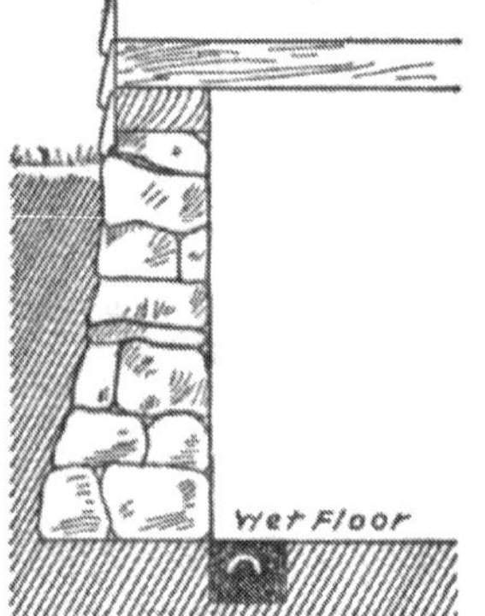

To insure a dry house cellar "cut a square-sided trench around the inside of the cellar wall and put a layer of small stones or gravel along the bottom. On this lay three-inch field tile, giving the proper slope all the way around to the outlet or cellar drain. Fill in the sides and top of the trench all about the tile with small stones or gravel, as suggested by the cross-section shown in cut, and all water that enters the cellar will be quickly carried off."

Arbor Seat

The cut does not give even a faint idea of the beauty of this arbor-seat, for when it is wholly covered by woodbine, or other vine, its appearance is very beautiful indeed. Anyone with a little gumption can build the framework and plant the woodbine. Nature will "do the rest." Such affairs about the place give it an air of refinement, and afford much comfort.

A Surface Drain in the Cellar

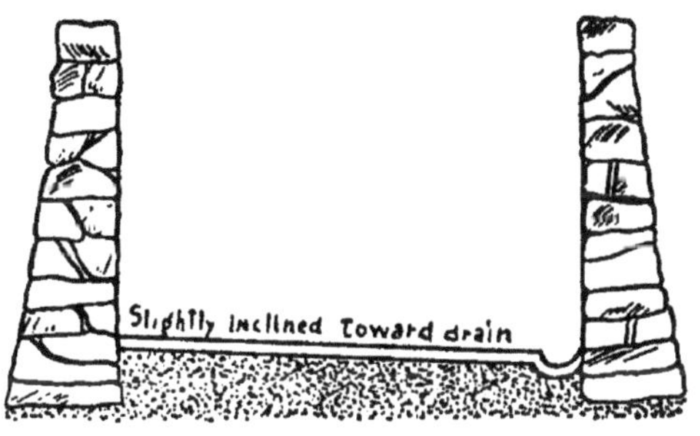

A damp cellar is an abomination and a menace to health. Cement it yourself; it need cost you only a few dollars for cement. Once experienced, you wouldn't part with this great comfort and convenience. Smooth the cellar floor, inclining it slightly toward one side and one end, if the cellar drain is at one corner. Along this side and end make a shallow rounded trench. Lay from an inch to an inch and a half of cement over the floor, making the open drain at side and end as shown in the cut. Any water that now gets into the cellar is at once carried by the open drain to the outlet drain, and there is no mud in the cellar.

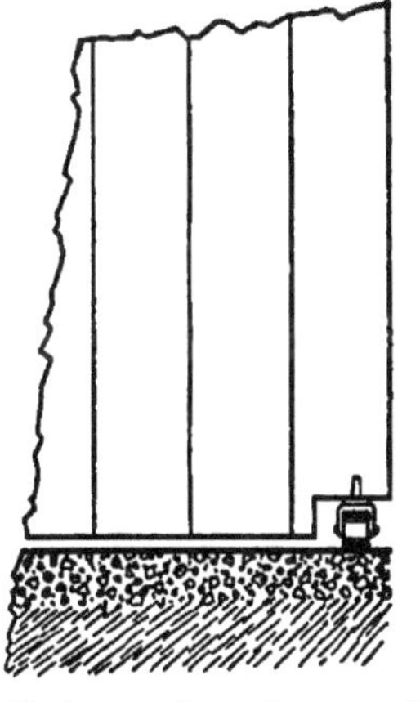

Caster for Large Swinging Doors

A sagging door is an abomination. And it seems almost impossible to keep heavy doors from sagging. I was visiting a farmer friend the other day who could say "Let 'em sag!" for he had placed casters on the doors of his new garage, as shown in the illustration. Of course this would not work on a common dirt floor, but he had a cement floor with a cement approach, and the plan worked like a charm. When you build your next farm structure put the idea into use.

Keeping the Door From Sagging

To keep a barn door or shutter from sagging is usually very difficult, but if the plan of bracing shown in the illustration is followed, and the boards are well nailed, doors or shutters of ordinary weight will not sag. The hinges, of course, should be placed on the left of the door as shown in the drawing.

497

Hinges That Prevent Sagging

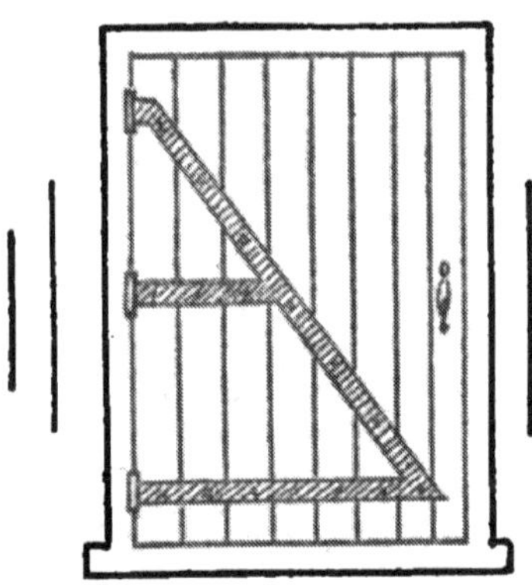

The hinges shown in the cut serve the double purpose of swinging the door and of holding it from sagging, the latter difficulty being completely overcome by this plan. Any blacksmith can make such a set of hinges from flat iron, and if they appear too prominent in the black state they can be painted the color of the doors.

Wind-Proof Door Joint

Either sliding or hinged double doors usually have a large crack between them. Screw an overlapping piece of board to the outside of one door and to the inside of the other. The doors will then shut with a wind-proof joint, as shown by the sectional views at the top of the cut.

Small Door in Big One

Needless effort is constantly being used in opening and shutting big barn doors for the entrance and exit of a single individual. Cut a small door in one of the big ones, as shown, and hinge it, putting a latch at the other edge. One can thus go in and out with small effort, and in winter when the big doors are drifted in, the small door, opening inwards, gives ready access to the barn without shoveling away the drifts.

A Screen-Door Storm-Door

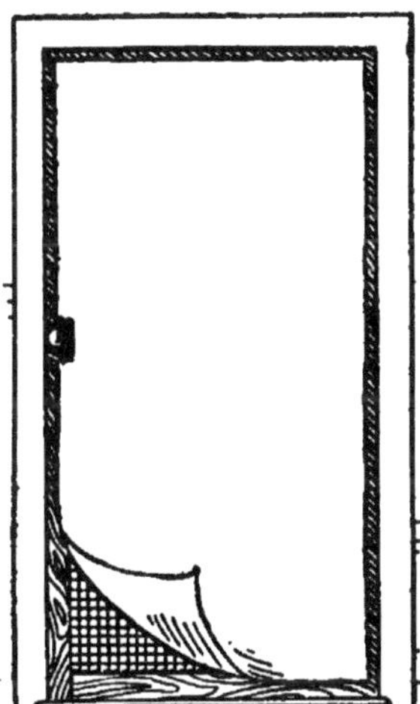

The winds will find their way into a house if there are any cracks about the doors. If you have no regular storm door, try the expedient shown in the cut. Put on the summer screen door and tack over the frame a strip of cotton cloth. When summer comes the cloth can be removed and the screen door is already in place, ready for fly time.

A Door-and a-Half

It often is the case that a door way must be closed against animals, but ventilation be still desired. The double door here shown will fill the bill. It is a half door hinged to the regular door as shown. When the half door is not desired it can be buttoned up against the full door and the whole used as one.

Glass in the Kitchen Door

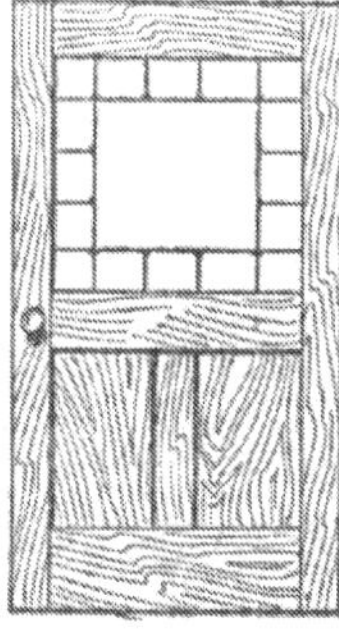

Fig. 1   Fig. 2

Many kitchens are dark and gloomy, particularly on stormy days. The outside door is usually made in the shape shown in Fig. 1. It is the work of but a few moments to remove the two upper panels, and the center rail. Have a window frame made to fit this opening, and the light and sunshine will make the dark kitchen bright and attractive, and much easier to work in. A pleasing sash is shown in Fig. 2. The center pane is clear, white glass, the small surrounding panes are of pale yellow glass, which even on a dark day gives the effect of sunshine.

Pivot Door

The cut shows the manner of making and operating a small door on pig and sheep pens or calf stables. It is pivoted on a bolt at the lower right hand corner. Gates for small stock may be made in fences on the same principle. Even large gates for horses and cattle may be operated by leaving off the top guard and letting the gate drop between post at each side, which will hold it securely.

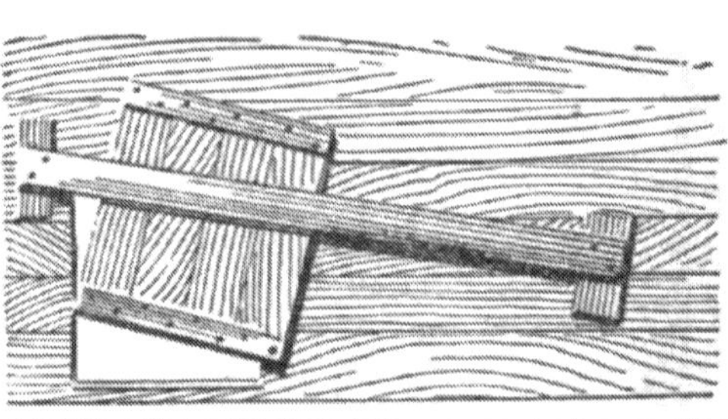

Saving Diagonal Bracing on the Door

The cut shows a pair of barn doors with the boards put on diagonally, which gives a much smaller chance for sagging than the up and down plan of nailing on the boards, as commonly seen. The boards could also be put on with the opposite diagonal slant; either way will give the bracing effect, since in the one case the long boards run from the top hinge to the opposite side at the bottom, and in the other case lift from the bottom hinge to the opposite upper corner.

Simple Self-Closing Latch

A wooden lock for a bull pen or box stall in the barn is very convenient. The one shown is quickly and easily made. Dotted lines show the position of the lock when door is being opened.

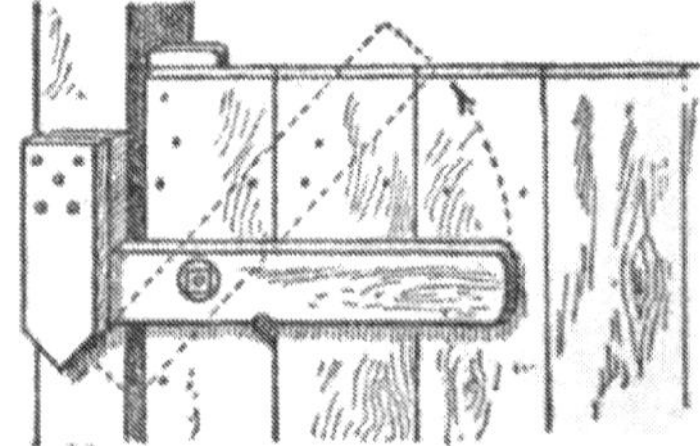

Dutch Double Door

Dutch doors are just the thing for box stalls and other places where ventilation and light are needed, but some restraint. In the doors figured, the upper part can be opened at pleasure, while the union of the edges of the two parts is made in such a way that when the two parts are shut together and the button turned, the two parts make one solid door and can be used as such. These are fine for the house as well as for the barn.

Double Latch

The cut shows a double latch for large doors that is both strong and easy to manipulate. It catches in grooved slots in the sill and in the beam above. The two pieces of the latch are bolted to the cleats of the door at top and bottom and at the center slide freely behind a flat iron strap, fastened to the middle cleat. The dotted lines show the latches drawn back to release the door. They may be held open or shut by a small iron or wooden pin, which should be attached to a cord hung on the door.

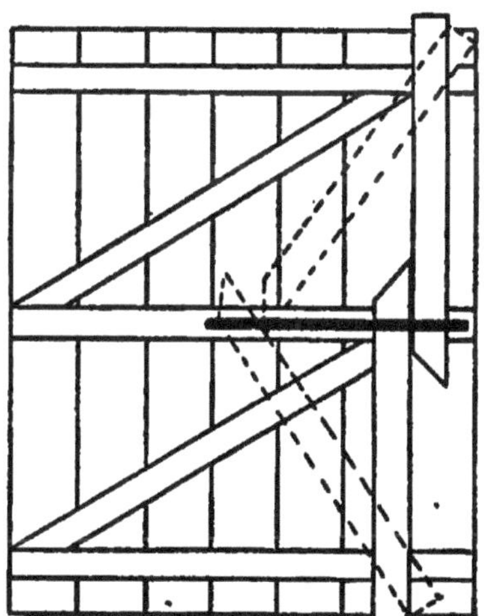

Barn-Door Fastenings

The old-fashioned swinging barn door still swings on many farms, and not infrequently is destitute of convenient and secure fastenings. Here are two good fastenings, and both are inexpensive and easily made. The first is made of a single piece of tough wood extending two inches above and below the door, the ends being let into grooves in the floor below and the cross beam above. It swings on a single strong bolt through its center. Dotted lines show how to release it. The second is not quite so simple, but is easily made and operated. As the cut shows, it is made of three pieces of hard wood bolted together where they touch each other. A third bolt through the lever between the two uprights, and through the door, enables the lever to move the two uprights in opposite directions whenever it is moved. The door is now fastened, but by raising lever to dotted lines the lower piece is lifted and the upper drawn down and the door released.

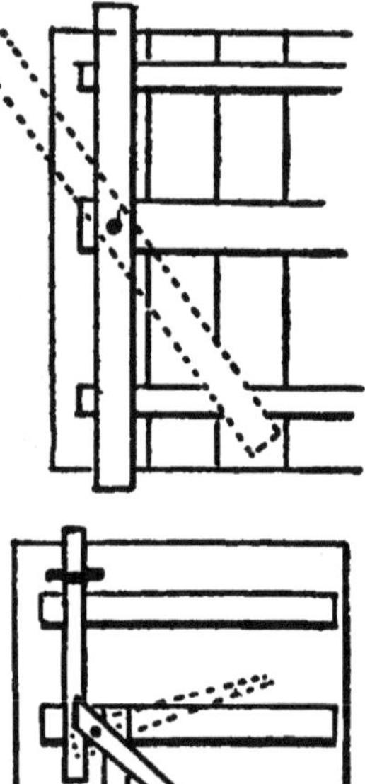

Bar for Open Door

The illustration shows how to arrange a drop-bar across a doorway

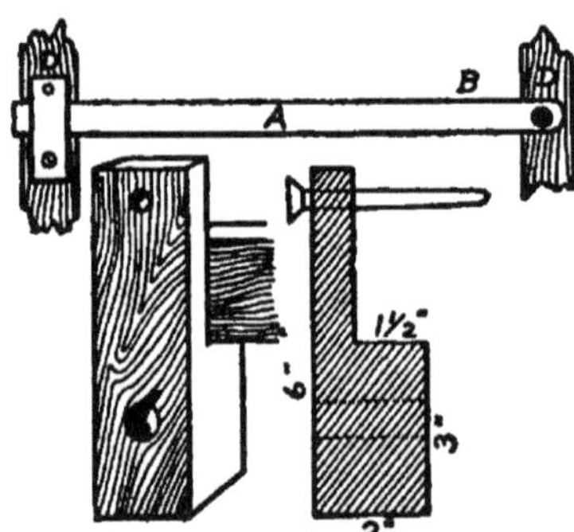

to keep stock in or out when door is open. The bar A is 2 x 1¼-inch material, fastened loosely at one end by a screw bolt, so it can be let down out of the way or raised and dropped into the block bolted to the opposite door post. This block and its construction is shown by the two small cuts below the bar. A hole is bored through upper part of block and into the post. A pin in this prevents stock from raising the bar.

Lock for Sliding Door

The drawing shows a lock that will be self-locking, and yet need no key to open, and at the same time that can not be opened by stock. Have a string go through the wall to raise the catch from the inside.

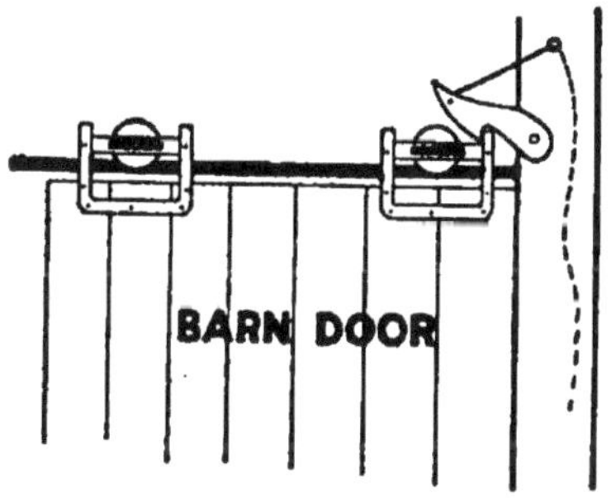

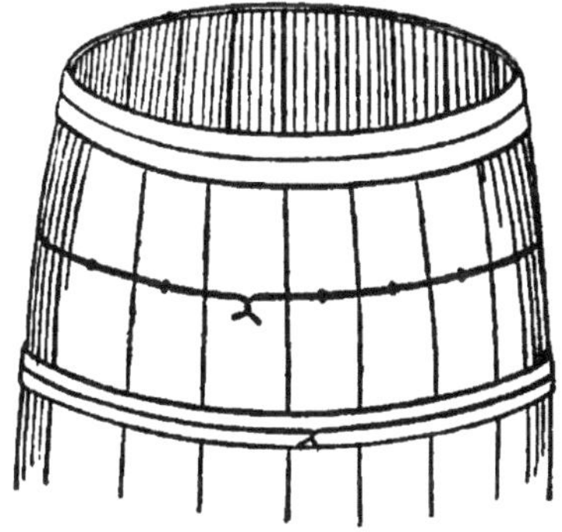

Extra Barrel Hoop

Hoops on apple, potato and other barrels will break or loosen. Take a a piece of No. 12 galvanized iron wire and make a wire hoop, twisting the wire tight with a pair of pincers. The wire may be kept in place with a few double-headed tacks. Where

a hoop is flat on the surface, it may be strengthened, as shown in the cut, by twisting a piece of this wire or even a piece considerably smaller, around the middle of it. A knowledge of this device may save much trouble.

Holder for Unweighted Windows

Take a stiff bit of hoop iron and break off a piece two inches long by half inch wide at one end, tapering to quarter inch at the other, thus making one end the heavier; bore a hole in the middle large enough to allow it to turn easily on the screw or nail used. With the heavier end up nail the article about three-quarters inch from the upper part of the sash. The corner (a) can be made to hold it up and (b) to hold it down, so it can't be opened.

Roof Scaffold for Shingling

Three boards are nailed together as shown in cut, and to these two strips are securely nailed. The upper ends of these strips are nailed to the boarding of the roof, and carried up as the work progresses. When patching a shingle roof two long strips could be used, the upper ends being nailed to the cap boards at the ridge. The "stage" could then be moved up on these strips, being securely nailed to the strips in each new situation.

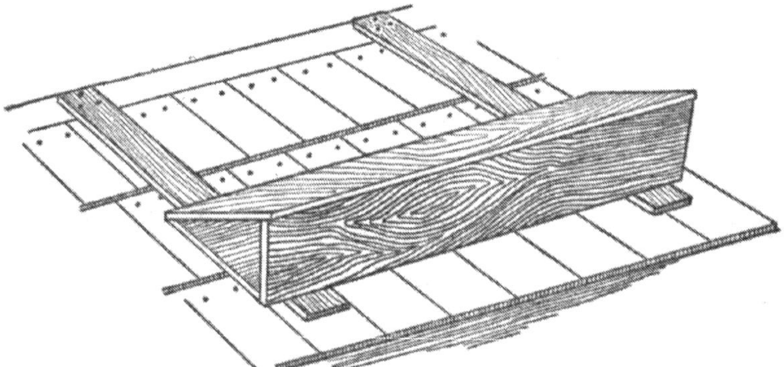

A Still Simpler Roof Scaffold

One of the most useful things ever invented is a roof scaffold, that once constructed serves indefinitely on any sloping roof. A triangular piece of one and a half inch plank is firmly nailed to each end and the middle of a rough board. Through each end of each piece of plank a four-inch lag screw is made to project half an inch or more. This is all. The scaffold is merely laid upon the roof and the men go upon it with impunity. The sharp points of the screws set down into the boards or shingles and never slip. They are never screwed down. A beginner in the use of these scaffolds will be a little suspicious of them the first hour, but soon feels as safe as upon the awkward, time-taking, nailed-down scaffold.

Wood Eaves Gutters

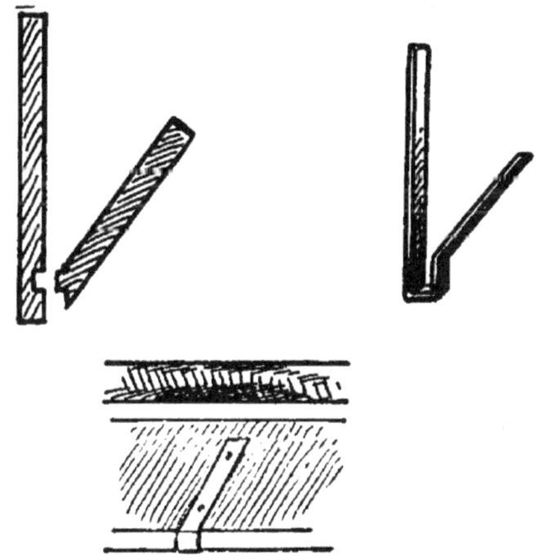

Farm buildings should be well supplied with eaves-troughs, or gutters, especially barns that have manure thrown out in heaps at one side of them. When you get tired of replacing rusted-out iron gutters, try this wooden one. The cuts show the very best way to make a simple eaves-trough. Two tongued and grooved boards are put together when the joint has been covered with white lead. It is then nailed, and an iron clamp (which any blacksmith can make in a few moments), is put on every eight feet. Such a gutter will "stay put."

Pen Partition

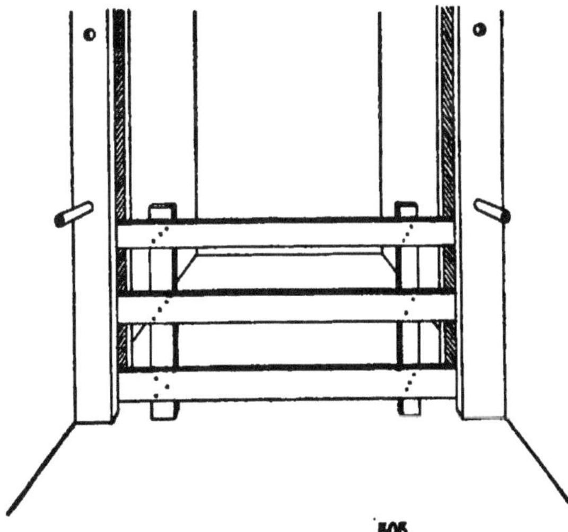

It is often desirable to separate the flock, but this is not always feasible where room is scarce. In such a case a part of the general pen can be cut off at will by the arrangement here figured. The panel of the fence slides up and down between two upright strips at either end, with pins and holes to keep the fence up out of the way, or to let it down to the bottom of the pen when desired.

Combination Tool-Box Saw-Horse

The cut shows a "horse" such as carpenters use, but modified so as to afford a tool box at the top, with nails, screws, etc., and a larger box below, where saws, planes, and other large tools may be placed. Such a device as this is very convenient to pick up and take to any farm building where a small job is to be done. The top is almost a work bench, and all the necessary tools are at hand. It is made entirely of short strips of board, and can be put together with very little labor, after the manner shown.

Homemade Warming Oven

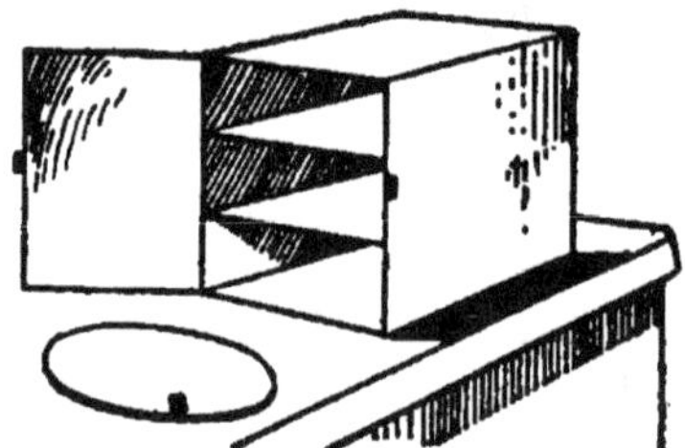

If one will try to procure a square cracker tin from the grocer he can provide a very excellent warming oven in the manner shown in the cut. A button of iron is riveted to each corner to raise the oven a bit above the stove top. Two sheet-iron slides are fitted to the interior so that dishes can be inserted at once, and the work is done.

A Roof of Boards

A roof of boards with battens over the cracks is the cheapest form of roof. But made in the ordinary way the rain drives under the battens and enters the building. The cut shows a simple way to obviate the difficulty. Each roof board is grooved at each edge (as shown in the cross-section diagram) with a tongue-and-groove plane before being nailed to the roof. The batten is then put on. The rain may drive under the edge as before, but falls into the groove and runs down the roof.

Truss for Roof

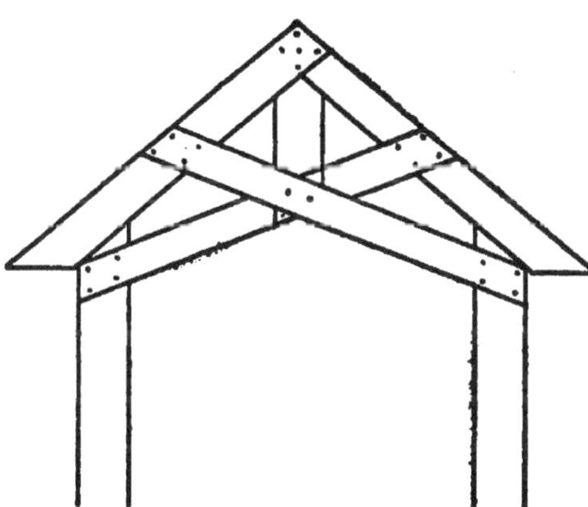

Sometimes we are puzzled to know how to do a piece of carpenter work without a carpenter. Here is the way to truss a roof that seems weak. It is thorough and permanent. Such boards are used as shown, being nailed solidly in place and where they cross. They should not be less than eight inches wide.

Ventilating Flue for Forge

Many farm repair shops are equipped with portable forges, and the smoke and gasses are very often found to be a nuisance. To remedy this, get an old dishpan, the larger the better. Cut a six-inch round hole in the center of the bottom. Then take a joint of six-inch stovepipe and trim the lower end so that three flanges may be bent as in Fig. 1. Punch a rivet hole in each flange and insert pipe through hole in dishpan. Rivet flanges to bottom of pan. Connect this joint of pipe with the other necessary joints and your smokehood is complete. Fig. 2. Suspend hood eighteen or twenty inches above the forge.

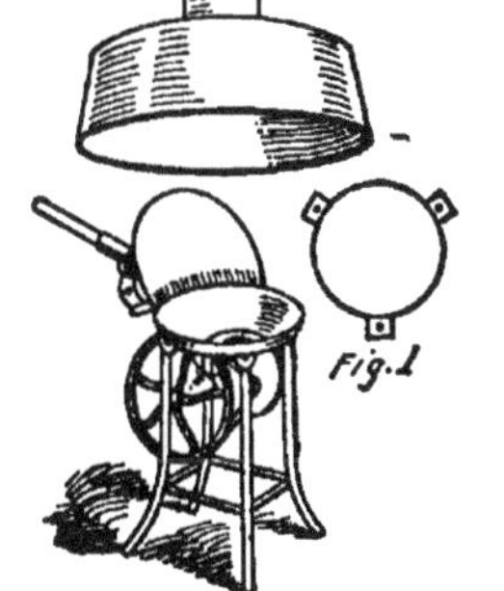

An Emergency Wrench

The time a person needs a wrench the most is when he does not have one. A device which can be used for a wrench is shown in the illustration. Take a bolt and two burrs, and adjust the distance between the two burrs so that they hold the burr which is to be turned. Use the bolt as a handle.

Home-made Lock Washer

I frequently have trouble in keeping unslotted nuts tight and in place. A lost nut is no good. The accompanying sketch shows a simple and effective way of keeping nuts in place. A piece of stiff tin is cut in the shape shown at the left, with a central hole the same size as that of the bolt. This is placed over the bolt before the nut is screwed on, and when the nut is tight the prongs of the tin are bent in 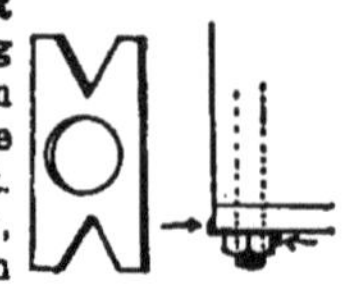 the shape shown at the right. The nut will usually hold if all the prongs are bent up, but when two are bent up and two down, they are sure to hold.

Adjustable Wooden Trestles

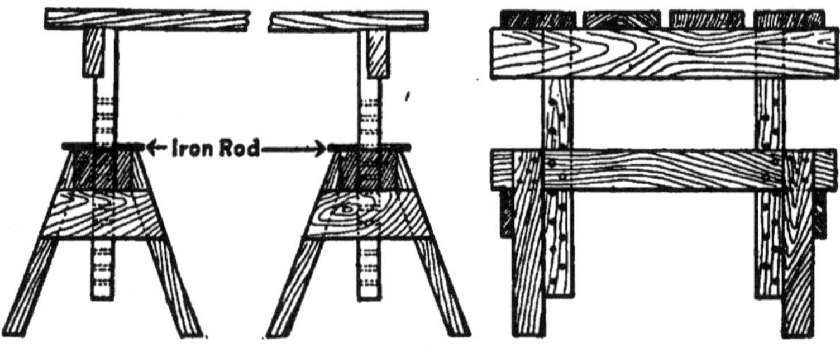

One of the biggest drawbacks to ordinary trestles is that they can be arranged for only one height. The illustrations show how to make some adjustable trestles which will carry a platform at various heights. A bolt or an iron rod is pushed through a hole in each of the sliding pieces of 2 x 4. The bolts hold the extension part firmly. Trestles of this kind are very handy in the house when painting the ceiling, when hanging paper, or when doing any one of a number of odd jobs which require something to stand on.

Pieces of 2x4 are used for legs, and for the top of the lower part of each trestle. The crosspieces on the extension parts, which bear the heavy planks, are of 2 x 6. The planks used for placing on the trestles should be at least two inches thick, to hold the person working on them. They do not need to be nailed in place.

Work Bench

This work bench is of the ordinary sort, so far as the top is concerned, but it has the whole front furnished with shelves, upon which the tools are placed, the glue-pot, the sandpaper, the nails and screws, and the thousand and one things needed in repairing, and so convenient to have right there at hand. Every farm ought to have a carpenter's bench, and such a device as this should go with it.

Another Work Bench

Four strips of board, or better, two of board and two of plank for the top, with two empty grocery boxes will quickly make a work bench that will have enough compartments for keeping all kinds of tools. The board shelf is for the long and short "planes," the small box for the small tools and the high box for hanging up saws. etc.

Leveling With a Square

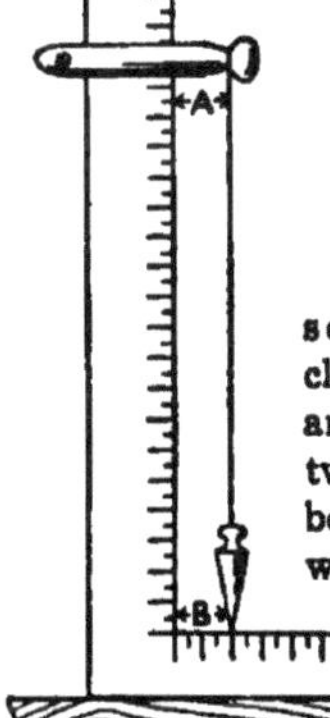

To level with a square, fasten a clamp to the vertical arm and attach a plumb-bob to the clamp. When the distance between the string and the vertical arm at B is equal to the distance between the string and the vertical arm at A, the surface upon which the lower arm of the square rests is level. This bit of knowledge may help sometime when the level is broken. Tuck it away where you will not lose it.

Metal Punch for Shop Work

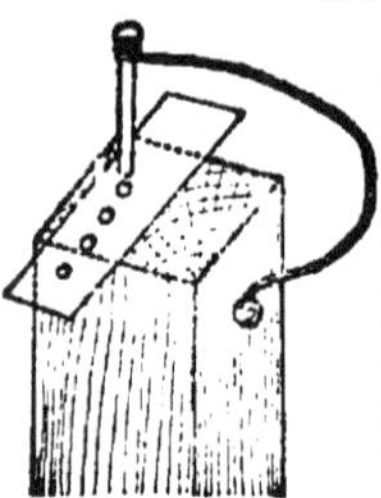

A good sheet-metal punch for ordinary shop work may be made as shown in the sketch. A bolt of the desired diameter with a tempered punch point is suspended over a hardwood block by a stiff piece of spring wire. A hole is bored in the block to register with the bolt and act as a die. When a piece of sheet metal is placed under the bolt and the bolt hit with a hammer, a clean-cut hole will be made. If the top of the block is covered with a plate of iron having a hole to correspond with the bolt, better and cleaner holes will be punched.

Fastening Bolts in Concrete

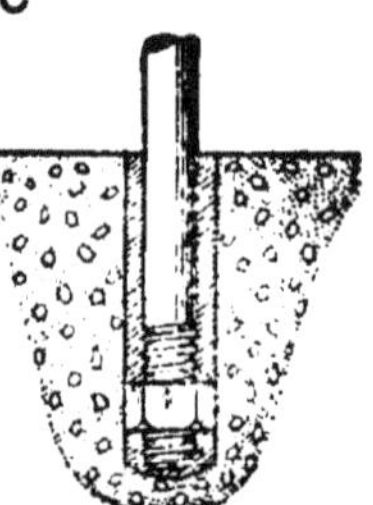

It is easy to fasten machinery to concrete floors if this plan is used: With a cold chisel make a hole in the concrete the right size to receive the nut of the bolt to be used. If the hole is a little larger at the bottom, so much the better. Put the bolt with nut in place into the hole, as shown, and fill the space with melted babbitt. Wrap the parts of the bolt exposed to the babbitt with thin oiled paper, and when the babbitt has hardened remove the bolt. When the machinery is removed the holes can be corked, leaving no projections on the floor.

Hanging Board for Tools

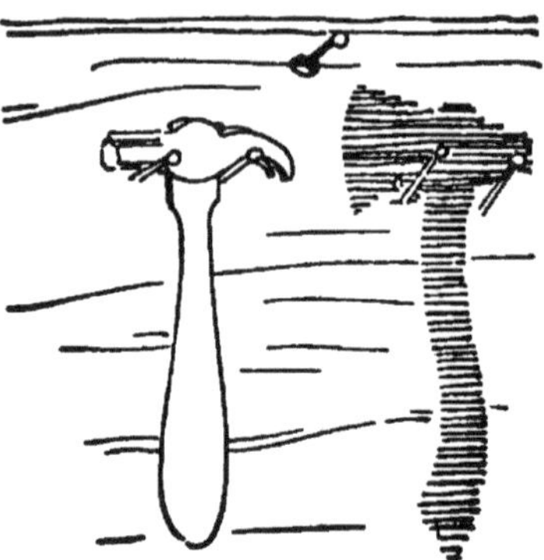

Tools are more easily found and more likely to be kept in place if they are hung up than when all are thrown together in chest or drawer. Have a large board with a hole in the top to hang it up by. Hold up each tool against it and drive in stout nails in the place most convenient to hang it up. With hatchet or hammer this would mean two nails under the head with the handle allowed to drop between them. Then draw or paint upon the board the exact outline of each tool in its place. You will then know just where to hang it, and if one is missing you can ask with assurance, "Who has taken my hammer?" or, "Where is my gimlet?"

Simple Holder for Bits

Bits lying around anywhere are apt to get dull and spoiled. Take a piece of wood an inch and one-half wide and an inch thick, bore holes through it as shown in the accompanying cut, and nail it up over the work bench. Put the bits stem end down into these holes and keep them there when not in use, and save your bits and money.

Bit and Drill Holder

Take a smooth three-quarter-inch board and bore a hole through it with each bit in your workshop. Hang it up near your brace and bits; then whenever you want to find the proper bit for a bolt, just see what hole the bolt fits and you'll know what size bit to use. It not only saves time, but gives the correct bit with absolute certainty.

Removing Broken Screws

Many times screws are inserted too tightly and the heads break off. The methods shown above for removing such screws are very practical. If the screw is not too small a hole can be drilled in the broken part as shown in Fig. 1. The square end of a file is then inserted as shown in Fig. 2, and given a few sharp blows with a hammer. By turning the file the screw may be taken out. Another method, which can be used best with large screws only, is to drill small holes, as shown in Fig. 3, and chisel a slot or groove, as shown in Fig. 4 and Fig. 5. Then a screw driver can be used to remove the broken part.

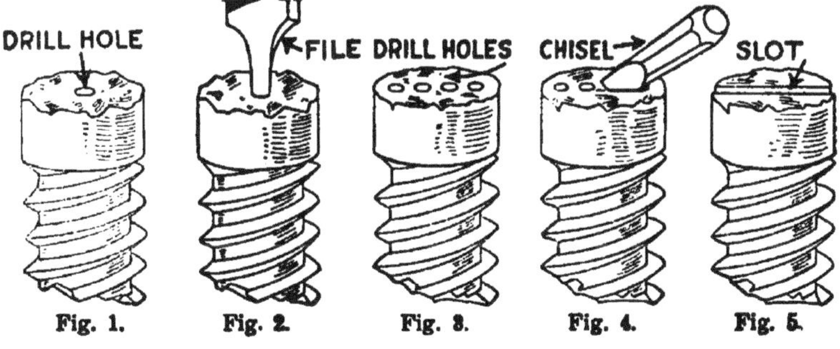

How to Hold Nuts in Place on Machinery

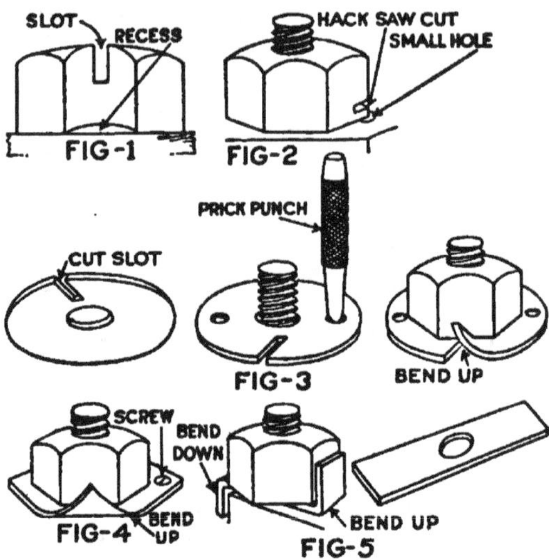

When machinery has been used a short time the paint wears off around the nuts, the nuts begin to loosen and will in time drop off unless there is some device to hold them tight. The loss of one nut may hold up work in the field for several hours or a longer time. If you have ever been delayed that way, you will welcome these ways to make the nuts hold.

The plan shown in Fig. 1 is easy to use. The nut is slotted with a hack-saw for about half its depth, and the under side of the nut is recessed parallel to the slot by filling. The tighter the nut is turned the more it will bind. In Fig. 2 a small slot is cut in the side of the nut, as indicated, near the bottom. This forms a lip. A small hole is drilled in the part of the machine where the corner of the nut comes, and when the nut is tightened this lip is bent down into the hole, thus locking the nut.

In Fig. 2 a common washer is used, a slot being cut part way into one side. A hole. say one-quarter inch, is drilled

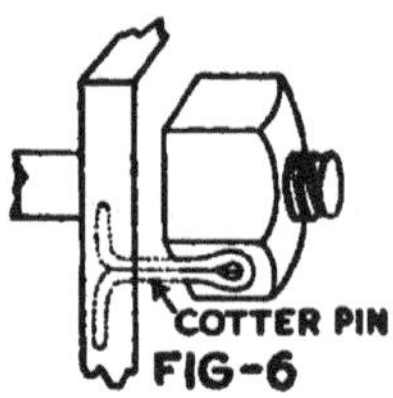

near the outer edge, and opposite from it an impression is made with a sharp prick-punch. This throws up a burr on the under side of the washer. The washer is put on and with a prick-punch held in the hole, an impression is made in the machine by tapping the punch with a hammer. Then the washer is given a half turn so that the cone or burr will set into this, the nut is tightened and the edge of the washer at the slot is turned up.

Fig. 4 shows a piece of sheet iron held in place on a wooden surface by means of a screw. Fig. 5 shows another way of using a strip of strap iron beneath the nut. Fig. 6 shows a cotter-pin used for holding a nut.

Belt Tightener

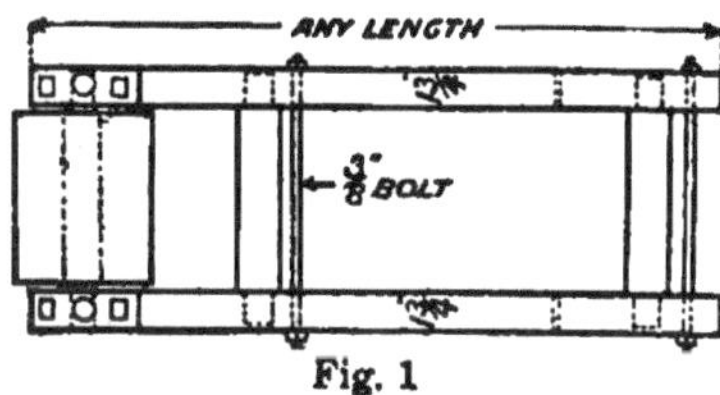

Fig. 1

Do your belts slip? A simple belt tightener is shown which can be made in your shop and fastened to the ceiling, floor or post. Pieces of hardwood one and three-quarter inches thick are used for the frame, a surface view of which is shown in F g. 1. The length will vary, depending on where the tightener is attached. The width varies with the width of the belt. The bolts holding the frame together are three-eighths of an inch in diameter. Here is a side view of the tightener fastened to the ceiling with lag-screws. The weight of the idler pulley makes the belt tight. When not in use the tightener can be hooked up to the ceiling by means of a small rope. The frame to which the tightener is fastened is braced with one and one quarter inch scrap iron.

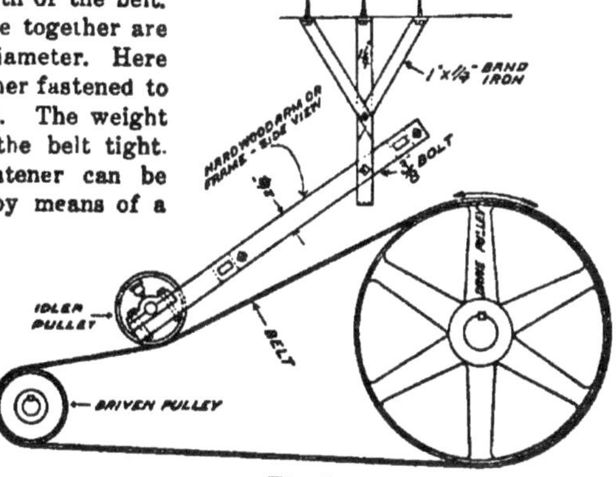

Fig. 2

Workshop Supply Cabinet

On every farm there is (or should be) a good assortment of bolts, washers, nuts, rivets, chain links, mowing-machine blades, nails, clevises, single-tree iron, screws, and various farm machine repairs which get hopelessly mixed unless some plan is adopted to keep them properly assorted. For this purpose, a cabinet like the one shown in the illustration is indispensable.

This cabinet can be made of almost any kind of rough wood, if carefully dressed. and drawers fitted into the upper part may be filled with the smaller farm supplies, while lower shelves will hold the heavier articles.

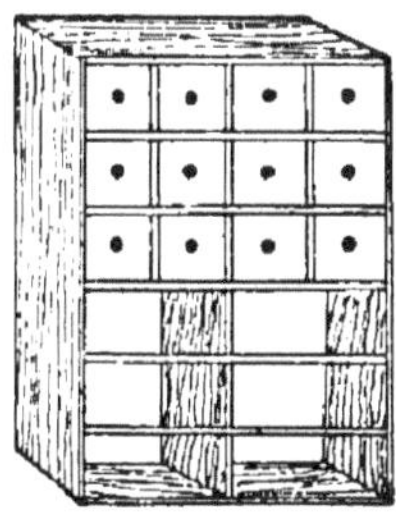

513

Chair for Kitchen Work

The ordinary wooden kitchen chair can be made very much more convenient by the additions shown in the cut. Make the folding arms as wide as you please, and when you sit down to prepare fruit or vegetables for cooking, with a pan on either arm and one in your lap, you will appreciate the convenience. That husband or boy who is handy with tools ought to be able to fix up a chair from the plain showing made by the cut. The brackets to hold up the shelves are hinged to the upright and fold under the arm when not in use.

Kitchen Cabinet

Make a kitchen cabinet by hanging a box over a table, as shown in the drawing. Add shelves and hinged doors, and put a row of cup hooks along one side. Paint the woodwork and cover the top of the table with zinc. It will then be ready for the labeled jars, boxes and bottles containing cereals, spices and extracts, while spoons, forks and other small articles are hung on the hooks. The bottom of the larger space is used for cake and pie pans.

Kitchen Table From Box

Note how easily and quickly this kitchen table can be made by any one who can saw a board in two. Two boxes (shredded wheat boxes are about the right shape) are placed, one upon the other, against the side wall. A projecting board top is nailed on, a curtain hung before the opening, and the table is complete, with splendid closet room below. If desired, the ends of the boxes can be hung, loosely, with the same material used for the curtain. Tack the edges.

Wood Box and Carrier

This is a wood carrier and a wood box combined, and can be made by any one handy with tools. Select handsomely grained wood, either oak, ash or some other that is at hand, and fashion in the form shown. The decoration in this case is etched on the side with a red-hot iron point, but this can be omitted. The handle may be of ash, as this is most easily bent. This device can be filled with wood in the shed, carried to the sitting room and left beside the stove until the wood is consumed. Such a carrier saves dirt and one's clothes.

Another Wood-Box and Carrier

A great convenience is a combined wood-box and carrier, such as is shown in the cut. Any ingenious boy can make one from green sticks with the bark left on, by boring holes for the insertion of ends where practicable, and elsewhere using round-headed screws. Such a carrier can be filled with wood at the pile, carried to the stove, and allowed to set beside it until the wood is used.

A Low Trunk

A low trunk to slide under the bed, out of the way, and to hold a man's clothing, is now to be had. Here in the cut is shown an improved form of a bed trunk, which can hold a man's possessions of the clothing sort, or can be used as a partnership affair, one side for his coats, waistcoats and trousers, and the other for madame's dresses. If the trunk is made long enough, the dresses can be laid in without folding or wrinkling—a great advantage. The form shown in the illustration is intended to be nearly as long as the bed, and nearly as wide, with handles and castors so it can be pulled out on either side, the covers opening on each side half way in for this purpose. Where closet room is limited, such a bed trunk will be found a very great convenience.

High Chair for Baby

The chair for the baby need not be made as fancy as the one shown, but use the ideas embodied in the cut. The table can be raised and lowered into place again when the baby is in the seat. Now he can not fall out. Under the chair are a shelf and a closet for playthings which the baby can use when placed upon the floor. Such a chair in a plain form is not beyond the home workshop, while it would be far better to hire a carpenter to make the chair than go without such a convenience.

Making Window Screens

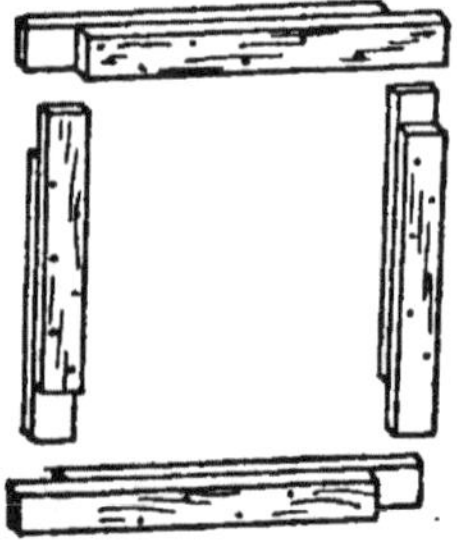

Farm homes, as a rule, are insufficiently supplied with window screens, possibly because it seems a difficult matter to make them, especially the framing of the corners. A way is shown in the cut that does away with all mortise and tenon joints, so that the frame can be put together in a few moments. Get a half-inch board and cut into two-inch strips. Cut strips into lengths just two inches shorter than needed for a frame, then place in pairs as shown in the drawing—one piece in each pair slipped two inches ahead of the other. Stretch the wire screening or netting between the two sets of strips, and tack fast to one set. Now brad them together and clinch the brads. Your screen is now finished and looks like above drawing.

Home-Made Bookcase

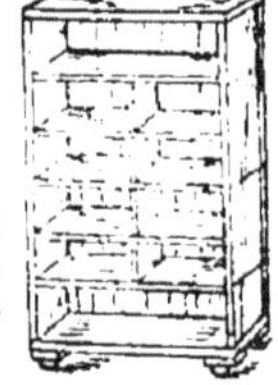

A bookcase that will meet ordinary requirements is shown in the drawing. It may be made as large as desired by using empty boxes of various sizes. The top and bottom boxes should be of the same length, the others of such dimensions as not to project beyond them. Plenty of screws should be used between partitions to hold the boxes together, and braces should be screwed to the back. Nail a block of wood underneath each corner and attach domes of silence to these, so that the bookcase may be easily and noiselessly moved.

Handy Devices   **House Furnishings**

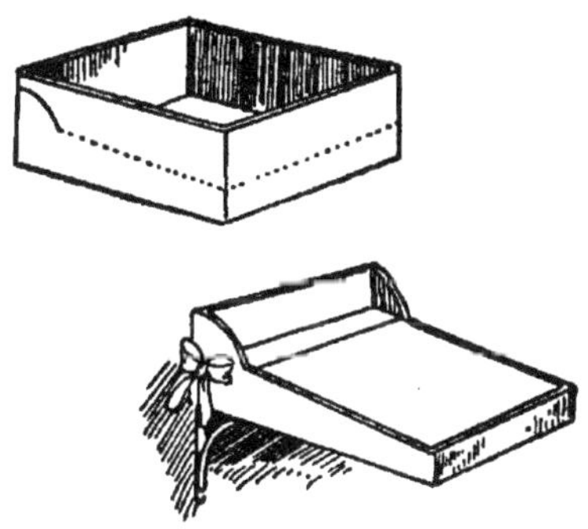

Making a Wall Desk

The first cut shows an empty grocery box, with dotted lines that show how it is to be cut down. A cover is now to be put on with hinges, and the whole covered tightly and smoothly with turkey red or other pretty cloth. Gimp tacks or round-headed brass ones, can be used to secure the edges. A bow of red ribbon at one side will add to the effect. A pair of stout bronze brackets to support the desk against the wall completes the affair. Here's a chance for a boy to get a pretty desk by making it himself.

Leather Magazine Cover

The cut shows a leather magazine cover that will be very serviceable and attractive as a gift. Get a shoemaker to put two eyelet holes, rimmed, at each end of the back, and press a crease at either edge of the back with the back of a dinner knife. The strings pass between leaves in two parts of the magazine and tie at the back, as shown. A monogram can be stamped upon one side by laying the leather on a flat board and forming the letters with a mallet and bit of hard wood, cut to a blunt edge, like the end of a screw-driver.

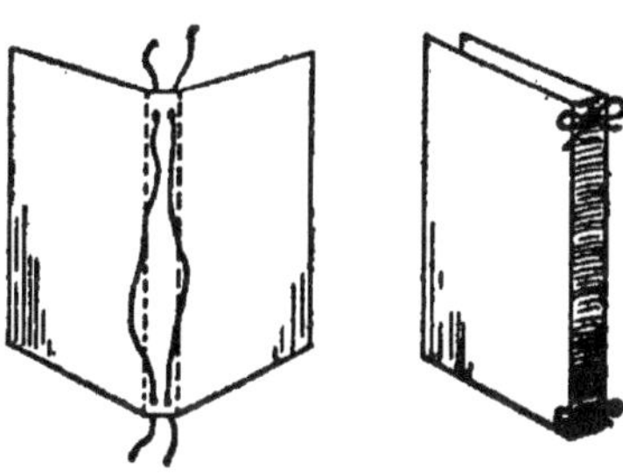

Handy Closet Device

Four wooden poles are secured in place near the top of the closet, as shown. On these, coat hangers are suspended for coats that would be stretched out of shape if hung on hooks in the usual way. When placed in position on the hanger, the latter is shoved along to the far side of the closet out of the way, yet each suit or coat is instantly available, and does not have to be looked for under other clothing.

Home-Made Sewing Table

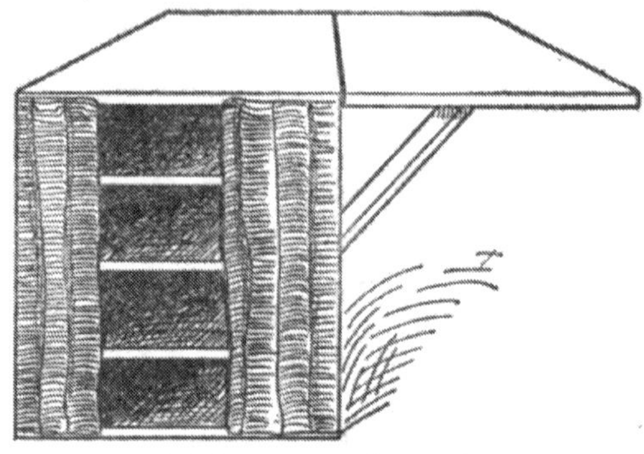

A home-made work-table was devised from a dry-goods box cut down to a suitable size. Shelves were inserted, and an extension leaf was hinged on to the top, thirty inches long and as wide as the top of the table. This was fitted with a brace which supported it when in use, the brace fitting into a slot near the base of the table, as shown in the cut. It was painted, a curtain was added in front, and a yard tape measure tacked with brass-headed tacks along the front edge. The cost was trifling; the result a very handy adjunct of the sewing room.

Cocoanut Twine Holder

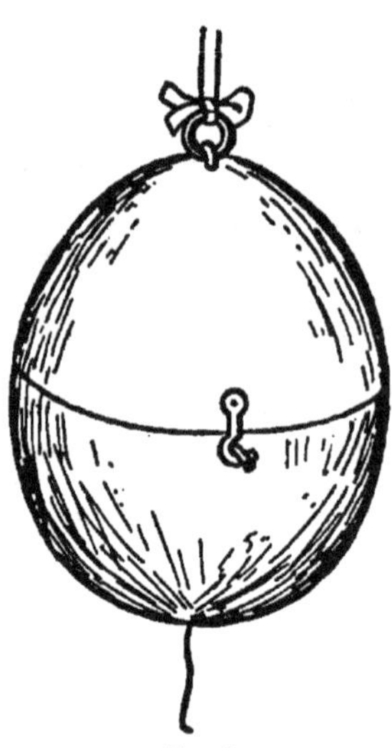

Here is a twine holder almost made to order by Nature herself. A cocoanut is evenly sawed through the middle. The inside is removed and the two parts are then hinged together by two sets of holes, and two bits of wire wound through them, as shown in Fig. 1. A little catch from an old box is placed on the opposite side. (One could be made from a bit of firm wire bent into the proper shape, together with two gimp tacks.) Then with a hole through the bottom for the twine to pass, and a ring fastened to the top by a bit of wire passing through the shell, the holder is complete.

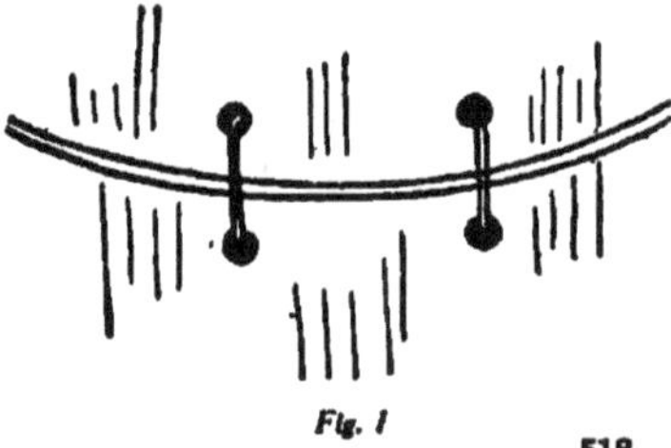

Fig. 1

Fig. 2

Drawer for China Closet

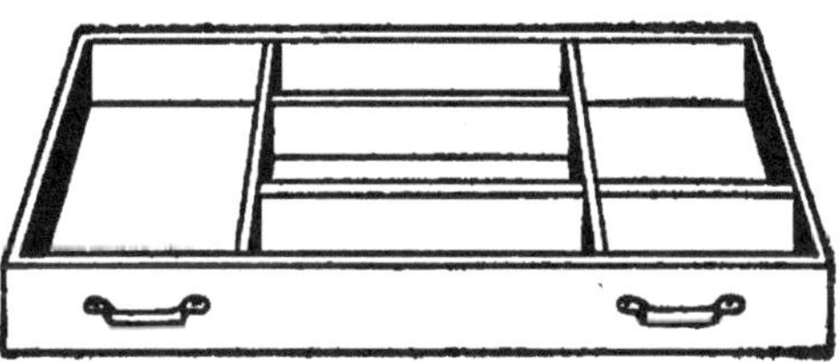

No matter if the china closet has no drawers at present, one like that shown in the cut can be made and fitted to slide in under one of the shelves, where it will be found the most convenient part of the whole closet. It is fitted with compartments for keeping silver knives and forks by themselves, teaspoons and tablespoons by themselves, with compartments provided for clean napkins and for those that are in use at the table and removed three times daily with the napkin rings upon them. The compartments for silver may well be lined with soft cloth, canton flannel or plush being useful for this purpose. Where knives, forks and spoons are heaped together promiscuously, there is much scratching of the bright surfaces that causes a quick loss of beauty.

Home-Made Bracket

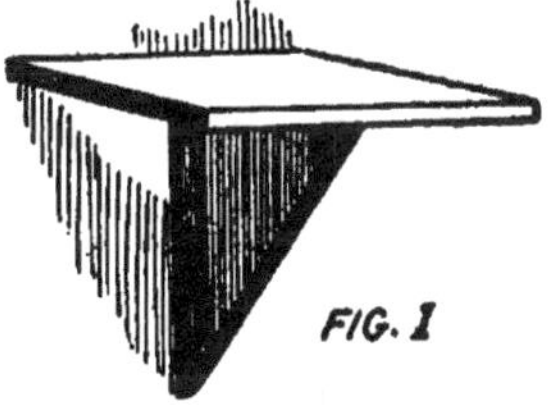
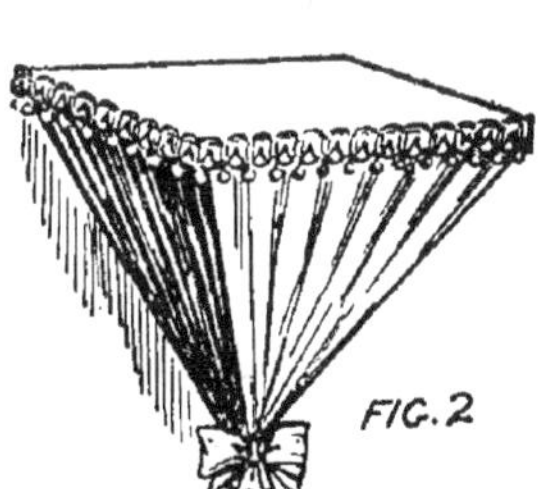

Figs. 1 and 2 show how a very pretty home-made bracket can be constructed. Two simple pieces of board are nailed together, and to the wall, as shown in Fig. 1. This is covered with a bit of soft remnant of silk, with a silk fringe finish about the edge, and a bow of harmoniously colored ribbon at the bottom. A handsome lamp, or vase, on such a bracket will add much to the furnishing of a room.

Umbrella Rack

In a house where space is at a premium a good place for an umbrella rack is on the inside of the closet door fastened across the bottom. It occupies space not otherwise used and is very convenient.

Convenient Chest Seat

Here is a grocery or dry goods box doing good service. A piece of the side is cut out and a drawer is made, as shown. A bottom for the upper part is put in where the dotted line shows. The cover is hinged and padded on top, and the whole is covered neatly with cloth. Then it is put on the veranda for a seat and for a place to keep the tennis or croquet set, or the boy's bat and ball, while in one end of the drawer (there can be a partition through the middle) can be kept the rubbers of all the family, ready to put on when going out and to receive when coming in, with no tracking of mud through the house.

Home-Made Vapor Cabinet

A circular cabinet bath that answers the purpose of a vapor bath can be made by taking three and a half yards of extra width floor oilcloth, cut exactly to the shape of a half disc, or circle, with half a circle cut in the middle diameter edge for the head. Fold like a cone, weight the edges down to the floor with a weight tied to. A common flat-iron will do.

Shelves for Children's Toys

To hold children's toys a set of shelves made of three horizontal boards nailed to two upright ones and braced by two strong sticks at the back is shown in the drawing. It is made three feet long, two feet high and one foot deep. You may paint it any color desired, and a neat curtain attached to the front will hide the toys from view.

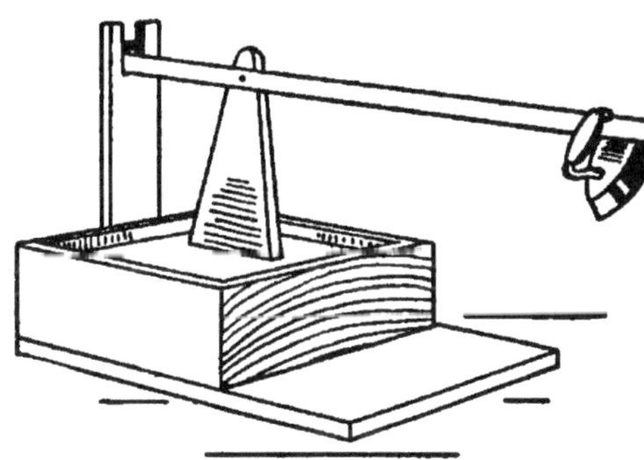

Home Meat Press

Housewives press cold corn beef, chicken, lamb, etc., in a round tin dish, with a plate over the meat. A better way is to make a square box of hard wood and fit it with the appurtenances shown in the cut. Then the meat can be pressed into a square loaf, where all the slices will be attractive. The "men folks" will surely be able to get up such a contrivance as this and present it to the "women folks."

Wrench for Fruit Cans and Jars

A handy home-made wrench for opening syrup cans, oil cans and small boxes and bottles with tops that screw on, is made as follows: Take two pieces of soft wood about a foot long and three-fourths of an inch thick and cut a shallow notch near the end of each, large enough to fit the top of a syrup can about an inch across. Tie these two pieces together at the ends loosely, but with a stout cord, as shown in drawing. Make small notches for the cord so that it will not slip off. The wrench will, of course, screw the tops on as well as off. A second notch, wider than the first, may be added for a fruit jar cover wrench.

Hot-Air Fruit Drier

The cut shows a fruit dryer capable of holding several bushels at once. The fireplace is of brick. Over the fire is a sheet of sheet iron. A course of bricks set on edge comes next; then another sheet of the iron. This last is perforated with inch-round holes. Air is admitted to this hot-air chamber through the bricks as shown. When heated it passes up through the drawers of fruit, the drawers having wire cloth bottoms. The whole top, supported on the wooden corners, can be lifted off and stored when not in use.

No dried fruit tastes so good as that dried by the sun in the open air. But the old fashioned method of exposing the fruit on open tables in the back-yard is not cleanly—flies simply make it their feeding ground. Use the device shown herewith, small frames three feet square, with wire netting bottoms. One can be arranged above another in the manner shown, and all can be covered with a sheet of cotton mosquito-netting. The air can then circulate in every direction through the trays, but insects cannot soil the fruit.

Sanitary Fruit Drier

Glass-Front Fruit Drier

A unique fruit drier is made of a big grocery box fitted with a hinged sash, and slat or wire gauze shelves, as shown. Have an entrance for air at the bottom and exits for air at the top. Have the sash side look toward the south. The sun-warmed air will rise through the shelves of fruit and pass out the top, keeping a constant current passing through the box.

To keep material neat, free from dust and unaffected by the air it is often desired to close tumblers, bowls and such dishes. An excellent method is shown in the cut. Press a circle of paper down over the top and hold it in place with a rubber band. This device is specially advised when sealing glasses of jelly, if paraffin is not available. Wet the circle of paper in white of egg. Press it down over the edge of the glass and slip a band on. When dry take the band off.

Making Tumblers Air-Tight

Those who have not yet put in a modern sink but must get along with an old-fashioned pump in the kitchen, will add greatly to their conveniences by the addition of a station or drop-leaf table as shown in the engraving. It may be as wide or even six inches wider than the sink, and about three and a half feet in length; if made from several boards they should be firmly held in place by cleats at the bottom and may be hinged so as to fold up or down as desired using a support to hold in a horizontal position. However, when once located it will be found of such great utility it will be very seldom disturbed. The support may be hinged to the table or secured by a hook or other means as thought best.

Movable Sink

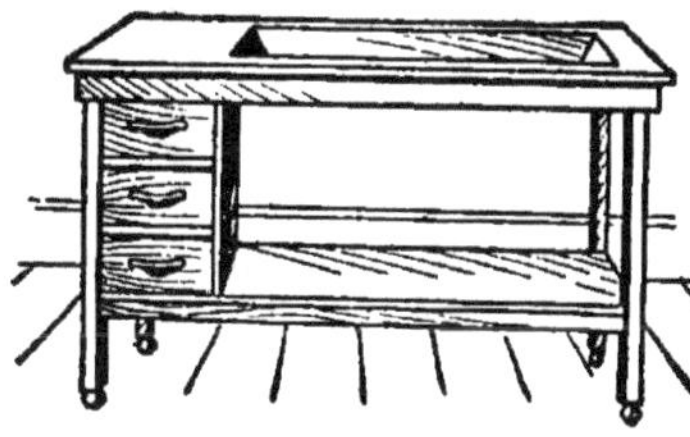

Jelly Press

Squeezing a hot jelly bag with the hands is uncomfortable and hard work. Make a simple little press of boards, like that shown in the cut, and put the jelly bag—a stout one—in it, as suggested. Make the lines to convey the juice to the pan nearer the fulcrum, if desired; or make the top board longer, if more power is wanted.

Folding Table for Sink

The lack of a suitable outlet sometimes makes it impossible to use a wet sink in a kitchen. Moreover on cold days in winter it is a comfort to take the dish washing near to the stove. A movable sink makes this possible. The framework is light and the sink is made at the tinman's of galvanized iron, with an outlet into a pail that may stand upon the bottom shelf. Such a sink, if light and small, can be pushed to the dining-table and the dirty dishes placed on it and pushed to the kitchen, saving many steps. Use big castors that turn easily.

Bins for the Kitchen Dresser

The cut shows a set of V-shaped bins placed under the broad shelf of the pantry dresser, in which flour, sugar, Graham and Indian meal, and often bulky articles may be kept. Each bin swings on a screw pivot at each side in the lowest point of the V. The advantage and convenience of such receptacles are too apparent to need comment.

Handy Kitchen Table

The accompanying drawing shows how a drop-leaf table can be easily remodeled so as to make a handy table for the kitchen. To make as shown, remove the entire top from the original table, being careful not to injure the leaves. Then make and adjust the drawer. Be careful not to

make the drawer too deep, as it would weaken the frame. The top can be made from the leaves. Place on the frame and adjust them so as to allow them to extend over the frame equally on all sides. When placed to suit, nail the one on the back part fast to the frame. The front one can be nailed fast to the frame or hinged to the other leaf, as best suits the individual. This one is hinged as shown.

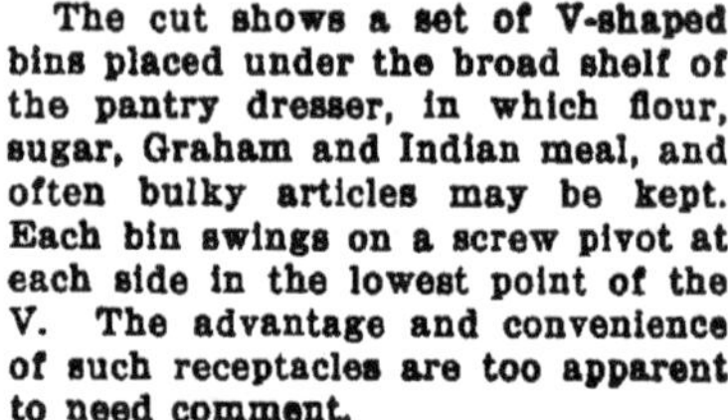

Simple Coffee Percolator

To make coffee for a small family the percolator shown in the drawing may be suspended in an ordinary coffee pot, with good results. A small baking powder can, a piece of wire, a center punch (or even a wire nail) and a hammer are the materials and tools required.

Heat Box for Raising Bread

The cut shows an arrangement that exactly reverses the operation of a regrigerator. The pan of dough is placed at the top of the space instead of the ice, and the hot water jug is put at the bottom where the food would be placed in a refrigerator. The heat rises all night long from the jug and surrounds the pan of dough. The box should be perfectly tight, woolen strip being used for the door to shut against. In very cold weather a blanket can be thrown over the box at night.

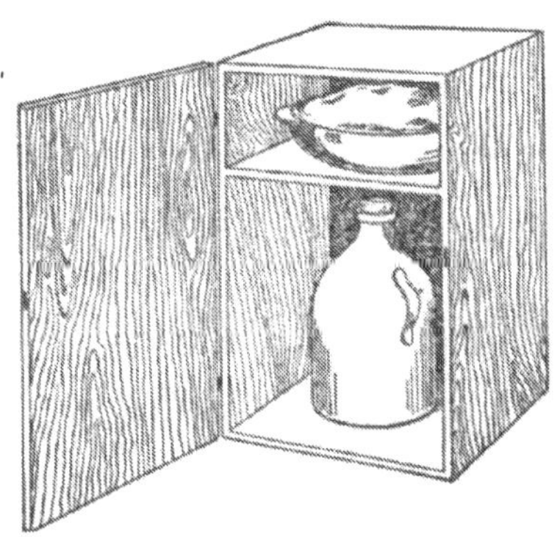

Fig. 1

Fig. 2

Bread and Cake Board

A bread and cake board should be made all in one piece as shown in Fig. 1, end pieces being put on to keep the board from warping. Then saw the board in two in the middle (as along the center line) and hinge the two parts together. The board can then be folded together, keeping the cutting surface always clean, and away from the dust.

Warning For Bake Oven

The notice on this oven door is to call the housekeeper's attention to what she has put in the oven and is in danger of forgetting as she bustles about doing other work meanwhile. This notice on the oven door, in big letters, is sure to keep the oven's contents in her mind as she passes back and forth through the kitchen.

Flour-Barrel Kneading-Board

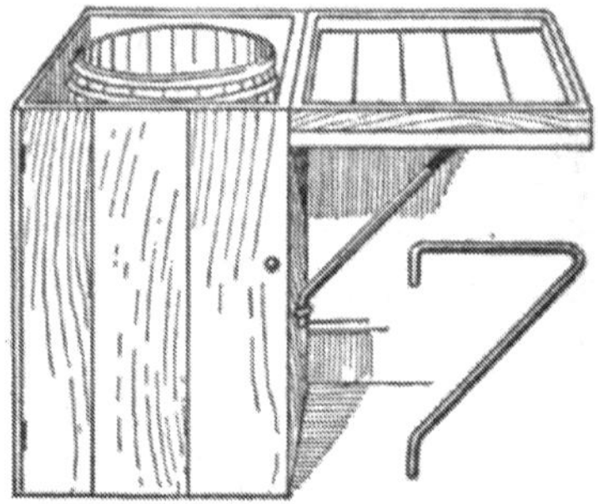

There are few housekeepers who would not appreciate such an arrangement as that figured in the illustration. The flour barrel is enclosed in a cabinet, the top of which is hinged at one side, and on being folded back becomes a kneading-board. This is supported by a swinging bracket made out of half-inch round iron, fitted to sockets. The kneading-board will always be kept perfectly clean, because when not in use it is upside down and shut in from the dust.

Oil Stove Stand

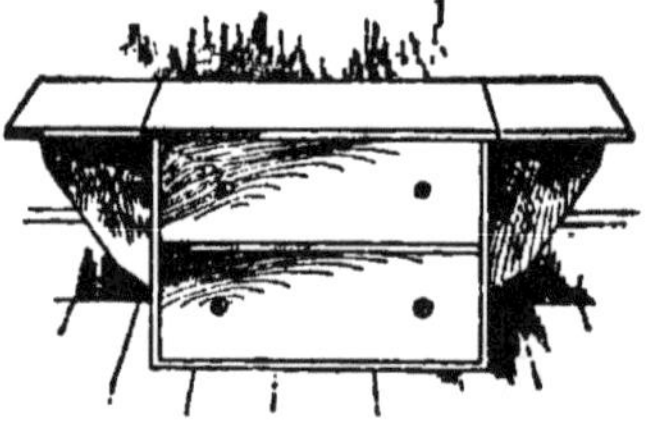

Oil stoves have now come to be a common fixture in the kitchens of the land, especially for summer use. Those not already provided with stands need to be raised two feet or so from the floor. Instead of turning a grocery box upside down, turn it on its side and fit a couple of drawers within. Hinge a shelf at each end, with hinged supports to turn back against the box when the shelves are to be put down. These shelves are uncommonly handy for holding kettles, dishes, etc., when cooking.

Home-Made Steamer

Even a large kettle is too small to put a couple of fowls in to boil. The best arrangement is shown in the cut. One milk pan is inverted over another and the fowls are really steamed, for but little water is needed. Lay three or four wooden skewers on the bottom of the pan and lay the fowls on these to keep them from burning to the tin. Turn the fowls once or twice while cooking. It is to be remembered that steam is as hot as boiling water.

To Keep Food From Burning

To keep the contents of tin dishes from burning upon the stove, we have various commercial articles, such as asbestos mats, asbestos-bottomed tins, etc. A device of this sort that can be made at home in five minutes is shown in the sketch. No. 12 galvanized wire is bent into the shape suggested, when it is ready to be placed under tin dishes when 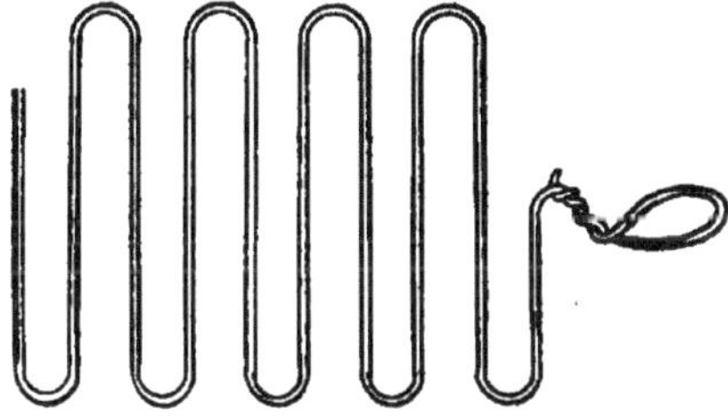 set upon the stove, or to have bread placed upon it for toasting on the top of the stove. A stouter wire can be used if preferred.

Drained Dish-Pan

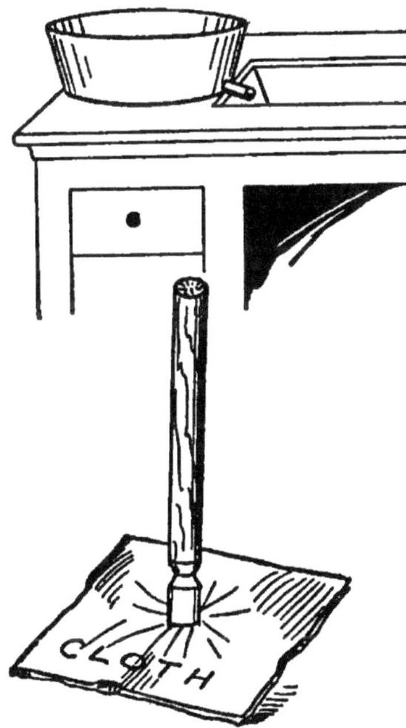

Oblong dish pans can be bought, which fit the sink and have a drain in the bottom. The cut shows how one woman has improved her dish pan along the same lines. The tinsmith soldered in a small tin tube at the bottom, and into the end of this she fitted a cork. After washing her dishes she piles them all up in the dish pan, pours hot water over them and leaves them to drain without wiping. In this way much labor is saved and a gloss given to the dishes that could not be obtained by the use of a dish towel.

Home-Made Dish-Mop

Take any smooth, even stick about foot in length. Notch it one inch from the bottom. Take several pieces of good cotton cloth, six or seven inches square. Place end of stick in center of cloth, as shown in the drawing, then bring cloth up about stick and tie in notch; next turn cloth down and tie about bottom of stick with stout twine. This makes a serviceable dish mop at practically no expense.

527

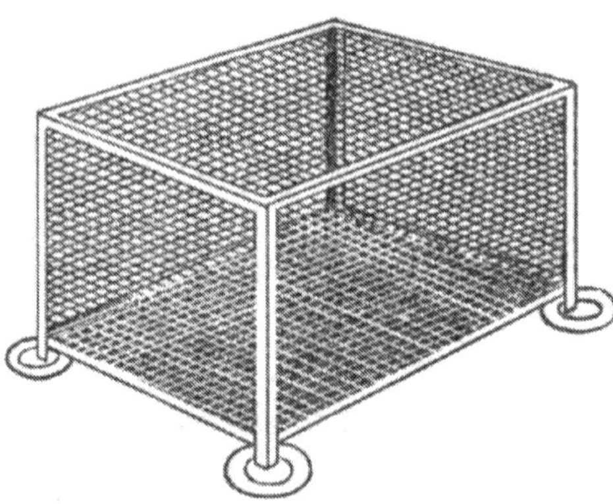

Wire Feed Box

If you are troubled by ants, flies and other insects getting into the food kept in the pantry make a box of wire netting with legs on it, which can be set in saucers of water, and thus keep all ants away, while the netting disposes of flies, moth millers, etc.   The top is on hinges, and the bottom is of boards; but the whole is so light that it may be moved to any place desired.

Home-Made Double Boiler

Two dishes already at hand can be utilized for a double boiler at almost no expense whatever.   Let the larger dish hold the hot water.   For the top of this cut a disc from tin or sheet iron, with an opening in the center just large enough to hold the smaller dish from touching the bottom of the lower vessel.   Cheap tin dishes can thus be combined in many ways that will save the purchase of expensive articles that will do no better work.

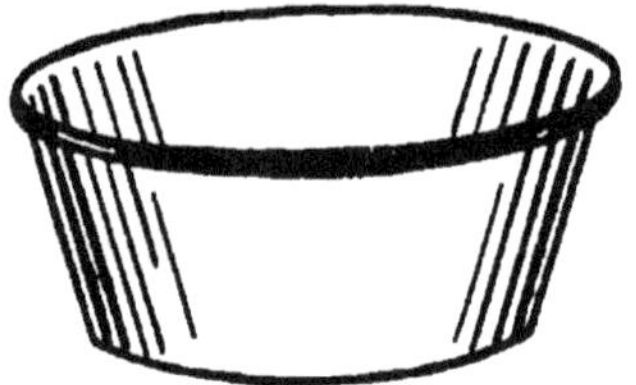

Hanger for Pot-Holders

Sew the metal clasps from old stocking supporters on to your pot-holders as shown in the drawing. They are just right to hang on a hook behind the stove.

Scraper For Sweet Corn

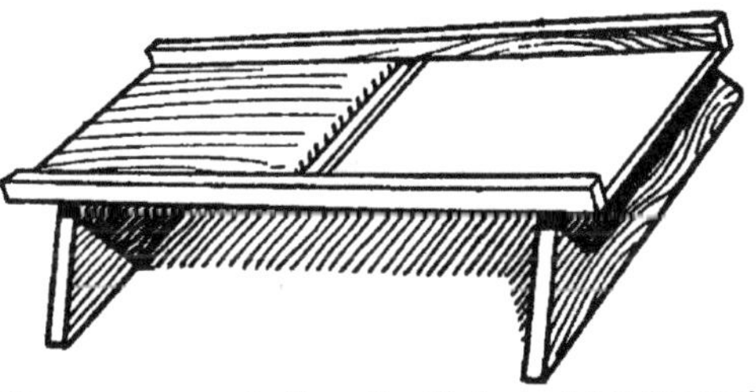

This simple little contrivance is a great labor saver where corn is to be made into puddings, fritters, etc. It is set on supports high enough for a dish to be slipped underneath. After the husk and silk are removed the ear is drawn over the sharp points which score the grains, and they are scraped off on the blade and fall through an opening into the dish below without any waste, or the spattering of hands, face and other surroundings as in the usual way. A dozen ears can be prepared in a minute or two, or it may be used for boiled corn by those who prefer it removed from the cob before eating.

Adjusting the Height of the Kitchen Table

The height of any working surface should, of course, vary with the height of the worker and whether or not she sits or stands at her work, but in general it should be about thirty-two inches for a woman of average height (five feet, five inches). It is wise for each housewife to perform the same task at different heights and see at which one she exerts least effort. The height of the table may be raised by using a block of wood with metal strips and screws or nails for fastening, as shown in Fig. 1; or it may be done by boring a hole in the block and inserting a bolt or wooden pin in the hole, as shown in the second illustration.

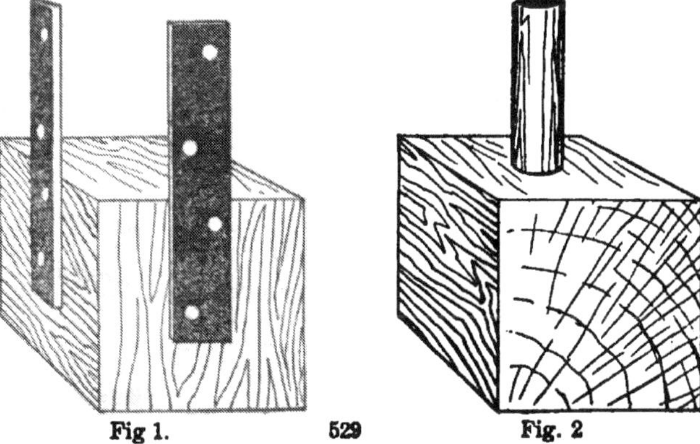

Fig 1.   529   Fig. 2

Keeping Food Dishes Hot

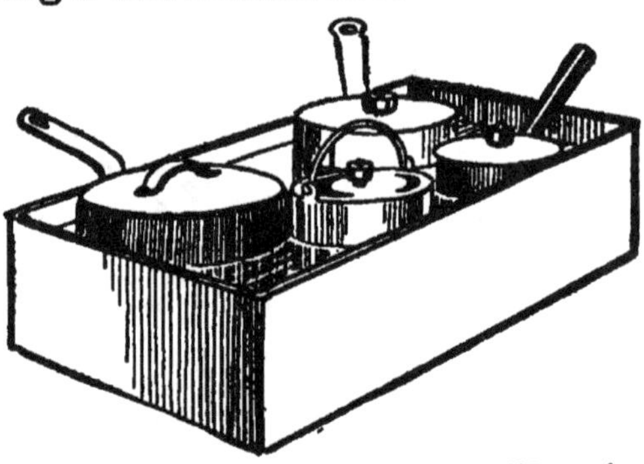

This is an open vessel (called by the French a Bain-Marie), intended to be kept on the back of the stove filled with hot water. Several stew-pans or large tin cups with covers and handles are fitted in, which are intended to hold cooked dishes desired to be kept hot. There is no better way of keeping vegetables, etc., warm and at the same time preserving the flavor. The cups or saucepans with tightly fitting lids set in a dripping pan filled with hot water would answer the purpose.

Chopping Knife

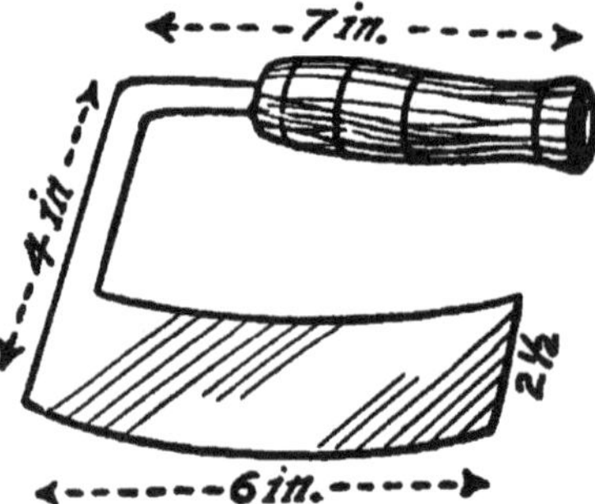

The chopping knife, shown in the drawing, appealed to us as being a very good device. It was made from a heavy piece of old scythe or cradle blade, which had been heated and properly bent by the blacksmith. The shape of the knife differs from the usual form. More work with less effort can be accomplished with this chopper than with the ordinary kind. One gets the slashing, or lengthwise, as well as the downward cut, with each stroke. It also makes a very fine implement for pounding or hacking steak.

A Home-Made Coal Hod

A grocery box fitted with a handle, and two sloping inside partitions, will give great satisfaction as a coal hod, since the sloping sides permit the coal to be shoveled up readily, as the shovel can always be run under the coal along the sloping boards.

Lifting the Clothes Line

This cut shows how a clothes-line full of clothes can be drawn up out of the way, between two buildings, two trees, or two very tall posts. When putting the clothes upon the line, it can be pulled down and held by two hooks, as shown in the cut. Then loosened, one end at a time, and drawn up taut out of the way. A clothes-line, high up, escapes being run into or dragging down into the dirt.

The Same Idea For Carpet Beating

People are often seen laboriously lifting a carpet upon a clothes line, only to have its edges dragging upon the ground. Rig a line and a pulley high up, as shown in the cut. The carpet can then be raised clear of the ground with small effort, while the subsequent "beating" process will be much easier. A cleat is used to fasten the rope.

Correct Notch for Clothes Prop

Do not neglect to make an auger hole at the bottom of the notch in the clothes prop, for the rope to slide down into, so it will not escape over the top of the prop and let the clothes down in the mud. Smooth the auger hole with your knife so it will not wear the rope. The prop can then be slipped along the line where needed.

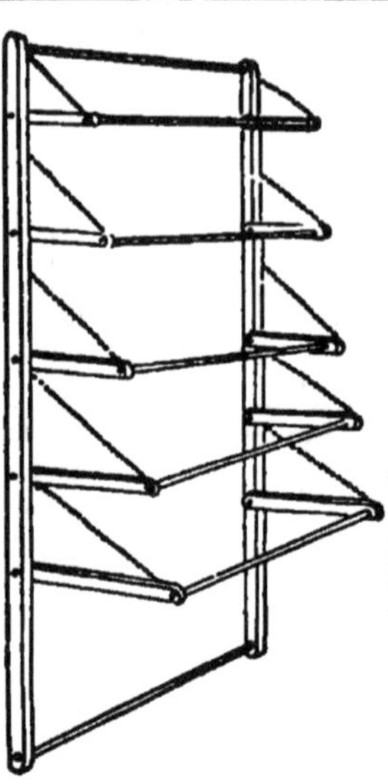

Clothes Drier

An excellent kitchen clothes drier is shown in the cut. It can be hung against the kitchen wall, and removed when not in use. The holders all fold up so that the whole thing occupies but little room if left on the wall. The clothes on each holder, beginning at the top, fall within, and so do not touch those on the next holder below.

Heat Saver for Heating Irons

First take an old bread pan, or anything that will cover two or three flat-irons. Take a piece of she‧t-iron and bend it as seen in the cut and use two rivets to make the handle. Use wood for the handles. Cover the irons and you can save half the fuel or wood. One large ironing will save more fuel than a new pan made to order would cost.

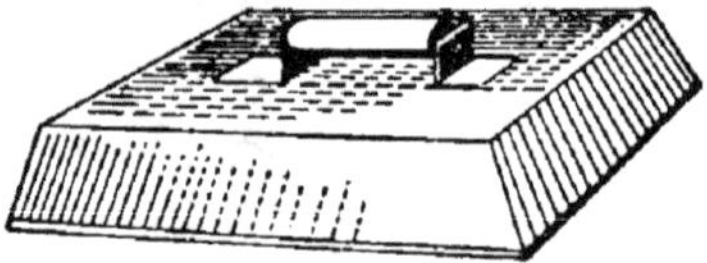

Another Kind of Clothes Drier

The cut shows how clothes driers can be made and attached to the kitchen wall, if desired. Light strips of wood are used, and these are supported by wire, attached to the outer ends and to the hooks in the wall. The inner end construction is also shown. The stout bit of wire fits into a screw eye in the wall. The hooks above are screw eyes opened a little, as shown. These driers can be made longer if desired, and can be turned against the wall when not in use.

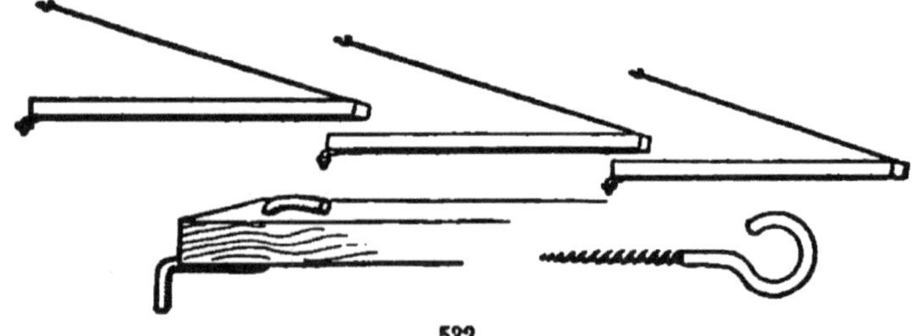

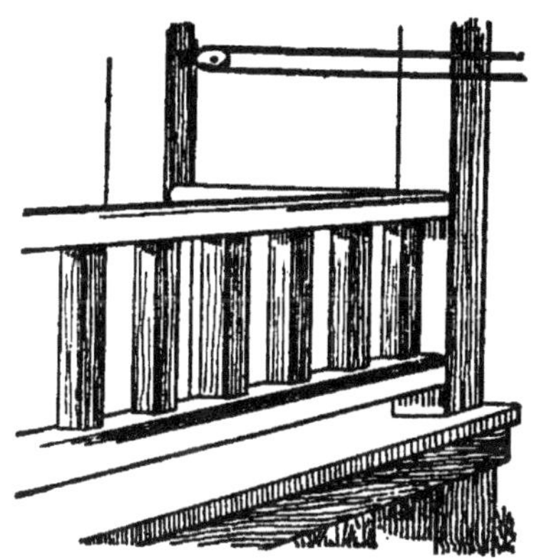

Endless Clothesline

If one end of the clothesline can be attached to a piazza or porch one need not step out upon the ground either to hang out or take in clothes. Have the line double, passing at either end about a pulley. Put the first garment upon the lower line, move it along and go on until all the clothes are pinned on, reversing the operation when removing them. This will be found a great convenience in winter, when heavy snows lie upon the ground.

Carpet Stretcher

Any old rake that has lost its teeth stapled to a bit of board, having broad ends projecting on its under side, forms an excellent carpet stretcher. Round the lower edges of the head and all edges where the staples touch the rake, so that the head will "play" a little. The long handle permits one to stand far back, so that his weight will not interfere with stretching the carpet.

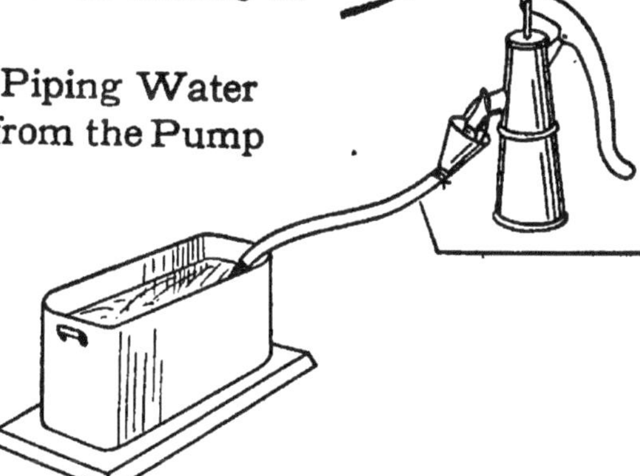

Piping Water from the Pump

It's hard work to carry water from pump or spigot on washdays. A much easier plan would be to tie a piec rubber hose around the neck of a funnel, and punch holes through each side of the funnel, and put a piece of wire through to make a handle by which to hang it up on pump spout or spigot. Have the hose long enough to reach to the stove or wash bench or wherever else you want much water.

538

Household | **Handy Devices**

Flower Box for Porch Rail

This sketch of a flower box for a porch railing tells its own story. The box can be made any length to fit the space where it is to be used. The irons, made by a blacksmith, not only hold the box to the rail, but also hold the sides from bulging out from the pressure of earth, and prevent warping.

Hanging Crate

Squashes need a cool place to keep well in winter. Piled in a heap on the bottom of the cellar is the worst possible arrangement, but one of the most common. A slatted crate, like that shown in the cut, to hang from the cellar ceiling, if the cellar be dry and airy, gives the best possible condition, as the air can thus circulate all around each specimen, while none are in contact with anything damp.

Flower Box Supports

Flower boxes on the porch look nice, but are apt to rot the boards underneath. Save the porch floor by placing the boxes on cross sections of logs cut about six inches in height, selecting pieces which are nicely covered with bark.

Scouring Knife Box

Here is a scouring knife box that the boys can make. It has a drawer which may contain the scouring material, cloth, etc.

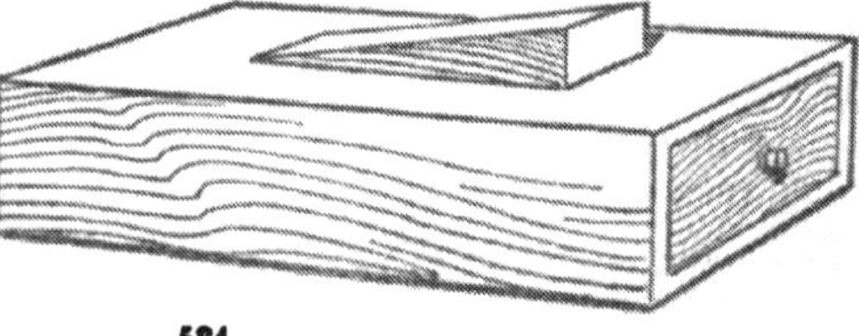

Polishing Board for Knives

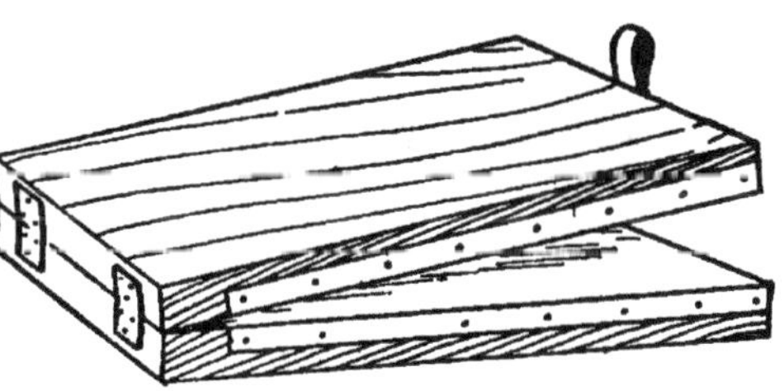

For a polishing board for cutlery, take two pieces of smooth pine board, fastened together at one end, as shown, by soft leather hinges. Leather loops tacked on opposite ends to lift upper board and also to hang it up when not in use. Stretch very fine sandpaper across inner surfaces of boards, bending it over and tacking it on the edges, as shown. To clean a knife raise the upper board, lay the blade on lower sheet, press down on upper board and draw knife blade out quickly. Fine emery cloth would wear better than the sandpaper.

Safe Ash Cans

The drawing shows how to make ash cans to be used about the farm instead of barrels. Take an old forty-quart milk can, which can be purchased cheaply from any milkman, cut the top off at the rim just above the handles, and then take a file and smooth down the rough edges, and lo! an ash can that is safe from hot embers or live coals, which have caused so many fires by being put into barrels in the shed.

Ash Sifter

The cut shows an ash sifter made of a large store box. The illustration shows plainly how it is constructed. Sifted ashes are used on the dropping boards to catch the manure. As the coal ashes are gathered from the stoves in the house they are at once sifted, the fine part used as mentioned, and the coarse part utilized on the walks.

Water Cooler

One end of the box is partitioned off (as shown in cut) for two pails of water, with holes in the top and bottom of the partition for communication with the ice space behind. This box is not double walled, nor lined. Wrap the ice in an old cloth if it is desirable to economize on ice. If not, this precaution need not be taken. Have the lids all fit tightly. The end need not be hinged, but can be nailed up tightly, if preferred. The pails can be put in through the top opening alone, very easily.

Portable Water Cooler

Water is usually carried into the field by farmers in a tin can, or a stone jug. In either case, it soon becomes warm and unpalatable. Make a box, with basket handle, of quarter-inch stuff, and fit the can into it as shown, packing the space all about the can with chaff. The weight is thus increased but a trifle, while the water can thus be carried more easily, and kept cool very much longer than by the old plan.

Iceless Water Cooler

Where a supply of ice is not at hand the drinking water soon becomes warm and unpalatable in summer, after being drawn from the well. To keep the water cold as long as possible, set a small pail of water into a large wooden pail, and have double covers arranged as shown. A few moments' work thus makes a diminutive refrigerator of the water pail.

Earth Refrigerator

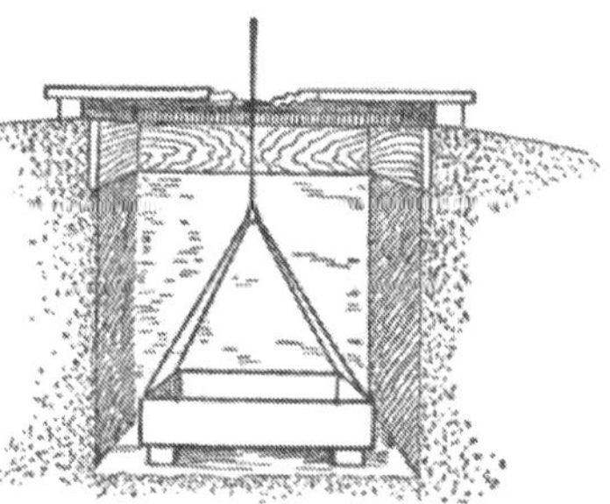

Some farmhouses are sadly in need of accommodations for keeping perishable food in summer. Ice they have none, and the cellars "keep things poorly." Sometimes meat, milk, etc., are hung in the well, but this is objectionable, since food or milk may be spilled into the water. Besides drilled wells cannot be thus used. A couple of hours work will sink a five-foot shaft in the shade of a tree, or on the shady side of a building. If the soil is firm, a lining of one board about the top will be enough. A tight platform, a hinged door, and a shallow box with ropes, as shown, will provide a ground refrigerator that will give coolness at small cost. Refrigerators should be thoroughly cleaned twice a week, sunned and aired, and the doors left open until all is dry inside.

Home-Made Dumb Waiter

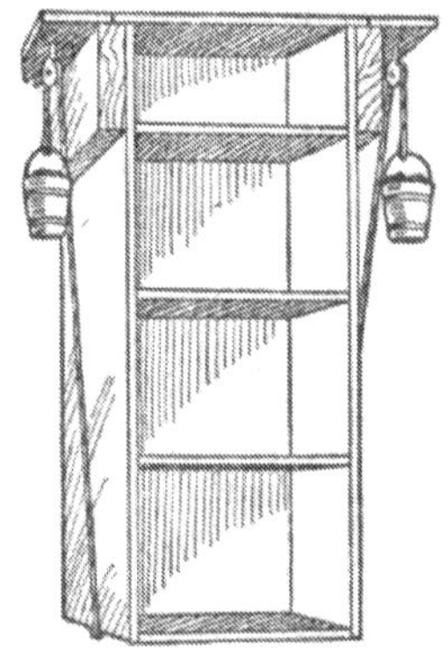

This home-made dumb waiter is not intended for the woman who is fond of running up and down stairs. If a man has a wife who likes to save steps he can make her one like it. There must, of course, be a cellar under the place where the contrivance is needed.

Make a plain cupboard, as illustrated, about four feet high and just wide enough to go between the joists of the floor. The top board runs out one inch beyond the sides to catch on to the joists. Find out where the good wife wants it and then cut the floor out to admit it.

You will have to cut with a chisel one inch of the flooring to let the top board drop on the joists flush with the floor, so that when the waiter is down there will be no sign of its presence except a bit of new floor. Fasten two little pulleys on the joist below the floor, then to a cleat across the bottom of the waiter attach strong cords on either side, run through the pulleys and tie to weights on either side, as shown in cut. The weights should just balance the waiter, just as the weights balance the window sash. When convenient have the waiter next to a wall. Have a ring or handle screwed on top to raise it by a cord or hood to hold it up while it is being loaded.

New Type of Dumb Waiter

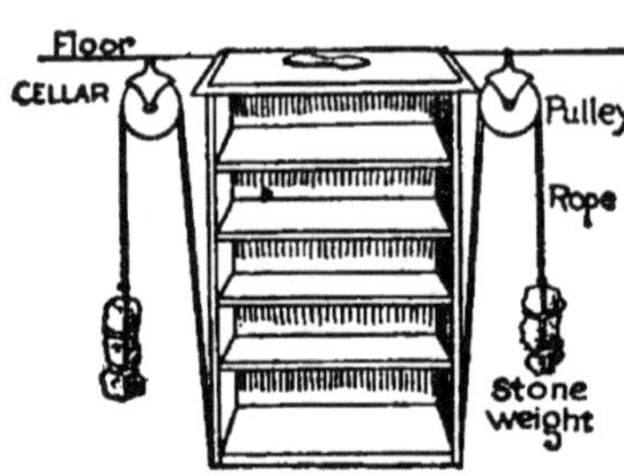

This is a device to overcome the tramping up and down cellar stairs for every tumbler of jelly or jar of preserves needed in a hurry. It is a dumb waiter built right in the pantry floor, and it goes down into the cellar. It is sixteen inches square on top with five shelves in it, like the sketch. The weights should be of the proper size to make it slide up and down easily. The handle is an old scrubbing brush.

Settee With Box Seat

A shady seat that serves a practical end is shown herewith. It is a lawn "settee" that has a hinged seat and a generous sized box beneath it, so that the young people may store their croquet mallets and balls or their tennis racquets, balls and net, safe from the weather, right by the playground. The seat will be an excellent point from which grandfather can view the game, and is so simple in construction that any boy ought to be able to make it.

Household Two-Step Stool

One of the handiest articles a housekeeper can have in the home is a two-step stool. The picture shows clearly how this convenience is made, as well as indicating the dimensions. The four legs may be fitted with rubbers such as are used for crutches and chairs. Besides making the stool noiseless, they will prevent it from sliding from under you when standing upon it.

Best Trap for Sink Drain

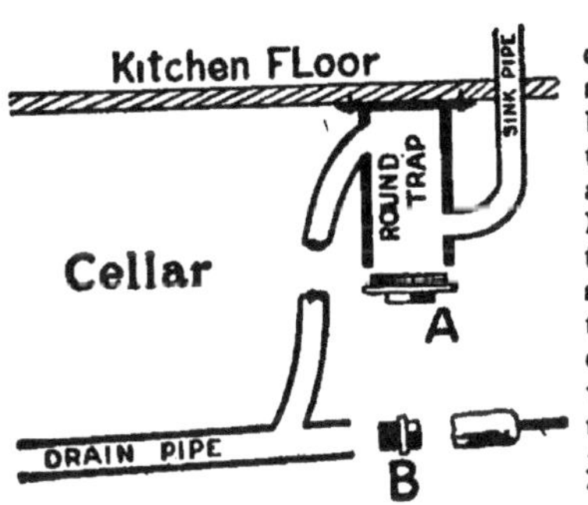

The old day of open sink drains is, or ought to be, past. The writer's sink drain empties into a cesspool, but it is to the kind of plumbing that attention is called. Do not use an S-trap—they will syphon out and leave the pipe open. The "round" trap, shown in the cut, is perfectly safe, and has the merit of a bottom that will unscrew, when grit and other material clog it. Likewise, when having plumbing done insist on the arrangement shown at B, at all joints. The cap can be unscrewed, a "plunger" inserted and the pipe cleaned in two minutes. With a closed pipe how will you get at an obstruction?

Trapping the Grease From the Sink

A grease-trap to catch and hold the grease from a kitchen sink, and prevent it from getting into the sewer or septic tank and closing it up: A, A, concrete cover and bottom; B, grease floating on water; C, water level; E, E, glazed sewer tile twenty-two inches in diameter; F, F, four-inch glazed sewer tile; G, pipe from kitchen sink; H, outlet to sewer or septic tank; J, house wall. By this arrangement the water and grease are separated, and when enough has accumulated the cover can be raised and the grease removed.

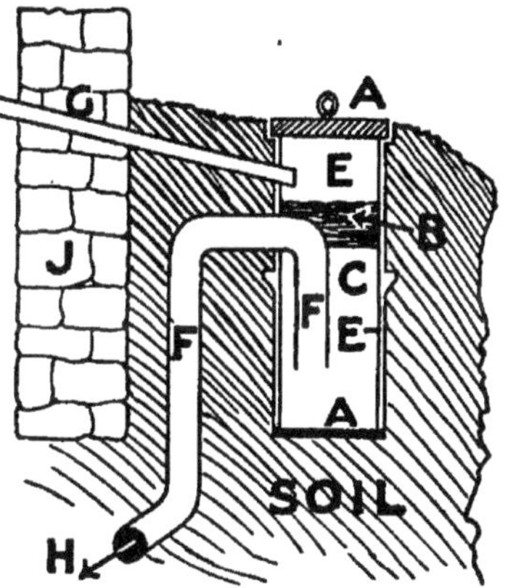

Convenient Wood Box

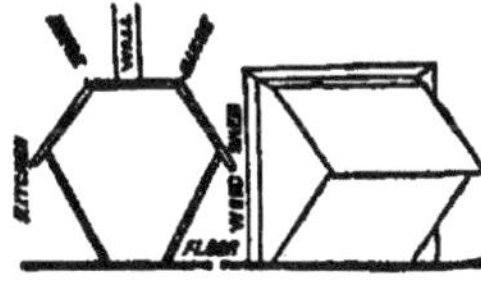

This wood box is filled from the outside and emptied from the inside of the house. The left-hand picture shows a sectional view; the right-hand cut shows one side of the same double box. Easy to make—simple to operate.

Wood Box and Carrier

The box shown in the cut is both a wood box and a wood carrier. It is just long enough to take in a stick of wood and big enough to hold an armful. It can be filled in the shed, and then taken in and left by the stove until the wood is used. The handle swings down out of the way when not in use. Make the box neat and attractive, and it will be ornamental as well as exceedingly convenient.

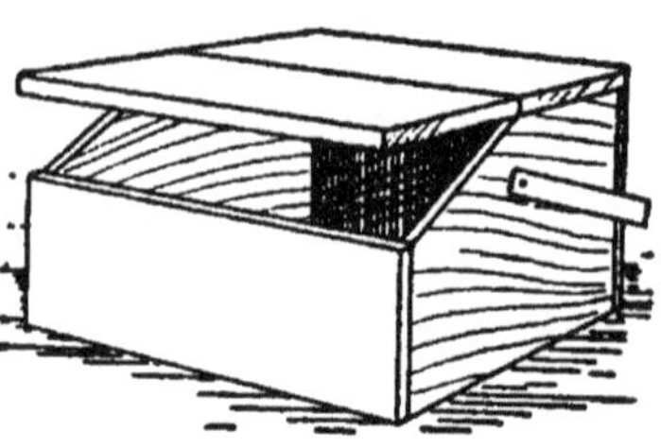

Wood Closet

Wood can be kept in the shed and litter out of the kitchen, by cutting an opening in the wall between the kitchen and the shed and adding a hinged door. The elimination of the wood box is an excellent feature for a kitchen having a limited floor space.

Fly Trap

To make one, take an ordinary metal fruit-jar lid, and cut an inch hole through the center of it. Then make a funnel out of common wire fly-screening, and put the small end through the hole in the fruit-jar lid (as shown) and solder it fast. Next screw the lid on a two-quart glass jar, set the trap on a board, put some bait under the funnel part—and watch the flies go in. When done eating they will go toward the light above. Several small notches should be made in the lower part of the funnel, so the flies can get at the bait.

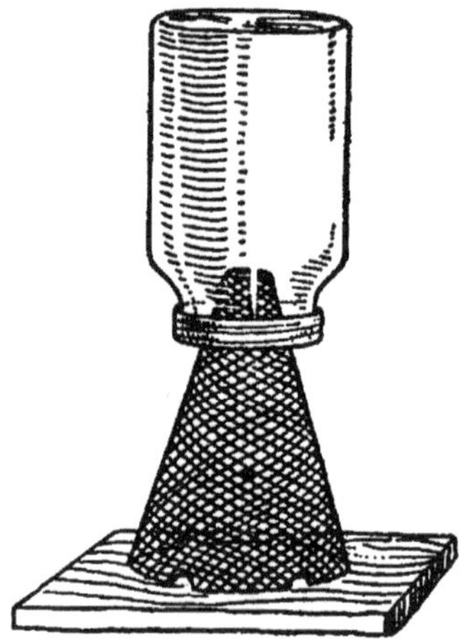

The Simplest of All Mouse Traps

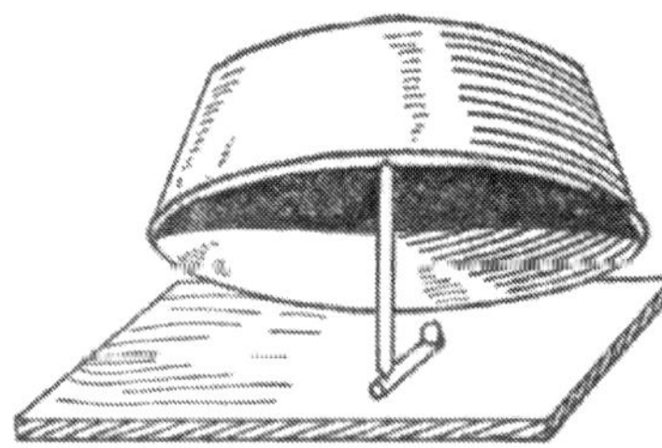

Many places are badly infested with mice. Here is a home-made trap that can be gotten ready in an instant and at no expense. Tilt a pan on one edge on top of a movable board and place a small stick under the open side—this stick resting on a rounded stick lying on the board and extending in under the pan. Place a tempting bit of bait on inner end.

Broom Hanger

A handy way to dispose of the broom without breaking the straw is to hang it on two large spools. Spools on which crochet cotton comes are satisfactory. Drive wire nails through the holes into the place you wish to hang the broom. Place the spools just far enough apart so that the handle of the broom will slip easily between them (see picture). Rest broom. end up, on these spools. This keeps it in shape and out of the way.

Window Lock

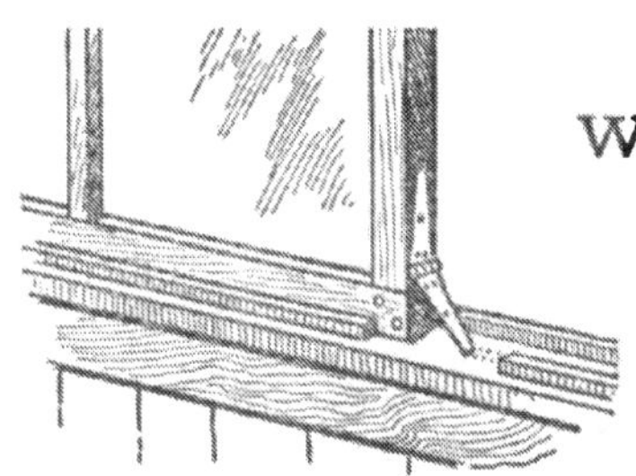

To lock windows which slide open horizontally, take an old hinge and attach as shown. If the free end is sharpened with a file it will hold more securely. To open the window lift the free end of the hinge.

Window Ventilator

In the winter time many folks live and sleep in unventilated quarters. The drawing shows a good way to let in fresh air without causing drafts. Simply take an old window screen, stretch some thin muslin across it and tack it in place, and then put it in a window as you would a fly-screen. Have one or more for each room where needed.

Tray for Invalid

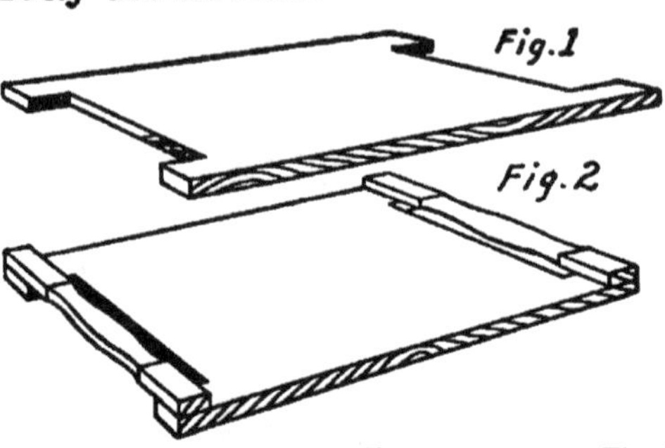

A tray on which to place an invalid's dinner, or one on which food can be carried to the dining table, or on which dirty dishes can be placed for removing to the kitchen sink, is a most useful household article. A home-made one is shown in the cuts. Cut out the ends from a piece of board about fourteen by twenty-five inches, as shown in Fig. 1. From one-square stuff make two handles, as shown in Fig. 2, and screw them to the ends of the board, as suggested. If made of handsome wood and smoothly finished this tray will be ornamental as well as useful.

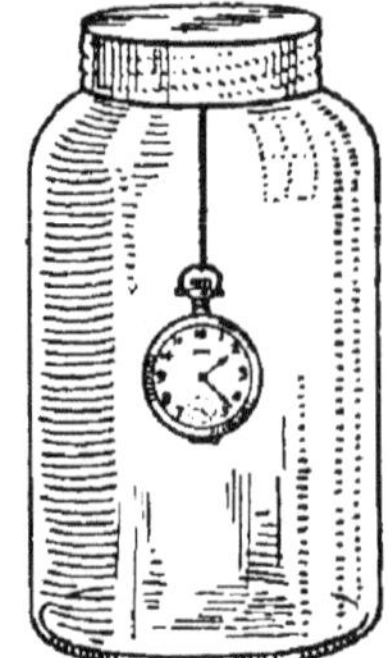

Silencing the Invalid's Watch

The ticking of a watch or clock is sometimes objectionable in the bedroom of a sick or nervous person. It can be muffled in the following manner: Procure a large clear-glass jar with the neck big enough to admit a watch or small clock. Then by means of a rubber band suspend the watch from the cover midway in the jar. The time can readily be seen without removing the watch.

Bed Table for Invalid

The cut shows a handy attachment for the bed of an invalid—a little table on which a breakfast or dinner may be placed and then swung around before the occupant of the bed. At other times it can hold the glass of water, the medicine or the book that is to be read a few moments at a time. It can always be put into a convenient position for use without the necessity of holding it, and pushed back out of the way.

Medicine Timer

This is made by cutting a piece of cardboard slightly larger than the top of a tumbler. Mark the figures, one to twelve, and opposite each hour mark cut a notch. Tie a knot in one end of a short piece of cord, pass it through the cardboard and fasten a button to the other end. When medicine is taken move the cord into the notch opposite the hour when it is to be taken again.

Fitting the Cork to the Bottle

Whittling a large cork down to make it fit a smaller medicine bottle is usually an unsatisfactory job. Here is a better way: Cut two clefts at right angles, as shown, and then you can neatly and easily force the cork into the small bottle. If properly done, the cork will fit into its place tightly and well. Try it.

Memorandum Pad

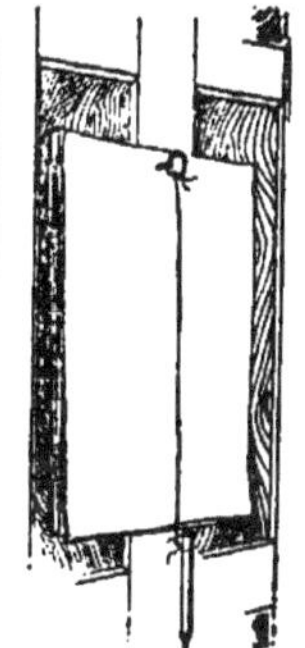

Take the sheets of brown paper that come from the village stores and cut into sheets of convenient size. Tie a pencil to a screw eye, such as is used in picture frames, and screw the sheets to the wall in a body, as shown. Write down wants as they occur, then a memorandum will be already and fully written out, when someone of the family goes to the village.

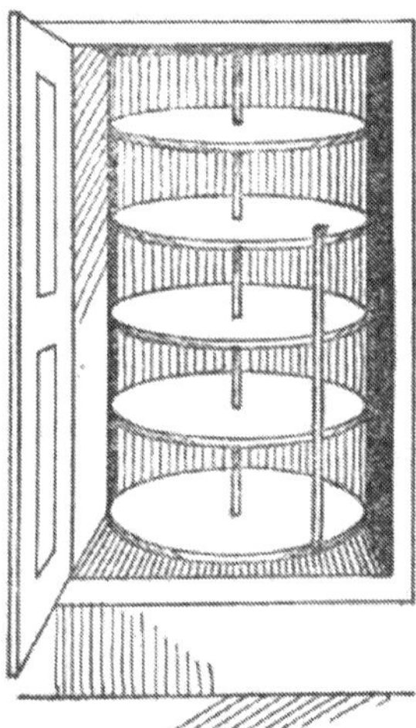

Circular Closet Shelves

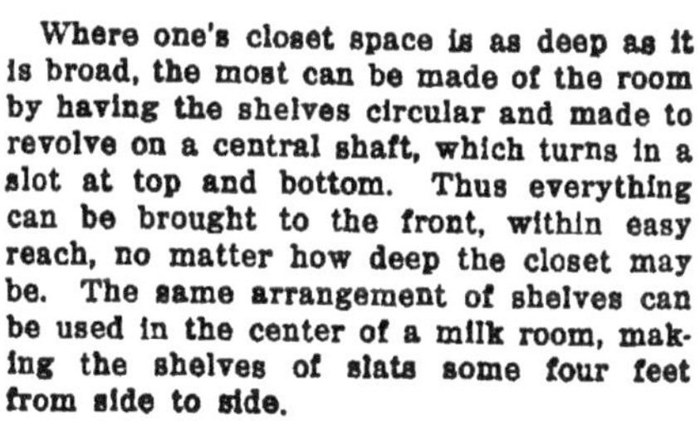

Where one's closet space is as deep as it is broad, the most can be made of the room by having the shelves circular and made to revolve on a central shaft, which turns in a slot at top and bottom. Thus everything can be brought to the front, within easy reach, no matter how deep the closet may be. The same arrangement of shelves can be used in the center of a milk room, making the shelves of slats some four feet from side to side.

String Holder

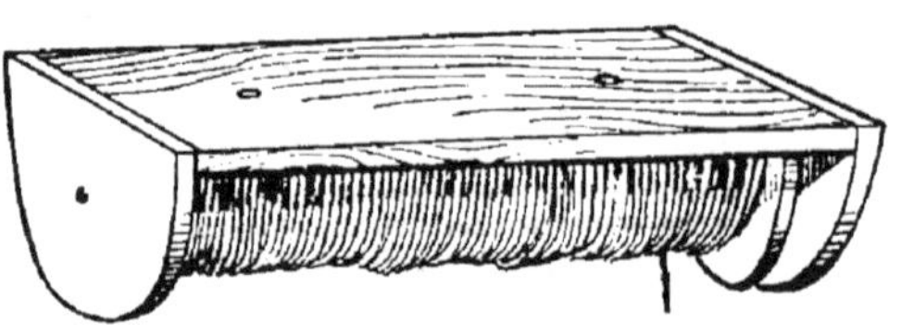

This is a roller with a wheel on each end that revolves in a frame screwed to the under side of the kitchen shelf where it is out of the way. In leisure moments the cord is taken out of the tangle, tied together, and wound on the roller, where it is always ready for use.

Keeping in the Baby

When there is a baby in the house one often sees a wide board across the door, slipped down between cleats at the sides that the door may be kept open. This is an inconvenient plan—much inferior to that shown in the cut. A piece of ticking or canvas is tacked at one side and hooked as shown at the other side. To step out one has but to unhook the upper ring. To unhook and hook again is but the work of an instant, while one must either climb over the wide board or laboriously lift it out of the cleats and then return it to them.

Holder for Stove Holders

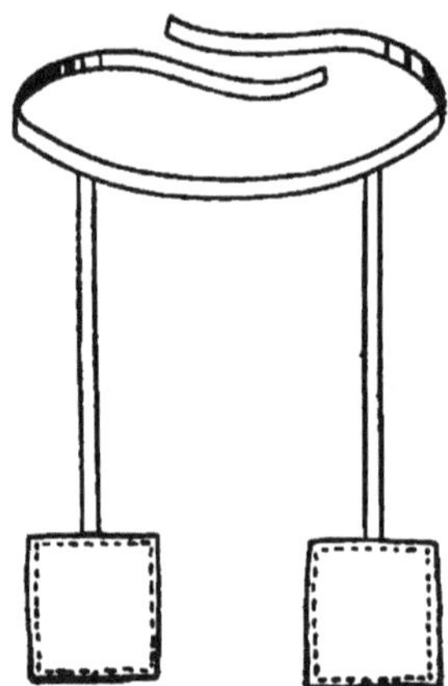

Here are a couple of holders that will always be at hand when the housewife is at work about her cooking. At such a time stove holders have a way of being uncomfortably mislaid—even if they were in use a moment before! But attached to a girdle, and that pinned about the waist, they cannot be otherwise than always ready at hand. The belt may have button and buttonhole if preferred. Such a device, daintily made, will serve as a very useful Christmas gift.

An Awning of Vines

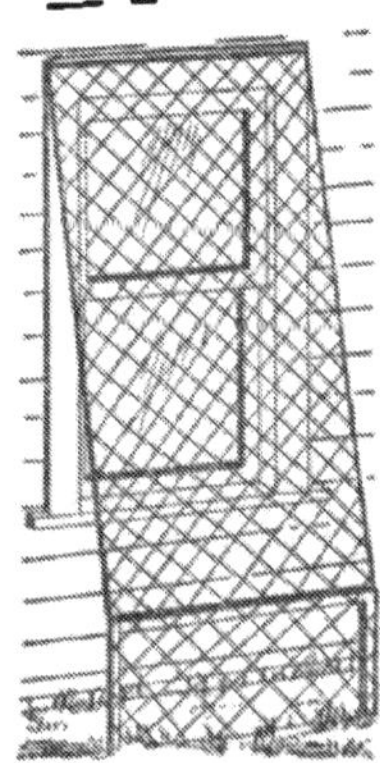

If vines grow closely over a window they shut out almost all light and air. Put up a strip of poultry netting in the manner shown and plant quick-growing vines to run over it, and space will be left for quite an amount of light and air to enter, while shutting out the hot sunlight.

Inexpensive Porch for Door

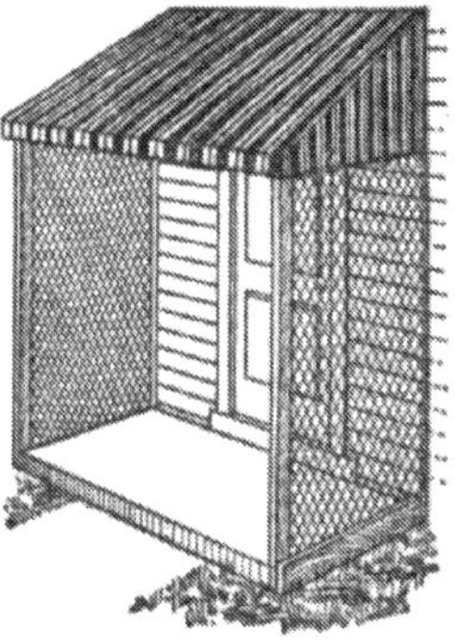

The outside kitchen door mustneed to be open in the summer to keep the kitchen from becoming unbearably hot. But if doorway is unshaded the sun pours in with a great heat. A covered porch like that shown in the cut is easily made, and will prove a great comfort. The awning has a frame work of strips of board and laths. The awning cloth is tacked to this. Put up wire netting at the sides and train morning-glory or other vines over it. It will be attractive, too, as well as comfort-giving.

Another Awning of Vines

When a vine is wanted over a doorway have it simulate an awning as shown in Fig. 1. To do this make a framework, preferably of iron rods or the frame of an old wagon or buggy top, and stretch over the slanted top a width of wire fencing. The vine reaching up to this must be gradually trimmed of offshoots until it reaches the netting, where it may be allowed to branch and blossom at will. A Virginia creeper would be admirable for the purpose.

Fig. 1   Fig. 2

Some iron bars or old pipes and a few feet of poultry netting will make the support for vines shown in Fig. 2, which will shut out an unpleasant view or shade a sunny window.

Keeping the Children in Bed

Anxious moments might be spared many a mother if the father of the family would hinge to the side of the bed a planed board that could be turned up at night and buttoned, as shown in the cut. With the other side of the bed against the wall, there will be no danger of the small boy or girl rolling out of bed in his or her sleep. By day the board can be turned down beside the bed, out of the way.

Shaving Mirror

No mirror is so good for shaving as one placed in the middle of a window, for the face gets all the light which the mirror reflects. Hence, the man of the house will welcome the device shown here. One pane of the bathroom window is removed and a pane cut from a broken mirror is put in its place. The original pane is put in behind it to protect the quicksilver, and both puttied in.

An old flat file, although an apparently useless thing, can be made very helpful to the housewife by driving it in beside the door and sinking it in a stake driven into the ground at right angles, as shown in the cut. The combination forms an indispensable foot scraper and labor saver, which can be made by anybody in two minutes. If the end of the file projects an inch or two as shown, it will be very useful to scrape the boot between the sole and upper. An old broom standing besides the door will complete the outfit.

Foot Scraper

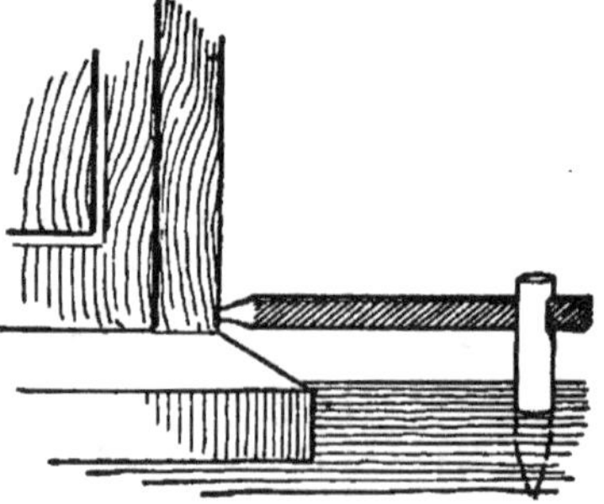

Kitchen Apron

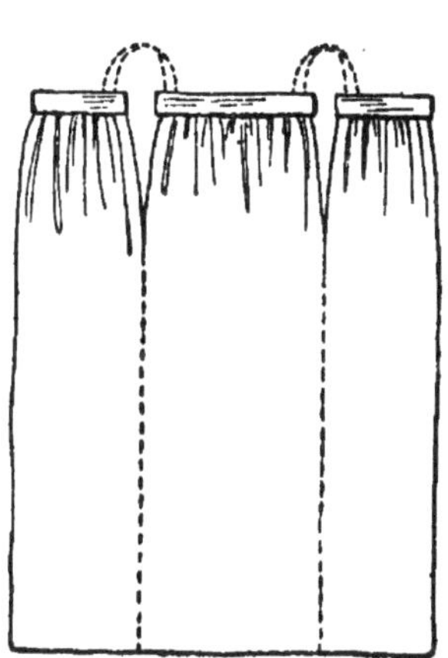

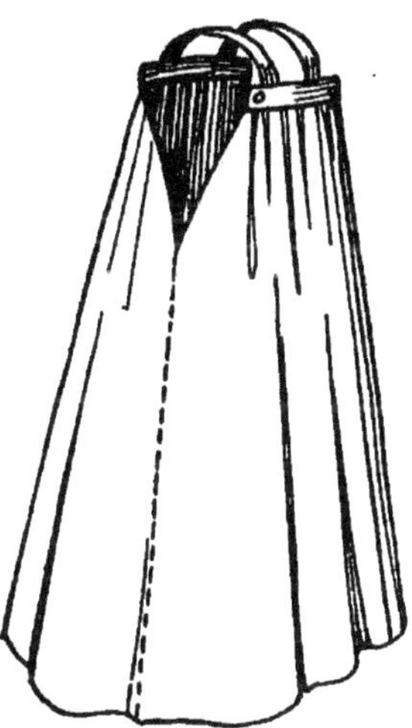

This is the simplest way in the world to make a kitchen apron. Take two breadths of cloth and sew them together as suggested — the whole breadth and the other divided into halves for the back. Sew up the seams about two-thirds of the distance, leaving the rest for " armsizes." Put plain bands at the top and " gather" the top of the cloth to them. Two shoulder straps go where the dotted lines appear. A button and button hole, or hook and eye, are placed at the outer ends of the short bands. The finished apron is shown at the right. These are just the aprons to make for girls, either to work or play in.

Household Truck

Women who are not strong find a scrub pail or a mop pail too heavy to move easily. They should make or have made a little truck. Pieces of wood about a foot square will do, nailed together so that the grain will run in opposite directions (this prevents warping and splitting). Then screw a cheap castor roller on each corner, as shown in picture. This truck is suprisingly useful, not only for pails of water, but also to move bookcases, etc. Raise one end of bookcase and run truck under; then raise up the other end, and shove the bookcase anywhere desired. By putting a board across the truck, even a stove can be moved in the same way.

Simple Food Tray

A cheese-box lid with part of a barrel hoop for a handle makes a convenient tray for carrying food to and from the cellar, or from kitchen to dining-room. The drawing shows how the handle is attached; it should be made very secure, and if the tray is finished with a coat of enamel paint it can easily be kept clean.

Signal Device

To call the "men folks" from distant fields to their meals is often beyond the power of an old-fashioned horn—especially if the wind is blowing the wrong way. The cut shows a good way to signal. The box is painted black, and can mean several things, according as it is fully elevated, or raised at other elevations.

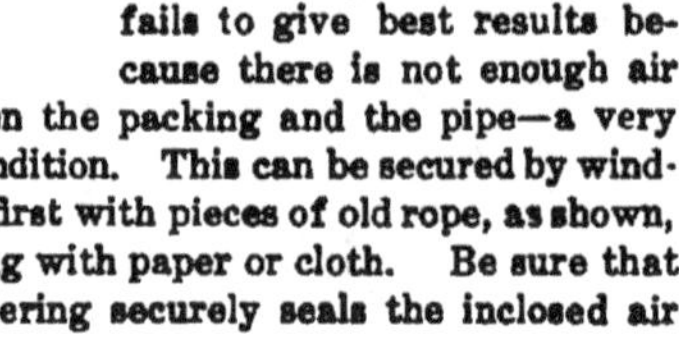

Keeping Pipes from Freezing

Water pipes are a source of trouble during cold weather. Prevention is not only better, but much cheaper, than cure. Winding pipes with old papers or cloth helps somewhat, but fails to give best results because there is not enough air space between the packing and the pipe—a very necessary condition. This can be secured by winding the pipe first with pieces of old rope, as shown, then wrapping with paper or cloth. Be sure that the outer covering securely seals the inclosed air space.

Protecting the Pump from Freezing

A pump with up-and-down handle is of such small diameter when the handle is forced down against it that it can be easily covered at night to keep it from "freezing up." The hoops and uprights here shown can be nailed together in a few moments. Pull on three grain sacks over one end. Rip open the bottoms of three others and pull them on over the other end. The cover can then be slipped over the pump, protecting it very thoroughly.

File Index for Clippings

This is better than a scrap-book, because the clippings are often printed on both sides of the paper, which makes it impossible to paste them in a book. This box has pasteboard partitions, and in each compartment is placed the envelope holding the clipping that has a subject beginning with a particular letter. An article on "Success with Strawberries," for instance, is put into its own envelope, the title written across one end, and then the envelope is placed in the "S" compartment. It's a fine way to save information that one wishes to refer to quickly again.

Hoist for Butchering

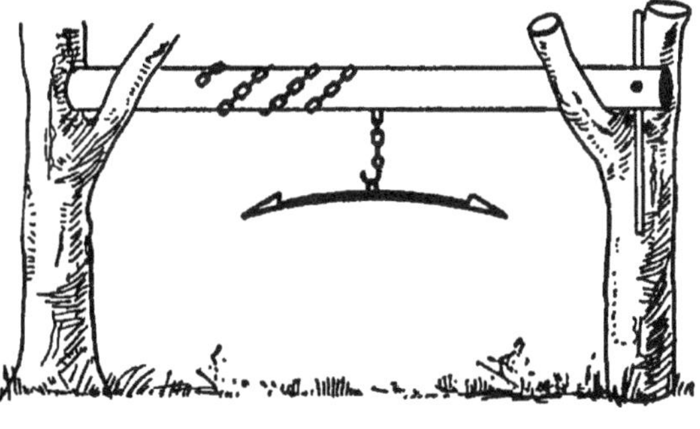

There has been considerable inquiry for a hoist on which to raise the carcasses at butchering time. For common farm use where it is employed but once or twice a year, it is doubtful if anything is cheaper or better than the old-fashioned contrivance illustrated herewith. A six-inch pole that is sound is placed in the crotches of two heavy poles well set, or in the branches of two nearby trees. To these the gambrels are fastened by chains, and this improvised cylinder is made to revolve and lift the pork by rolling it by means of a crowbar or strong stick which fits into holes bored into the pole at right angles. This is prevented from unwinding by a pin thrust into a hole bored in the post. Of course, pulleys and ropes are better, but these are not always owned.

Outdoor Kettle Crane

A stout post is set in the ground, and drain tiles slipped down over the wood, iron "eye" pieces being driven into the post between the two sections of the tiling. Have the blacksmith make a crane from round iron to fit these eye-pieces.   With this ar-

Kettle Hanger

Set posts in the ground far enough apart to hang the necessary number of kettles; put across a piece of timber with mortise through for the hooks, which can be made of old wagon tire with an iron bolt to fasten at the proper height.

rangement the kettle can be pulled over the fire and drawn away from it for examination or emptying. The tiling protects the post from the fire, an air space being left between the wood and the tiling.

Scalding Barrels

Here is a drawing of a home-made device for cleaning vinegar or other barrels by steam. An ordinary kettle is the boiler. The cover is a piece of sheetiron or anything suitable; in the center is fitted a piece of pipe ten inches long. Wooden blocks hold the barrel in place, so that the pipe goes through the bung, as shown. The kettle is filled one-third full of water and a fire built underneath.

Shoe Cleaner

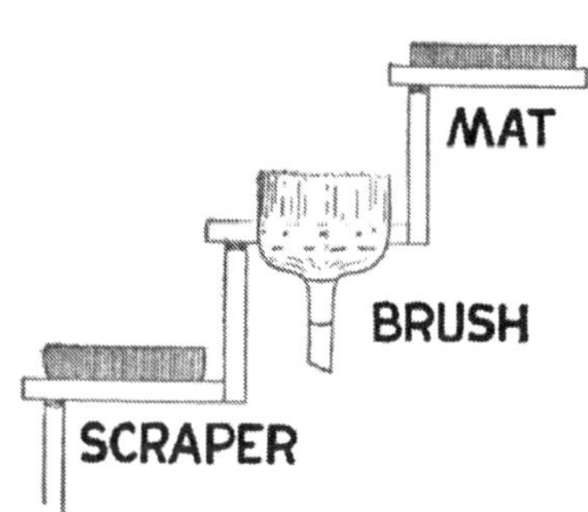

Everyone knows how hard it is to clean the bottom of shoes after a visit to the barnyard and to stand bow-legged and knock-kneed, and almost sprain the ankles while waving a six-foot broom in a vain endeavor to prepare to enter the house without carrying the barnyard in at the same time. An old broom can be taken to the chopping block and the "splints" cut down about two inches from the nearest row of string. Then cut the handle to a four-inch length, and nail to the side of the doorstep, bristles upward. This can be used in connection with the scraper and the door-mat (see drawing).

Foot Warmer

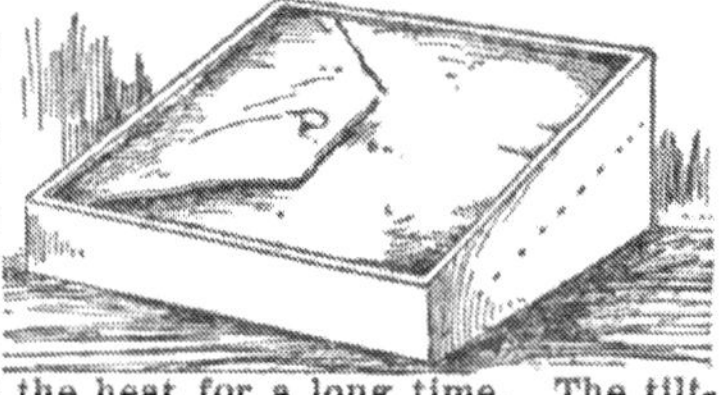

If the feet are kept warm during long rides on cold days in winter, it is surprising how comfortable the rest of the body will be. The usual slab of soapstone can be made to hold its heat for a long time by having a thick cloth bag in which to slip it. Put the stone into a shallow box as shown. The wood being a poor conductor, will help retain the heat for a long time. The tilting of the stone also makes it fit the feet better.

Another Foot Warmer

Farmers who have to take long rides in cold weather need a foot warmer, for with the feet warm the whole body will be at least measurably comfortable Take an empty varnish can and box it on all sides but one, as shown in the cut. Pack wool about the sides and ends, and have a flap of felt to cover over the top on which the feet are to rest. Fill with boiling water, and the heat will be kept in for many hours.

Carrying Freight on a Sleigh

The cut tells its own story. By fitting this device to the back of a sleigh, the latter becomes almost as convenient as a pung. A trunk, bags of grain, a barrel of flour—almost any article can thus be carried behind. Put on with bolts and nuts the device can be removed in a moment when freight accommodations are not needed. This saves the purchase of a pung, where one is not already at hand.

Winter Driving Hood

For those who must ride or work out-of-doors on windy and zero days, the cloth head covering shown in the cut will be found altogether comfortable. Any housewife can cut out such an article from cloth, pinning together first a paper try-pattern to get the size right. The flaring cape part at the bottom goes under the coat collar, leaving no crack for the cold air to enter. Fine for husking corn.

Snow Packer

Roads about farm buildings can be nicely broken out after each snow-storm by the use of the device shown here. Boards are nailed or bolted to two wood or iron curved pieces. If of iron, the upper ends can terminate in such a fitting as will hold a pair of thills taken from a wagon not in use. If the curved pieces are of 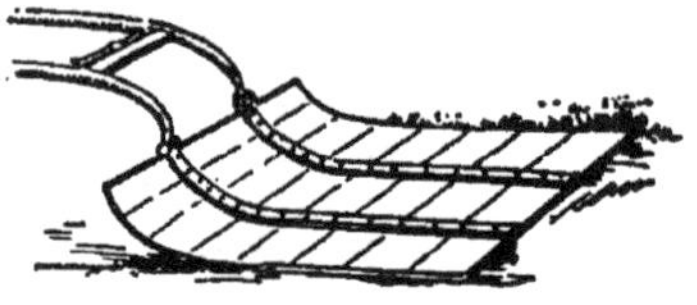 wood, have a blacksmith fit two pieces of iron to screw on the ends, to hold the thills. The driver stands upon the boards and by his weight the snow is packed down, making a hard road. This plan is better in north latitudes than pressing the snow aside; a trench soon fills with drifting snow.

Snow Scoop

Cut a drygoods box in half, as shown in the cut. Two side strips strengthen the device and afford two handles. Two loops of rope at the lower points allow a horse to be hitched to the scraper, when a path, or road, about the farm buildings can be cleared of snow in short order.

Using the Lantern as a Stove

It is not well enough known that a lighted lantern carried beneath the robes will keep the occupants of the sleigh or automobile warm, even on a bitter cold day. But a lantern in such a position is a source of danger, as it is quite easy to upset it. Secure it to a piece of plank, as shown in the cut, and it will ride without any danger of capsizing. The plank can be kept for use whenever the lantern is used as a stove.

Converting the Carriage into a Sleigh

Take one-fourth-inch buggy tire iron and short hardwood runners of same width, about three feet long, carved out to admit wheel to rest in groove. Two bolts are put through both runners and wheels, and runners are drawn tight to wheel. The iron part of runners slants to height of axle on the wheel and is held to wheel by clips and a bolt. A brace at rear, fastened to wheels, prevents the runners from twisting off in turning. To prevent runners from turning a leather strap is run from wheel to shaft of buggy.

A Little Water Power

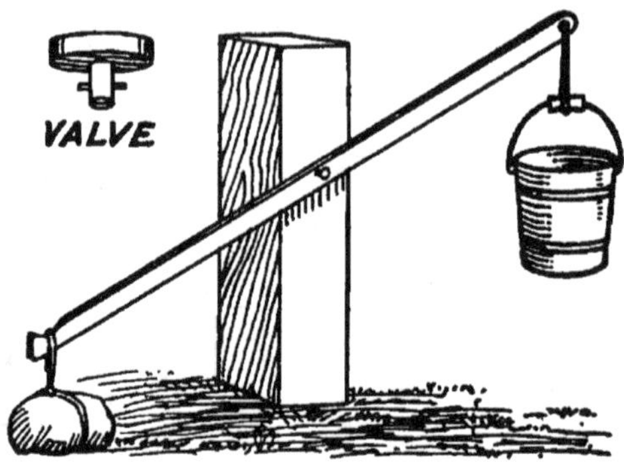

A friend of ours has a u s e f u l contrivance by which water is utilized as a power for light mechanical work, like pumping, stirring a vat of cream, scaring away birds, etc. It consists of a pole balanced upon a pivot, or bolt, to one end of which is suspended a weight and to the other a water bucket. This bucket has a large auger hole in its bottom, forming a rough valve, the stem of which projects an inch or two through the bottom. (This valve is shown in the upper part of cut.)

The motion is that of a walking beam. First the weight draws up the bucket. Water running into the bucket fills it, and its weight carries it down, lifting the weighted end. When it strikes the ground the valve is forced up, letting the water out. Relieved of its weight the bucket rises, only to fill and descend again.

The valve in the cut shows the loose plug with a head that completes the bucket valve. It has a pin or spike through its lower end so it cannot rise too high and float away. It is held down by water until forced up from below. This little power is quickly constructed and is quite effective.

Water Gate on a Cable

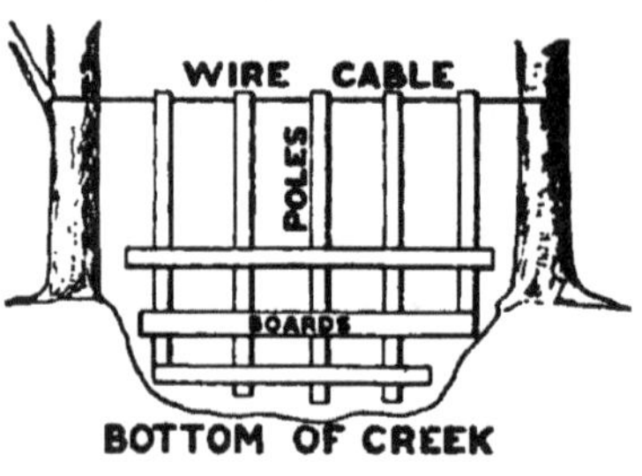

Take eight or ten strands of No. 11 wire and fasten them to a tree or post, on one side of creek, about eight or ten feet from the ground. Cut several poles long enough to reach from bottom of creek to the wire above. Flatten the large or upper ends and bore an inch and a half hole through each one. Now run the first-named wire through these holes, and fasten the ends to a tree or post on the other side of creek. Distribute these poles evenly on the cable and nail on boards as indicated in the cut. If high water takes off the boards, the poles still remain, as the water never gets them.

Windmill Power Transmission

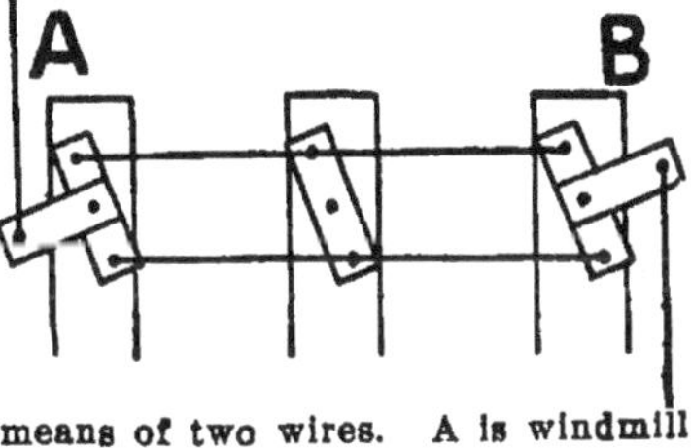

The other day I saw a windmill pumping water from a well some twenty rods distant. With every stroke, the mill had to lift about a hundred pounds of stones wired to the pump's piston. The lifting was done by a single wire. It reminded me of the old Dutchman who went to the mill with grain in one end of his sack and stone in the other end to make it balance across the horse's back. Here is a better way by means of two wires. A is windmill shaft and B pump piston.

Note—The artist has represented the levers as bolted to the tops of the posts, whereas, they should be on the side, near the top.—Editor.

Flood Gate in Fence

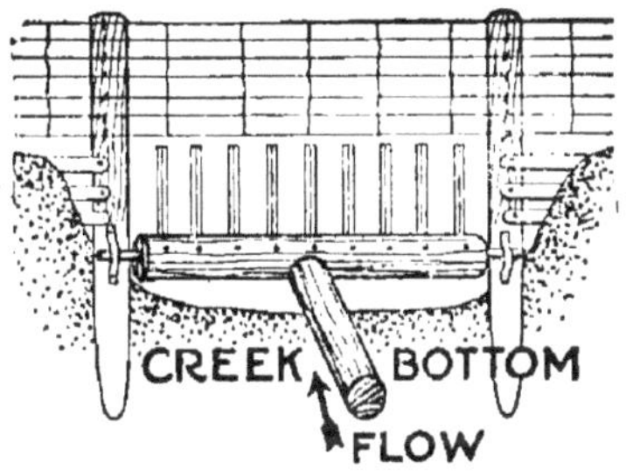

Set a heavy post deep into the ground on each side of the stream, as shown. Take a round log of good diameter and notch it at either end, as for a windlass. Loosely fit over each end a large clip made of old wagon-tire iron and bolt these clips fast to the bottoms of post near creek bottom (see cut). Now bore large auger holes in top of log and drive a solid stake three or four feet long into each hole and secure with a spike. Up stream mortise into the log a moderately heavy post, this post to lie flat upon the bottom of the creek. When freshets come this floodgate will swing and allow the heavy current to pass over it, and as the current recedes the post-weight will automatically bring the gate back into its proper position.

One-Man Saw Handle

A double handle for a cross-cut saw is easily made, as shown, from a forked branch of a tree. Such a handle gives the workman better control over the saw. One man alone can use the cross-cut with a handle of this kind, and do nearly as much work as with a helper, which is worth considering, since help is so scarce. If the saw-handle is made from a well-seasoned branch it will last much longer than one made from a green branch. Make one to use in cutting winter wood.

Improved Saw-Horse

The front, lower cross-bar is left out, because it is always in the way of one's feet, and iron straps are placed across the ends to keep these from spreading. The dotted line shows a light chain that is attached to the middle bar and is brought over the log to be sawed, which is held in place by the operator's foot placed in the ring at the end of the chain. The chain can be adjusted to suit any size of wood. The hardest part of wood sawing is the holding of the wood in place with one's knee or foot, in the old way.

Another Saw-Horse Holder

To hold a stick of wood firmly on a sawbuck and prevent it from turning or tilting up, bend a piece of strap iron and attach to a board, as shown in the picture. Put the iron hook over the pole between the legs of the saw-horse, and stand with the left foot on the board. This is much easier than the old method of holding the pole steady.

Saw-Horse Attachment

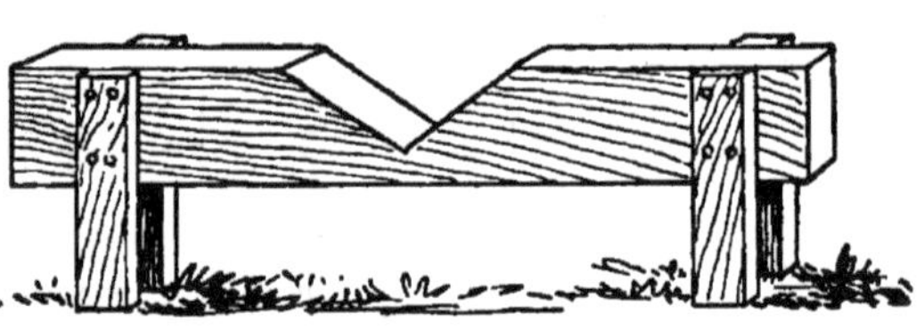

Having been at a loss for some handy place to saw wood, especially large poles, I thought of this simple device to be placed on top of a carpenter's saw horse. I have used it for some time and find it very useful. The upright pieces are to be nailed on as shown in the illustration, so they will just fit over the saw-horse. If necessary to prevent slipping, a nail can be driven in the bottom of the notch and allowed to project about a quarter of an inch on the under side.

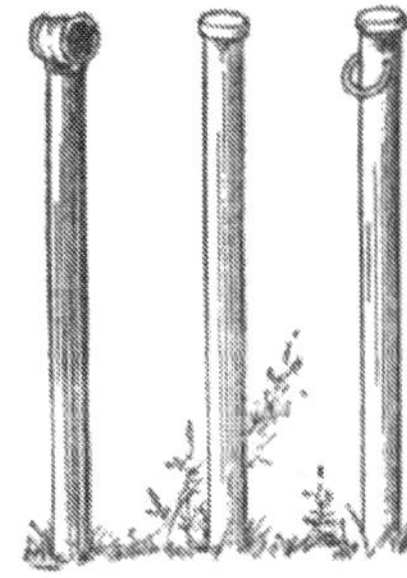

Cheap Hitching Posts

Here is the way to secure durable hitching posts. They are simply two-inch, or two-and-a-half-inch iron piping, with either a coupling or cap screwed to the top, either of which can be bought with a bit of pipe. A ring can be inserted below the cap by a blacksmith, if desired, drilling a hole through the walls of the tube. The cuts tell the whole story. Set the post in its hole and pour concrete around it, ramming it down hard.

Rapid Hitching Post

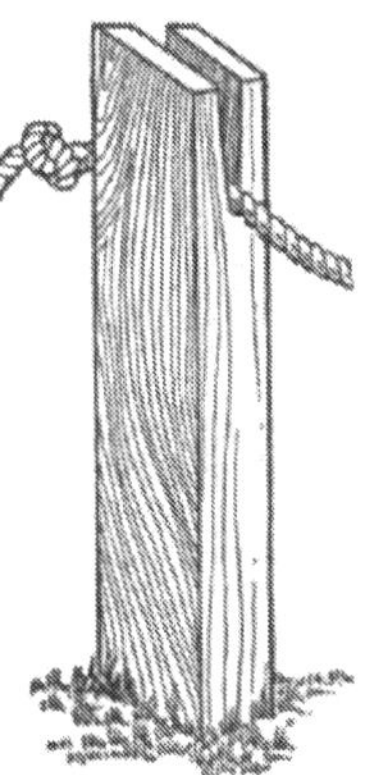

This illustration gives a good idea of a hitching post that has been used for several years with perfect satisfaction, and which saves the trouble of tying the horse while at the water trough or tank. Cut a slit in the top of a strong hardwood post, as shown in the sketch. The slit should be made one and a half inches wide at the top, and only half an inch wide at the lower end, so that it will hold the halter rope securely when placed in it. If there is a knot at the end of the rope the horse can not pull it out.

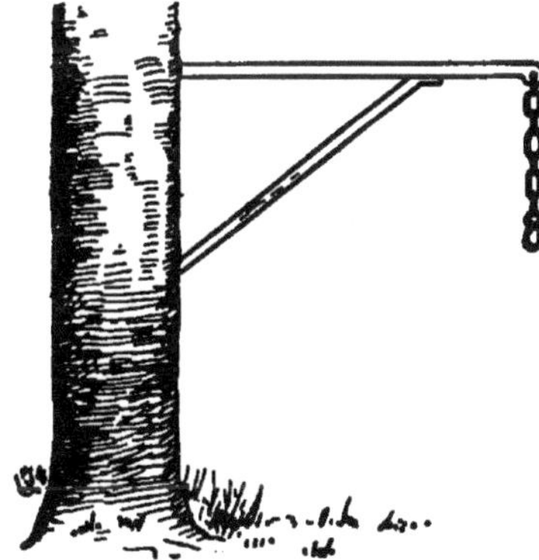

Hitching Post for Trees

Where trees are in a position to be used as hitching posts, it is wise to make some provision like that shown in the cut. Many a fine tree has been ruined by horses gnawing the bark, but with an iron arm, and a short hitch chain at the end, that danger is avoided. A blacksmith can make such a device in half an hour.

Rope Hitch on Hook

To hitch a rope into a hook quickly, and so that it will not draw tight and stick, use a half-hitch as shown in the cut; the strain comes on the upper rope. A hitch of this kind will not slip and it may be taken off very easily and quickly.

Stand for Tools

A small and stout box with corner up-rights and a top like that shown in the cut can be made in a few moments, and will be found a great convenience in many farm buildings. It can be "stood up" the hoes, forks, crowbars and other tools, that otherwise would be tumbling about in corners and upon the floor. One or two of these boxes made on a rainy day will save many a hunt for tools on pleasant days.

Cess-Pool

For the disposal of sewage nothing is better than a well-constructed cess-pool. Locate it far away from and lower than the well, digging the opening in sandy, or gravelly, soil if possible. Make the opening round, or oval, and stone up with loose rocks as a well is stoned. Dig seven or more feet in depth, according

to the nature of the soil, capacity required, etc. In sand, or gravel, a small cess-pool will answer all purposes. Cover in manner suggested in cut, leaving trap-door in plank cover for convenience in cleaning out the interior when necessary. Cover with earth and sod. Have the joints of pipe leading to the cess-pool cemented, and the pipe securely "trapped" in the cellar, or below the sink.

Barrel Rat Traps

Here are two very easy and cheap ways of making barrel rat-traps—Fig. 1 and Fig. 2. The first method involves luring the rats to the paper-covered barrel top for several nights; and then, when they feel quite safe, cutting a cross-shaped slit in the stiff paper, through which they fall into the barrel. The second method consists of a barrel with a hinged cover, as shown. "A" represents stop to keep cover from swinging backward; "B" bait.

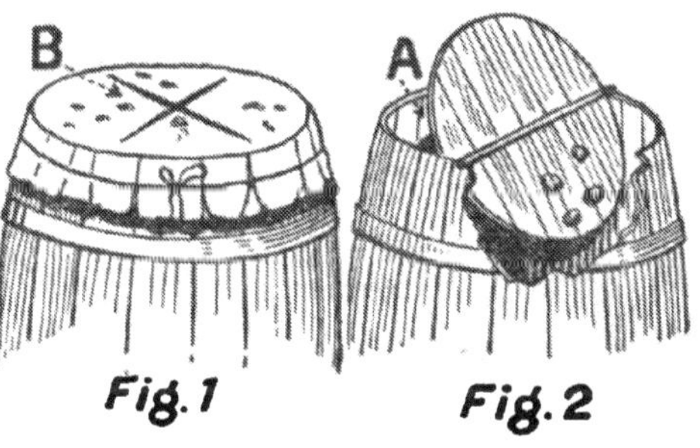

Runway for Animals

Many stables have side doors high up from the ground that would make the most convenient outlet for cattle. A run-way can readily be constructed as shown in the cut. It can be made less steep than the one shown if desired. Earth is filled inside the rock wall, and firmly trodden down. A loose stone wall only is required— easy to build if flat rocks are at hand. Such a run-way may also be built inside the barn to permit cattle to go from the first floor to the basement for water, or other purposes.

Making a Needle for Twine

Take an ordinary opener-key, such as often comes with canned goods, put it in the vise, give it a few strokes with a file, and lo!— it becomes a needle with a large eye for twine. It's just the thing for mending horse blankets, grain bags, baskets, etc.

Butter Carrier

The cut shows a very serviceable butter carrier for hot weather that anyone can make. Beneath the top is screwed a galvanized iron pan with open ends—only the lower part of the ends being left in place to keep the

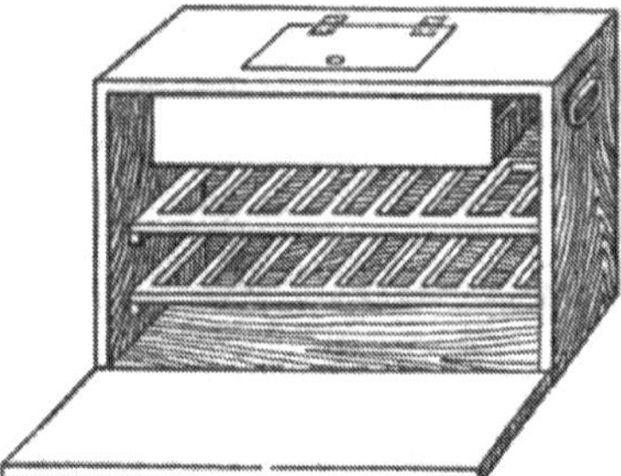

water from running out and down upon the butter. Cracked ice is put into the pan through the hinged door on top. There is an opening in the rear to let out the water as the ice melts. The butter is laid on the racks. The air over the ice is cooled and sinks, forcing up the warmer air.

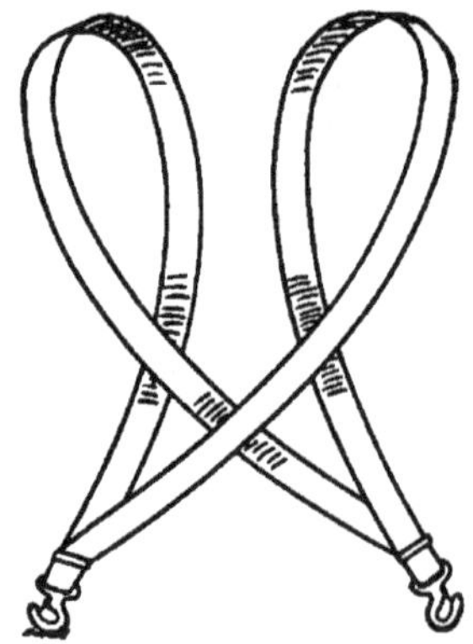

Utilizing Old Pails

When the bottom of a tin pail becomes useless the pail need not be thrown away as also useless. Unsolder the bottom, scribe a circle of the required diameter on a bit of board and cut it, or saw it out. This

nailed into a pail, as shown in the cut, will make the pail serviceable for grain, vegetables, fruits, etc. Success is often the sum of little economies.

Harness for Basket Carrying

Where one has to carry scores of baskets full of corn, apples, potatoes, or other farm products, from one building to another, or up or down stairs, a harness to put over the shoulders will be found a great relief on the muscles of the arms. With this device part of the load is supported directly by the muscles of the back. Make it of a couple of broad straps of leather.

Box for Produce

When one carries to customers, or to the general market, a variety of farm products, as butter, eggs, fresh berries, etc., it is desirable to have a box that gives separate space for each variety. The box here shown fills this need, and is made of such light material that it can be carried easily, even when well filled. The handles are arranged to fall down over the ends that the cover may be raised. The shelves can be movable and made either so'id or of slate. Make the ends of three-quarter inch stuff, and the rest of three-eighth inch pine.

Barrel Carrier

Barrels of apples, or potatoes, are unhandy things to carry, but with the two devices like these shown in the cuts they can be moved with ease. Three-eighths inch round iron is bent by a blacksmith into the form shown. Two men with two of these affairs can carry full barrels of fruit or vegetables with ease as well as safety.

Replacing Basket Handles

When handles have been broken from baskets new ones can be placed on them in the manner shown in the cut. Handles of broken-down washtubs can be taken, or two old bails from broken pails can be cut down and fitted to the basket in a few moments in the way shown.

Rope Sling for Barrel

The cut plainly shows the right method of attaching a rope to a barrel keg or other round object. It is very simple, holds securely and has proved satisfactory in practice. There is no untying to do—as soon as the rope ends are loosened the loops slide down and off.

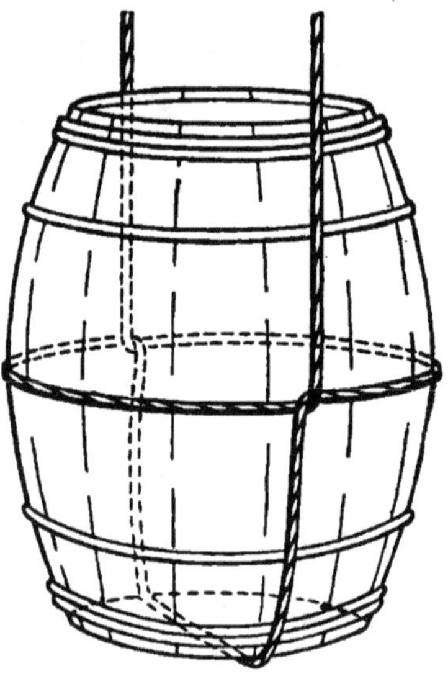

If you have any trap doors in the barn loft, have them so made as to avoid accidents. This cut shows one way to do it. B is a plank door. A, A are pieces of plank hinged to partition and that swing back out of way when not in use. These come up against battens of door when it is raised and are hooked firmly to it. No danger of falling into this trap, even in a dark night.

Baling Old Paper

By using a good-sized box with a slot cut in the bottom you can bale your waste paper. Arrange a cord and a lining of old cloth or very tough paper as shown in cut. **Pack**

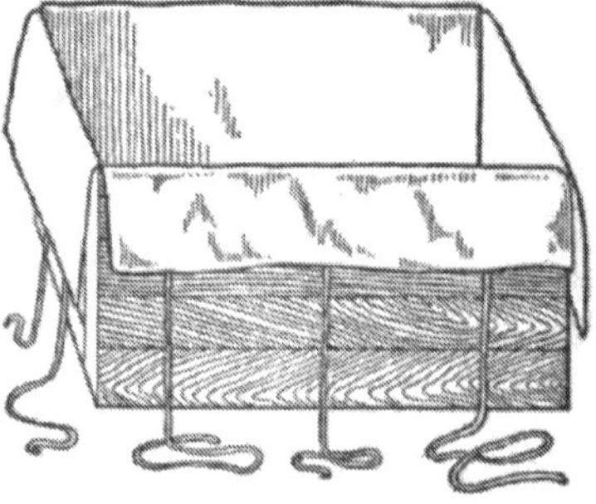

the paper down hard, fold the lining over the top and tie securely. Turn the box over push through the slot and your bale will slip out ready to weigh and sell.

Protecting the Trap Door

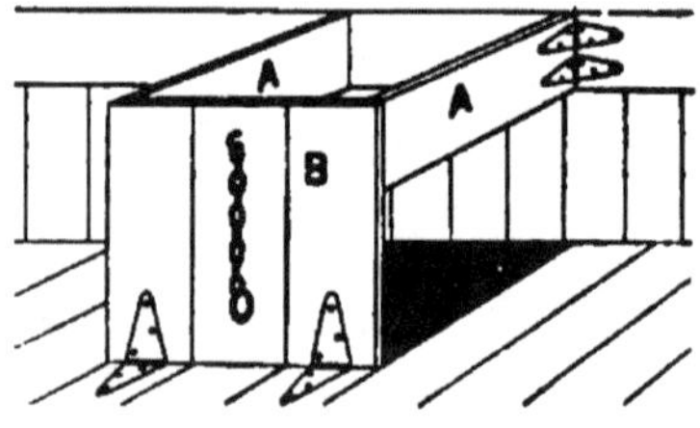

Pulley Hoist

The cut shown speaks for itself. It shows one of the finest arrangements for taking heavy articles down from chambers, or for raising them up to any floor. The supports being at either side of the window, give the whole opening through which the weight may be swung in when raised to a position opposite the window. The supports are fastened inside at the floor to keep them from slipping.

Buried Rain Barrel

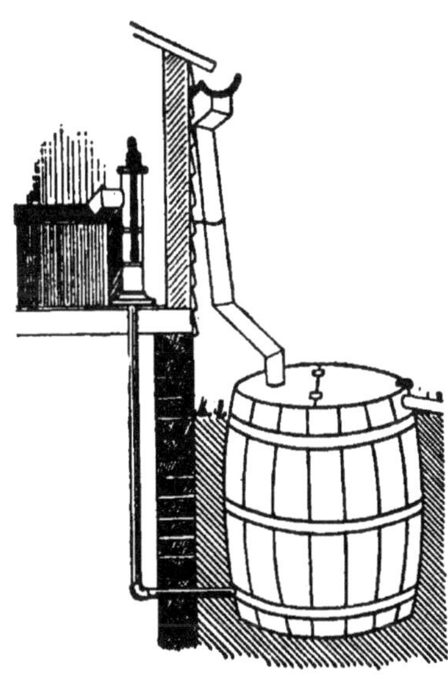

If the house is not equipped with wind-mill or engine, tanks and such like, it may be necessary to utilize the rain water from the roof. A large hogshead is buried to within a few inches of the top at the side of the kitchen. One-half the cover is nailed on and the other half hinged as shown in cut and locked. The conductor from the roof passes through the lid, a wire screen being fastened underneath to keep out leaves and dirt. An overflow pipe is placed in front about two inches from the top. A one and a quarter inch pipe is laid from the cellar into the tank near the bottom and carried up into the kitchen sink. All who require soft water for washing or other purpose will find this arrangement cheap and useful.

Simple Rope Ladder

There is always danger of fire catching the roof of farmhouse or barn. Quick work will often save much loss. But if one happens to be alone, it is almost impossible to raise long, heavy ladders, especially the one needed to reach any place on the roof. A very convenient ladder to keep on hand is one made of light ropes, as shown in the cuts. A staple over the rope keeps the latter from slipping off the "rounds." At one end are two hooks and a connecting cross-bar which any blacksmith can make in a few moments. To the center of the cross-bar is attached a long cord with a bit of weight fastened to its outer end. It is but an instant's work to throw this cord over the ridge-pole of a building and having secured the weighted end upon the other side, to draw the ladder up over the roof until the hooks cross the ridge. Then pull the ladder back into place with the hooks biting into the wood of the saddle boards, and one can go anywhere upon the roof. This ladder should be long enough to reach the ground from any building. One person alone can raise a wooden ladder to the eaves, and with this rope ladder he has command of the situation. The "rounds" of the rope ladder need not extend to the ground, but the ropes should, that the ladder may be manipulated from the ground.

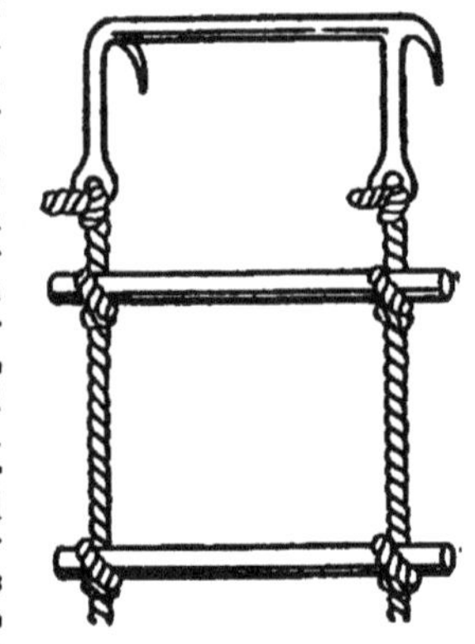

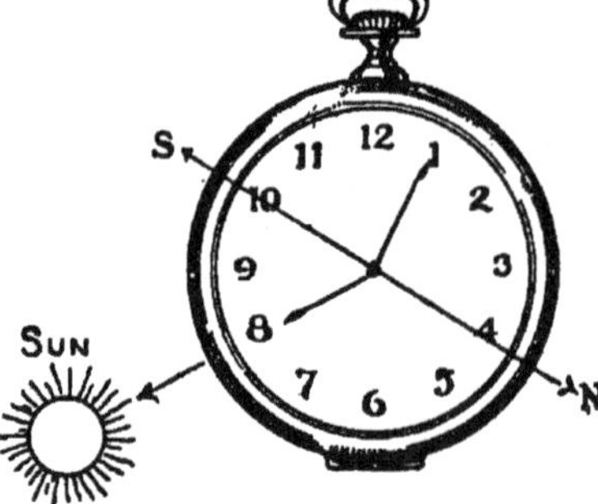

Watch Compass

Point the hour hand at the sun and lay a pencil or stick exactly half way between the hour hand and the 12 (as shown by the arrow in the accompanying sketch); then the end between the hour and the 12 always points due south. Of course, a correction must be made if "Daylight Saving" is in effect.

Cross-Roads Cream Station

Many patrons of a creamery wagon live on cross-roads and are obliged to meet the wagon at the junction of the main and cross-roads. This often necessitates a long and vexatious wait, with valuable time lost. Have a closet built at the junction of the roads as suggested in the cut, with lock and key. Let the creamery man have a key to fit the lock, and the cream can be set in and left for the driver to collect when he comes along. Leave openings in the rear to ventilate the closet that, being closed, it may not be overheated by the sun. Several neighbors on a cross-roads can unite in the use of such a closet, each one putting his name on his can, taking turns in carrying the cream.

Chopping Block

Here is a good holder for chopping old boards. The block is a four-foot stick eight inches in diameter, and two pieces of scantling nailed on one and one-half inches apart, as shown. Be sure the cooks have plenty of kindling this s u m m e r,

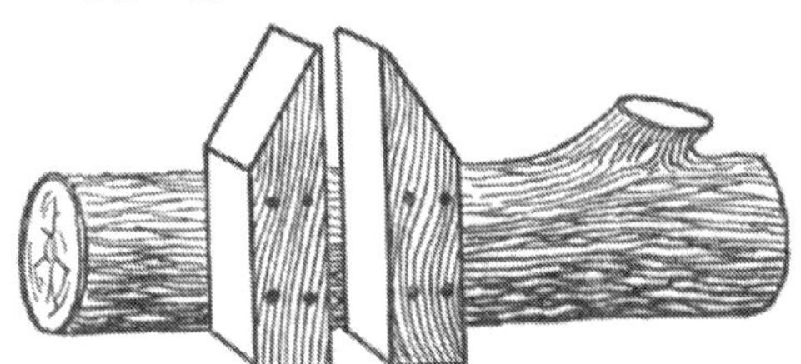

whatever the method you use in cutting it. The old-fashioned saw and sawbuck will make nice kindling out of old boards by the application of plenty of elbow grease.

Telephone Index Chart

For a telephone chart use a circular piece of cardboard about six inches wide. In the center cut a hole just large enough to get it over the mouthpiece on the telephone. Divide the cardboard into sections and in the sections put the names of the people in your neighborhood, with the telephone number after each name. Arrange the names alphabetically.

Under Shoe Brake

If you live in a hilly country it is a good idea to adopt the wooden shoe shown here for use on wagon wheels when carting heavy loads down a steep incline. It saves the tire from wear and the whole wheel from being strained. To make it, take a piece of hard wood shaped like the letter J, two feet long, four inches thick and six inches wide. Make a groove in top two inches deep and a little wider than the tire. Mortise hole through front end large enough to admit a stout chain. To use it, set shoe in front of hind wheel, drive on and put the chain through shoe and around felloe of wheel and fasten the other end of chain to coupling pole and drive ahead.

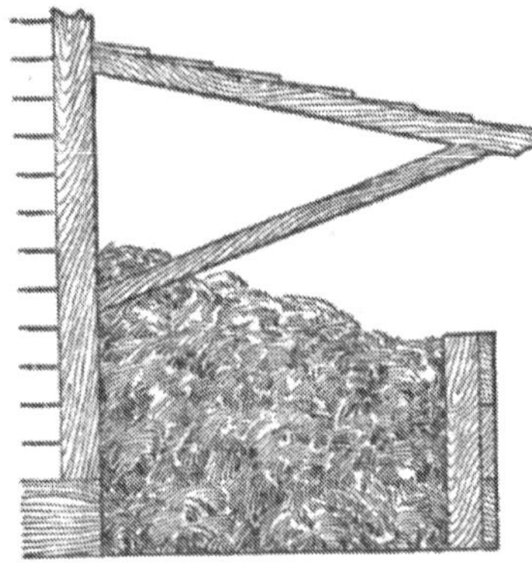

Manure Shed

On many farms the manure is thrown out of the small tie-up windows into piles that have a "shed" roof above them, as shown in this diagram. This works well, provided a front fence is used to keep the manure from rolling down under the eaves, and to protect it from driving rain storms. The roof and the front protection keep out most of the rain. The tail end of the cart can be backed in under the roof when the manure is to be loaded.

Sodding

If you have sodding to do go to the meadow and cut a strip a foot wide, and roll it up as in the picture and haul to where wanted. In this way a large space can be covered in a short time. This is often preferable to seeding.

Wagon Wheel Holder for R. F. D. Mail Boxes

A clever and useful community mailbox holder for country districts is shown in the picture. In the construction of this an ordinary worn - out wagon wheel has been used. Wheels of this type may be found around most any farm place. First a 2 x 4 post was driven in the ground and in the center of the upper end a large spike was driven. After the several boxes were fastened to the wheel by means of wires, the hub was placed over the spike in such a manner that the hub rested upon the top of the post. When the farmer wishing to secure his mail comes to the wheel, he does not have to get out of the vehicle, but simply turns the wheel around until his mail holder comes to him.

Bee Smoker Fuel

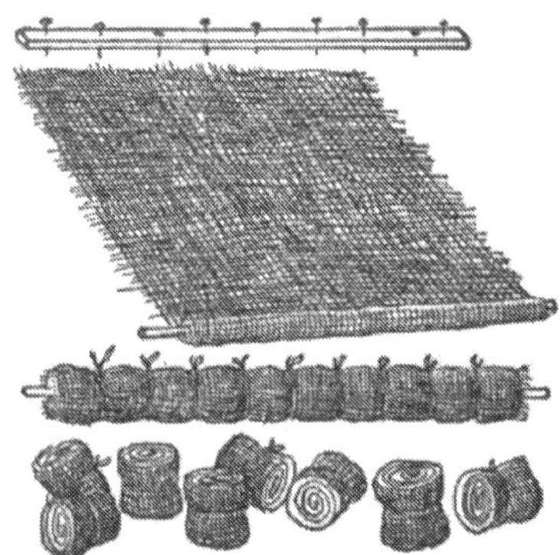

The cut shows how an experienced bee-keeper makes his smoker fuel. Old, rotten phosphate sacks are ripped apart and wound up, not too tight, on a light stick, as shown, and then tied every four inches with soft cotton twine. Next the roll is cut with a sharp ax between the strings. Now dissolve a pound or two of saltpetre in water, making it very strong, and dip one end of each wad in the solution and throw them out in a pile to dry. Throw a dash of red lead in the pan so that the end of the wad dipped in this solution will be marked. This is the end to light. This fuel lights instantly, makes no sparks and lasts for a long while.

An Outside Cellar Door Lift

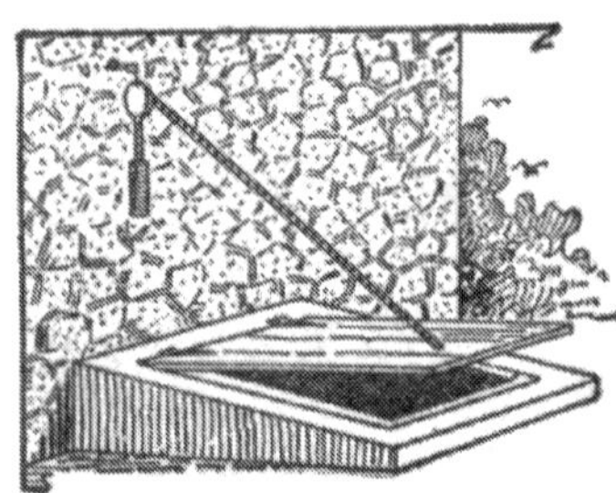

Attach the hinges to the upper end, and near the lower end of the door bore a hole through which a rope may be run and knotted on the under side. Carry the other end of the rope up to a small pulley wheel fastened to the house; run it over and to the end tie a weight or sack of stone just sufficient to hold up the door when given the least lift. With weight of the right proportions, the door goes up at a touch and stays until lowered. When part way open it looks like the accompanying sketch.

Stairway That Lifts Up

Where space is limited, the stairway to the stable loft can be arranged to swing up out of the way when not in use. It is strongly hinged at the top, and has a cord passing from the bottom up over a pulley, with a balance weight on the other end of the cord. It is within reach of the hand when raised, and so can be pulled down whenever needed.

Protecting the Roof from Painter's Hooks

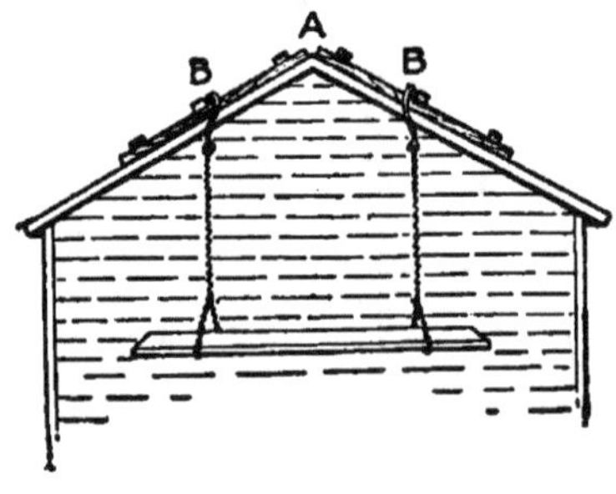

This device is merely two sixteen-foot boards connected by a heavy strap hinge at A. At intervals cleats are nailed to them, so that when put over the ridge of a building a person may walk up or down the roof and place painter's hooks where they are wanted without piercing the roof, for the hooks, B, set into the boards instead of into the roof material. When not in use the boards close together and are laid away with the rest of the scaffolding.

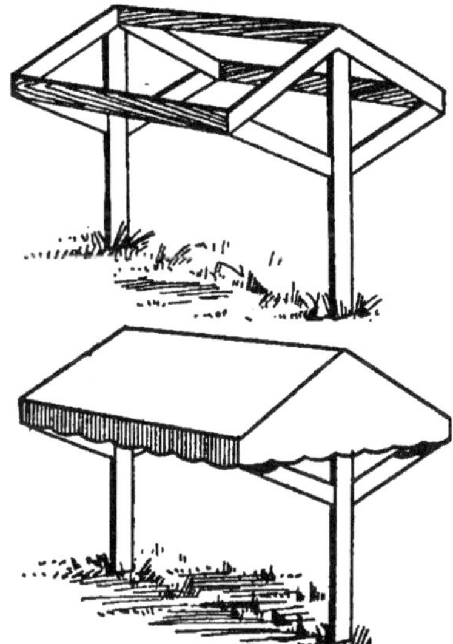

Lawn Awning

Many farm houses have neither piazzas nor trees about them. Set out some trees this spring. But while waiting for the trees to grow, make an awning like that shown in the cuts, under which two or three chairs can be placed during the heat of the summer. It is an unfortunate house that has no place out of doors where one can find a shady seat for a few moments' rest and refreshment. The framework shown in the cut is covered with awning cloth, or even white cotton.

Keeping Surface Water from the Well

Hundreds of wells ought to have the wall raised and the earth graded in a slope away from the top, to keep out surface water. Take an old cart-wheel tire (if not big enough a blacksmith can make it so), and support it above the opening of the well. With rocks and cement build the wall up inside the iron tire, and imbed the latter in the cement, as the work goes on. This iron band helps to get a true circular wall, and, in particular, keeps the wall from cracking and breaking apart. Now fill in, and tamp down good clay from the wall out for a dozen feet, as shown by the dotted line, and the wall will be secure against impure surface water.

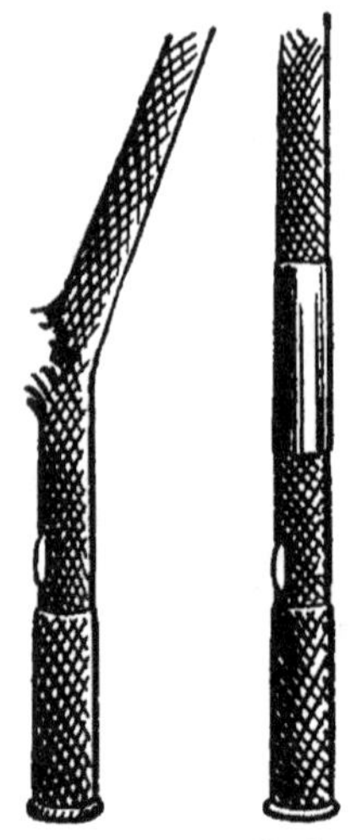

Repairing the Whip

The best of whips are not over tough, and will frequently snap in two. Instead of throwing a broken whip away as useless, take it into a tinsmith's shop and have him clasp a bit of tin around the break and solder it after the fashion shown in the illustration. He can roll the tin tightly about the break, and hold it thus with pincers while applying the solder. Thus mended the whip will be as strong as before it was broken.

One-Man Lifting Device

This is one practical lifting device where one man has to do his work alone. It will lift bags of grain, boxes and pick up big stones and swing them around upon a drag or stone boat. The height to which this device will lift an object depends on the height of the standard and the length of the lever. Make it of a size to suit your needs, and you will find it saving you many a straining lift.

Picnic Lunch Box

When the family goes on a picnic, as it ought to occasionally, the box shown in the cut is just the thing to take the dinner in. Get a nice corner-mortised box at the grocer's, and put two handles upon it as shown. These can be made from ash. Put in a bottom about half way down, and give access to the lower half by a sliding or hinged cover at the side. In this lower space can be put provision that must not be crushed. The housewife will appreciate this feature.

Woodbine Hedge

Drive stakes two feet apart and connect each pair by a bent rod. Over successive pairs of these stretch poultry netting tightly and neatly, as shown in the drawing. Plant woodbine along either side, and in two or three years you will have as handsome a hedge as grows out of doors. Woodbine delights in moist, loose and very rich soil. It will pay grandly for all of these conditions.

Raised Flower Bed

A raised flower bed shows off its contents to better advantage, and can be more easily worked, as one need not greatly stoop in caring for the flowers. Make the edge of cobble stones, laid in cement (any one can do it who has gumption!) and the flower bed will be doubly attractive. A concrete edging will be more durable, but not so handsome. The cut shows how the bed will look, minus the flowers.

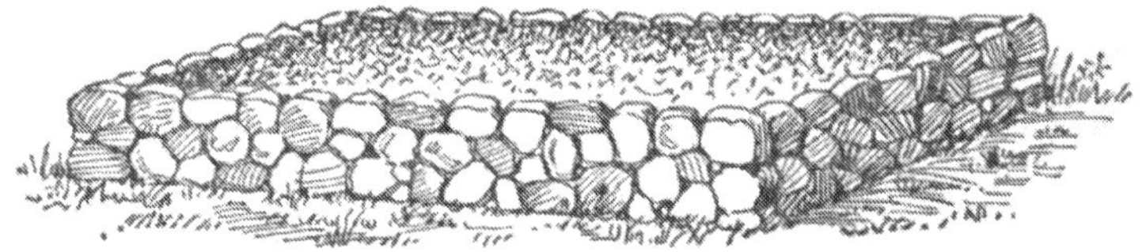

Convenient Wheel for Ladders

Did you ever try to carry a long ladder alone? It is hard, and often exasperating work. Make a box, and screw wooden wheels to the sides, as shown in the cut. Make the box wide enough to take in any ladder; then, when needed, slip one end of the ladder into the box, take the other end in the hands, like wheelbarrow handles, and wheel the ladder to the place where it is to be used.

Chasing the Robins

Robins bothering the strawberry patch in the garden? Well, here is one way to protect them. Pussy has been busy catching mice all winter, and it won't hurt her to have a brief change of occupation. Surely she won't object if we put a nice collar around her neck, fasten her to a sliding tether attached to a taut overhead wire—and "turn her loose" to scare away the birds in that little berry patch. Not likely she will catch any, but she will give 'em a good scare.

Making An Ox Yoke

The act of making yokes for oxen has been all but lost. Here is the method: Take a gnarlly stick of elm or maple, five feet long, ten inches deep and seven inches wide. Bore bow holes before the yoke is shaped, at exact right angles to the stick, and two inches in diameter. From outside to outside of the holes should be about eleven inches; and from the end of the timber to the edge of the first hole, one-half as far. The yoke is worked down so that when done it is only

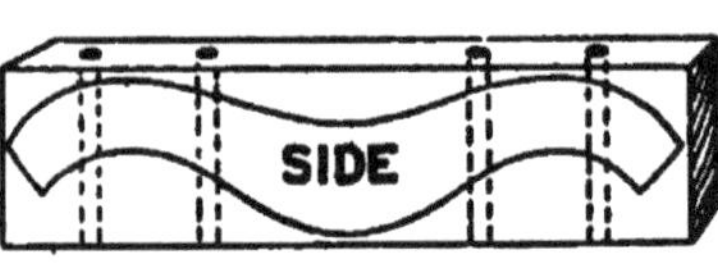

fifty-three inches long. Of course, a staple must hold the ring at the exact middle of the yoke, and this point must hang an inch or two below the ends. The bows should be made of the best seasoned hickory, shaped roughly and thoroughly steamed before bending. They vary in length, from loop to top, from two feet to two and one-half feet.